# Microbiome in Inflammatory Lung Diseases

Gaurav Gupta • Brian G. Oliver •
Kamal Dua • Alisha Singh •
Ronan MacLoughlin
Editors

# Microbiome in Inflammatory Lung Diseases

Springer

*Editors*
Gaurav Gupta
School of Pharmacy
Suresh Gyan Vihar University
Jaipur, India

Brian G. Oliver
School of Life Sciences
University of Technology Sydney
Sydney, NSW, Australia

Kamal Dua
Discipline of Pharmacy, Graduate School of Health
University of Technology Sydney
Sydney, NSW, Australia

Alisha Singh
School of Pharmacy
Suresh Gyan Vihar University
Jaipur, India

Ronan MacLoughlin
R&D Science and Emerging Technologies
Aerogen Limited
Dangan, Ireland

ISBN 978-981-16-8956-7 ISBN 978-981-16-8957-4 (eBook)
https://doi.org/10.1007/978-981-16-8957-4



This Springer imprint is published by the registered company Springer Nature Singapore Pte Ltd.
The registered company address is: 152 Beach Road, #21-01/04 Gateway East, Singapore 189721, Singapore

# Preface

Worldwide, respiratory disorders are one of the most prevalent non-communicable diseases and are also a major source of illness and mortality globally, imposing a significant personal and socioeconomic cost on the health care system. Although each respiratory illness has its own set of trigger events and pathogenic processes, there are certain characteristics among them, such as epithelium injury or dysfunction, airway inflammation, and airway remodeling. Most respiratory illness therapies now only give short relief from symptoms and do not provide effective prevention. Furthermore, poor illness control reduces people's quality of life, but traditional pathogenic mechanisms in respiratory disorders cannot adequately explain the occurrence and progression of the diseases. Respiratory illnesses have become a major public health concern in all countries across the world, particularly in emerging countries and underdeveloped populations. Respiratory disorders are becoming more prevalent worldwide, particularly affecting the elderly and children. Acute respiratory distress syndrome (ARDS), asthma, chronic obstructive pulmonary disease (COPD), and pulmonary fibrosis are only a few of the respiratory disorders that have claimed over 354 million lives worldwide. COPD (3.9% of global prevalence) and asthma (3.6% of global prevalence) are the most prevalent respiratory disease. It is reasonable to predict an increase in SARS-CoV-2 and more serious COVID-19 infections in individuals with respiratory diseases, particularly in the case of asthma and COPD.

The human microbiota consists of commensal, symbiotic, and pathogenic microorganisms (protozoa, fungi, microbes, and viruses) that coexist in our bodies and form organ-specific microbial communities. The structure and size of the microbiome will differ between body parts since they are influenced by the host and environmental factors. The ability to access an organism's DNA, metabolites, RNA organisms, and proteins are part of the microbiota, which confirms the microbiome's essential function in human health and disease. The human microbiome plays an important function in physiology, and certain microbiome species are thought to be helpful. However, certain microbiome elements are thought to be particularly harmful to human health. Several experiments have shown that bacteria and viruses are linked to chronic inflammation, which increases the risk of lung disease. The function of the lung microbiome has been widely overlooked, and it was once thought that healthy lungs were sterile, perhaps due to the difficulty of accessing the lower airways without invasive procedures. Recent findings have shown that healthy lungs have a rich and complex microbiota, as well as the understanding that its modifications affect lung disease, implying that the microbiome plays a role in the development and progression of lung disease. There is a lot of detail about the microbiota of the gastrointestinal (GI) tract compared to the microbiome of the lungs. This would be the first book to address the role of the microbiome in lung disease.

This book, *Microbiome in Inflammatory Lung Diseases*, will address lung microbiome transition, namely the reduction of probiotic species and a possible increase in pathogenic bacteria, which tends to be a central element in the resistance, chronization, and evolution of respiratory diseases. This book would present developments in lung microbiome research and show the ability to use the microbiome as a pathway to inhibit inflammatory lung diseases and modulate therapeutic techniques, implying the microbiome as a valuable solution in inflammatory lung diseases patients. This book attracts a range of audiences including clinical researchers working in the field of respiratory diseases, undergraduate and postgraduate students from various disciplines such as pharmacy, microbiology, immunology, pharmacology, biotechnology, and health sciences. Moreover, the listed editors from three different nations (Australia, the UK, and India) hold extensive experience working with gut microbiota and various inflammatory diseases that bring an extra edge to the book in its compilation.

The editors of this book would like to express their sincere gratitude to all the authors for their time and their valuable contributions to the production of this book.

| | |
|---|---|
| Jaipur, India | Gaurav Gupta |
| Sydney, NSW, Australia | Brian G. Oliver |
| Sydney, NSW, Australia | Kamal Dua |
| Jaipur, India | Alisha Singh |
| Dangan, Ireland | Ronan MacLoughlin |

# Acknowledgement

The publication of this book was finalized during the coronavirus (COVID-19) pandemic. We would like to dedicate this book to all those who were affected by the pandemic and, in particular, to our health workforce around the world for their dedication and care during this difficult time.

# Contents

1 **Introduction to Lung Disease** . . . . . . . . . . 1
Waleed Hassan Almalki

2 **Introduction to Microbiome** . . . . . . . . . . 13
Shivkanya Fuloria, Vetriselvan Subramaniyan, Mahendran Sekar, Yuan Seng Wu, Srikumar Chakravarthi, Rusli Bin Nordin, Pradeep Kumar Sharma, Dhanalekshmi Unnikrishnan Meenakshi, Ajay Mendiratta, and Neeraj Kumar Fuloria

3 **Role of Microbiome in Inflammation During Tuberculosis** . . . . . . . 29
Kuldeepak Sharma, Mateja Erdani Kreft, Mateja Škufca Sterle, and Darko Vasic

4 **Interplay of Microbiome, Inflammation, and Immunity in Inflammatory Lung Diseases** . . . . . . . . . . 43
Hitesh Malhotra, Anjoo Kamboj, Peeyush Kaushik, and Rupesh K. Gautam

5 **Microbiome in Asthma** . . . . . . . . . . 65
Khalid Saad Alharbi, Sattam Khulaif Alenezi, and Sulaiman Mohammed Alnasser

6 **Microbiome in Chronic Obstructive Pulmonary Disease (COPD)** . . . . . . . . . . 79
C. Sarath Chandran, Anitha Jose Subin, Alan Raj, K. K. Swathy, and Indu Raghunath

7 **Microbiome in Asthma-COPD Overlap (ACO)** . . . . . . . . . . 103
Shibi Muralidar, Gayathri Gopal, and Senthil Visaga Ambi

8 **Microbiome in Acute Respiratory Distress Syndrome (ARDS)** . . . . . . . . . . 117
Gayathri Gopal, Shibi Muralidar, Abishek Kamalakkannan, and Senthil Visaga Ambi

9 **Role of Brain–Gut–Microbiome Axis in Depression Comorbid with Asthma** . . . . . 135
Shvetank Bhatt, K. Sreedhara R. Pai, C. R. Patil, S. N. Manjula, and S. Mohana Lakshmi

10 **Understanding the Impact of the Microbiome on Lung Cancer** . . . . 153
Anindita Goswami, Sanchita Chandra, Sarmistha Adhikari, and Paramita Mandal

11 **Microbiome in Pulmonary Tuberculosis** . . . . . 167
Arnab Rakshit, Aarti Verma, Saloni Verma, Gurjit Kaur Bhatti, Amit Khurana, Jasvinder Singh Bhatti, Snehal Sainath Jawalekar, and Umashanker Navik

12 **Lung Microbiome: Friend or Foe of *Mycobacterium tuberculosis*** . . 207
Summaya Perveen and Rashmi Sharma

13 **Microbiome in Idiopathic Pulmonary Fibrosis** . . . . . 227
Sachchidanand Pathak, Anurag Mishra, Gaurav Gupta, Abhay Raizaday, Santosh Kumar Singh, Pramod Kumar, Sachin Kumar Singh, Neeraj Kumar Jha, Dinesh Kumar Chellappan, and Kamal Dua

14 **SARS-CoV-2 and Microbiota** . . . . . 241
Edda Russo, Lavinia Curini, Alessio Fabbrizzi, and Amedeo Amedei

15 **Microbiome in SARS-CoV-2 (Covid-19)** . . . . . 281
Subha Manoharan, Lakshmi Thangavelu, Mallineni Sreekanth Kumar, Gaurav Gupta, Kamal Dua, and Dinesh Kumar Chellappan

16 **Microbiome in Influenza-A Virus Infection** . . . . . 295
Suhas Suresh Awati, Santosh Kumar Singh, Abhay Raizaday, Pramod Kumar, Yogendra Singh, Mohammad Arshad Javed Shaikh, and Gaurav Gupta

17 **Microbiome in Upper Respiratory Tract Infections** . . . . . 309
Piyush Mittal, Manjari Mittal, Ujjawal Rawat, and Ambika

18 **Challenges in Understanding the Lung Microbiota** . . . . . 327
Olorunfemi R. Molehin, Olusola O. Elekofehinti, Adeniyi S. Ohunayo, and Oluwatosin A. Adetuyi

19 **Microbiome in Inflammatory Lung Diseases: Challenges and Future Prospects** . . . . . 339
Nitin Verma, Komal Thapa, and Kamal Dua

**20 Microbiota Targeted Via Nanotechnology for Lung Cancer Therapy: Challenges and Future Perspectives** . . . . . . . . . . . . . . . . . . 359
Monika Yadav and Anita Kamra Verma

**Correction to: Interplay of Microbiome, Inflammation, and Immunity in Inflammatory Lung Diseases** . . . . . . . . . . . . . . . . . . . . . . . . . . . . . . . . C1
Hitesh Malhotra, Anjoo Kamboj, Peeyush Kaushik, and Rupesh K. Gautam

# About the Editors

**Gaurav Gupta** is an Associate Professor in the School of Pharmacy at the Suresh Gyan Vihar University (SGUV), Jaipur, Rajasthan, India. He has more than 11 years of experience in molecular and biochemical pharmacology, including respiratory diseases and cancer biology by employing experimental animal models to understand the cellular and molecular mechanism. Dr. Gupta is dedicated to improving outcomes in healthcare through his initiatives in pharmacology and phytochemistry research and effective teaching in the field of Pharmaceutical Sciences. Dr. Gupta has more than 300 research and review articles in peer-reviewed international journals and is co-author of one book. He is a member of various national and international societies.

**Brian G. Oliver** is currently the co-director of the Respiratory, Sleep, Environmental and Occupational Health Clinical Academic Group of Maridulu Budyari Gumal, the Sydney Partnership for Health, Education, Research and Enterprise (SPHERE), A NHMRC AHRTC. He is a translational researcher who aims to identify and develop new ways of treating respiratory diseases. His scientific training began at the National Heart and Lung Institute, UK, where he mastered the isolation and in vitro culture of several types of human lung cells. He then had further training in both molecular biology (the University of Leeds) and then respiratory virology at Imperial College, UK, before commencing his Ph.D. at the University of Sydney. He now leads the Respiratory Cellular and Molecular Biology Group with laboratory facilities located at both UTS and the Woolcock Institute. The work from his group is recognized to be among the best in the world, evidenced by selection for presentation at symposia at both national and large international conferences as well as resulting in prestigious publications.

**Kamal Dua** is a Senior Lecturer in the Discipline of Pharmacy at the Graduate School of Health, University of Technology Sydney (UTS), Australia. He has research experience of over 12 years in the field of drug delivery systems targeting inflammatory diseases. Dr. Dua is also a Node Leader of Drug Delivery Research in the Centre for Inflammation at Centenary Institute/UTS, where the targets identified from the research projects are pursued to develop novel formulations as the first step towards translation into clinics. Dr. Dua researches in two complementary areas, drug delivery and immunology, specifically addressing how these disciplines can advance one another, helping the community to live longer and healthier. This is evidenced by his extensive publication record in reputed journals. Dr. Dua's research interests focus on harnessing the pharmaceutical potential of modulating critical regulators such as interleukins and microRNAs and developing new and effective drug delivery formulations for the management of chronic airway diseases. He has published more than 80 research articles in peer-reviewed international journals and authored or co-authored four books. He is an active member of many national and international professional societies.

**Alisha Singh** is an Assistant Professor in the School of Pharmacy, Suresh Gyan University, Jaipur, Rajasthan, India. She has been a part of multiple funded projects as a Research Intern at TNMC and BYL Nair Ch. Hospital, Mumbai. Her total work experience is of 4 years which includes working with Cipla Ltd. as a Medical Advisor. Her research interests include cost analysis studies, health economics, and drug utilization studies.

**Ronan MacLoughlin** is currently Associate Director of R&D, Science and Emerging Technologies in Aerogen Ltd. Dr. MacLoughlin has more than 20 years of experience in Respiratory Drug Delivery with several nebulizers and accessory product launches over that time. He has the responsibility of new product development, establishing and exploiting the science underpinning respiratory drug delivery, and identifying new and potentially disruptive emerging technologies. To this end, he has developed multiple technologies and products with several patents granted and pending that cover the range of drug, device, drug/device combination products, patient interventions, and patient interfaces. Dr. MacLoughlin currently serves as chair of the Industry Representative Group in CURAM, the Science Foundation Ireland, center for the development of the next generation of smart medical devices, and sits on the board, (previously chair) of the Medical and Engineering Technologies Center (MET), the Enterprise Ireland Technology Gateway. Additionally, he is currently chair of the Paediatric and Cystic Fibrosis working group within the International Society for Aerosols in Medicine (ISAM). Finally, Dr. MacLoughlin holds the position of adjunct Associate Professor in Trinity College Dublin (School of Pharmacy and Pharmaceutical Sciences) and Honorary Senior Lecturer in the Royal College of Surgeons, Ireland (School of Pharmacy and Biomolecular Sciences).

# Introduction to Lung Disease

1

Waleed Hassan Almalki

**Abstract**

The global incidence of lung disease (LD) affecting children and adults is steadily increasing. The source of mortality and morbidity of lung diseases is unknown. However, current data from the WHO and other institutions show that there are approximately 400 million people worldwide suffering from mild to severe COPD and asthma. Lung diseases can be classified as non-infectious (asthma, chronic obstructive pulmonary disease (COPD), lung cancer, cystic fibrosis, and idiopathic pulmonary fibrosis (IPF)) or infectious (tuberculosis, influenza and COVID-19) disease and method transfer. Lung diseases have a huge impact on a global scale and are becoming more common due to the ageing population and the lack of appropriate interventions to minimise the risk factors that lead to the development of these diseases. Asthma, COPD, fibrosis, COVID-19, and influenza-like lung diseases have become life-threatening and life-threatening, effective treatments and appropriate preventive measures have become challenges for researchers.

**Keywords**

Lung diseases · Asthma · COPD · COVID-19 · Fibrosis

W. H. Almalki (✉)
Department of Pharmacology and Toxicology, College of Pharmacy, Umm Al-Qura University, Makkah, Saudi Arabia
e-mail: Whmalki@uqu.edu.sa com



G. Gupta et al. (eds.), *Microbiome in Inflammatory Lung Diseases*,
https://doi.org/10.1007/978-981-16-8957-4_1

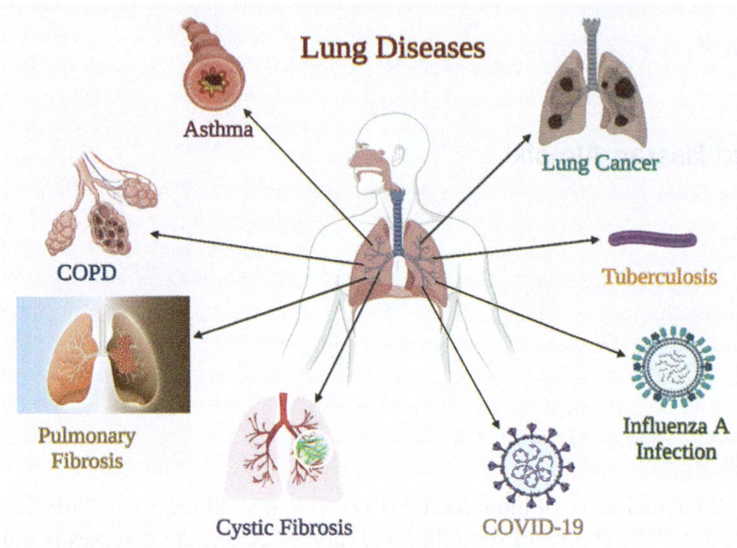


**Fig. 1.1** Lung Diseases

## 1.1 Introduction

The enhancement of worldwide incidence of Lung Disease (LD), which affects both children and adults, is steadily rising. The source of pulmonary illness fatality and morbidity is unknown although current WHO and other agency figures indicate that about 400 million individuals worldwide suffer from mild to severe COPD and asthma alone. Furthermore, *Haemophilus influenzae* infection in the lower respiratory tract causes between 250,000 and 500,000 fatalities each year [1–3]. In 2015, *Mycobacterium* TB infection of the lower respiratory tract affected 10.4 million people globally, killing 14% of those infected. Other non-communicable illnesses, such as lung cancer induced by cigarettes smoking or exposure to environmental toxins, claim the lives of 1.6 million people each year and are on the rise. Lung disorders can be classified as non-communicable (asthma, chronic obstructive pulmonary disease (COPD), lung cancer, cystic fibrosis, and idiopathic pulmonary fibrosis (IPF)) or communicable (tuberculosis, influenza, and COVID-19) depending on disease etiopathology and method of transfer (Fig. 1.1) [4, 5].

## 1.2 Overview of Lung Diseases

Here we briefly discussed communicable and non-communicable lung diseases.

### 1.2.1 Asthma

Asthma is a diverse and complicated lung illness marked by varied airflow restriction, bronchial hyperactivity, and, most critically, elevated airway inflammation. Asthma impacts around 10% of the adult population in nearly every country, totalling around 300 million people worldwide. Furthermore, asthma is projected to be the main cause of 383,000 fatalities globally, with low- and middle-income nations accounting for nearly 80% of asthma-related deaths [6, 7]. Asthma is also a significant financial burden, costing up to $USD 3100 per person each year. Asthma is a prevalent but misunderstood respiratory condition that can strike anybody at any age. House dust mites, pollen, moulds, cigarette smoke, environmental exposures to harmful chemicals, and air pollution are all risk factors for asthma. Asthma is a complicated condition that can present as disease "episodicity", or times in which symptoms arise and then disappear after therapy. In addition, the illness may be "chronic" in people, as evidenced by the persistence of typical asthmatic clinical signs. Wheezing, breathlessness, rapid breathing, and coughing are all common asthma signs. Multiple factors, like contact to allergens, irritants, pulmonary tract infections with bacteria or virus, sinusitis, physical activity, thunderstorms, and extreme cold, might exacerbate symptoms. Asthma has been classified into phenotypic and/or endotypes based on current advances in asthma pathogenesis [8–10]. This is critical for effective asthma therapy, as advised by a new Lancet panel that describes the discovery of "treatable characteristics" in asthma patients and then precisely addresses these qualities for illness control.

According to studies, the most of asthma episodes are caused by Th2 activation, in which type 2 T-helper cells are mobilised into the airways in reaction to an outside or endogenous stimulus and release high levels of cytokines including IL-13, IL-9, IL-5, and IL-4. IL-4 is engaged in the transformation of B-cell IgE to immunoglobulin E that leads to the secretion of inflammatory intermediaries like cysteinyl leukotrienes and catecholamines, whereas IL-5 is only engaged in the intake of eosinophils that leads to the advancement of the upper airway's allergic rhinitis [11–13]. IL-4 and IL-13, in combination with inflammatory markers, produce contraction of airway muscle, leading in bronchospasm, overproduction of mucus, and enhance influx of immune cell, leading in hyperreactivity of airway and reduction in airway dimension in the lower respiratory tract, reducing airflow. The airway epithelium has been discovered to have a significant function in regulating Th2 responses by generating master moderators such as IL-33, IL25, or thymic stromal lymphopoietin, which govern the production of Th2 mediators and induce asthma to develop early in childhood [14, 15]. Wheezing and airway hyperactivity to nonspecific stimulation characterise the early phases of asthma; nevertheless, later phases (extreme forms) of asthma contribute in airway remodelling with successive recurrences due to enhanced inflammation assisted by systemic variables or other local (infections of bacteria or virus) [16, 17].

### 1.2.2 COPD

Emphysema, small airway degradation, chronic bronchitis, and chronic asthma are all examples of COPD, which is a unified name for a set of progressing, disruptive, incurable lung diseases. As per the WHO's total prevalence of illness research, there were 251 million COPD patients worldwide, with 90% of them coming from low- and middle-income nations. In 2015, an estimated 3.17 million people died, accounting for 5% of all fatalities, a rise of 11.6% from 1990. In contrast to death, the Centers for Disease Control and Prevention projected a financial impact of USD $32.1 billion in 2010 for health expenses and missed work days due to COPD in the U.S., which is expected to rise to USD $49.0 billion by 2020 [18, 19]. Moreover, the actual figure of COPD cases in the world is already a debatable point, based on the reality that so many asthma incidences in the older people are frequently misdiagnosed as COPD, and the lack of information sets from underdeveloped or developing Middle Eastern and Asian countries, which could bring up the fatality score by several millions. Individuals with any type of COPD encounter a broad variety of complaints, the most common of which is dyspnoea or breathlessness throughout daily routines, which increases with time, whereas people with severe COPD have repeated complications and ER visits throughout the year. This is due to the partial alveoli destruction (emphysema) or the aggregation of inflammatory cells and large amount of mucus in bronchioles (chronic bronchitis), which reduces the gaseous exchange abilities of the lungs and induces blockage to the flow of air, resulting in hypoxemia and consequent failure of organ, particularly in cigarette smokers and exsmoker having to suffer from COPD. Furthermore, individuals with any form of COPD may experience typical symptoms such as persistent cough (dry or wet cough), fatigue, wheeze, and tightness of chest, such symptoms are frequently misinterpreted as age-related [20, 21]. Some people may not show symptoms until the disease has progressed to the point where it is life-threatening. After years of investigation, there is no treatment for regenerating destroyed tissue and restoring pulmonary functioning. Furthermore, COPD is a chronic condition that worsens with age. Current therapies are intended to halt the course of etiopathogenesis and give short term relief to patients, but they are incapable of recovering affected areas' impaired functionality [22].

Tobacco smoking was discovered to be the single largest prevalent risk factor for COPD, as per current databases. Lengthy contact to non-cigarette smoking irritants (e.g., airborne grit particulates, anthropogenic particulates, and metal pollutants) has been linked in aggravation (smokers) or the establishment (nonsmokers) of COPD, according to recent findings, which is still relatively understudied. Furthermore, research suggests the significance of genetic susceptibility, such as alpha-1-antitrypsin (AAT) insufficiency, in the progression of COPD; however, the specific fundamental processes remain unknown. AAT deficient individuals, on the other hand, are more susceptible to pulmonary infections, and hereditary factors account for just 1% of all COPD occurrences, underlining tobacco smoking and air pollution as important contributors [23, 24]. It is worth noting that not all smokers or exsmokers acquire the condition; about 20–30% of smokers or exsmokers suffer from the disease throughout the course of their lives. Passive smoking was even found among the risk variables for COPD (51.2%, $n = 87$) in the research; however, the specific fundamental processes are yet unknown. COPD is caused by a blockage of airflow and an inflow of inflammatory cells, particularly $CD8^+$ T lymphocytes, neutrophils, and macrophages, into the alveolar and peripheral regions as a result of cigarette smoke or atmospheric particulates/gases. The immune system's reaction to various types of COPD, though, was discovered to be varied [25, 26].

Over mucus generation, elevated inflammatory cells, increased MUC5AC gene expression in responding to secreted serine proteases, higher ROS levels from inhaled smoke, or triggered macrophages characterise chronic bronchitis, which inhibits the air space and causes destruction to adjacent cells, culminating in remodelling (fibrosis) of the respiratory tract and deterioration of pulmonary elasticity, while the emphysema is caused by cigarette smoke. Inhaled smoke increases inflammation in alveolar sacs and the bronchioles, resulting in narrower airway walls and the progressive deterioration of alveolar sacs, resulting in, function recoil, and alveolar structure loss [27, 28]. It is unclear how COPD patients' adaptive and innate immune defenses are activated. Moreover, immune cell MMP, IL8, and CXC overexpression as well as the mediators of proinflammatory secretion such as transforming growth factor beta (TGF-β), leukotriene B4, IL1 (Th1 responses), and TNF-α cause local fibrosis and a disequilibrium of oxidant-antioxidant proportion (ROS/RNS) and are thought to be important variables in disease worsening [29, 30].

### 1.2.3 Lung Cancer

Lung cancer-related fatality is primarily caused by late diagnosis and ineffective therapeutic approaches in 70% of lung carcinoma patients, who are usually in later stages of the illness (stage III or IV). It is a highly aggressive, quickly metastasizing cancer that affects both men and women. Lung cancer fatality is greater than the cumulative death rate of the other four main types of carcinoma in the U.S., according to statistics (pancreas, colon, breast, and prostate). Smoking histories of 20 years or more appear to be related with a higher risk of progression and death. Tobacco-induced lung carcinoma susceptibility is thought to be largely reliant on competing gene-enzyme connections at the level of procarcinogens, as well as the resulting amount of DNA destruction [31, 32]. As a result, lung carcinoma is thought to be usually avoidable through quitting smoking and prevention. To minimise the unavoidable growth in pulmonary malignancies in nations where smoking has risen, community awareness and support are necessary to limit or eliminate smoking tobacco. Lung carcinoma is the leading source of cancer associated mortality in both women and men throughout the world. According to a research by scientists, roughly 1.8 million new instances of lung carcinoma were diagnosed in 2012, account for 12.9% of all new cancer occurrences. As per the Global Burden of Disease Study 2020, lung carcinoma caused a significant amount of health impact and expenditure throughout the world. According to one investigation, men's cancer deaths are unrelated to their economic status. Interestingly, the research found that a country's economic progress level is linked to lung cancer deaths in women. Lung carcinoma has a complicated diversity due to its genesis in many sites in the bronchial tree and the varying manifestations of patient signs as well as indications depending on the kind and anatomic site [33, 34]. Lung tumour is conventionally divided into two types: non-small-cell lung carcinoma (NSCLC) (85% of all lung cancers) and small-cell lung carcinoma (SCLC) (15% of all lung malignancies). Giant cell carcinoma, squamous cell carcinoma, and adenocarcinoma are the three types of NSCLCs. Certain histology features and accurate immunohistochemical biomarkers were added to this classification of lung cancer, allowing for a convincing differentiation among preinvasive tumours and aggressive adenocarcinomas. Additionally, the development of molecular characterisation of lung tumours and the ever-expanding arsenal of effective treatments has had a significant impact on how lung carcinoma is categorised today. Even in the same histopathological subtype, findings suggest that lung carcinoma is a collection of molecularly and histologically diverse illnesses [35, 36].

### 1.2.4 Cystic Fibrosis

The cystic fibrosis (CF) is the greatest prevalent autosomal recessive illness, affecting around 1/3500 births. The majority of individuals show signs and symptoms at birth or shortly after delivery, with respiratory illnesses and low weight growth being the most common. Persistent pulmonary infections and pancreatic failure should lead to a diagnosis of CF. Before to CF newborn testing, however, a clinical odyssey with a sweat test generally followed the ultimate diagnosis. A sweat chloride content of more than 60 mmol/L is considered diagnostic for CF. High salt loss with perspiration and male sterility are two more common illness symptoms. Chronic pulmonary infections caused by particular microorganisms, as well as severe inflammation, can result to bronchiectasis, decreased pulmonary functioning, and finally pulmonary failure [37, 38]. *Pseudomonas aeruginosa* and *Staphylococcus aureus* are common CF pathogens, but some patients will develop infections with more uncommon and difficult-to-treat infectious organisms including *Stenotrophomonas maltophilia*, *Achromobacter xylosoxidans*, *Burkholderia cepacia,* and *Mycobacterium* further in the illness. Diseases can affect almost all organ and worsen with age, involving allergic bronchopulmonary aspergillosis, haemoptysis, gastrointestinal blockages, nasal polyps, CF-related hyperglycaemia, and liver illness, among others [39, 40].

Autosomal recessive illness is caused by mutations in the CF trans membrane conductance regulator (CFTR) gene, which is situated on the long arm of chromosome 7. The Cystic Fibrosis Mutation Database has found and published over 1400 individual variations, rendering population testing purely using genetic methods unfeasible. Although lung symptoms are the most common cause of morbidity and death, the typical CF phenotype is extremely complicated, encompassing numerous epithelium lined organs. In recent decades, substantial advancement has been achieved toward a better knowledge of the route that connects CFTR gene alterations to clinical symptoms of CF, especially the processes that underpin the obvious failure of lung defense [41, 42].

### 1.2.5 Idiopathic Pulmonary Fibrosis

Idiopathic pulmonary fibrosis (IPF) is amongst the very dangerous types of idiopathic intermittent pneumonias, with persistent increasing fibrosis, inexorable decrease in pulmonary functioning, increasing respiratory insufficiency, and a high death rate. Appropriate diagnosis is critical for prognosis and therapy choices. In North America and Europe, a comprehensive assessment of the worldwide prevalence of IPF indicated a rate of 2.8–9.3 per 100,000 annually, with substantially reduced rates in Asia and South America. Among nations, there is significant regional heterogeneity, which might be due to exposure to ecological or professional risk variables. Depending on previous statistics, IPF has a significant fatality rate, with a projected median survival of 2–3 years after diagnosis. Recent data suggests that survivability has not improved. Fatality rates seem to be growing as well, however this might be due to better detection and diagnosis [43, 44]. IPF is characterised by UIP, which is a histological marker. Temporal and regionally variable fibrosis, clustering of fibroblasts and myofibroblasts, and extensive accumulation of unorganised collagen and extracellular matrix, with or without honeycomb cyst development, are all characteristics. While the exact triggers for these activities are uncertain, present theories imply that IPF is the result of an abnormal healing mechanism in reaction to complex interplay among hosts and atmosphere. The "multiple strike theory" proposes that IPF is produced by the combination of a hereditary propensity to abnormal epithelial cell regulation and environmental stressors. Fibrosis is caused by the long-term effects of fibrotic diseases with a known aetiology and causes such as asbestos, immune complexes, medications, or radiation ingested [45, 46].

### 1.2.6 Tuberculosis

The acid-fast bacterial strains M. tuberculosis causes TB in humans, with the animal-adapted strain M. bovis accounting for a lower number of zoonotic cases (143,000 in 2018). M. tuberculosis is extremely infectious when transferred in aerosols from the airways of patients with active TB by coughing, spitting, or sneezing, despite missing many of the traditional pathogenic elements seen in other pathogenic bacteria such as exotoxins. The pathogenic organism inhalation into the alveoli of the lower airways occurs when a susceptible individual is exposed to droplets harbouring the bacteria. Local macrophages, which are typically the first immune cells to interact with M. tuberculosis in the lungs, internalise the bacteria [47]. Inhibition of a set of pathogenic genes in the bacteria leads in a loss of virulence in an experimental model of tuberculosis, but not a reduction of mycobacterial proliferation in the lack of stress or famine under optimal in vitro circumstances. Protein kinases, metal transporter, proteases, gene controllers, macrophage activity inhibitors, cellular membrane proteins, lipids metabolism enzymes, and proteins of unknown activity, such as PE and PE PGRS proteins, are among the pathogenic determinants of M. tuberculosis. M. tuberculosis can resist RNS and ROS activity, as well as lysosomal fusion and phagosome acidification, after being phagocytosed by alveolar macrophages [48]. These activities are important for the pathogen's survival in the host in latent TB, as well as for bacterial multiplication, tissue dispersion, and destruction in active TB patients, as well as downstream person-to-person spread. The parenchymal destruction, traction bronchiectasis, bronchostenosis, cavitation, and fibrosis are examples of architectural lung destruction that can occur with respiratory TB [49, 50].

### 1.2.7 Influenza A Virus Infection

Influenza is a contagious respiratory illness caused by the influenza A and influenza B viruses in humans. The Centers for Disease Control and Prevention (CDC) estimates that influenza virus infection caused 9.2 million to 35.6 million infections and 140,000–710,000 hospitalised in the U. S. among 2010 and 2017. Influenza A viruses produce pandemic seasonal illnesses that kill around 500,000 people each year throughout the world, according to the most current estimates of 291,243–645,832 fatalities each year. Clinical signs of influenza virus infection range from a mild upper respiratory infection with tiredness, muscle aches, headache, coughing, runny nose, sore throat, and fever to serious and, in some instances, lethal pneumonia caused by the influenza virus or secondary bacterial infection of the lower airway [51, 52]. In certain circumstances, influenza virus infection can cause a variety of non-respiratory problems, including heart, central nervous system, and other organ systems. While yearly seasonal epidemics are the norm, rare and unexpected worldwide pandemic outbreak involving nonhuman influenza A virus subtypes do happen. Every 10–50 years, a pandemic influenza outbreak occurs, defined by the addition of a new influenza strain. A viral strain that is antigenically distinct from formerly circulating strains; in humans, the absence of pre-existing protection is frequently linked to the intensity of illness and increased fatality [53].

Human influenza viruses are spread by the pulmonary route, but avian influenza viruses are spread via the faecal–respiratory pathways, faecal–oral, faecal–faecal, or in wild birds. Based on the mode of propagation, the virus infects and replicates in epithelial cells of the pulmonary or digestive tract. Furthermore, human infections of the eye and conjunctivitis have been linked to several avian influenza A viruses, particularly those of the H7 subtype (inflammation of the conjunctiva). In humans, the intensity of infection is linked to viral multiplication in the lower airways, which is followed by significant inflammation caused by immune cell infiltration [54, 55].

### 1.2.8 COVID-19

A new coronavirus known as SARS-CoV-2 appeared in the Chinese city of Wuhan at the end of 2019 and triggered an epidemic of atypical viral pneumonia. This new coronavirus illness, also known as COVID-19, has spread rapidly throughout the globe due to its high transmissibility. In regard of both the numbers of sick persons and the geographic scope of epidemic locations, it has massively exceeded MERS and SARS. COVID-19 is still spreading across the world, posing a serious risk to human health [56, 57]. The first individual with SARS-CoV-2 infection was detected with pneumonia of unknown aetiology, with signs identical to infections of SARS-CoV and MERS-CoV and was hospitalised and died. Additionally, patients who required ICU admission had greater TNF-α, MIP-1A, MCP-1, IP-10, and G-CSF, levels than others who did not, suggesting that the cytokine outburst was connected to illness intensity [58–60].

SARS-CoV-2 infection tends to affect people of various ages, with the average age of infection being about 50 years old. Clinical symptoms, on the other hand, vary with age. Most young individuals and adolescents have relatively minor illnesses (moderate pneumonia or non-pneumonia) or are asymptomatic, but elderly men (>60 years old) with co-morbidities are more prone to suffer serious lung infections that needs hospitalisation or even death. Pregnant women did not have a greater risk of illness than non-pregnant women [61, 62]. But it was an isolated incident, confirmation of SARS-CoV-2 transplacental transfer from an affected mother to a newborn was described. Fever, tiredness, and a dry coughing are the main typical signs of infection. In investigations of patients in China, less typical signs include chest discomfort, diarrhoea, vomiting, nausea, haemoptysis, headaches, sore throat, sputum secretion, hunger, and fever. Patients in Italy also experienced self-reported olfactory and taste abnormalities. Following an incubation period of 1–14 days (most often around 5 days), most patients displayed symptoms of sickness, pneumonia, and difficulties in breathing occurred within a median of 8 days following disease start [63, 64]. COVID-19 indications that are serious, like acute respiratory distress syndrome (ARDS) and severe pneumonia, are connected to the virus's activation and secretion of cytokines and chemokines, which results in a "cytokine storm" which induces destruction and inflammation, especially in the lungs. COVID-19 produces IL-1, LTs, IL-2, TNF-α, IL-6, GM-CSF, IL-12, and other chemokines due to NF-κB expression in many cells such as the gastrointestinal system, lungs, kidney, liver, central nervous system, and cardiovascular system. Risks associated with T, such as a greater death rate [65, 66].

### 1.2.9 Conclusion

Lung diseases have a huge impact worldwide and are becoming more common due to the ageing population and the lack of appropriate interventions to minimise the risk factors that contribute to the progression of these diseases. Asthma, COPD, fibrosis, COVID-19, and influenza-like lung diseases are becoming life-threatening and dangerous, so effective treatment approaches and adequate prevention are becoming a challenge for researchers.

## References

1. Alba MA, Flores-Suárez LF, Henderson AG, Xiao H, Hu P, Nachman PH et al (2017) Interstitial lung disease in ANCA vasculitis. Autoimmun Rev 16(7):722–729
2. Albtoush OM, Al-Mnayyis A, Christopher K, Werner S, Jürgen H, Horger M (2018) [Differential diagnosis of cystic lung diseases]. Rofo 190(12):1103–1107
3. Ali MS, Ghori UK, Musani AI (2019) Orphan lung diseases. Med Clin North Am 103(3): 503–515
4. Azadeh N, Moua T, Baqir M, Ryu JH (2018) Treatment of acute exacerbations of interstitial lung disease. Expert Rev Respir Med 12(4):309–313
5. Barba T, Mainbourg S, Nasser M, Lega JC, Cottin V (2019) Lung diseases in inflammatory myopathies. Semin Respir Crit Care Med 40(2):255–270
6. Bittmann I (2021) [Drug-induced interstitial lung diseases]. Pathologe 42(1):11–16
7. Boateng E, Krauss-Etschmann S (2020) miRNAs in lung development and diseases. Int J Mol Sci. 21(8):2765
8. Britto CJ, Brady V, Lee S, Dela Cruz CS (2017) Respiratory viral infections in chronic lung diseases. Clin Chest Med 38(1):87–96
9. Chen J, Jin Y, Yang Y, Wu Z, Wu G (2020) Epithelial dysfunction in lung diseases: effects of amino acids and potential mechanisms. Adv Exp Med Biol 1265:57–70
10. Cinetto F, Scarpa R, Rattazzi M, Agostini C (2018) The broad spectrum of lung diseases in primary antibody deficiencies. Eur Respir Rev. 27(149):180019
11. Collins BF, Raghu G (2019) Antifibrotic therapy for fibrotic lung disease beyond idiopathic pulmonary fibrosis. Eur Respir Rev. 28(153):190022
12. Cottin V (2016) Eosinophilic lung diseases. Clin Chest Med 37(3):535–556
13. Cottin V, Hirani NA, Hotchkin DL, Nambiar AM, Ogura T, Otaola M et al (2018) Presentation, diagnosis and clinical course of the spectrum of progressive-fibrosing interstitial lung diseases. Eur Respir Rev 27(150):180076
14. Cottin V, Valenzuela C (2020) Diagnostic approach of fibrosing interstitial lung diseases of unknown origin. Presse Med 49(2):104021
15. Cottin V, Wollin L, Fischer A, Quaresma M, Stowasser S, Harari S (2019) Fibrosing interstitial lung diseases: knowns and unknowns. Eur Res Rev. 28(151):180100
16. Dawod YT, Cook NE, Graham WB, Madhani-Lovely F, Thao C (2020) Smoking-associated interstitial lung disease: update and review. Expert Rev Respir Med 14(8):825–834
17. Fazen LE, Linde B, Redlich CA (2020) Occupational lung diseases in the 21st century: the changing landscape and future challenges. Curr Opin Pulm Med 26(2):142–148
18. Ferro F, Delle SA (2018) The use of ultrasound for assessing interstitial lung involvement in connective tissue diseases. Clin Exp Rheumatol 36 Suppl 114(5):165–170
19. George PM, Spagnolo P, Kreuter M, Altinisik G, Bonifazi M, Martinez FJ et al (2020) Progressive fibrosing interstitial lung disease: clinical uncertainties, consensus recommendations, and research priorities. Lancet Respir Med 8(9):925–934
20. Griese M (2018) Chronic interstitial lung disease in children. Eur Respir Rev. 27(147):170100
21. Gupta N, Ryu JH (2020) Controversies and evolving concepts in orphan lung diseases. Semin Respir Crit Care Med 41(2):175–176
22. Ha YJ, Lee YJ, Kang EH (2018) Lung involvements in rheumatic diseases: update on the epidemiology, pathogenesis, clinical features, and treatment. Biomed Res Int 2018:6930297
23. Harris EJA, Musk A, de Klerk N, Reid A, Franklin P, Brims FJH (2019) Diagnosis of asbestos-related lung diseases. Expert Rev Respir Med 13(3):241–249
24. Hoy RF, Brims F (2017) Occupational lung diseases in Australia. Med J Aust 207(10):443–448

25. Jacobs K, Kligerman S (2019) Immune-Mediated Lung Diseases. Semin Ultrasound CT MR 40(3):213–228
26. Jiang J, Xiao K, Chen P (2017) NOTCH signaling in lung diseases. Exp Lung Res 43(4–5): 217–228
27. Kanne JP (2019) Smoking-related diffuse lung diseases. Semin Roentgenol 54(1):30–36
28. Khateeb J, Fuchs E, Khamaisi M (2019) Diabetes and lung disease: a neglected relationship. Rev Diabet Stud 15:1–15
29. Khoor A, Colby TV (2017) Amyloidosis of the lung. Arch Pathol Lab Med 141(2):247–254
30. Kolb M, Bondue B, Pesci A, Miyazaki Y, Song JW, Bhatt NY et al (2018) Acute exacerbations of progressive-fibrosing interstitial lung diseases. Eur Respir Rev. 27(150):180071
31. Kolb M, Vašáková M (2019) The natural history of progressive fibrosing interstitial lung diseases. Respir Res 20(1):57
32. Kumar A, Cherian SV, Vassallo R, Yi ES, Ryu JH (2018) Current concepts in pathogenesis, diagnosis, and management of smoking-related interstitial lung diseases. Chest 154(2):394–408
33. Kylhammar D, Rådegran G (2017) [Pulmonary hypertension due to lung diseases]. Lakartidningen 114:562–569
34. Lahousse L (2019) Epigenetic targets for lung diseases. EBioMedicine 43:24–25
35. Latshang TD, Schoch OD (2017) [Lung diseases and altitude mountaineering]. Ther Umsch 74(10):555–562
36. Lenga Ma Bonda W, Iochmann S, Magnen M, Courty Y, Reverdiau P (2018) Kallikrein-related peptidases in lung diseases. Biol Chem 399(9):959–971
37. Liu PP, Yang SN, Dai HP, Wang C (2020) [The role of exosome in the lung diseases]. Zhonghua Jie He He Hu Xi Za Zhi 43(8):692–7
38. Liu Y, Gao H, Wang X, Zeng Y (2020) Methylation of inflammatory cells in lung diseases. Adv Exp Med Biol 1255:63–72
39. Maher TM, Wuyts W (2019) Management of fibrosing interstitial lung diseases. Adv Ther 36(7):1518–1531
40. Margaritopoulos GA, Antoniou KM, Wells AU (2017) Comorbidities in interstitial lung diseases. Eur Respir Rev. 26(143):160027
41. Mira-Avendano I, Abril A, Burger CD, Dellaripa PF, Fischer A, Gotway MB et al (2019) Interstitial lung disease and other pulmonary manifestations in connective tissue diseases. Mayo Clin Proc 94(2):309–325
42. Montesi SB, Fisher JH, Martinez FJ, Selman M, Pardo A, Johannson KA (2020) Update in interstitial lung disease 2019. Am J Respir Crit Care Med 202(4):500–507
43. Nathan N, Berdah L, Borensztajn K, Clement A (2018) Chronic interstitial lung diseases in children: diagnosis approaches. Expert Rev Respir Med 12(12):1051–1060
44. Nathan N, Berdah L, Delestrain C, Sileo C, Clement A (2020) Interstitial lung diseases in children. Presse Med. 49(2):103909
45. Nogee LM (2017) Interstitial lung disease in newborns. Semin Fetal Neonatal Med 22(4): 227–233
46. Olson AL, Gifford AH, Inase N, Fernández Pérez ER, Suda T (2018) The epidemiology of idiopathic pulmonary fibrosis and interstitial lung diseases at risk of a progressive-fibrosing phenotype. Eur Respir Rev. 27(150):180077
47. Palla J, Sockrider MM (2019) Congenital lung malformations. Pediatr Ann 48(4):e169–ee74
48. Perelas A, Silver RM, Arrossi AV, Highland KB (2020) Systemic sclerosis-associated interstitial lung disease. Lancet Respir Med 8(3):304–320
49. Perlman DM, Maier LA (2019) Occupational lung disease. Med Clin North Am 103(3): 535–548
50. Ray A, Jaiswal A, Dutta J, Singh S, Mabalirajan U (2020) A looming role of mitochondrial calcium in dictating the lung epithelial integrity and pathophysiology of lung diseases. Mitochondrion 55:111–121
51. Reinero C (2019) Interstitial lung diseases in dogs and cats part II: known cause and other discrete forms. Vet J. 243:55–64

52. Richeldi L, Varone F, Bergna M, de Andrade J, Falk J, Hallowell R et al (2018) Pharmacological management of progressive-fibrosing interstitial lung diseases: a review of the current evidence. Eur Respir Rev. 27(150):180074
53. Rivera-Ortega P, Molina-Molina M (2019) Interstitial lung diseases in developing countries. Ann Glob Health. 85(1):4
54. Rout-Pitt N, Farrow N, Parsons D, Donnelley M (2018) Epithelial mesenchymal transition (EMT): a universal process in lung diseases with implications for cystic fibrosis pathophysiology. Respir Res 19(1):136
55. Salinas M, Florenzano M, Wolff V, Rodríguez JC, Valenzuela H, Fernández C et al (2019) [Update on interstitial lung diseases]. Rev Med Chil 147(11):1458–1467
56. Aljabali AA, Bakshi HA, Satija S, Metha M, Prasher P, Ennab RM et al (2020) COVID-19: underpinning research for detection, therapeutics, and vaccines development. Pharm Nanotechnol 8(4):323–353
57. Anand K, Vadivalagan C, Joseph JS, Singh SK, Gulati M, Shahbaaz M et al (2021) A novel nano therapeutic using convalescent plasma derived exosomal (CPExo) for COVID-19: a combined hyperactive immune modulation and diagnostics. Chem Biol Interact 344:109497
58. Khan T, Agnihotri K, Tripathi A, Mukherjee S, Agnihotri N, Gupta G (2020) COVID-19: a worldwide, zoonotic, pandemic outbreak. Altern Ther Health Med 26:56–64
59. Mehta M, Prasher P, Sharma M, Shastri MD, Khurana N, Vyas M et al (2020) Advanced drug delivery systems can assist in targeting coronavirus disease (COVID-19): a hypothesis. Med Hypotheses 144:110254
60. Prasher P, Sharma M, Gupta G, Chellappan DK, Dua K (2020) Are medicinal plants an alternative strategy to combat COVID-19? Altern Ther Health Med 26:92–93
61. Rawat S, Dhramshaktu IS, Pathak S, Singh SK, Singh H, Mishra A et al (2020) The impact of COVID-19 pandemic infection in patients admitted to the hospital for reasons other than COVID-19 infection. Altern Ther Health Med 26:108–111
62. Satija S, Mehta M, Sharma M, Prasher P, Gupta G, Chellappan DK et al (2020) Vesicular drug delivery systems as theranostics in COVID-19. Newlands Press Ltd, London
63. Shahcheraghi SH, Ayatollahi J, Aljabali AA, Shastri MD, Shukla SD, Chellappan DK et al (2021) An overview of vaccine development for COVID-19. Ther Deliv 12(3):235–244
64. Sharma M, Prasher P, Mehta M, Zacconi FC, Singh Y, Kapoor DN et al (2020) Probing 3CL protease: rationally designed chemical moieties for COVID-19. Drug Dev Res. https://doi.org/10.1002/ddr.21724
65. Singh Y, Gupta G, Mishra A, Chellappan DK, Dua K (2020) Gender and age differences reveal risk patterns in COVID-19 outbreak. Altern Ther Health Med 26:54–55
66. Sunkara K, Allam VR, Shukla SD, Chellappan DK, Gupta G, MacLoughlin R et al (2021) COVID-19 in underlying COPD patients. EXCLI J 20:248

# 2 Introduction to Microbiome

Shivkanya Fuloria, Vetriselvan Subramaniyan, Mahendran Sekar, Yuan Seng Wu, Srikumar Chakravarthi, Rusli Bin Nordin, Pradeep Kumar Sharma, Dhanalekshmi Unnikrishnan Meenakshi, Ajay Mendiratta, and Neeraj Kumar Fuloria

**Abstract**

The microbiome is the indigenous microbial population (microbiota) and the host environment in which it lives, and it is revolutionising how doctors think about germs in human health and illness. The understanding that most microbes in human bodies perform vital ecosystem functions that benefit the whole microbial host system is perhaps the most basic development. The microbiome is a

S. Fuloria · N. K. Fuloria (✉)
Faculty of Pharmacy & Centre of Excellence for Biomaterials Engineering, AIMST University, Bedong, Malaysia
e-mail: shivkanya_fuloria@aimst.edu.my; neerajkumar@aimst.edu.my

V. Subramaniyan · S. Chakravarthi · R. B. Nordin
Faculty of Medicine, Bioscience and Nursing, MAHSA University, Jenjarom, Selangor, Malaysia
e-mail: drvetriselvan@mahsa.edu.my; srikumar@mahsa.edu.my; rusli@mahsa.edu.my

M. Sekar
Department of Pharmaceutical Chemistry, Faculty of Pharmacy and Health Sciences, Royal College of Medicine Perak, Universiti Kuala Lumpur, Ipoh, Perak, Malaysia
e-mail: mahendransekar@unikl.edu.my

Y. S. Wu
Centre for Virus and Vaccine Research, & Department of Biological Sciences, School of Medical and Life Sciences, Sunway University, Subang Jaya, Selangor, Malaysia

P. K. Sharma
Accurate College of Pharmacy, Greater Noida, Uttar Pradesh, India

D. U. Meenakshi
College of Pharmacy, National University of Science and Technology, Muscat, Oman
e-mail: dhanalekshmi@nu.edu.om

A. Mendiratta
Department of Pharmacy, SMAS, Galgotias University, Greater Noida, India
e-mail: ajay.20smas3020002@galgotiasuniversity.edu.in



G. Gupta et al. (eds.), *Microbiome in Inflammatory Lung Diseases*,
https://doi.org/10.1007/978-981-16-8957-4_2

collection of varied and numerous bacteria that live in the gastrointestinal system. Generally, this ecosystem comprises billions of microbial cells that play a vital role in human health control. Immunity, nutrition absorption, digestion, and metabolism have all been linked to the microbiome. Researchers have discovered that changes in the microbiome are linked to the development of diseases including obesity, inflammatory lung disease, and CVS diseases, carcinoma in recent times. A change in the microbial population of the intestine has a big impact on human health and disease aetiology. These changes are caused by a combination of factors, including lifestyle and the existence of an underlying illness. Dysbiosis makes the host more susceptible to infection, the type of which varies depending on the anatomical location. The distinct metabolic processes and roles of these bacteria inside each bodily location are accounted for by the inherent variety of the human microbiota. As a result, it is critical to comprehend the human microbiome's microbial makeup and behaviours as they relate to health and illness.

**Keywords**

Microbiome · Microbiota · Pathogens · Factors · Liver · Lung

## 2.1 Introduction

The name "microbiome" is not totally new: it combines the word "microbe" with the suffix "-ome", a latinized form of the Greek suffix "-" that means "total", "aggregate", or "collective action". As a result, the microbiome is the total of all microorganisms that share a common point, generally a common place/habitat throughout a given time frame, as described in a long-forgotten microbiome vision. This is a unique microbial communities that lives in a rather well-defined environment and has specialised physio-chemical characteristics. As a result, the word not only relates to the microbes engaged, but also to the environment in which they operate [1, 2]. The term "microbiome" has been used to characterise the combined genome of our endogenous microorganisms (microflora), with the notion that a full genetic perspective of Homo sapiens as an existence must encompass the genes in our microbiome [3–5]. Besides being woefully stringent because it focusses on colonising microbiota and, worse yet, colonisers solely of Homo sapiens, and confusingly unidisciplinary because it only contains a genetic/genomic component, this perspective is also both ambiguous and incorrect in its qualitative aspect. Because the suffix "-ome" refers to collectivity and completeness, there is no letter in the term "microbiome" to witness to the genetic portion of "genome".

The gut microbiome is made up of the cumulative genomes of bacterium, fungus, viruses, and archaea that live in the gut. The amount of information available on the microorganisms that live in our intestines is rapidly increasing. Until recently, the different allelic variants of genes were blamed for human population heterogeneity. The human gut is home to trillions of bacteria, each with a genome larger than all of

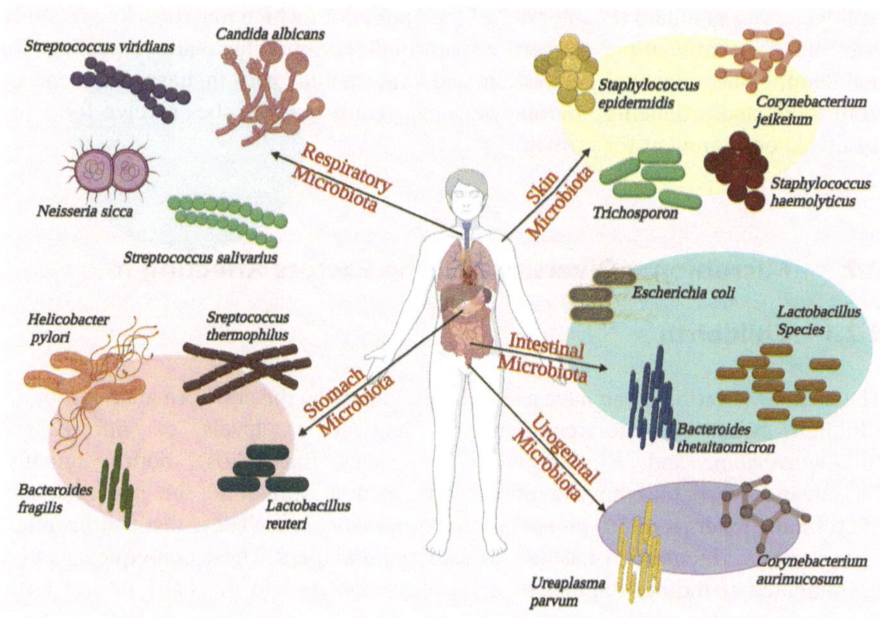


**Fig. 2.1** Human Microbiota

the human cells in the body. In the gut, bacteria are distributed spatially, with the colon having the greatest variety and abundance [6, 7]. Due to its closeness to the environment, the colon contains higher aerobes than the small intestine. It is proven tough to cultivate commensals since they are anaerobic, especially in the upper intestine. Advances in omics-based methods have contributed to a better knowledge of the gut ecology and the numerous variables that influence its microbial composition. This technique has paved the way for a slew of new studies on the role of gut microbiota in immunological equilibrium, which has ramifications for both wellness and illness. The National Institutes of Health (NIH) launched the human microbiome study in 2007, which discovered bacteria on the human body's different surfaces. *Actinobacteria*, *Proteobacteria*, *Bacteroidetes*, and *Firmicutes* are the most common phyla in humans, with *Bacteroidetes* and *Firmicutes* dominating the gut. The gut colonisation process begins at birth and is influenced by whether the baby is delivered vaginally or through C-section [8–10]. Through development, the microbiota, on the other hand, alters as a result of several environmental variables. Each person's microbiota is distinct, much like their genetic fingerprint, however around a third of the species are shared by most people (Fig. 2.1) [11, 12].

Gut microbiota is influenced by a variety of variables, such as cleanliness, nutrition, geographical region, and human genotype. Furthermore, researches in people and animals have shown that sex hormones and age have a role in defining the microbial makeup of the intestinal tract. Commensal bacteria and humans have coevolved and share a symbiotic association. Intestinal microbes out-compete the pathogens and maintain the integrity of the epithelium which may be a key factor in preventing inflammation. Diverse microbial communities are essential for maintaining the intestinal ecosystem and play a vital role in harvesting energy from foods and producing micronutrients. In return the microbes receive food and a suitable environment for growth [13, 14].

## 2.2 Microbiome Diversity and the Factors Affecting It

### 2.2.1 Childbirth

The method of birthing can have a significant impact on the newborn gut microbiota. Children born by caesarean surgery had lower levels of *Bacteroides*, *Bifidobacterium*, and *Escherichia coli* species than those born vaginally. Caesarean-born babies have a microbiome that is similar to the maternal skin microbiome in terms of *Streptococcus* and *Staphylococcus*. These distinctions relate to an increased incidence of asthma and allergy disorders. These consequences may be mitigated if mother vaginal bacteria are transferred to the baby during birth. Preterm babies have been linked to changes in the gut microbiota, but not to atopic sensitisation [15–17].

### 2.2.2 Metabolic Components of the Microbiota and Diet

Infant nutrition, particularly breastfeeding, was being found to enhance colonisation by *Lactobacilli* and *Bifidobacteria*, which is another important variable determining gut microbiome biodiversity. Maternal milk includes oligosaccharides and a variety of fatty acids, which influence the gut microbiota and its ability to generate metabolites that safeguard versus asthma and allergies by promoting the formation of Treg cells. This impact is also caused by raw milk consumption in the first year of birth, which is most likely due to increased amounts of proteins in the serum component as well as unsaturated omega-3 fatty acids. Polyphenols and fish oils, among other nutritional ingredients, are crucial for microbiome diversity [18, 19]. By fermenting different nutritional polysaccharides, bacteria like *Ruminococcaceae* and *Lachnospiraceae* can impact the gut microbiota by generating SCFAs like acetate, butyrate, and propionate. Notably, these acids have anti-inflammatory effects on the immune system that can attenuate FA and lung diseases by enhancing epithelial barrier purpose and triggering DC analogues, Treg cells, and synthesis of IL-10, in addition to acting as a vital source of energy for bowel colonocytes [20, 21].

### 2.2.3 The Significance of the Microbiome in Childhood

Early childhood exposure is critical for the microbiota, and there is mounting evidence that gut microbiota dysbiosis has a significant impact on immune system functioning. Perinatal food intake of the mother or infant, antibiotic usage, and interaction with older siblings are all possible causes. The highest inter-individual microbial diversity is seen in the first 3 years of life, according to evidence from various populations. The presence of bacteria in tiny levels in the meconium and placenta implies that interactions with the microbiome may start before birth. Bacterial exposure in the first months of life can trigger the innate immune system in a number of ways, affecting FA. Early vaccination with class IV and XIV Clostridium spores and other bacteria will reduce the IgE levels circulating in adulthood [22, 23]. In contrast, 3-week-old babies with a greater *Clostridium difficile* faecal load and a higher *Bifidobacterium* to *C. difficile* ratio had a higher number of skin test positive responses to aeroallergens and food. Similarly, high levels of faecal *E. coli* in babies' faeces are linked to IgE-mediated eczema in their first month. Surprisingly, the same colonisation pattern might have various outcomes depending on the age. Colonisation of *M. catarrhalis*, *H. influenzae*, or *S. pneumoniae* in the first month of birth, for example, raises the chance of asthma by causing a rise in atopic indicators such as serum IgE and eosinophils, but not at 12 months [24, 25].

### 2.2.4 The Significance of Antibiotic Exposure

The use of antibiotics in the 1950s has been linked to an increase in allergy cases. This is considered to be caused by antibiotics causing dysbiosis, which has been linked to the onset of asthma and Alzheimer's disease. Because maternal antibiotic use during gestation enhances the incidence of allergies in children, and antibiotics usage in the first month of birth has been linked to cow's milk allergies, the age of early introduction may be crucial. Antibiotics used intrapartum have been found to alter the microbiota of children aged 3–12 months. Antibiotics have been shown to have an impact on the microbiota of elderly people, according to research. In newborns, antibiotic treatment is linked to chronic allergic airway inflammation, but not in adults [26]. Antibiotics, even at modest dosages, can alter microbiome composition; nevertheless, the links among antibiotic use and allergy disorders get stronger as the number of antibiotics administered increases, and various drug families have distinct impacts. According to several research, beta-lactam antibiotics are largest prevalent cause of FA in children under the age of two, but macrolides are linked to FA in children beyond the age of two. In the case of asthma, additional research is needed to determine if the infection or the medicines themselves enhances susceptibility [27, 28].

## 2.3 Disease and the Human Microbiome

### 2.3.1 Infectious Diseases

In medical college, C difficile infection was presented as an example of a disease in which disturbance of the normal microbiota serves a significant part in pathogenesis. While the link among antibiotic use and the establishment of C difficile infection has long been recognised, more recent research has begun to identify the processes that underpin it. Substantial research was done on the microbial activities expressed by the indigenous microbiota that regulate C difficile colonisation resistance. The function of the gut microbiota in bile acid and bile salt metabolism is one field where research is ongoing. Microbes that can execute de-conjugation and transformation processes transform entangled bile salts into unconjugated main and secondary bile acids when they are released by the liver into the gastrointestinal tract. Some of these molecular species help C difficile spores germinate, while others stop the organism's vegetative form from growing. The discovery of innovative therapies has resulted from a better knowledge of molecular mechanisms [29, 30]. Microbiota replacement treatment, particularly faecal microbiota transplant, is an important field of research since this pathophysiology of C difficile infection, particularly persistent infection, is linked to a lack of natural microbial communities and functionality. Numerous additional infections and inflammatory disorders can be influenced by the gut microbiota. The microbiota condition of patients receiving allogeneic transplant of stem cell is linked to the likelihood of acquiring bacteraemia. The presence of gastrointestinal microorganisms in the airways of individuals with sepsis and ARDS appears to promote the lung chronic inflammation. Ultimately, the gut microbiota's makeup may have a significant impact on the repair of surgical intestinal anastomoses. These findings might affect therapy, as well as prognosis and outcome [31, 32].

### 2.3.2 Cardiovascular (CVS) Diseases

The gut microbiota produces compounds that not only impact the gut but also have systemic effects. Some GI bacteria may produce trimethylamine N-oxide (TMAO) metabolites, which may be linked to heart disease. Liver flavins comprising monooxygenase help the gut microbiota convert trimethylamine from phosphatidylcholine, choline, and l-carnitine-rich diets to TMAO. In experimental animals, TMAO disrupts lipid transport and causes the secretion of progenitors that induce foam cell development and artery stiffness. It has been revealed that intestinal dysbiosis is linked to CVS disease. A clinical research was conducted on two groups of people: those who had a low risk of CVS disease and those who had a high risk of CVS disease. A disturbed gut flora was linked to a greater risk of CVS disease, according to their results. CVS illnesses have been linked to the predominance of particular species. In germ-free mice animal studies, faeces transplantation from hypertension patients with overexpressed *Klebsiella* and *Prevotella* raised blood pressure. Additionally, the *Firmicutes* to *Bacteroidetes* ratio in the faeces microbiota of hypertensive mice increased significantly [33–35].

### 2.3.3 Metabolic Disease and Obesity

The intricate metabolic interaction among the endogenous microbiota of the intestine and the host has prompted researchers to look into the microbiome's possible involvement in metabolic disorders including hyperglycaemia and obesity. Decade ago, ground-breaking research found a link among obesity and the gut microbiota in both human and animal illness models. The use of leptin defective mice in a study looking at the function of the microbiota and obesity showed the strong relationship among host and microbial variables in the complicated pathophysiology of diseases like obesity. Although these researches a complete grasp of the processes that underpin this link remains difficult. Moreover, according to a recent meta-analysis of numerous research, the direct link among the microbiome and obesity might be less than formerly thought. Whatever the magnitude of the impact, it is apparent that the microbiota may alter the digestive tract's nutrition processing [36, 37]. The expression of key metabolic regulating peptides like glucagon-like peptide YY and peptide 1 can be influenced by microbially generated products including bile acids and short chain fatty acids. Some of the ways through which the microbiota might impact human energy metabolism have been elicited in recent research. Other research has found that altering the host diet has an impact on the gut microbiota, establishing a complicated system in which extrinsic and intrinsic microbiome connections can affect host metabolism. How accidental modification of the microbiota—for example, by antimicrobial therapy disturb the natural equilibrium and tilt towards the genesis of the metabolic syndrome and obesity is an intriguing area of study that has attracted a lot of interest. The impacts of microbial metabolism on different organ systems have been studied recently [38, 39].

### 2.3.4 Cancer

The gut microbiota has a big influence on its host's health. According to research on the interactions between microbial populations and their hosts, these organisms engage in biochemical processes that influence tumorigenesis, neoplasm growth, and immune treatment reaction. Persistent intra-abdominal infections, antibiotic medications, or both may raise the colorectal cancer risk, based on a well-studied model on variables that could promote to dysbiosis in the gut. In addition, final products produced by the gut microbiota have an effect on intestinal cell coverage, either promoting or inhibiting tumorigenesis. Besides from colorectal cancer, the microbiome of the intestine has been demonstrated to have a contribution in extra - intestinal cancers like hepatic carcinoma by allowing organisms to spread to other regions of the body. H. pylori also increases the risk of stomach cancer in humans. *Clostridium* and *Fusobacterium* are highly represented in people with stomach cancer, according to new research on the human microbiota and cancer [40, 41]. In the context of breast cancer, ecological and host variables have a direct impact on the disease's development. Bacterial populations, on the other hand, have the potential to cause breast cancer. When compared to healthy people, those with breast cancer have more *Staphylococcus*, *Enterobacteriaceae,* and *Bacillus* bacteria in their breast tissue. In addition, bacteria such as *Staphylococcus* and *Escherichia coli* epidermidis isolated from cancer patients caused a double-stranded break in HeLa cells' DNA. *Lactobacillus* spp., which has a variety of health advantages, was not identified in the breast tissue of breast cancer patients. A greater abundance of *Bacteroides massiliensis* has been linked to prostate cancer. The intricate connections among cancer and the human microbiota have been attributed to a change in the human microbiome [42, 43].

### 2.3.5 Lung Disease

The investigation of microbial populations has sparked a re-examination of previously thought-to-be microbe-free areas, like the upper and lower pulmonary tracts. Despite the fact that the lungs were once believed to be a sterile environment, the use of culture-independent techniques shows that the lungs are home to a small biomass of microorganisms that are rather varied. Initial research questioned the significance of this tiny community of bacteria in a healthy lung, however most current research suggests that the makeup of the lung microbiota might influence baseline inflammatory tone in healthy persons [44, 45]. In addition, it is known that microbial populations are prevalent and physiologically relevant in particular pulmonary illness conditions. Although it has generally been established that several cystic fibrosis patients get persistently colonised by harmful organisms, the airways of these patients have lately been discovered to contain a far more varied ecosystem than initially assumed. Although the significance of this discovery for the aetiology of lung illness and cystic fibrosis has yet to be determined, it is fair to suppose that microbe–microbe interaction in this milieu is just as significant as those in the GI tract. The significance of microbial populations in the development of lung illnesses including chronic obstructive pulmonary disease (COPD) and asthma is a hot topic of research right now. Several of the previous research indicate correlation instead of causation, however most current research is looking at how the lung microbiome could promote the inflammatory processes that are so important in COPD pathogenesis. Further research is anticipated to shed more light on the function of changed microbial populations in the development of various lung illnesses [46, 47]. Polymicrobial relationships in chronic and acute rhinosinusitis have been studied in the upper airway. Pathogens and other microorganisms have been studied in respect of their capacity to alter host physiology, much as they have been in the lower airways. Some microorganisms are known to be rich in sinusitis. According to previous research, those with sinusitis showed Corynebacterium pseudotuberculosis. The possible pathogenicity of this bacterium has been demonstrated in an animal model of sinusitis. Other members of the native sinus population modulate resistance to colonisation by this organism, according to a study of the native microbiome of the upper airway in individuals either with or without sinus illness [48, 49].

### 2.3.6 Allergic Diseases

It has been discovered that the human microbiota may have a role in allergy disorders. However, very less is known about the impact of the pulmonary microbiome on pulmonary tract immune modulation. A healthy microbiota, on the other hand, influences the mucosa of the lungs and shapes the pulmonary tract. By microaspiration, a dysbiotic flora immediately impacts the microbiota of the lungs, increasing the development of pulmonary illnesses in people. This was demonstrated in germ-free mice by the researchers. The absence of an immunological regulation system in experimental mice resulted in pulmonary and allergy disorders. Caesarean (CS) delivery of neonates has also been identified as a risk factor for allergic diseases [50, 51]. Children are predisposed to such illnesses because of the absence of normal mother flora during CS. Children who get CS have reduced numbers of good flora (Bacteroidetes) in their stomach, according to molecular research. This decreases Bacteroidetes' anti-inflammatory activity and promotes to local tissue inflammation (allergic rhinitis and asthma) caused by environmental and genetic factors. Current empirical studies found a link among the generation of allergy antigen and dysbiotic gut flora in children, leading to airway illness [52, 53]. At the age of four, children with reduced *Faecalibacterium*, *Akkermansia*, and *Bifidobacterium* microbiota diversity were more vulnerable to various allergen pulmonary hypersensitivity, which may relate to asthma [54, 55]. Mice that were not exposed to germs were more vulnerable to allergic airway illness. The susceptibility was restored after microbial colonisation, and there was a reduction in allergen sensitivity. Clinical investigations of allergy prevalence in Europe revealed that farming settings with varied microbial ecosystems had a reduced occurrence of airway allergies. The stimulation of the innate immune response in pulmonary epithelial cells has been related to the cause of this condition. Contact to agricultural dust including *Lactococcus lactis* G121 and *Acinetobacter lwoffii* F78 microbes diversity has been shown to decrease pulmonary inflammation in mouse [56, 57].

## 2.4 The Microbiome as a Therapeutic Target

The microbiota could perform a contribution in a number of illnesses, such as when a microbiome lacks a useful function or when harmful microbial activity is present. As a result, it is appealing to believe that restoring a favourable microbe function and structure may be a unique therapy for some illnesses. To achieve this, a number of different techniques have been presented. While the effectiveness of this innovative technique to illness protection and management has been limited to a few diseases and treatments to yet, the potential of what the future may contain justifies a consideration of what the future may hold [58, 59].

### 2.4.1 Antibiotics

Antibiotic-mediated alterations in microbial flora modify the disease-related microbiota and can assist recover health, but ancillary disruption to indigenous gut flora produced by therapeutic antibiotics serves a significant part in the genesis of C difficile infection. This technique was employed long before the microbiome became a popular topic of discussion. Antibiotic therapy studies, for example, were used to treat pouchitis, irritable bowel syndrome, and liver encephalopathy in individuals who had a colon resection for ulcerative colitis. It was thought that there were no hidden typical microbial pathogens involved in such initial treatment trials. Bacterial overpopulation or microbiota imbalance were suspected, and the antibiotic was administered in the hopes of correcting the problem. This method has the apparent drawback of being empirical in nature. We still do not know how a specific antibiotic regimen will affect a particular microbial population [60, 61]. For the avoidance of repeated C difficile infection, a variation of this strategy has been recommended. Some of the most modern antibiotics produced for the therapy of C difficile infection are meant to be more tightly limited to the pathogen with the goal of reducing collateral harm to the endogenous microbiome, which is linked with recurring illness. Fidaxomicin, which has a lesser propensity for altering the microbiome, has similarly been linked to a decreased risk of recurring illness while retaining strong pathogen effectiveness. This technique is confined to treating C difficile infection, however when managing a known bacterial disease, the use of wide range antibiotics must be restricted to preserve the microbiome. As a result, proper antibiotic management can assist to restrict the growth or choice of antibiotic-resistant organisms while also protecting the native microbiome [62, 63]. Using bacteriophage treatments to treat infections is another option that is considered to have little impact on the microbiome. Bacteriophage treatments have been designed to treat particular bacterial infections, and they are improbable to have off-target impacts on other microbiome members due to the unique characteristics. While bacteriophage is recognised to choose for resistant bacterial variations, these resistant bacteria frequently have changed surface features that, in addition to phage resistance, reduce pathogenicity in the host. Although much more research is required until bacteriophage can be turned into therapeutic agents, there is a surge of attention in developing new treatments to reduce microbiome disruption [64].

### 2.4.2 Microbial Biotherapies and Probiotics

Since most microbiota-related diseases are considered to be caused by a lack of helpful organisms, replacing "missing" microbiome components is an approach that precedes contemporary interest in the microbiota. The WHO defines probiotics as "alive microbes that, when supplied in appropriate numbers, impart a health advantage on the host". Considering this, numerous potential probiotics have yet to be produced or verified to meet the criteria. Also when research is done to demonstrate prospective health advantages, the mechanistic foundation is frequently ignored. As a result, some have said that the probiotic area has a "non-scientific" component to it. Additionally, regulatory authorities like the US FDA have permitted many probiotics to be classified as nutritional supplements as long as they are not "designed to diagnostic, heal, ameliorate, cure, or protect a human illness [65, 66]". As a result, new terminology for live biotherapeutics intended for use as medicines has been proposed. However, if the official WHO definition of probiotics is followed, the new definition may not be needed because it requires formal controls and verification of new drugs. However, many studies have used probiotics in the treatment of various diseases. In fact, ielie Metchnikoff, who won the Nobel Prize for research on phagocytosis in 1908, suggested that microorganisms could have beneficial and detrimental effects on the host. He suggested that the intake of fermented milk products could be beneficial to health, which led to the development of members of *Lactobacillus* and *Bifidobacterium* as possible probiotics. These organisms are generally given as therapeutic foods, such as kefir and yoghurt, which are fermented milk products. Children's acute gastroenteritis can be prevented and treated with these medicines, according to studies. Results include limiting the development of antibiotic-related diarrhoea and preventing C. difficile infection. Previous small trials using typical probiotics have had mixed results, with most published guidelines suggesting their use. A major double-blind, multicentre, randomised, placebo-controlled study in older adults subsequently found that a probiotic combination of lactobacilli and bifidobacteria sought to stop antibiotic-associated diarrhoea or C difficile infection [67]. This adds to the evidence that probiotics should not be used to avoid these diseases on a regular basis. Many of the classic probiotic microbes were obtained from fermented food items, as previously stated. They were picked for criteria other than theory - based or empirically proven modes of action as a result. Emerging microbiome research, particularly those that look at microbial activity, has resulted in the design and animal evaluation of organisms that might be utilised clinically for certain conditions. Turning to C difficile infection, the relevance of metabolism of bile acid in pathophysiology has led clinical studies with bile acids, its analogues, and organisms that might possibly change metabolism of bile acid inside the GIT. Even though this therapy is currently in the early stages of research, the concept of generating live biotherapeutics based on intelligently determined modes of action will ideally become a key method for probiotic development in the future [68, 69].

## 2.5 Prebiotics and Diet Therapy

Altering the microbiome's ambient circumstances to give nutrients that promote the development and preponderance of helpful bacteria and their activities is another method for beneficially altering the native microbiome. This technique has mostly been used to alter the gut microbiota by changing the diet. At its most fundamental level, this strategy is providing a particular source of food that is intended to promote the growth of beneficial bacteria or microbial activities. Prebiotics are non-digestible polysaccharides that are digested by certain bacteria to help them develop. Many techniques are being developed to boost the synthesis of butyrate as well as other free fatty acids, due to the positive benefits of microbial fermentation products like butyrate. Since similar therapies assume the presence of the necessary bacteria, a variant is to give a "synbiotic" that contains both the probiotic carbohydrate and the prebiotic organism [70]. While concentrating on a particular food has proved beneficial, larger dietary modifications that rely at least in part on changing the native microbiome have also been employed. Exclusive enteral nutritional (EEN) treatment has been shown to be effective in treating children with IBD, especially Crohn's disease. This comprises a well-defined liquid diet that is the sole source of nourishment. Although it has a high rate of success in producing recovery in these youngsters, lengthy compliance to this diet is challenging. Current research on the impact of EEN on the gut microbiota has found that it has a statistically significant impact on the structure and activity of the microbiota. These alterations are linked to functional variations in the microbiota, which go against what prior research suggests would be helpful, highlighting how little we understand regarding medicinal microbiome modification [71].

## 2.6 Microbial Restoration

A logical implication of the probiotic method is the substitution or repair of an unhealthy population. There are a few distinctions, though. Microbiota transplant, which involves transferring an entire microbiome from a healthy person to a person suffering from a microbiota-related illness, has piqued curiosity. Such therapies, notably the transplant of entire faeces or material generated from faeces, have a long history. New attention in faecal microbiota transplant (FMT) as a therapy for recurring C difficile infection has prompted a number of investigations into this type of microbiota treatment. In 1958, FMT was initially used to treat antibiotic-associated pseudomembranous colitis, which was thought to be caused by C difficile infection. Human studies of FMT for recurring C difficile infection have subsequently employed several faeces formulations and evaluated various administration methods. The patient's own faeces were given to the placebo group in one placebo-controlled experiment [21, 35]. The positive development of all types of FMT in treating recurring C difficile infection has sparked speculation that microbiota substitution may be utilised to treat other illnesses. Unfortunately, this result has yet to be reproduced in other diseases like obesity and inflammatory bowel disease (IBD). Small experiments have had contradictory results. This therapy may not be immediately applicable to other disorders, according to some. The spore-forming portion of the gut microbiota appears to be the source of FMT's therapeutic impact in recurring C difficile. FMT as produced for C difficile, which favours the injection of spore-forming organisms, may not always be effective in other situations [18, 45].

## 2.7 Conclusion

The investigation of the human microbiome is crucial because it provides an in-depth understanding of the interactions among humans and their microbiota. This information will be useful in future research investigations aimed at improving these organisms' ability to resist life-threatening illnesses. It is worth noting that long-term usage of broad-spectrum antibiotics has the potential to alter the human microbiome. As a result, the indigenous microbial population becomes unbalanced, allowing invading diseases to thrive. Treatments that include pre and probiotics, on the other hand, should be recommended.

## References

1. Aleman FDD, Valenzano DR (2019) Microbiome evolution during host aging. PLoS Pathog 15(7):e1007727
2. Barko PC, McMichael MA, Swanson KS, Williams DA (2018) The gastrointestinal microbiome: a review. J Vet Intern Med 32(1):9–25
3. Belizário JE, Faintuch J (2018) Microbiome and gut dysbiosis. Exp Suppl (2012) 109:459–476
4. Berg G, Rybakova D, Fischer D, Cernava T, Vergès MC, Charles T et al (2020) Microbiome definition re-visited: old concepts and new challenges. Microbiome 8(1):103
5. Bilen M (2020) Strategies and advancements in human microbiome description and the importance of culturomics. Microb Pathog 149:104460
6. Bradley CA (2019) Tumour microbiome defines outcomes. Nat Rev Gastroenterol Hepatol 16(11):649
7. Brusselaers N (2019) Prescribed drugs and the microbiome. Gastroenterol Clin N Am 48(3): 331–342
8. Callewaert C, Ravard Helffer K, Lebaron P (2020) Skin microbiome and its interplay with the environment. Am J Clin Dermatol 21(Suppl 1):4–11
9. Cussotto S, Clarke G, Dinan TG, Cryan JF (2019) Psychotropics and the microbiome: a chamber of secrets.... Psychopharmacology. 236(5):1411–1432
10. Daeschlein G, Hinz P, Kiefer T, Jünger M (2019) [Role of the microbiome in chronic wounds]. Hautarzt 70(6):422–431
11. Dannenberg L, Zikeli D, Benkhoff M, Ahlbrecht S, Kelm M, Levkau B et al (2020) Targeting the human microbiome and its metabolite TMAO in cardiovascular prevention and therapy. Pharmacol Ther 213:107584
12. Davenport ER, Sanders JG, Song SJ, Amato KR, Clark AG, Knight R (2017) The human microbiome in evolution. BMC Biol 15(1):127
13. Davidson GL, Raulo A, Knowles SCL (2020) Identifying microbiome-mediated behaviour in wild vertebrates. Trends Ecol Evol 35(11):972–980
14. Dickson I (2020) Microbiome signatures for cirrhosis and diabetes. Nat Rev Gastroenterol Hepatol 17(9):532
15. Dominguez-Bello MG, Godoy-Vitorino F, Knight R, Blaser MJ (2019) Role of the microbiome in human development. Gut 68(6):1108–1114
16. Douglas AE (2019) Simple animal models for microbiome research. Nat Rev Microbiol 17(12): 764–775
17. Dréno B (2019) The microbiome, a new target for ecobiology in dermatology. Eur J Dermatol 29(S1):15–18
18. Ferrer M, Méndez-García C, Rojo D, Barbas C, Moya A (2017) Antibiotic use and microbiome function. Biochem Pharmacol 134:114–126

19. Finlay BB, Pettersson S, Melby MK, Bosch TCG (2019) The microbiome mediates environmental effects on aging. BioEssays 41(10):e1800257
20. FitzGerald MJ, Spek EJ (2020) Microbiome therapeutics and patent protection. Nat Biotechnol 38(7):806–810
21. Garud NR, Pollard KS (2020) Population genetics in the human microbiome. Trends Genet 36(1):53–67
22. Giles EM, Couper J (2020) Microbiome in health and disease. J Paediatr Child Health 56(11): 1735–1738
23. Gould AL, Zhang V, Lamberti L, Jones EW, Obadia B, Korasidis N et al (2018) Microbiome interactions shape host fitness. Proc Natl Acad Sci U S A 115(51):E11951–E11e60
24. Greenhill C (2020) Gut microbiome influences exercise response. Nat Rev Endocrinol 16(2): 68–69
25. Gulati M, Plosky B (2020) As the microbiome moves on toward mechanism. Mol Cell 78(4): 567
26. Haque MM, Mande SS (2019) Decoding the microbiome for the development of translational applications: overview, challenges and pitfalls. J Biosci 44(5):118
27. Hintze G (2019) [Microbiome – growing importance in medicine]. Deutsch Med Wochenschr 144(14):929
28. Hitchings R, Kelly L (2019) Predicting and understanding the human microbiome's impact on pharmacology. Trends Pharmacol Sci 40(7):495–505
29. Jagodzinski A, Zielinska E, Laczmanski L, Hirnle L (2019) The early years of life. Are they influenced by our microbiome? Ginekol Pol 90(4):228–232
30. Johnson D, Letchumanan V, Thurairajasingam S, Lee LH (2020) A revolutionizing approach to autism spectrum disorder using the microbiome. Nutrients. 12(7):1983
31. King KC, Stevens E, Drew GC (2020) Microbiome: evolution in a world of interaction. Curr Biol 30(6):R265–R2r7
32. Klassen JL (2018) Defining microbiome function. Nat Microbiol 3(8):864–869
33. Koontz JM, Dancy BCR, Horton CL, Stallings JD, DiVito VT, Lewis JA (2019) The role of the human microbiome in chemical toxicity. Int J Toxicol 38(4):251–264
34. Kumar M, Ji B, Zengler K, Nielsen J (2019) Modelling approaches for studying the microbiome. Nat Microbiol 4(8):1253–1267
35. Marijnissen GM, Zwittink RD, Kuijper EJ, van Furth EF (2020) Microbiome and psychiatry: autism as an example. Tijdschr Psychiatr 62(2):131–140
36. Maruvada P, Leone V, Kaplan LM, Chang EB (2017) The human microbiome and obesity: moving beyond associations. Cell Host Microbe 22(5):589–599
37. Mills S, Stanton C, Lane JA, Smith GJ, Ross RP (2019) Precision nutrition and the microbiome, part I: current state of the science. Nutrients. 11(4):923
38. Mitchell AB (2019) The lung microbiome and transplantation. Curr Opin Organ Transplant 24(3):305–310
39. Mohajeri MH, Brummer RJM, Rastall RA, Weersma RK, Harmsen HJM, Faas M et al (2018) The role of the microbiome for human health: from basic science to clinical applications. Eur J Nutr 57(Suppl 1):1–14
40. Myers B, Brownstone N, Reddy V, Chan S, Thibodeaux Q, Truong A et al (2019) The gut microbiome in psoriasis and psoriatic arthritis. Best Pract Res Clin Rheumatol 33(6):101494
41. Neugent ML, Hulyalkar NV, Nguyen VH, Zimmern PE, De Nisco NJ (2020) Advances in understanding the human urinary microbiome and its potential role in urinary tract infection. mBio. 11(2):e00218–e00220
42. Neuman H, Koren O (2017) The pregnancy microbiome. Nestle Nutr Inst Workshop Ser 88:1–9
43. Paoli A, Mancin L, Bianco A, Thomas E, Mota JF, Piccini F (2019) Ketogenic diet and microbiota: friends or enemies? Genes. 10(7):534
44. Paxton RJ (2020) A microbiome silver bullet for honey bees. Science. 367(6477):504–506
45. Peirce JM, Alviña K (2019) The role of inflammation and the gut microbiome in depression and anxiety. J Neurosci Res 97(10):1223–1241

46. Perez-Muñoz ME, Arrieta MC, Ramer-Tait AE, Walter J (2017) A critical assessment of the "sterile womb" and "in utero colonization" hypotheses: implications for research on the pioneer infant microbiome. Microbiome. 5(1):48
47. Peterson SN, Bradley LM, Ronai ZA (2020) The gut microbiome: an unexpected player in cancer immunity. Curr Opin Neurobiol 62:48–52
48. Picardo SL, Coburn B, Hansen AR (2019) The microbiome and cancer for clinicians. Crit Rev Oncol Hematol 141:1–12
49. Rabb H, Pluznick J, Noel S (2018) The microbiome and acute kidney injury. Nephron 140(2): 120–123
50. Rajagopala SV, Vashee S, Oldfield LM, Suzuki Y, Venter JC, Telenti A et al (2017) The human microbiome and cancer. Cancer Prev Res. 10(4):226–234
51. Robinson CD, Bohannan BJ, Britton RA (2019) Scales of persistence: transmission and the microbiome. Curr Opin Microbiol 50:42–49
52. Chellappan DK, Ning QLS, Min SKS, Bin SY, Chern PJ, Shi TP et al (2019) Interactions between microbiome and lungs: paving new paths for microbiome based bio-engineered drug delivery systems in chronic respiratory diseases. Chem Biol Interact 310:108732
53. Singhvi G, Girdhar V, Patil S, Gupta G, Hansbro PM, Dua K (2018) Microbiome as therapeutics in vesicular delivery. Biomed Pharmacother 104:738–741
54. Salazar N, González S, Nogacka AM, Rios-Covián D, Arboleya S, Gueimonde M et al (2020) Microbiome: effects of ageing and diet. Curr Issues Mol Biol 36:33–62
55. Salvucci E (2019) The human-microbiome superorganism and its modulation to restore health. Int J Food Sci Nutr 70(7):781–795
56. Schwartz DJ, Langdon AE, Dantas G (2020) Understanding the impact of antibiotic perturbation on the human microbiome. Genome Med 12(1):82
57. Selway CA, Eisenhofer R, Weyrich LS (2020) Microbiome applications for pathology: challenges of low microbial biomass samples during diagnostic testing. J Pathol Clin Res 6(2):97–106
58. Silbergeld EK (2017) The microbiome. Toxicol Pathol 45(1):190–194
59. Skowron KB, Shogan BD, Rubin DT, Hyman NH (2018) The new frontier: the intestinal microbiome and surgery. J Gastrointest Surg 22(7):1277–1285
60. Suraya R, Nagano T, Kobayashi K, Nishimura Y (2020) Microbiome as a target for cancer therapy. Integr Cancer Ther 19:1534735420920721
61. Thahir AIA, Gordon A, Salam A (2020) Does gut microbiome associate with the growth of infants? A review of the literature. Enferm Clin 30(Suppl 4):66–70
62. Ticinesi A, Lauretani F, Tana C, Nouvenne A, Ridolo E, Meschi T (2019) Exercise and immune system as modulators of intestinal microbiome: implications for the gut-muscle axis hypothesis. Exerc Immunol Rev 25:84–95
63. Unger MM, Becker A, Keller A, Schäfer KH, Schwiertz A, Oertel WH (2020) The role of the gut microbiome in idiopathic Parkinson's disease. Nervenarzt 91(12):1085–1095
64. van der Goot E, van Spronsen FJ, Falcão Salles J, van der Zee EA (2020) A microbial community ecology perspective on the gut-microbiome-brain axis. Front Endocrinol 11:611
65. van Tilburg BE, Gutierrez MW, Arrieta MC (2020) The fungal microbiome and asthma. Front Cell Infect Microbiol 10:583418
66. Vorholt JA, Vogel C, Carlström CI, Müller DB (2017) Establishing causality: opportunities of synthetic communities for plant microbiome research. Cell Host Microbe 22(2):142–155
67. Walter J, Armet AM, Finlay BB, Shanahan F (2020) Establishing or exaggerating causality for the gut microbiome: lessons from human microbiota-associated rodents. Cell 180(2):221–232

68. Wheeler KM, Liss MA (2019) The microbiome and prostate cancer risk. Curr Urol Rep 20(10): 66
69. Williams CL, Garcia-Reyero N, Martyniuk CJ, Tubbs CW, Bisesi JH Jr (2020) Regulation of endocrine systems by the microbiome: perspectives from comparative animal models. Gen Comp Endocrinol 292:113437
70. Young VB (2017) The role of the microbiome in human health and disease: an introduction for clinicians. BMJ. 356:j831
71. Zhang X, Li L, Butcher J, Stintzi A, Figeys D (2019) Advancing functional and translational microbiome research using meta-omics approaches. Microbiome. 7(1):154

# Role of Microbiome in Inflammation During Tuberculosis

# 3

Kuldeepak Sharma, Mateja Erdani Kreft, Mateja Škufca Sterle, and Darko Vasic

**Abstract**

The correlation between TB, which is one of the primary causes of chronic inflammatory disease and healthy microbiota in the community, is not sufficiently well known. The gut microbiota is critical for the regulation of inflammatory and immunological responses, and it has a substantial impact on the patient's wellness. It was revealed that individuals' lung mediated secretion and faecal samples had distinct tuberculosis-associated microorganisms, both prior and after the misleading effect of antibiotics. Numerous studies have demonstrated that isolated microbiota from patients' faeces predicted the activation of pro-inflammatory immunological mechanisms, therefore establishing the value of the intestinal flora in tuberculosis therapy and prevention. The purpose of this study was to examine the microbiome and human defence system of TB patients. Collective data shows in this study that some intestinal microbiotas including anaerobes may influence the host, but additional mechanistic investigations are necessary. Therefore, we study the function of intestinal microbiota with the TB pathogenesis and may govern the immunity and inflammatory environment of host microbiota and their metabolites, as well as the pathways underlying host immunity.

K. Sharma (✉)
Institute of Pharmacology and Toxicology, Faculty of Medicine, University of Ljubljana, Ljubljana, Slovenia

M. E. Kreft
Institute of Cell Biology, Faculty of Medicine, University of Ljubljana, Ljubljana, Slovenia

M. Š. Sterle · D. Vasic
Prehospital Emergency Medical Service, Health Centre Ljubljana, Ljubljana, Slovenia



G. Gupta et al. (eds.), *Microbiome in Inflammatory Lung Diseases*,
https://doi.org/10.1007/978-981-16-8957-4_3

**Keywords**

Tuberculosis · Microbiome · Inflammation · SCF · Immunity

## 3.1 Introduction

Human beings are almost sterile but get colonised quickly and dynamically throughout the body during their childbirth and early development [1]. These mostly lie in the intestines, including bacteria, viruses, and archaea and eukaryotic microorganisms to a lower degree [2, 3]. The human microbiota is made up of around 10,000 bacterial species that assist humans in performing regular physiological functions [4]. The microbiota contributes to the establishment of an optimum underlying ecosystem that is demonstrated by activities such as energy extraction from nutrition, synthesis of supplementary growth factors, and activation of both the natural and adoptive immune systems [5]. As a result, the involvement of bacteria in humans offers important biological assistance [6].

Inflammation is particularly sensitive to gut microbiota, and experiments in dextran sulphate that distort the mucous membrane and trigger intestinal inflammatory process in rodents which demonstrates the impacts of inflammation in microbiota [7]. The effect of inflammation is a sequence of molecular and cellular processes which always significantly impact bacterial communities or generate substantial environmental stress. In retrospective studies, inflammatory disorders including bowl, arthritic, and celiac disease are one of the biggest sets of known human diseases that alter the microbiome. Other aspects of interdependent microbial stimulation to host defense in intestinal illnesses and conditioning of systemic immunology are poorly understood although they undoubtedly contribute to the onset, maintenance, and resolution of inflammatory disorders.

## 3.2 Altered Gut Microbiome Diversity Associated Inflammation During Diseases

The human gut microbiota, comprising a complex microbial ecology, comprises hundreds of generally stable groups of species in a healthy person but its composition may rapidly vary due to disease, age, nutrition, antibiotic usage, host genes, and inflammation. Host inflammatory illnesses may be regulated by the gut microbiota and its metabolites. Many studies have connected the intestinal microbiome with inflammatory conditions. Forbes and colleagues [8] have demonstrated that inflammatory illnesses mediated by the immunological signalling, such as Crohn's disease, rheumatoid arthritis, and ulcerative colitis change the balance of intestinal microbiome. Moreover, the significance of intestinal dysbiosis in the development of inflammatory illnesses such as lung diseases (tuberculosis, COPD, asthma etc.), colorectal cancer, major types of diabetes mellitus, metabolic syndrome, obesity is reported in several cases [9–13] (Fig. 3.1).

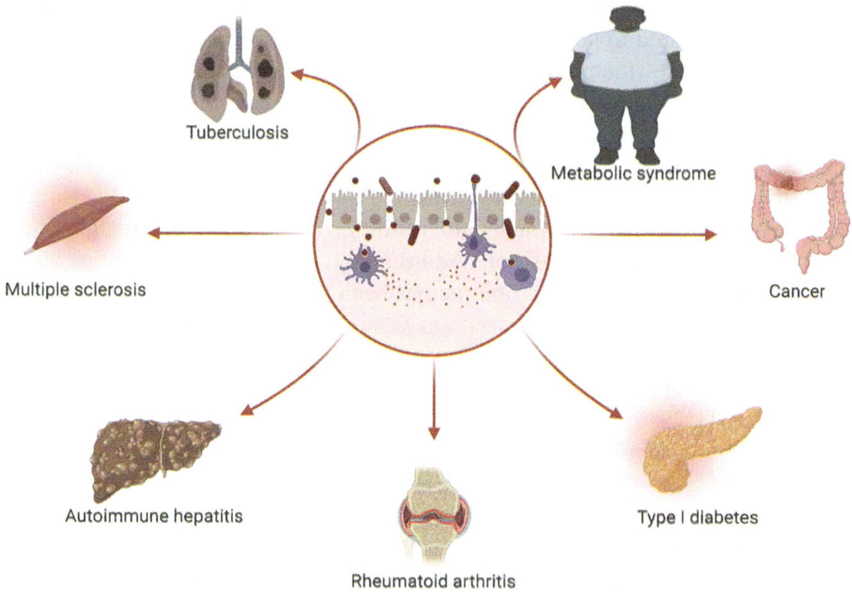


**Fig. 3.1** Dysbiosis mediated various disease

The microbiome may be involved in TB infection, according to new findings [14, 15]. TB is amongst the most deadly infectious illnesses in the world. In 2019, according to the WHO 2020 report, about 10 million individuals were infected with and acquired TB (adult men, adult females, and kids account for 5.6, 3.2, and 1.2 million, respectively), and 1.4 million were fatalities [16]. It is caused by the Mycobacterium tuberculosis complex and has remained a significant source of illness and mortality across the world [17]. M. tuberculosis can infect not just the lungs, but also other organs, namely the brain and spine [16]. The microbiota and its associated metabolic products are thought to contribute to tuberculosis infection vulnerability, disease progression, and aggressiveness. Alteration in the population dynamics of the microbiome may have an influence on immunological signalling pathway. Learning as much as possible about the complex processes that regulate the link between the microbiome and tuberculosis inflammation may reveal significant variables that predict and govern TB disease development, aggressiveness, clinical prognosis, as well as the possibility of relapse.

## 3.3 Tuberculosis and Its Therapy Relationship with Intestinal Dysbiosis

The intestinal microbiome may be dramatically altered as a result of lung infection with influenza virus via a process involving type I interferons [18]. Initial research indicated the association of TB with intestinal microbiota abnormalities. A large number of studies have revealed significant differences in the intestinal microbiota of TB patients compared to healthy control participants. In particularly in comparison to nutritious subjects, the microbiota biodiversity in faecal matter from TB patients has been considerably decreased. Hu et al. [19] conducted a cross-sectional study in which they evaluated the gut microbiota composition of tuberculosis patients from China and discovered that TB infection resulted in a loss of variety, which was mostly due to changes in Bacteroides relative abundance. Additionally, Hu et al. discovered that anti-TB treatment can result in quick, significant alteration in microbial diversity and composition. Additionally, the effect of antibiotics used to treat tuberculosis on the gut microbiome has been studied. During anti-TB therapy, the relative abundance of the Clostridium genus fell considerably, but the relative abundance of many Bacteroides genus members rose, including Bacteroides fragilis and Bacteroides OTU230. Furthermore, during 1 week of tuberculosis therapy, OTU8 and OTU2972 belonging to the family Erysipelotrichaceae grew significantly, but the remainder of the Erysipelotrichaceae family decreased. Similarly, Negi and colleagues [20] employed an in vivo mouse model to show that wide-spectrum antibiotics might lead to major changes in the gut microbiota's composition, reduction in the amount of commensal bacteria Campylobacter, Lactobacillus, and Bifidobacterium, and rise in Enterococcus and Bacteroides. In addition to, studies have suggested that changed microbiome balance may impair the effectiveness of anti-TB medicines and therapy. Negi et al. [20] continued on to explore the effects of intestinal microbial dysbiosis on the efficacy of isoniazid (INH) against tuberculosis pathogens in the in vivo mouse model. They found that reduced numbers of Campylobacter, Bifidobacterium, and Lactobacillus might lead to poor immunological response to INH and to more severe granulomatous diseases by antibiotic pre-treatment. This study also indicates that the damage of innate immunity and intestinal defence was affected by differences in the microbiota during INH therapy resulting in lower levels of antimicrobial peptides RegIII and pro-inflammatory cytokines TNF-$\alpha$ and IFN-$\gamma$, but higher levels of IL-10 anti-inflammatory cytokine. Previously, similar findings were discovered by Khan and Mendonca [21] and reported to disrupt alveolar macrophage immunological function against Mycobacterial TB Infections with intestinal modifications linked with higher Bacteroides and Verrucomicrobiaceae abundance and significant reduced Lachnospiraceae densities. Furthermore, there were also 31 competent participants from a Chinese study group and 46 tuberculosis patients, with considerable reduction in the variety and number of microbiological products, defined by the notable decrease in bacteria that are generating short chains of fatty acids (SCFAs). The study also connected microbial alterations to inflammatory response variations during TB in phenotypes. In comparison with their healthy close contacts, the authors discovered the distinct gut microbiome profiles in the TB infected with Erysipelotrichaceae, Blautia, anaerostipes in their faeces, and inferred mechanism modulating the immune response of short chains of fatty acids [22]. In numerous illnesses, including cancer, the SCFAs microbial inflammatory cascade was described and provides fresh routes for the research of TB biomarkers and microbiome-oriented disease-prevention or supportive treatment. Earlier,

Namasivayam and team found that the animal model shifted the order Clostridiales phylum Firmicutes and certain species of the phylum Tenericutes [15]. In addition, in Bacteroides vulgatus, single nucleotide polymorphisms (SNPs) were substantially different for TB patients in comparison with heathy controls [23]. Liu, Luo Research Group [24] separate individual patients into fresh and recurring TB patients (RTB). According to their findings, Proteobacteria and Actinobacteria were the substantially more elevated, whereas in the RTB patients' faecal sample Bacteroides with a range of valuable commensal bacteria were decreased. Moreover, in comparison with the healthy group Lachnospira and Prevotella were significantly less in newly RTB. Similar results have been found in teenagers experiencing TB. A cases-controlled investigation has shown a decreased diversity of microbiota, increased Prevotella, enterococcus abundance, decreased and reduced abundance for host health benefits in paediatric patients with Ruminococcaceae, Faecalibacterium prausnitzii, and Bifidobacteriaceae [25]. Furthermore, an Indian study group using 16S rRNA genotype to discrimination between TB patients and healthy controls found that Actinobacteria and Firmicutes in TB patients had statistically greater [26] (Table 3.1).

The studies have revealed that the intestinal flora patterns across nutritionally balanced individuals and TB participants varied and that microbiome fingerprints may also be detected at different stages of TB development. Additionally, these

**Table 3.1** Exhibiting gut microbiome mediated various mechanism and functions in host cell

| Intestinal Microbiota | Function | Mechanism |
|---|---|---|
| *Lactobacillus* | Boost the anti-viral reaction [27] | Modify TNF-α and IL-12 release in dendritic cells |
| *Faecalibacterium prausnitzii, Bacteroides thetaiotaomicron* | Encourage differentiation of the cell goblet [28] | Facilitates production of mucous |
| *Bifidobacterium* | Optimise the responsiveness to humoral immunity [29]; ease gut inflammation [30]; accelerate differentiation among Th cells [31] | Trigger interleukin expression in particular 1 and 6; Minimise the pro-inflammatory and lipopolysaccharide levels |
| *Clostridium* | Provoke Tregs to replicate and discriminate [32] | Significantly raise Foxp3 transcription and boost TGF-β signalling |
| Firmicutes | Promote epithelial barrier function by: (1) repressing claudin-2 protein expression in response to IL-10RA [33]; (2) increasing NLRC3 protein expression [34]; and (3) controlling intestinal macrophage growth [35]. | Generation of butyric acid |
| *Bacteroides fragilis* | Impacts Th1's immune response [36] and accelerates Treg cell differentiation [37] | Polysaccharide A production |

commensals generate metabolites that have the potential to influence immunological signalling. Dumas et al. illustrated that a reduction in Firmicutes and Bacteroidetes and an increase in Proteobacteria were associated with an increase in subsequent airway colonisation by tuberculosis pathogens in antibiotic-treated mice, implying an importance for intestinal flora in initial defence, presumably by maintaining the processes of mucosal-associated invariant T cells. The above results indicate that intestinal microbiome can represent a significant impact towards TB treatment. This demonstrates some of the fundamental distinctions between people and mice, but also some of the common outcomes associated with tuberculosis and the microbiota. However, it is also important to further examine if the causative link and the changed intestinal microbial ecology with reduced bacterial diversity promote TB susceptibility (Fig. 3.2).

## 3.4 Gut Microbiota and Inflammatory Immune Activity in Tuberculosis

When there is an inflammatory event in the lung, it is conceivable that the intestinal microbiome may be altered, and this will affect the lung–gut axis [38]. It is generally accepted that microbes and by-products produced by the lung microbiota are transported through circulation to the intestines and vice versa [39]. They can then go to the liver, stimulating inflammatory mechanism units like neutrophils and macrophages [40], which are particularly essential during Mtb infection [41]. The balance of the gut microbiota promotes systemic and lung immunity by modulating T cell development, immune cell migration and death, through triggering TLR signalling, and inflammatory tone suppression [42]. Additionally, the lung microbiome plays a key role in the defence mechanism against infections by stimulating the Th1/Th2 immune reaction, which results in the management of inflammatory process [43]. Thus, the human microbiota can be critical in the defense against tuberculosis, and alterations in the microbiota may contribute to the TB spectrum and interpretation [44].

According to current scientific understanding, the microflora contributes the majority of its function to host immune response and inflammation through its ability to generate enzymatic activities that metabolise nutrients, resulting in SCFAs (primarily carbohydrates) with anti-inflammatory and immunomodulatory features [45]. Additionally, the by-products and other substance of microorganisms, such as carbohydrates of the bacterial capsule, endotoxins & lipopolysaccharides that are generated to preserve the integrity of the intestinal wall, as well as upregulation or downregulation of particular crucial immune system signalling ligands significantly impact vitamin production, energy modulation, and intestinal endocrine hormone regulation [46]. Thus, the tight association in between intestinal microbiota and the immunological transduction pathway may influence the inflammatory reaction, since this connection can result in either anti-inflammatory or

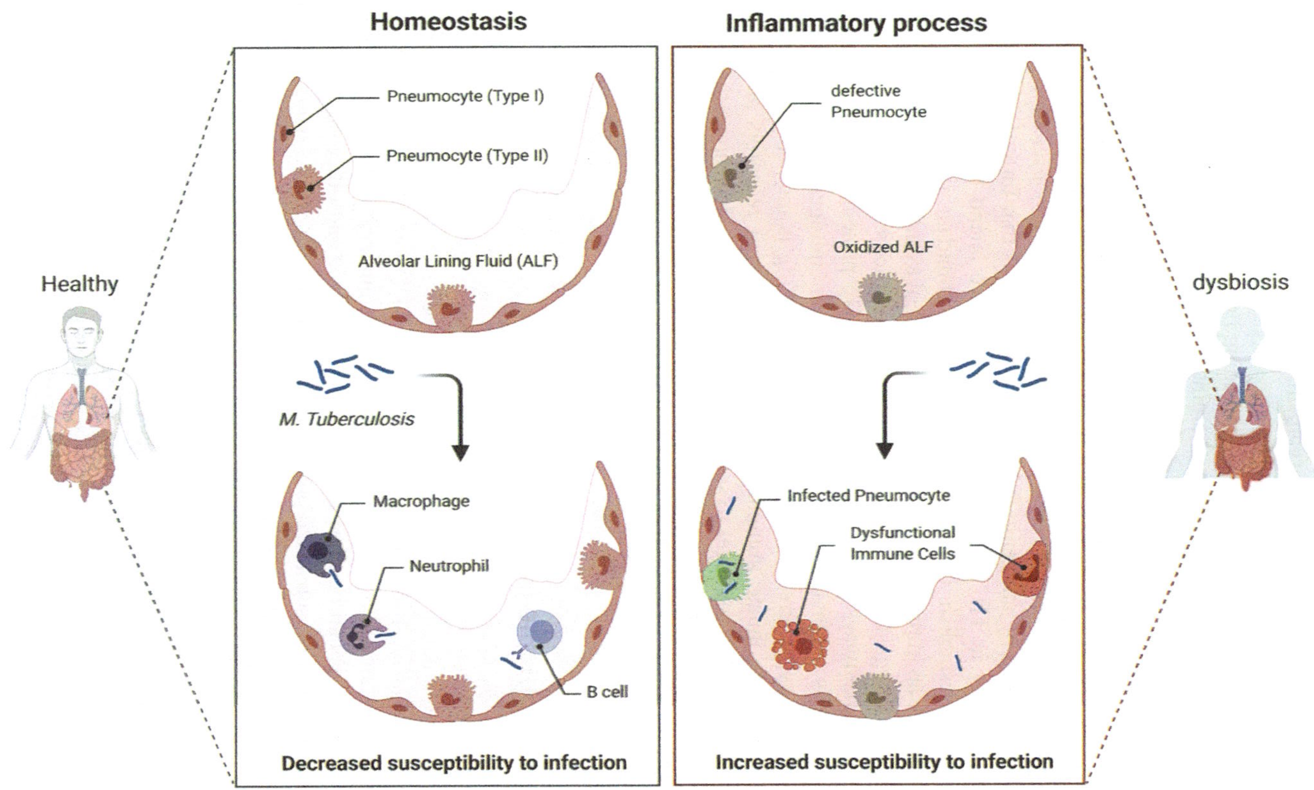


**Fig. 3.2** Illustration of gut–lung axis defect associated immune-inflammation response mediated increase susceptibility towards M. tuberculosis

pro-inflammatory pathways via regulatory cell activation. On the other side, the intestinal flora has the ability to impact drug biotransformation and function as a barricade against infectious microorganisms through colonisation and resource competition.

The gut microbiota produces SCFAs including propionate, butyrate, acetate, and lactate. Recently published research examines the role of butyrate in facilitating communication in between microbial communities and the immunological regulatory framework. In addition to, by inhibiting histone deacetylases, butyrate regulates the production of certain cytokines in T Cells and macrophages, and also the stimulation of Treg Cells. However, increased butyrate levels may be harmful in tuberculosis (TB) because they decrease pro-inflammatory responses (interleukin-1/17, TNF-$\alpha$) elicited by antigenic epitopes of tuberculosis that further consequently produce mortality in TB. Therefore, in this study, we intend to find mechanistic exploration to establish if SCFAs generated by abundant intestinal anaerobic organisms have an impact on peripheral immunological events [22]. Thus, numerous investigations suggest that SCFAs stimulate immunological activities via signal transduction pathways by interacting with G protein-coupled receptors. Pro-inflammatory signalling is mediated by the mitogen-activated kinases (MAPK) pathway, whereas anti-inflammatory signalling is mediated by the beta-arrestin-2 route. In innate immunity, SCFAs bind to their receptors and activate dendritic cells and macrophages to secrete IL-10. It has been also shown that SCFAs can promote the development of B cells and Treg cells in the gut–lung axis [45]. As a result, SCFAs are responsible for balancing and regulating the immune response overall.

## 3.5 Microbiome-immune Crosstalk

Interactions between the host and bacteria are crucial in regulating tissue homeostasis at both proximal and peripheral sites. Nonetheless, microbial dysbiosis and compromised barrier function have been linked to inflammation and metabolic disruption in distant areas, for example, during inflammatory processes after antibiotic treatment. This can be done in part by macrophages and other innate immune cells triggered through substance produced by bacteria such as LPS generated from gut microbial species. Macrophages are the primary host cells for intracellularly available TB pathogen, and pathogenic infiltration of macrophages with antimicrobial defence mechanisms promotes intracellular multiplication and preservation [47]. In particular during infection with TB, the respiratory microbial community was engaged in determining alveolar macrophages activity. Infection of mice with Mycobacterium tuberculosis following an 8-week course of the TB medicines isoniazid and pyrazinamide resulted in a modestly increased lung bacterial load. This was related to changes in alveolar macrophage phänotypes, including reduced synthesis of ATP and cellular respiration, MHCII expression, interleukin-1, and TNF-alpha generation [21]. Control of TB infection depends crucially on the presence of functional IFN-gamma and interleukin-12 signalling. Additionally, protection mediated by IFN-gamma occurs largely as a result of adoptive T cells activation. It has also been demonstrated more recently that the involvement of innate and innate-like lymphoid cells (ILC) is significant [48]. The ILC is categorised according to its cytokine expression patterns into three categories: Class 1; NK and noncytotoxic type 1 (Interferon, Tumour necrosis factor), Class 2; including interleukine-4/5/13, and Class 3; includes the Interleukin-17/22 [49]. The accumulation of IFNγ-expressing Nk cells in the pleural fluid in individuals with TB pleurisy has been documented [50]. Individuals with latent TB infection had a higher number of intravascular natural killer cells in their blood plasma and an enhanced capability for cytotoxicity, which was related with increased granzyme B, perforin expression [51] and CD27+ NK lung cell accumulation were also associated with latent tuberculosis in non-human primates [52]. Simultaneously, individuals with active TB had significantly less natural killer cells spreading in their peripheral circulation. Natural killer cells have been shown to contribute to the response of CD8+T cells, and to lysis of monocytes, macrophages, and regulatory T cells cultured with mycobacterial antigens [53].

CD4+ T cells are, on the other hand, crucial for controlling Mycobacterium TB illness with increasingly identified CD8+ T and B contributing lymphocytes. While IFN, IL-12, and TNF are important in regulating Mycobacterium TB-associated inflammation, strict management of all specific immune effector pathways, including Treg cells and IL-10, is necessary to prevent unacceptable pathology and maintain TB microorganism in control [48]. A study exhibits the presence of increased lung anaerobes (Haemophilus, Veillonella, and Prevotella) is associated with lower CD4+ lymphocyte counts in healthy HIV-infected people who eventually evolve aggressive TB [54]. Conversely, the presence of myeloid, type I and II interferon, and inflammatory genes is associated with higher CD4+ lymphocyte counts in active tuberculosis patients [55] (Fig. 3.3).

## 3.6 Conclusion

It has now been well known that the variety of inflammatory metabolic pathologies associated with intestinal microbiota alterations considerably differs between TB infected persons and healthy controls. The uneven quantity of intestinal microbiota can be used as a biomarker of microbiome, which distinguishes the active disease from the healthy, or as a therapeutic target for supplementary immunomodulator treatment or as a nutritional approach for the prevention or control of TB or lung problems. All viable candidates for such therapies include probiotics, prebiotics, and techniques for gut microbiota transfer.

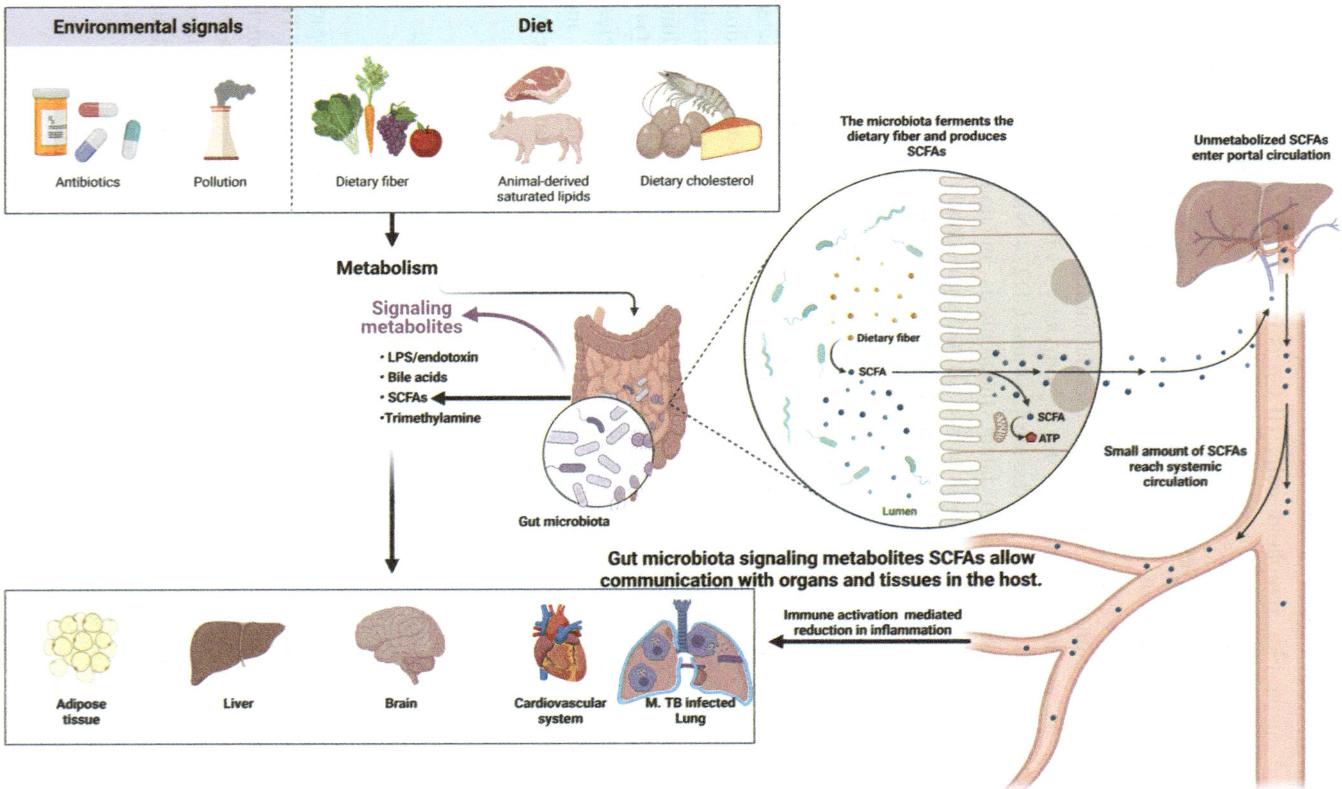


**Fig. 3.3** Exhibits gut microbiome signalling metabolite SCFAs mediated immune modulation mediated suppression of lungs inflammatory process during tuberculosis

## References


1. Sharon I, Morowitz MJ, Thomas BC, Costello EK, Relman DA, Banfield JF (2013) Time series community genomics analysis reveals rapid shifts in bacterial species, strains, and phage during infant gut colonization. Genome Res 23(1):111–120
2. Qin J, Li R, Raes J, Arumugam M, Burgdorf KS, Manichanh C et al (2010) A human gut microbial gene catalogue established by metagenomic sequencing. Nature 464(7285):59–65
3. Human Microbiome Project C (2012) Structure, function and diversity of the healthy human microbiome. Nature 486(7402):207–214
4. Croucher R, Dahiya M, Gowda KK (2013) Contents and price of vendor assembled paan quid with tobacco in five London localities: a cross-sectional study. Tob Control 22(2):141–143
5. Blaser MJ, Falkow S (2009) What are the consequences of the disappearing human microbiota? Nat Rev Microbiol 7(12):887–894
6. Sommer F, Bäckhed F (2013) The gut microbiota – masters of host development and physiology. Nat Rev Microbiol 11(4):227–238
7. Perše M, Cerar A (2012) Dextran sodium sulphate colitis mouse model: traps and tricks. J Biomed Biotechnol 2012:718617
8. Forbes JD, Chen C-Y, Knox NC, Marrie R-A, El-Gabalawy H, de Kievit T et al (2018) A comparative study of the gut microbiota in immune-mediated inflammatory diseases—does a common dysbiosis exist? Microbiome. 6(1):221
9. Arrieta MC, Stiemsma LT, Dimitriu PA, Thorson L, Russell S, Yurist-Doutsch S et al (2015) Early infancy microbial and metabolic alterations affect risk of childhood asthma. Sci Trans Med 7(307):307ra152
10. Knip M, Siljander H (2016) The role of the intestinal microbiota in type 1 diabetes mellitus. Nat Rev Endocrinol 12(3):154–167
11. Meijnikman AS, Gerdes VE, Nieuwdorp M, Herrema H (2018) Evaluating causality of gut microbiota in obesity and diabetes in humans. Endocr Rev 39(2):133–153
12. Delzenne NM, Neyrinck AM, Bäckhed F, Cani PD (2011) Targeting gut microbiota in obesity: effects of prebiotics and probiotics. Nat Rev Endocrinol 7(11):639–646
13. Arthur JC, Perez-Chanona E, Mühlbauer M, Tomkovich S, Uronis JM, Fan TJ et al (2012) Intestinal inflammation targets cancer-inducing activity of the microbiota. Science. 338(6103): 120–123
14. Mori G, Morrison M, Blumenthal A (2021) Microbiome-immune interactions in tuberculosis. PLoS Pathog 17(4):e1009377
15. Namasivayam S, Maiga M, Yuan W, Thovarai V, Costa DL, Mittereder LR et al (2017) Longitudinal profiling reveals a persistent intestinal dysbiosis triggered by conventional anti-tuberculosis therapy. Microbiome 5(1):71
16. Liu Y, Wang J, Wu C (2021) Microbiota and tuberculosis: a potential role of probiotics, and postbiotics. Front Nutr. 8(191):626254
17. Kone B, Somboro AM, Holl JL, Baya B, Togo AA, Sarro YDS et al (2020) Exploring the usefulness of molecular epidemiology of tuberculosis in Africa: a systematic review. Int J Mol Epidemiol Genet 11(1):1–15
18. Yildiz S, Mazel-Sanchez B, Kandasamy M, Manicassamy B, Schmolke M (2018) Influenza A virus infection impacts systemic microbiota dynamics and causes quantitative enteric dysbiosis. Microbiome. 6(1):9
19. Hu Y, Yang Q, Liu B, Dong J, Sun L, Zhu Y et al (2019) Gut microbiota associated with pulmonary tuberculosis and dysbiosis caused by anti-tuberculosis drugs. J Infect 78(4):317–322
20. Negi S, Pahari S, Bashir H, Agrewala JN (2020) Intestinal microbiota disruption limits the isoniazid mediated clearance of Mycobacterium tuberculosis in mice. Eur J Immunol 50(12): 1976–1987
21. Khan N, Mendonca L, Dhariwal A, Fontes G, Menzies D, Xia J et al (2019) Intestinal dysbiosis compromises alveolar macrophage immunity to Mycobacterium tuberculosis. Mucosal Immunol 12(3):772–783

22. Naidoo CC, Nyawo GR, Sulaiman I, Wu BG, Turner CT, Bu K et al (2021) Anaerobe-enriched gut microbiota predicts pro-inflammatory responses in pulmonary tuberculosis. EBioMedicine 67:103374
23. Hu Y, Feng Y, Wu J, Liu F, Zhang Z, Hao Y et al (2019) The gut microbiome signatures discriminate healthy from pulmonary tuberculosis patients. Front Cell Infect Microbiol 9:90
24. Luo M, Liu Y, Wu P, Luo DX, Sun Q, Zheng H et al (2017) Alternation of gut microbiota in patients with pulmonary tuberculosis. Front Physiol 8:822
25. Li W, Zhu Y, Liao Q, Wang Z, Wan C (2019) Characterization of gut microbiota in children with pulmonary tuberculosis. BMC Pediatr 19(1):445
26. Krishna P, Jain A, Bisen PS (2016) Microbiome diversity in the sputum of patients with pulmonary tuberculosis. Eur J Clin Microbiol Infect Dis 35(7):1205–1210
27. Drakes M, Blanchard T, Czinn S (2004) Bacterial probiotic modulation of dendritic cells. Infect Immun 72(6):3299–3309
28. Wrzosek L, Miquel S, Noordine M-L, Bouet S, Chevalier-Curt MJ, Robert V et al (2013) Bacteroides thetaiotaomicron and Faecalibacterium prausnitzii influence the production of mucus glycans and the development of goblet cells in the colonic epithelium of a gnotobiotic model rodent. BMC Biol 11(1):61
29. Zhu G, Liu X, Fang Y, Zhai B, Xu R, Han G et al (2018) Increased mTOR cancels out the effect of reduced Xbp-1 on antibody secretion in IL-1α-deficient B cells. Cell Immunol 328:9–17
30. Xue J, Ajuwon KM, Fang R (2020) Mechanistic insight into the gut microbiome and its interaction with host immunity and inflammation. Anim Nutr 6(4):421–428
31. Yan S, Yang B, Zhao J, Zhao J, Stanton C, Ross RP et al (2019) A ropy exopolysaccharide producing strain Bifidobacterium longum subsp. longum YS108R alleviates DSS-induced colitis by maintenance of the mucosal barrier and gut microbiota modulation. Food Funct 10(3):1595–1608
32. Tinoco-Veras CM, Santos A, Stipursky J, Meloni M, Araujo APB, Foschetti DA et al (2017) Transforming growth factor β1/SMAD signaling pathway activation protects the intestinal epithelium from clostridium difficile toxin A-induced damage. Infect Immun. 85(10):e00430–e00417
33. Zheng L, Kelly CJ, Battista KD, Schaefer R, Lanis JM, Alexeev EE et al (2017) Microbial-derived butyrate promotes epithelial barrier function through IL-10 receptor–dependent repression of claudin-2. J Immunol 199(8):2976–2984
34. Cheng D, Xu J-H, Li J-Y, Wang S-Y, Wu T-F, Chen Q-K et al (2018) Butyrate ameliorated-NLRC3 protects the intestinal barrier in a GPR43-dependent manner. Exp Cell Res 368(1): 101–110
35. Koh A, De Vadder F, Kovatcheva-Datchary P, Bäckhed F (2016) From dietary fiber to host physiology: short-chain fatty acids as key bacterial metabolites. Cell 165(6):1332–1345
36. Horai R, Sen HN, Caspi RR (2017) Commensal microbiota as a potential trigger of autoimmune uveitis. Expert Rev Clin Immunol 13(4):291–293
37. Round JL, Lee SM, Li J, Tran G, Jabri B, Chatila TA et al (2011) The toll-like receptor 2 pathway establishes colonization by a commensal of the human microbiota. Science. 332(6032):974–977
38. Sze MA, Tsuruta M, Yang S-WJ OY, Man SFP, Hogg JC et al (2014) Changes in the bacterial microbiota in gut, blood, and lungs following acute LPS instillation into mice lungs. PLoS One 9(10):e111228
39. He Y, Wen Q, Yao F, Xu D, Huang Y, Wang J (2017) Gut–lung axis: the microbial contributions and clinical implications. Crit Rev Microbiol 43(1):81–95
40. Young RP, Hopkins RJ, Marsland B (2016) The gut–liver–lung axis. Modulation of the innate immune response and its possible role in chronic obstructive pulmonary disease. Am J Respir Cell Mol Biol 54(2):161–169
41. Eum S-Y, Kong J-H, Hong M-S, Lee Y-J, Kim J-H, Hwang S-H et al (2010) Neutrophils are the predominant infected phagocytic cells in the airways of patients with active pulmonary TB. Chest 137(1):122–128

42. Samuelson DR, Welsh DA, Shellito JE (2015) Regulation of lung immunity and host defense by the intestinal microbiota. Front Microbiol 6:1085
43. Balcells ME, Yokobori N, Hong BY, Corbett J, Cervantes J (2019) The lung microbiome, vitamin D, and the tuberculous granuloma: a balance triangle. Microb Pathog 131:158–163
44. Tarashi S, Ahmadi Badi S, Moshiri A, Nasehi M, Fateh A, Vaziri F et al (2018) The human microbiota in pulmonary tuberculosis: Not so innocent bystanders. Tuberculosis (Edinb). 113: 215–221
45. Eribo OA, du Plessis N, Ozturk M, Guler R, Walzl G, Chegou NN (2020) The gut microbiome in tuberculosis susceptibility and treatment response: guilty or not guilty? Cell Mol Life Sci 77(8):1497–1509
46. Parada Venegas D, De la Fuente MK, Landskron G, González MJ, Quera R, Dijkstra G et al (2019) Short chain fatty acids (SCFAs)-mediated gut epithelial and immune regulation and its relevance for inflammatory bowel diseases. Front Immunol 10:277
47. Upadhyay S, Mittal E, Philips JA (2018) Tuberculosis and the art of macrophage manipulation. Pathog Dis. 76(4):fty037
48. O'Garra A, Redford PS, McNab FW, Bloom CI, Wilkinson RJ, Berry MP (2013) The immune response in tuberculosis. Annu Rev Immunol 31:475–527
49. Vivier E, Artis D, Colonna M, Diefenbach A, Di Santo JP, Eberl G et al (2018) Innate lymphoid cells: 10 years on. Cell 174(5):1054–1066
50. Schierloh P, Yokobori N, Alemán M, Musella RM, Beigier-Bompadre M, Saab MA et al (2005) Increased susceptibility to apoptosis of CD56dimCD16+ NK cells induces the enrichment of IFN-gamma-producing CD56bright cells in tuberculous pleurisy. J Immunol. 175(10): 6852–6860
51. Roy Chowdhury R, Vallania F, Yang Q, Lopez Angel CJ, Darboe F, Penn-Nicholson A et al (2018) A multi-cohort study of the immune factors associated with M. tuberculosis infection outcomes. Nature 560(7720):644–648
52. Esaulova E, Das S, Singh DK, Choreño-Parra JA, Swain A, Arthur L et al (2021) The immune landscape in tuberculosis reveals populations linked to disease and latency. Cell Host Microbe. 29(2):165–78.e8
53. Roy S, Barnes PF, Garg A, Wu S, Cosman D, Vankayalapati R (2008) NK cells lyse T regulatory cells that expand in response to an intracellular pathogen. J Immunol. 180(3): 1729–1736
54. Segal LN, Clemente JC, Li Y, Ruan C, Cao J, Danckers M et al (2017) Anaerobic bacterial fermentation products increase tuberculosis risk in antiretroviral-drug-treated HIV patients. Cell Host Microbe. 21(4):530–7.e4
55. Cliff JM, Kaufmann SHE, McShane H, van Helden P, O'Garra A (2015) The human immune response to tuberculosis and its treatment: a view from the blood. Immunol Rev 264(1):88–102

# 4 Interplay of Microbiome, Inflammation, and Immunity in Inflammatory Lung Diseases

Hitesh Malhotra, Anjoo Kamboj, Peeyush Kaushik, and Rupesh K. Gautam

**Abstract**

The microbiome present in and on the human body encompasses a diverse community of bacteria that affects overall health and various physiological as well as the pathological mechanism of the host. For instance, metabolism, regulation of the immune system, inflammatory processes, carcinogenesis, and host defence mechanism. Not all, certain bacteria may interact with pro-inflammatory molecules to induce inflammatory reactions in the body by amplifying the cascade of inflammatory pathways involving the production of interleukins and cytokines. Similarly, the structural component of gut bacteria and metabolic derivatives suppresses the inflammatory reactions. Furthermore, the microbiota activates the immune system and thus induces a protective mechanism against the pathogens as well as regulates the pathways involved in the development of tolerance for antigens. The chapter highlights the relationship between the microbiota and the regulation of inflammatory and immune processes. Also, the mechanism behind microbiota mediated regulation of inflammatory molecules and immune cells in combating various pulmonary inflammatory conditions.

The original version of this chapter was revised by updating the affiliations of the authors. The correction to this chapter can be found at https://doi.org/10.1007/978-981-16-8957-4_21

H. Malhotra · P. Kaushik
Guru Gobind Singh College of Pharmacy, Yamunanagar, India

A. Kamboj
Chandigarh College of Pharmacy, Mohali, India

R. K. Gautam (✉)
MM School of Pharmacy, MM University, Sadopur-Ambala, India



G. Gupta et al. (eds.), *Microbiome in Inflammatory Lung Diseases*,
https://doi.org/10.1007/978-981-16-8957-4_4

**Keywords**

Asthma · Cytokines · Immunity · Inflammation · Lungs · Macrophages · Microbiome · Microbiota · Pro-inflammatory molecules

## 4.1 Introduction

Microbiota in the human body is composed of around $10^{13}$ different species with at least numerous times gene complexity as compared to our genome. Microbiota is considered an essential component of the gut which surpasses human cells by manifold [1]. The composition of microbes varies among individuals due to various reasons such as genes, demographic area, diet, pathological conditions, and lifestyle. Microbiota such as protozoa, fungi, and bacteria can be altered due to mutation, irrational use of antibiotics, and dietary changes [2, 3]. The phyla abundantly present in the gut are Actinobacteria, Firmicutes, Proteobacteria, and Bacteroidetes [4]. Firmicutes, lactic acid producing gram-positive microbe are widely marketed as probiotics. In addition, Collinsella and Bifidobacterium species of phyla Actinobacteria have also been used as a functional food for humans. In contrast, Bacteroidetes and Proteobacteria cause up-regulation of macrophages which in turn induce pro-inflammatory molecules in the body. The human microbiota contains friendly as well as pathogenic bacteria which exists in complex symbiosis. Thus microbiota shows a vital role in human physiology and ailment by forming a synergetic association with the host and regulation of energy equilibrium. Also, the bacteria act as an immune system booster and suppress the colonization of many pathogens. Such mutual balanced relation is required for maintaining energy metabolism as well as immune function in the host and disturbances in equilibrium lead to systemic and non-systemic diseases. Therefore, some researchers considered microbiota as a "forgotten organ" [5–7]. Altered microbiota composition results in chronic inflammation and metabolic dysfunction. Furthermore, metabolites from microbiota contain short-chain fatty acids (SCFAs) that induce inflammatory reactions by modulating the activity of monocytes and macrophages in humans [8, 9].

Inflammation is the biological reaction shown by the body after antigen encounter and produces two responses, i.e., self-regulating defense response which initiates wound healing and amplification of inflammatory response with consequences series of diseases like atherosclerosis and cancer [10, 11]. The word inflammation in Latin means a fire that is characterized by swelling, redness, pain, and impaired body function. Redness occurs due to increased blood flow while swelling and pain are due to the accumulation of fluid and release of pro-inflammatory molecules. Loss of functions appears because of any noxious agents like bacteria, fungi, viruses and their products, in the body which ultimately triggers the inflammatory processes in the body and initiates the healing process. The inflammatory process proceeds due to the initial recruitment of neutrophils and later lymphocytes, macrophages, and plasma cells, results in the development and progression of chronic diseases [12, 13]. Thus by regulating the inflammatory processes in the body one can easily control the mortality and morbidity due to inflammatory reactions [14]. For instance, macrophages in the liver are termed as Kupffer cells while in nervous system they are termed as microglials [15]. Inflammatory and immune cells respond instantly against noxious stimuli and trigger inflammatory reactions in the body which destroys cells. In addition, the immune reaction also leads to the activation of an inflammatory defense system [16]. On pathogenic exposure, toll-like receptors are over-expressed on Kupffer cells and help in removing foreign molecules and protect against infection. Macrophages further activate antigen-presenting cells, phagocytes, and the generation of various cytokines that regulates the immune system. Thus, inflammation can occur due to microorganism attacks, trauma such as accidents, surgery, autoimmune disorders, and allergy [17, 18].

The inflammatory cells usually involved are dendritic cells, macrophages, and mast cells. Activation of these inflammatory cells due to any harmful stimulus will lead to the production of inflammatory mediators and lead to clinical symptoms of inflammation. Generally speaking, inflammation includes four stages, that is, vasodilation leads to increased blood flow, increased vascular permeability, leukocyte recruitment, and metabolic disorders [19]. Acute inflammation involves the accumulation of neutrophils in affected tissues, while chronic inflammation involves macrophages, lymphocytes, and plasma cells accumulation. While during an allergic reaction, eosinophils and T-lymphocytes are accumulated at the site of inflammation. In acute inflammation, the concentration of amyloid A protein and C-reactive protein abruptly rises. In addition, the level of certain biomarkers like interleukins and interferons in serum moderately rises in acute inflammation. For instance, in cardiac disorders and obesity, the CRP is markedly increased as compared to normal individuals [20, 21].

In the intestine, a tightly regulated immune system provides protection against foreign particles like pathogens and all together suppresses excessive immune activity. Since the enteric system is the main site of pathogen entry, so a specialized system, Gut-Associated Lymphoid Tissue (GALT) is present which acts as a checkpoint for antigen entry and determinant for inflammatory responses [22]. Macrophages play three major roles in any inflammatory process, i.e., phagocytosis, antigen presentation, immune modulation by regulation of growth factors, and cytokines production. Macrophages through the production of cytokines and inflammatory biomarkers contribute to beneficial as well as harmful outcomes of inflammatory reactions by phagocytosing and destroy the foreign invader [23]. Thus, macrophages show a crucial part in inflammation initiation and progression. The activation of pro-inflammatory M1 macrophages by certain pro-inflammatory molecules like IL-1β and IFN-γ leads to the genesis of TNF-α, IFN-γ, IL-6, IL-12, and ROS which cause an inflammatory reaction [24]. On the other hand, when M2 macrophages are activated by IL4, IL13, IL10, they initiate the production of biomolecules involved in anti-inflammatory activity, such as TGFβ, IL1, and IL10receptor antagonists [25]. Thus, M2 macrophages block the inflammatory process and promote healing and type-II immunity. Thus, depending upon the subsets of macrophages activated modulation of inflammatory and immune reactions occur [26].

## 4.2 Human Microbiota

Microbiota is distributed in the entire GIT, starting from the buccal cavity till the end of the alimentary canal, i.e., rectum. Around 108–1000 CFU (colony forming unit) of bacteria are present in 1 g saliva. While in the stomach, duodenum, and jejunum the count is approximately 102–104 CFU/g of the content. But this count increases to 1010 CFU in the distinct part of intestine. Bacteria in these colonies belong to phyla Firmicutes, Bacteroidetes, Actinobacteria, Proteobacteria, and Verrumicrobia. Besides, the profile of microflora in an individual remains stable over time [27].

The studies reveal that the concentration of Firmicutes, Proteobacteria, Actinobacteria, and Fusobacteria is very high in the oral cavity as compared to other phyla [28]. The clinical studies reveal that human saliva contains more than 100 genera of bacteria such as Veillonella, Streptococcus, Neisseria, Porphyromonas, Prevotella, Rothia, Haemophilus, and Fusobacterium. The most prominent genus in the saliva is Streptococcus, which accounts for 23% of the total population [29, 30]. Moreover, individuals belonging to different geographical locations shown significant variation in the bacterial sequencing as compared to the individuals belong to the same geographical area. For instance, Enterobacter concentration is high in the Congo population while it is entirely lacking in people of China, California, Turkey, and Germany. Similarly, the people of Bolivia contain a high frequency of Serratia [31]. In the stomach, the frequency of microflora is very limited due to the corrosive effect of gastric juice containing hydrochloric acid and proteolytic enzymes. In spite, certain phyla like Actinobacteria, Firmicutes, Proteobacteria, Bacteroidetes, and Fusobacteria are abundantly present in the stomach. Besides the genera, Prevotella, Streptococcus, Haemophilus, Porphyromonas, and Neisseria were also present in gastric juice [32, 33]. Furthermore, microbiota in the jejunum and ileum of middle-aged subjects is mainly composed of Firmicutes, Proteobacteria, Bacteroidetes, Fusobacteria, and Actinobacteria. Out of these Streptococcus and Gammaproteobacteria are the dominant sequences [34]. The concentration of microbiota in the large intestine, i.e., in the colon and rectum is mainly dominated by Firmicutes and Bacteroidetes, while Proteobacteria (*Acinetobacter johnsonii*, *E. coli*, *Sutterella wadsworthensis*, and *N. subflava*), Verrucomicrobia, and *Fusobacterium varium* are also present. The clinical studies reveal that in old age Firmicutes and Proteobacteria are the dominant microflora in the colon and rectal segment. The former was represented by *Streptococcus salivarius* and *Butyrivibrio fibrisolvens* subgroups while the latter by Klebsiella and Escherichia subgroups [34, 35]. The quantity and quality of microbiota differ widely between individuals of different age groups and geography. Instead, *E. coli*, *Bacteroides vulgatus*, *Bacteroides uniformis*, and *Ruminococcus torques* are identified in every individual [36].

### 4.2.1 Gut Microbiota and Intestinal Mucosa Development

Microbiota in the body possesses protective, metabolic as well as immunological functions. Microbiota along with digestive enzymes, mucins, and the mucosal epithelial barrier is termed as a non-immune component of immunity [37]. Microbiota significantly influences the development of an immune system in the intestinal and extra-intestinal area mainly by preventing pathogen intrusion either directly or indirectly via activation of the immune system [38]. Comparative pre-clinical studies demonstrate that animals containing microbiota show sufficient immune response against antigen as compared to microbiota-free animals. Even microbiota in the host body is also responsible for the development of ultrastructure, for instance, villus capillaries and angiogenesis. Furthermore, the development and activation of lymphoid tissues like gut-associated lymphoid tissue (GALT), lymphocytes, and antibodies are profoundly influenced by the microbiome [39, 40].

## 4.3 Inflammation and Gut Microbiota

Microbiome with their metabolites regulates the inflammatory processes in the host body and mediates significant role in various immune-inflammatory disorders such as multiple sclerosis, rheumatoid arthritis, Crohn's disease, and ulcerative colitis [41, 42]. Furthermore, alteration in microbiota composition results in the genesis of inflammatory diseases like bronchial asthma, diabetes mellitus, and obesity. For instance, bacteria of the Enterobacteriaceae family are linked with an inflammatory disease like IBD [43]. Similarly, *E. coli* has been associated with inflammatory bowel diseases. In addition, alteration in Firmicutes and Proteobacteria population is linked with many inflammatory conditions like obesity and diabetes mellitus [44]. It has been observed that non-alcoholic steatohepatitis patients had a major content of Proteobacteria in the body. In addition, in patients with RA, although the intestinal flora contained high levels of Lactobacillus and Prevotellacopri, the number of Bifidobacterium and Bacteroides was found to be decreasing [45, 46].

Major metabolites of the gut microbiota, such as short-chain fatty acids, exhibit strong anti-inflammatory effects primarily via the G protein-coupled receptor 3 (GPR 3), which is present on the cell membranes of macrophage immune cells [47, 48]. For instance, butyrate, a major metabolite of Ruminococcaceae, Eubacterium, Clostridia, and Firmicutes, significantly antagonizes inflammation and modulates the energy requirements of the body. Butyrates suppress the NF-κB pathway and thus regulate the innate immune response and inflammatory processes. Also, butyrate ameliorates inflammation by targeting the interferon-gamma and Peroxisome Proliferator-Activated Receptor-γ (PPAR-γ) signaling pathway [49].

Macrophages are important cells involved in chronic inflammatory conditions that exhibit inflammatory and anti-inflammatory roles against cytokines and microbial products. Macrophages are classified into two groups according to their activity [14]. That is, the classically activated M1 phenotype is activated by IFN-γ, which exerts inflammatory activity, and the activated M2 phenotype is generated in

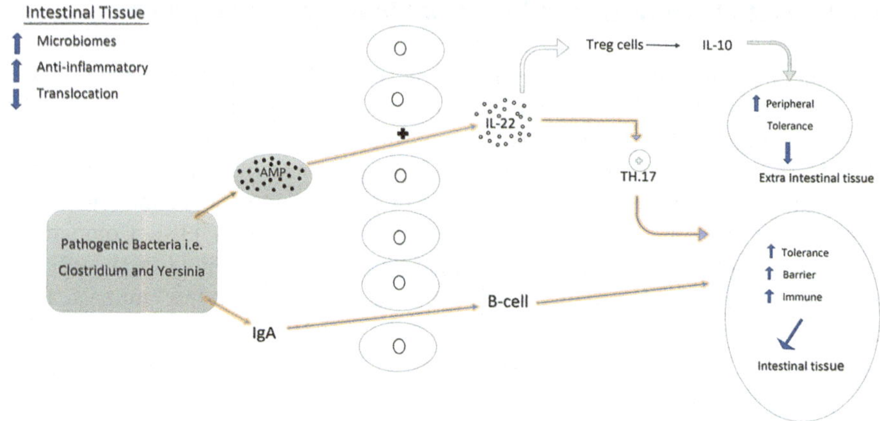


**Fig. 4.1** Regulation of inflammatory processes by microbiome [50]

response to IL-4 or IL-13 to exert anti-inflammatory action. Currently, researchers target macrophages phenotype for the management of various chronic inflammatory diseases such as atherosclerosis, pulmonary disorders. Various pre-clinical and clinical studies reveal that imbalance in M1 and M2 macrophages activity is linked with certain chronic conditions like rheumatoid arthritis. Thus, from the above discussion, it is clear that there is a strong association between macrophages and inflammatory diseases [16, 17] (Fig. 4.1).

### 4.3.1 Inflammation Driving Capacity

In certain conditions like chronic inflammatory diseases, the disease-causing agents sometimes disturbed the entire microbiota either by decreasing the bacterial diversity and/or overgrowth of bacteria responsible for inducing aggressive inflammatory and immune reactions such as *B. fragilis* and *E. coli* [51, 52]. Gram-negative bacteria possess lipopolysaccharide as the basic component of the outer cell membrane. Lipopolysaccharide comprises variable provinces of polysaccharide or oligosaccharide and lipid core area that is responsible for the shock-like condition in the host. Indeed, LPS interacts with macrophages and causes the release of inflammatory molecules such as TNF-α, IL-6, and IL-1 [53]. Aerobic *Escherichia coli* and anaerobic *B. fragilis* are gram-negative bacteria that contain high proportions of lipopolysaccharides and can induce immunity by stimulating the natural immune response [54]. Nutritional composition significantly affects the composition of the microflora. For example, a fat rich diet increases the fraction of gram-negative bacteria in the intestine and promotes the penetration of lipopolysaccharides through enterocyte membranes. For instance, in diabetes mellitus and obesity the high-fat diet results in increased lipopolysaccharide concentration which triggers systemic inflammatory reaction and insulin resistance. The fat diet also stimulates the release of bile juice in the duodenum which produces detergent action on the intestinal membrane and causes increased leakage of lipopolysaccharide. Besides gram-negative bacteria, many gram-positive bacteria like enterococcus are also involved in the alteration of microbiota and induce aggressive systemic inflammation [55, 56].

Pre-clinical and clinical studies also reveal that a combination of different taxa is a prerequisite for the development of a pathogenic state in the host. For example, in rodents for the onset of peritonitis, a combination of the systemic abscissa of anaerobic such as *B. fragilis* and *F. varium* and aerobic bacteria like *Enterococcus faecalis* and *E. coli* is required. Where *E. coli* is responsible for abscess development and high mortality rate and *B. fragilis* induces inflammatory processes, i.e., enhanced TNF-α production. In addition, *B. fragilis* also suppresses phagocytosis of *E. coli* and on the other hand, *E. coli* hinders intracellular killing of *B. fragilis* [57–59].

### 4.3.2 Bacterial Neutralization of Inflammation

Certain microbiota in the body are rarely involved in any inflammatory processes and even have the ability to counteract the inflammation. The mechanistic approach behind the anti-inflammatory potential of microbiota involves

1. Counteracting the bacteria that induce an aggressive inflammatory reaction.
2. Retards the penetration of lipopolysaccharides by improving the barrier effect.
3. Down-regulation of inflammatory and immune components.

Clinical studies reveal that some bacteria like *Faecalibacterium prausnitzii* (family Ruminococcaceae) combat systemic inflammation by modulating the release of IL-6 and IL-8 [60, 61]. Moreover, *F. prausnitzii* metabolites down-regulate the NF-κB pathway and IL-8 release. Furthermore, stimulation of mononuclear cells and regulation of anti-inflammatory molecules IL-10 and IL-12 lead to anti-inflammatory action. In humans, the most abundant strain responsible for the anti-inflammatory activity is Lactobacillus and Bifidobacterium. These strains suppress the undesired immune process such as allergic and autoimmune reactions [57].

## 4.4 Microbiota & Immune System

Furthermore, the intestinal immune system and gut microbiota work mutually and help the mucus defense system to discriminate the antigenic and non-antigenic mediators. On exposure to antigens such as bacteria and other pathogenic microorganisms, the intestinal immune system initiates and regulates innate and adaptive immune responses [62]. After antigen encounter, immune cells that react include lymphocytes, macrophages, and dendritic cells, through specialized membrane-bound receptors, known as toll-like receptors along with major histo-compatibility complexes I and II molecules. TLRs recognize pathogen-associated molecular patterns (PAMPs) such as lipids, polysaccharides, peptidoglycans, nucleic acids as well astragalin, and activate the immune system [63]. When enabled, the immune system initiates an intracellular signaling pathway, up-regulation of inflammation inducing genes and production of cytokines and interferon are responsible for inflammatory reactions. It also releases certain co-stimulatory molecules that activate the adaptive immune response [64, 65].

NOD-like receptors (NLRs) or nucleotide-binding domains are other family receptors resides in the cytoplasm and are responsible for detection of pathogens in mammalian cells. Around 20 different members are identified till now but NOD1 and NOD2 are most widely distributed in the body [66]. A large population of NLR is present in the body where TLRs are expressed at low concentrations in the GI tract. As the antigen starts invading the host cell, they interact with NLR present in the cytoplasm and thus defense mechanism will be activated [67]. NLR also regulates signaling pathways such as NF-κB and MAPK (mitogen-activated protein kinase) [68]. NOD1 can detect the peptidoglycan moiety present in the cell wall of gram-negative bacteria and NOD2 detects muramyl dipeptide that is present in a wide range of bacteria. Thus by regulating the inflammatory and immune signaling pathway, NLRs play a crucial role in various inflammatory diseases. Therefore, both NLR and TLR regulate inflammatory processes in the body against antigen encounters [69, 70].

In certain cases, inflammation occurs due to allergic or autoimmune diseases such as rheumatoid arthritis, diabetes mellitus-I, Crohn's disease, asthma, and multiple sclerosis. In addition, in certain metabolic alterations, mild systemic inflammation is also seen. Chronic inflammation leads to increased death rate and morbidity due to cancer, obesity, cardiovascular, respiratory and liver disorders. The microbiota in the body significantly regulates inflammation and the altered composition of microbiota can alleviate the inflammatory processes by enhancing the expression of pro-inflammatory cytokines [71].

### 4.4.1 Gut Microbiota and Innate Immune System

A mutual relationship between microflora and the innate immune system leads to normal growth, development, and regulation of inflammatory processes in the body. Such immune system can discriminate between the non-pathogenic and pathogenic microbial components by recognizing the specialized receptors such as "pattern recognition receptors" (PRRs) and TLRs as well as sequences present on the bacterial cell wall, for instance, pathogen-associated molecular patterns (PAMPs). These sequences are also termed microbe-associated molecular patterns (MAMP) because of associated molecules like lipopolysaccharides, peptidoglycans, flagellin, formylated peptides, etc. On mammalian innate immune cells such as macrophages, neutrophils, and dendritic cell TLR are present. NLR also modulates inflammatory signaling pathways such as nuclear factor kappa B (NF-κB), regulates immune activity by regulating TLR expression on the surface of immune sensory cells via the microbiota microbial recognition factor (MAMP), and finally nuclear factor kappa B (NF-κB) signaling which causes activation of inflammatory pathway and triggers co-stimulatory molecules of antigen-presenting cells that stimulate the production of inflammatory molecules as well as activation of T-cells [72–74].

Immune cells when encountered by any antigen initiate inflammatory reactions and specify immune processes. For example, activation of TLRs in dendritic cells after contact with bacterial components leads to activation of both the host's natural and adaptive immune responses. Perhaps epithelial cells are directly involved in immune cell activation mainly by transporting immunoglobulins secreted by plasma cells and its binding to specific receptors such as a receptor of polymeric immunoglobulins (pIgR) [75]. In humans, epithelial cells express the important LPS binding molecule CD 14 which along with TLR helps in the discrimination of self and non-self. In addition CD 14 may be involved in shaping the interaction between the immune system and microbiota [76, 77]. An epithelial cell also regulates other immune cells of the immune system and results in inflammatory reactions. Microbiota modulates the immune system by altering the production of mucin from goblet cells. Mucin by forming the complex with pathogenic bacteria directly suppresses the intestinal as well as systemic infections [78].

### 4.4.2 Gut Microbiota and Adaptative Immune System

In the bowel, the main immune compartment is lamina propria which constitutes dendritic cells, T-cells, and Ig-A secreting plasma cells. Gut-associated lymphoid tissue is primarily responsible for acquired immune response. Lamina propria contains certain regulatory cells responsible for maintaining tolerance to food and self-antigens [79]. In addition, CD 4 and CD 8 lymphocytes as well as Ig-A secreting plasma cells are also exist in the lamina propria. Antigen encountered is immediately taken up by dendritic cells in the lamina propria and transferred to MLN and lymphoid tissues. Furthermore, antigens are transported to antigen-presenting cells like macrophages. Dendritic cells are also responsible for the regulation of T-cells by continuing the balance between native T-cells and effector cells to maintain tolerance against antigens [80]. Besides, DC on antigen exposure expresses CD 103, which migrates to lymphoid tissue and generates effectors lymphocytes from native cells. Thus, dendritic cells remove antigen and pass away, leaving long-standing memory cells that can quickly respond to re-encounter with the same antigen. The entry of lymphocytes and dendritic cells into GALT depends upon the activation of adhesion molecules such as L-selectin, CCR7, and the interaction of CCL21. Through the secretion of IL-2, DCs are required for the generation of IFN-δ, which secretes a Th1 response, whereas the production of IL-1, IL-6, and IL-23 maintains the Th17 response. In addition, interactions with immature DCs and/or TGF-β and IL-10 lead to Treg cells involvement in anti-inflammatory activity. Native lymphocytes in lymphoid tissues alter the expression of adhesion molecules by suppressing the expression of L-selectin and CCR7, thereby inhibiting invasion into lymphoid tissues while enhancing the expression of novel adhesive agents that direct their migration to distinct tissues. This phenomenon is called "homing" [81].

## 4.5 Role of Microbiota in Allergic Ailments

The frequency of allergic diseases like eczema, asthma, hay fever increases globally in the last few decades, mainly in Western countries [82]. Genetic susceptibility and microbiome play a very important role in the prevalence of allergic diseases by regulating immune processes. In the human body, microflora develops immediately after birth but its concentration and type depend upon the type of delivery, breastfeeding, diet, and geography. As discussed earlier microflora influences the maturation of host immunity and its distortion may lead to the development of allergic disease. There are diverse reasons for this distortion such as dietary changes, irrational use of drugs like antibiotics, antacids, proton pump inhibitors, and NSAIDs [83, 84]. Strachan in 1989 proposed a hygiene theory which states that "a lack of microbial exposure during childhood results in a perturbation in gut microbiota composition and aberrant immune responses to innocuous antigens later in the life with the development of atopic diseases i.e. a chronic inflammatory disorders caused by aberrant T-helper 2 (Th2)-type immune responses against common innocuous environmental antigens (allergens) in susceptible individual" [85–87].

It has been suggested that a lack of microbial antigens may lead to immune aberrations from Th2 cytokine to Th1-type profile, resulting in enhanced Th2 cell responses to allergens. Th1-associated diseases include diabetes mellitus type-I, multiple sclerosis, and Crohn's disease while in chronic parasitic/worm infections high level of Th-2 responses is observed with a high risk of allergy. In contrast, regulatory T-cells by regulating the production of IL-10 and TGF-β as well as Th1 and Th2 responses reveal anti-inflammatory and immunosuppressive action. Thus Treg cells activity is linked with the development of atopic and autoimmune diseases [87, 88]. According to pre-clinical and clinical studies, the balance of Treg cells and TH2 cells is important for an allergic immune response to common environmental allergens. It was seen that individuals suffering from multiple food allergies were mainly linked with TGF-β producing Treg cells rather than Th2 cells as demonstrated in Fig. 4.2. Gut microbiota through the development of IgA antibody also involved in pathogen and allergen elimination [90, 91].

Gut microbiota modulates the immune response to even aeroallergens and leads to allergic airways manifestations. In pre-clinical studies, it has been seen that alteration in gut microbiota due to overuse of antibiotics may predispose to allergic airway disease. It was observed that treatment with *Lactobacillus reuteri* suppresses allergic airway manifestations. These studies support that host-microbiota communication may affect the systemic immune response and inflammatory process. Though, some inhaled antigens are also swallowed in the GI tract which induces tolerance through the sensors present in the GI tract. Besides, stimulated Treg cells migrate to different tissues throughout the body as mentioned in Table 4.1 [107, 108].

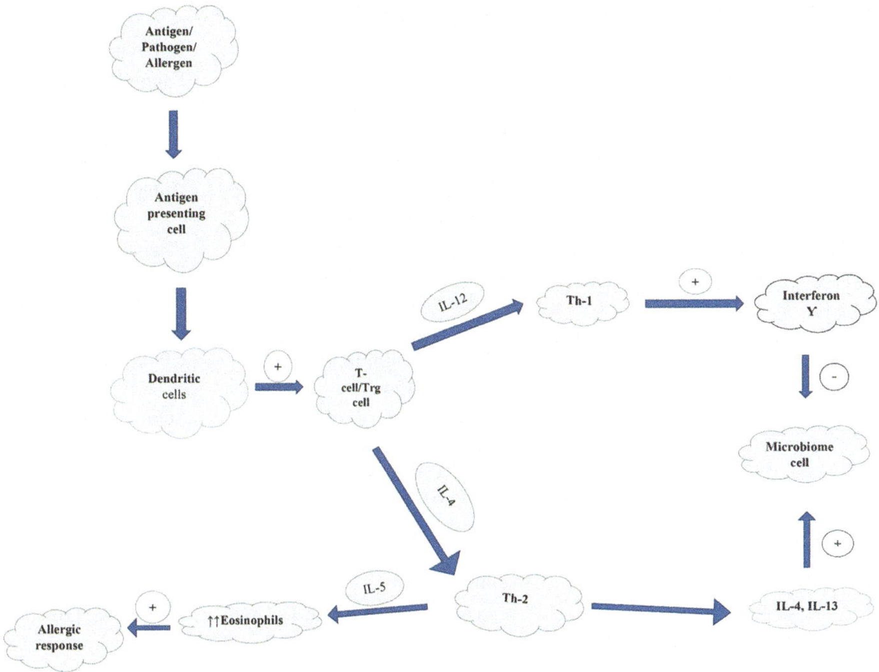


**Fig. 4.2** Role of microbiota and immune cells in atopic diseases [89]

## 4.6 Microbiota and Pulmonary System

Historically, the lung was considered sterile in the human body, but recent studies have shown that the lower respiratory tract contains different bacterial communities in both healthy and diseased states [109]. In the early days of the study, the total gut microbiota was mainly considered, but the field of total lung microbiological research is developing rapidly, and challenging observations are made on the relationship between lung microbiota and respiratory diseases. Studies of lung microbial communities can reveal new insights and methods of the pathogenic mechanism of lung infections [110, 111]. Lung infections have always been the leading cause of the global burden of disease, as have several lungs previously thought to be only indirectly related to the pathogenesis of microbial disease.

The lungs of fetus, just like intestines, have to be sterile and infant's lungs are probably to collect microbial groups after birth. Soon after delivery, the baby's mucosal floor hastily fills with microbes (intestinal and vaginal microbes in normal delivery, dermal microbiota in caesarean section) [112]. Initially microbiota in infants is uniform at various body sites but after some days or weeks the microbial diversity becomes site-specific [113]. The communities of microbiota in lung still now have no longer been studied in toddlers, however current longitudinal research

**Table 4.1** Influence of the microbiota on host tolerance to pathogens [92]

| S. no. | Pathogens | CFU | Results | References |
|---|---|---|---|---|
| 1 | *Klebsiella pneumoniae* | CFU Inc | Increase in blood neutrophils count and IL-10 level in lungs but, TNF-α level declines | Fagundes et al. [93] |
| 2 | *Pseudomonas aeruginosa* | CFU Inc | Low level of IL-10 and IL-1 in lungs and suppression of apoptotic mechanism in lungs | Fox et al. [94], Robak et al. [95] |
| 3 | *E. coli* | CFU Inc | Increased level of IL-6 and IL-1 with reduction inTNF-α. Modulates NF-κBsignaling pathway | Chen et al. [96] |
| 4 | *K. pneumoniae* | CFU Inc | Marked reduction in the level of IL-6 and TNF-α. | Clarke [97], Brown et al. [98] |
| 5 | *S. pneumoniae* | CFU Inc | Modulates the level of IL-17 and GM-CSF. Level of TNF-α decreases while IL-1 and 6 declines | Schuijt et al. [99] |
| 6 | *P. aeruginosa* | CFU Inc | Increase count of neutrophils, IL-6 and CXCL2 while IgA level declines | Robak et al. [95] |
| 7 | *Mycobacterium tuberculosis* | CFU Inc | Treg cell increases while Th1 cells decline | Khan et al. [100] |
| 8 | *Influenza virus* | – | Stimulation of natural killer cells and myeloid cells. Up-regulation of IL-1 and IL-33 while down-regulation of IFN-γ, MCP-1, TNF-α and IL-6. Level of IgA increases | Belkacem et al. [101], Youn et al. [102], Park et al. [103] |
| 9 | *P. aeruginosa* | CFU Inc | Marked improvement of histology. Serum level of IL-6 declines while the level of Il-10 increases. Treg cells increase in lungs. | Khailova et al. [104] |
| 10 | *S. pneumoniae* | – | Level of IgA and IgG increases. Serum level of IL-10 rises. | Racedo et al. [105] |
| 11 | *K. pneumoniae* | CFU Dec | Survival ↗; body weight loss | Vieira et al. [106] |

of toddlers with cystic fibrosis has proven that there may be consistency among the intestine microbiota and the breathing microbiota, and there may be proof that the gastrointestinal tract takes precedence [114]. Analysis of relative abundance of phylum reveals that the most widespread phyla seen are Bacteroides, Firmicutes, and Proteobacteria. A genus regarded in whole some human beings is Pronotera, Veillonella, Streptococcus, and Pseudomonas. Active tobacco intake appears to alternate the configuration of the microorganism of the higher airway. The variation in the microbiota of the individuals suffering from cystic fibrosis in comparison to the samples acquired from the western countries has been observed and the geographical variation in microbial content seen in the intestine but not reported in lungs [115–117].

### 4.6.1 Microbiota in Diseased Lung

**Cystic fibrosis**: Cystic fibrosis (CF) has been linked with persistent pulmonary infections, was first described in the 1930s, and is considered a valid reason high morbidity and deaths in patients suffering from respiratory disease. Common pathogens include *Staphylococcus aureus, H. influenza,* and *Pseudomonas aeruginosa*. Early eradicate able and intermittent infection occurs in immune compromised patients with pathogens like *Stenotrophomonas maltophilia* and *Burkholderia cepacia* complex. Analytical methods reveal that numerous microbiota is present in the airways of CF patient [118, 119]. Rogers et al. demonstrate that patients suffering from pulmonary infection possess diverse bacterial species [120, 121].

Follow-up studies have again demonstrated different microbial properties for colonization and respiratory infection in cystic fibrosis, including many species that could not be identified by culture techniques. This knowledge may partly explain the long-term difference between the indications for sputum culture and the resulting sensitivity and response of the patient to antibiotic therapy. The clinical significance of this observation indicates that the administration of antibiotics does not "suppress" the infection but changes the internal structure of the dynamic and heterogeneous microbial community. The cystic fibrosis lung microbiome contains several unknown or unrecognized bacterial species which are pathogenic and leads to inflammation and destruction of the airways in cystic fibrosis [122, 123]. Studies have confirmed that the diversity of the lung microbiota in cystic fibrosis varies with time or the severity of the disease. The diversity of bacterial communities decreased with age and severity of airway obstruction. Studies have shown that compared to heterozygous ΔF508 and non-ΔF508 patients, homozygous patients with ΔF508 mutations have less microbial diversity. Klepac-Ceraj et al. found that microbial multiplicity decreases with age, confirming the negative effects of diversity on respiratory and systemic antibiotics [124].

A recent study compared lung tissue, sputum, and throat samples obtained from subjects with chronic cystic fibrosis and reveals that sputum significantly represents the microbiome predominant in the lower respiratory tract cells that were relatively homogeneous, but over-represented complexity of atypical bacterial species. The majority of the lower respiratory tract contains one to three typical respiratory pathogens specimens. The clinical trial investigated whether enteral probiotic use affects the frequency and severity of lung injury in patients with cystic fibrosis. These studies showed that compared with placebo-treated patients and compared with subjects before and after probiotics, the frequency of lung deterioration was significantly reduced. Both studies hypothesized that reducing inflammation of the gut wall might have an indirect effect on lung inflammation, which could be beneficial [125].

**Asthma and Allergic Airways Disease**: Over the years, it has been recognized that there is a correlation between the decrease in the frequency of infections in children and initiation of allergic reaction and asthma. This has led to the "hygiene hypothesis" that reduced exposure to infections in early life will lead to impaired mucosal tolerance and self-increased immunopathology. A variety of research has found an affiliation among early adolescence exposure to antibiotics and initiation of bronchial allergies and allergies, which has led to the initiation and progression of various diseases due to alteration in microbiota [126]. Hilty et al., analyze the microbiome of oral and nasal samples from asthmatics and chronic obstructive pulmonary disease patients and healthy controls. In asthmatic patients, the authors found higher frequencies of Proteobacteria and lower frequencies of Bacteroides compared to control. This rise in Proteobacteria was caused by Neisseria, Moraxella, and Haemophilus [127]. Thereafter, Huang et al. compare lung microbial from 65 poorly controlled asthmatics with lung microbial plants from 10 controls patients and revealed an increased bacterial load and bacterial diversity in asthmatics [128]. They identified an increase in the concentration of Proteobacteria in asthmatics and correlated with the degree of bronchial hypersensitivity. There was a significant difference between the microbiome composition of the lungs and the functional response to clinical intervention among patients who benefited from bronchial hypersensitivity when administered with clarithromycin.

Besides microbiome in lungs plays a vital role in allergic conditions, a further hypothesis is that the perturbation of the microbiome because of poor diet and antibiotic use in the Western European region results in mucosal disruption. Data supporting this conjecture consisted of correlations between asthma/allergic responses and the use of antibiotics in developed countries and correlations with disturbed fecal microbiome and atopic diseases. Confirmation of this conjecture in antibiotic treated and sterile rodents supports an association between alteration in the gut microbiota and pulmonary allergic reactions. Few studies indicate that significant improvement observed when Lactobacillus was given orally in allergic pulmonary inflammation [129–131].

**Chronic Obstructive Pulmonary Disease**: It has been speculated that chronic obstructive pulmonary disease (COPD) is caused by chronic pulmonary infections and inflammation [132]. Studies have shown that the bronchioles and alveoli of patients suffering from COPD patients possess unique microbes, which is likely associated with the intermittent onset of chronic disease progression and infectious exacerbation [133]. Microbial strains from samples of COPD patients grouped with asthmatics and healthy controls showed similar increases in the relative presence of Proteobacteria and a decrease in Bacteroides. It is noteworthy that the COPD sample, compared to the control, contained a significant amount of Haemophilus, the most commonly cultured microorganism for exacerbation of COPD [134]. A study conducted by ErbDownward et al. analyzed the microbial communities and transplanted lung tissue in patients with severe COPD was compared to smokers and non-smokers show no evidence of lung disease or obstructive ventilator impairment. They found a similar bacterial load and significantly reduced microbial diversity in subjects with severe respiratory obstruction. Common genera in COPD patients are Haemophilus, Streptococcus, Pseudomonas, and Prevotella. The researchers analyzed tissues in several parts of the same lung to reveal significant regional heterogeneity of lung microbial communities from the lungs of critically ill patients [135]. Sze et al. also studied surgically obtained lung tissue samples and found that the number of bacterial cells in COPD lung tissue was the same and that there was no significant difference in microbial diversity compared to the control group. Diversity is greater in both COPD subjects and controls than in CF patients. Other microbial communities were found in COPD subjects compared to smoking and non-smoking controls. The concentration of Proteus and Bacteroides was similar, but Firmicutes augmented compared to the control group due to increase in Lactobacillus count [136].

A survey of respiratory microbiota in eight COPD patients found the same common doors (Proteobacteria, Bacteroides, Actinobacteria, Firmicutes) as in previous studies. Among them, Prevotella, Haemophilus, Fusobacterium, Streptococcus, Moraxella, Acinetobacter, and Neisseria account for a large proportion. The most common doors of COPD and control samples are actinomycetes, Firmicutes, and Proteobacteria. A significant contribution of this study is that the microbial community of the sample is categorized according to the subject's exposure to inhalation, bronchodilators, and/or inhaled corticosteroids, as well as disease status. The data from the above study along with clinical observations that inhaled corticosteroids may cause severe pneumonia [137, 138].

**Lung Transplantation**: The death ratio and graft malfunction rate for transplant patients are still greater than other organ transplants, with infection and bronchiolitis obliterans syndrome accounting for the majority of the morbidity (BOS). Multiple correlations between microbial invasion and BOS have been found, even though the etiology is unknown [139]. Lung transplantation can alter the lung microbiota by causing a variety of alterations in the host's respiratory defense system, such as airway deformation, interruption of normal lymphatic pathways, and immune suppression induced by anti-rejection medications. Even the prevalence of single species (such as *P. aeruginosa* and *S. aureus*) in certain cases overpowered the presence of single species, according to a newly published research by comparing the microbiota of samples from lung transplant patients. The author additionally looked into the occurrence of fungal species and discovered a lot of Candida and *Aspergillus niger* [140, 141].

**Idiopathic Interstitial Pneumonia**: The study characterizes the microflora of individuals with idiopathic interstitial pneumonia, which is a heterologous group of chronic pulmonary illnesses. Friaza et al. investigated the microbial communities of 20 individuals suffering from idiopathic pulmonary fibrosis, non-specific and acute interstitial pneumonia. Similar respiratory infections (for example, *Haemophilus influenza*) and previously unknown species were discovered. In the presence of pneumonia and bacterial load, negative relevance was seen, indicating an in vivo antagonistic relationship between Pneumocystis and bacterial species [142].

## 4.7 Conclusion

Intestinal microbial studies in humans and animal models reveal a large number of findings concerning the mechanism through which microorganism affects inflammation and tolerance against microbes as well as a mode of communication in microorganisms. The bidirectional connection between the gut microbial plexus surfaces and mucosal immunity is dynamic. Physical barriers, such as mucosal lining, the release of antimicrobial proteins by epithelium, and complex cell mediated immune response by IgA production, are among the guest's immunological mediators. Conversely, lymphatic maturation, epithelial repair through endotoxin signaling, initiates the balance of Th1/Th2 and the resistance to the mucous membrane of microorganisms can be promoted. However, it has been observed that abuse in enteric microorganism flora affects the sensitivity to allergic airway disease. Recent studies have also clarified a series of mechanisms that affect bacterial laminates, the gene expression of host epithelial cells, and activation of microbial toxic factors through small molecules similar to human hormones.

If the changed lung microbiome is linked to the disease's development, it will instantly pique the researcher's curiosity as a potential new treatment target. Through the use of probiotics, prebiotics, antibiotics, and molecular quorum-sensing inhibitors, the lung microbiome, like the microbiome in other compartments, may rectify ecological problems and restore "healthy" microbial populations. Targeting antibiotics to the narrower direct pathogens of the microbial spectrum without affecting the remaining members of the microbial community should be an aim for a better understanding of pulmonary microbiota diseases. Clinical investigations on the use of oral probiotics in gastrointestinal disorders have found that they are particularly effective in preventing antibiotic-associated diarrhea and treating acute infectious diarrhea. These findings also show that in lung diseases that involve a direct infectious component, such as pneumonia or bronchiectasis, the most appropriate first target of targeted microbial plant therapy may involve a direct infectious component. Oral probiotics are widely studied in the anticipation of respiratory infections, and there is evidence that they are useful. Pulmonary droplets of prebiotics or probiotics, enteral administration, have significantly varying effects on the lung microbial flora. Systemic and local inflammation, mucosal resistant airway reactivity, and microbial cooperation and antagonism are all likely to be affected by the production of true small molecule pharmacopeias by resident microorganisms.

## References

1. Zhang X, Shen D, Fang Z et al (2013) Human gut microbiota changes reveal the progression of glucose intolerance. PLoS One 8:e71108
2. Belkaid Y, Hand TW (2014) Role of the microbiota in immunity and inflammation. Cell 157:121–141
3. Pickard JM, Zeng MY, Caruso R et al (2017) Gut microbiota: role in pathogen colonization, immune responses, and inflammatory disease. Immunol Rev 279:70–89

4. Tap J, Mondot S, Levenez F et al (2009) Towards the human intestinal microbiota hylogenetic core. Environ Microbiol 11:2574–2584
5. Backhed F, Ding H, Wang T et al (2004) The gut microbiota as an environmental factor that regulates fat storage. Proc Natl Acad Sci U S A. 101:15718–15723
6. Mazmanian SK, Liu CH, Tzianabos AO et al (2005) An immunomodulatory molecule of symbiotic bacteria directs maturation of the host immune system. Cell 122:107–118
7. O'Hara AM, Shanahan F (2006) The gut flora as a forgotten organ. EMBO Rep 7:688–693
8. Correa-Oliveira R, Fachi JL, Vieira A et al (2016) Regulation of immune cell function by short-chain fatty acids. Clin Transl Immunol 5:e73
9. Sommer F, Backhed F (2013) The gut microbiota–masters of host development and physiology. Nat Rev Microbiol 11:227–238
10. Wellen KE, Hotamisligil GS (2005) Inflammation, stress, and diabetes. J Clin Invest 115:1111–1119
11. Xie W, Li M, Xu N et al (2013) MiR-181a regulates inflammation responses in monocytes and macrophages. PLoS One 8:e58639
12. Hakansson A, Molin G (2011) Gut microbiota and inflammation. Nutrients 3:637–682
13. Tracey KJ (2002) The inflammatory reflex. Nature 420:853–859
14. Hume DA, Ross IL, Himes SR et al (2002) The mononuclear phagocyte system revisited. J Leukoc Biol 72:621–627
15. Geissmann F, Manz MG, Jung S et al (2010) Development of monocytes, macrophages, and dendritic cells. Science 327:656–661
16. Yona S, Kim KW, Wolf Y et al (2013) Fate mapping reveals origins and dynamics of monocytes and tissue macrophages under homeostasis. Immunity 38:79–91
17. Fujihara M, Muroi M, Tanamoto K et al (2003) Molecular mechanisms of macrophage activation and deactivation by lipopolysaccharide: roles of the receptor complex. Pharmacol Ther 100:171–194
18. Gordon S (2002) Pattern recognition receptors: doubling up for the innate immune response. Cell 111:927–930
19. Molne J, Wold A (2007) Inflammation, 1st edn. Liber AB, Stockholm
20. Pepys MB, Baltz ML (1983) Acute phase proteins with special reference to C-reactive and related proteins (pentaxins) and serum amyloid A protein. Adv Immunol 34:141–212
21. Ridker PM (2003) Clinical application of C-reactive protein for cardiovascular disease detection and prevention. Circulation 107:363–369
22. Forchielli ML, Walker WA (2005) The role of gut-associated lymphoid tissues and mucosal defence. Br J Nutr 93:S41–S48
23. Palmblad J (1984) The role of granulocytes in inflammation. Scand J Rheumatol 13:163–172
24. Anderson CF, Mosser DM (2002) A novel phenotype for an activated macrophage: the type 2 activated macrophage. J Leukoc Biol 72:101–106
25. Gordon S (2003) Alternative activation of macrophages. Nat Rev Immunol 3:23–35
26. Mosser DM (2003) The many faces of macrophage activation. J Leukoc Biol 73:209–212
27. Eisen JA (2007) Environmental shotgun sequencing: its potential and challenges for studying the hidden world of microbes. PLoS Biol 5:e82
28. Costello EK, Lauber CL, Hamady C et al (2009) Bacterial community variation in human body habitats across space and time. Science 326:1694–1697
29. Huyghe A, Francois P, Charbonnier Y et al (2008) Novel microarray design strategy to study complex bacterial communities. Appl Environ Microbiol 74:1876–1885
30. Rheims H, Sproer C, Rainey FA et al (1996) Molecular biological evidence for the occurrence of uncultured members of the actinomycete line of descent in different environments and geographical locations. Microbiology 142:2863–2870
31. Nasidze I, Li J, Quinque D et al (2009) Global diversity in the human salivary microbiome. Genome Res 19:636–643
32. Bik EM, Eckburg PB, Gill SR et al (2006) Molecular analysis of the bacterial microbiota in the human stomach. Proc Natl Acad Sci U S A 103:732–737

33. Li XX, Wong GLH, To KF et al (2009) Bacterial microbiota profiling in gastritis without *Helicobacter pylori* infection or non-steroidal anti-inflammatory drug use. PLoS One 4:e7985
34. Wang M, Ahrné S, Jeppsson B et al (2005) Comparison of bacterial diversity along the human intestinal tract by direct cloning and sequencing of 16S rRNA genes. FEMS Microbiol Ecol 54:219–231
35. Hayashi H, Takahashi R, Nishi T et al (2005) Molecular analysis of jejunal, ileal, caecal and rectosigmoidal human colonic microbiota using 16S rRNA gene libraries and terminal restriction fragment length polymorphism. J Med Microbiol 54:1093–1101
36. Pettersson B, Ahrné S, Wang M et al (2003) The mucosa-associated bacteria from the sigmoid colon of nine healthy 60-year-old individuals, identified by bacterial 16S rDNA. Lund University Publications (LUP, Lund), Lund, Sweden
37. Liu Z, Li N, Neu J (2005) Tight junctions, leaky intestines, and pediatric diseases. Acta Pediatr 94:386–393
38. Srikanth CV, Mccormick BA (2008) Interactions of the intestinal epithelial with the pathogen and the indigenous microbiota: a three-way crosstalk. Interdiscip Perspect Infect Dis 2008:626827
39. Neu J (2007) Perinatal and neonatal manipulation of the intestinal microbiome: a note of caution. Nutr Rev 65(6 Pt 1):282–285
40. Rakoff-Nahoum S, Paglino J, Eslami-Varzaneh F et al (2004) Recognition of commensal microflora by toll-like receptors is required for intestinal homeostasis. Cell 118:229–241
41. Forbes JD, Chen CY, Knox NC et al (2018) A comparative study of the gut microbiota in immune-mediated inflammatory diseases-does a common dysbiosis exist? Microbiome 6:221
42. Yang HE, Li Y, Nishimura A et al (2017) Synthesized enone fatty acids resembling metabolites from gut microbiota suppress macrophage-mediated inflammation in adipocytes. Mol Nutr Food Res 61:1700064
43. Arrieta MC, Stiemsma LT, Dimitriu PA et al (2015) Early infancy microbial and metabolic alterations affect risk of childhood asthma. Sci Transl Med 7:307ra152
44. Matsuoka K, Kanai T (2015) The gut microbiota and inflammatory bowel disease. Semin Immunopathol 37:47–55
45. Liu X, Zou Q, Zeng B et al (2013) Analysis of fecal Lactobacillus community structure in patients with early rheumatoid arthritis. Curr Microbiol 67:170–176
46. Zhu L, Baker SS, Gill C et al (2013) Characterization of gut microbiomes in nonalcoholic steatohepatitis (NASH) patients: a connection between endogenous alcohol and NASH. Hepatology 57:601–609
47. Le Poul E, Loison C, Struyf S et al (2003) Functional characterization of human receptors for short chain fatty acids and their role in polymorphonuclear cell activation. J Biol Chem 278:25481–25489
48. Maslowski KM, Vieira AT, Ng A et al (2009) Regulation of inflammatory responses by gut microbiota and chemoattractant receptor GPR43. Nature 461:1282–1286
49. Kinoshita M, Suzuki Y, Saito Y (2002) Butyrate reduces colonic paracellular permeability by enhancing PPARgamma activation. Biochem Biophys Res Commun 293:827–831
50. Grigg JB, Sonnenberg GF (2021) Host-microbiota interactions shape local and systemic inflammatory diseases. J Immunol 198:564–571
51. Kleessen B, Kroesen AJ, Buhr HJ et al (2002) Mucosal and invading bacteria in patients with inflammatory bowel disease compared with controls. Scand J Gastroenterol 37:1034–1041
52. Swidsinski A, Weber J, Loening-Baucke V et al (2005) Spatial organization and composition of the mucosal flora in patients with inflammatory bowel disease. J Clin Microbiol 43:3380–3389
53. Gangloff SC, Hijiya N, Haziot A et al (1999) Lipopolysaccharide structure influences the macrophage response via CD14-independent and CD14-dependent pathways. Clin Infect Dis 28:491–496
54. Lindberg AA, Weintraub A, Zahringer U et al (1990) Structure-activity relationships in lipopolysaccharides of *Bacteroides fragilis*. Clin Infect Dis 12:S133–S141

55. Cani PD, Amar J, Iglesias MA et al (2007) Metabolic endotoxemia initiates obesity and insulin resistance. Diabetes 56:1761–1772
56. Cani PD, Bibiloni R, Knauf C et al (2008) Changes in gut microbiota control metabolic endotoxemia-induced inflammation in high-fat diet-induced obesity and diabetes in mice. Diabetes 57:1470–1481
57. Kim JM, Kim YJ, Cho YJ (2000) Synergy of *Bacteroides fragilis* and *Escherichia coli* in the induction of KC gene expression in mouse peritoneal tissues. Scand J Infect Dis 32:643–649
58. Onderdonk AB (2005) Animal models simulating anaerobic infections. Anaerobe 11:189–195
59. Onderdonk AB, Bartlett JG, Louie T et al (1976) Microbial synergy in experimental intra-abdominal abscess. Infect Immun 13:22–26
60. Biagi B, Nylund L, Candela M et al (2010) Through ageing, and beyond: gut microbiota and inflammatory status in seniors and centenarians. PLoS One 5:e10667
61. Collins MD, Lawson PA, Willems A et al (1994) The phylogeny of the genus *Clostridium*: proposal of five new genera and eleven new species combinations. Int J Syst Evol Bacteriol 44:812–826
62. Testro AG, Visvanathan K (2009) Toll-like receptors and their role in gastrointestinal disease. J Gastroenterol Hepatol 24:943–954
63. Zarember KA, Godowski PJ (2002) Tissue expression of human Toll-like receptors and differential regulation of Toll-like receptor mRNAs in leukocytes in response to microbes, their products, and cytokines. J Immunol 168:554–561
64. Poltorak A, He X, Smirnova I et al (1998) Defective LPS signaling in C3H/HeJ and C57BL/10ScCr mice: mutations in Tlr4 gene. Science 282:2085–2088
65. Schwandner R, Dziarski R, Wesche H et al (1999) Peptidoglycan- and lipoteichoic acid-induced cell activation is mediated by Toll-like receptor 2. J Biol Chem 274:17406–17409
66. Philpott DJ, Girardin SE, Sansonetti PJ (2001) Innate immune responses of epithelial cells following infection with bacterial pathogens. Curr Opin Immunol 13:410–416
67. Girardin SE, Tournebize R, Mavris M et al (2001) CARD4/Nod1 mediates NF-*kappa*B and JNK activation by invasive *Shigella flexneri*. EMBO Rep 2:736–742
68. Girardin SE, Boneca IG, Carneiro LA et al (2003) Nod1 detects a unique muropeptide from gram-negative bacterial peptidoglycan. Science 300:1584–1587
69. Chen G, Shaw MH, Kim YG et al (2009) NOD-like receptors: role in innate immunity and inflammatory disease. Annu Rev Pathol 4:365–398
70. Pickard KM, Bremner AR, Gordon JN et al (2004) Microbial gut interactions in health and disease. Immune responses. Best Pract Res Clin Gastroenterol 18:271–285
71. Cheroutre H (2004) Starting at the beginning: new perspectives on the biology of mucosal T cells. Annu Rev Immunol 22:217–246
72. Akira S, Hemmi H (2003) Recognition of pathogen-associatedmolecular patterns by TLR family. Immunol Lett 85:85–95
73. Medzhitov R, Janeway CJR (2000) Innate immune recognition: mechanisms and pathways. Immunol Rev 173:89–97
74. Ulevitch RJ (1999) Endotoxin opens the Tollgates to innateimmunity. Nat Med 5:144–145
75. Strober W, Fuss IJ, Blumberg RS (2002) The immunology of mucosal models of inflammation. Annu Rev Immunol 20:495–549
76. Bland PW (1998) Mucosal T cell–epithelial cell interactions. Chem Immunol 71:40–63
77. Funda DP, Tuèková L, Farre MA et al (2001) CD14 is expressed andreleased as soluble CD14 by human intestinal epithelialcells in vitro: lipopolysaccharide activationof epithelial cells revisited. Infect Immun 69:3772–3781
78. Linden SK, Florin TH, Mcguckin MA (2008) Mucin dynamicsin intestinal bacterial infection. PLoS One 3:e3952
79. Campbell DJ, Butcher EC (2002) Intestinal attraction:CCL25 functions in effector lymphocyte recruitmentto the small intestine. J Clin Investig 110:1079–1081
80. Macdonald TT, Monteleone G (2005) Immunity, inflammation, and allergy in the gut. Science 307(5717):1920–1925

81. Neutra MR, Mantis NJ, Kraehenbuhl JP (2001) Collaboration of epithelial cells with organized mucosal lymphoid tissues. Natl Immunol 2:1004–1009
82. Beasley R, Keil U, Von Mutius E et al (1998) World wide variation in prevalence of symptoms of asthma, allergicrhinoconjunctivitis, and atopic eczema: ISAAC. Lancet 351 (9111):1225–1232
83. Romagnani S (2007) Coming back to a missing immune deviation as the main explanatory mechanism for the hygiene hypothesis. J Allergy Immunol 119:1511–1513
84. Sepp E, Julge K, Vasar MC et al (1997) Intestinal microflora of Estonianand Swedish infants. Acta Paediatr 86:956–961
85. Palmer C, Bik EM, Digiulio DB et al (2007) Development of the human infant intestinalmicrobiota. PLoS Biol 5:e177
86. Strachan DP (1989) Hay fever, hygiene, and household size. Br Med J 299(6710):1259–1260
87. Romagnani S (2006) Regulatory T cells: which role in thepathogenesis and treatment of allergic disorders? Allergy 61:3–14
88. Romagnani S (2004) Immunologic influences on allergy and the TH1/TH2 balance. J Allergy Clin Immunol 113:395–400
89. Purchiaroni F, Tortora A, Gabrielli M et al (2013) The role of intestinal microbiota and the immune system. Eur Rev Med Pharmacol Sci 17:323–333
90. Akdis M, Verhagen J, Taylor A et al (2004) Immune responses in healthy and allergic individuals are characterized by a fine balance between allergen-specific T regulatory1 and T helper 2 cells. J Exp Med 199:1567–1575
91. Rautava S, Kalliomaki M, Isolauri E (2005) New therapeutic strategy for combating the increasing burdenof allergic disease: probiotics – A Nutrition, Allergy, Mucosal Immunology and Intestinal Microbiota (NAMI) Research Group report. J Allergy Clin Immunol 116:31–37
92. Dumas A, Bernard L, Poquet Y et al (2018) The role of the lung microbiota and the gut–lung axis in respiratory infectious diseases. Cell Microbiol 20:e12966
93. Fagundes CT, Amaral FA, Vieira AT et al (2012) Transient TLR activation restores inflammatory response and ability to control pulmonary bacterial infection in germfree mice. J Immunol 188:1411–1420
94. Fox AC, McConnell KW, Yoseph BP et al (2012) The endogenous bacteria alter gut epithelial apoptosis and decrease mortality following Pseudomonas aeruginosa pneumonia. Shock 38:508–514
95. Robak OH, Heimesaat MM, Kruglov AA et al (2018) Antibiotic treatment-induced secondary IgA deficiency enhances susceptibility to Pseudomonas aeruginosa pneumonia. J Clin Invest 128:3535–3545
96. Chen LW, Chen PH, Hsu CM (2011) Commensal microflora contribute to host defense against Escherichia coli pneumonia through Toll-like receptors. Shock 36:67–75
97. Clarke TB (2014) Early innate immunity to bacterial infection in the lung is regulated systemically by the commensal microbiota via NOD-like receptor ligands. Infect Immun 82:4596–4606
98. Brown RL, Sequeira RP, Clarke TB (2017) The microbiota protects against respiratory infection via GM-CSF signaling. Nat Commun 8:1512
99. Schuijt TJ, Lankelma JM, Scicluna BP et al (2016) The gut microbiota plays a protective role in the host defence against pneumococcal pneumonia. Gut 65:575–583
100. Khan N, Vidyarthi A, Nadeem S et al (2016) Alteration in the gut microbiota provokes susceptibility to tuberculosis. Front Immunol 7:529
101. Belkacem N, Serafini N, Wheeler R et al (2017) Lactobacillus paracasei feeding improves immune control of influenza infection in mice. PLoS One 12:e0184976
102. Youn HN, Lee DH, Lee YN et al (2012) Intranasal administration of live lactobacillus species facilitates protection against influenza virus infection in mice. Antivir Res 93:138–143
103. Park MK, Ngo V, Kwon YM et al (2013) Lactobacillus plantarum DK119 as a probiotic confers protection against influenza virus by modulating innate immunity. PLoS One 8: e75368

104. Khailova L, Baird CH, Rush AA et al (2013) Lactobacillus rhamnosus GG improves outcome in experimental Pseudomonas aeruginosa pneumonia: potential role of regulatory T cells. Shock 40:496–503
105. Racedo S, Villena J, Medina M et al (2006) Lactobacillus casei administration reduces lung injuries in a Streptococcus pneumoniae infection in mice. Microbes Infect 8:2359–2366
106. Vieira AT, Rocha VM, Tavares L et al (2016) Control of Klebsiella pneumoniae pulmonary infection and immunomodulation by oral treatment with the commensal probiotic Bifidobacterium longum 5(1A). Microbes Infect 18:180–189
107. Noverr MC, Huffnagle GB (2005) The microflora hypothesis of allergic diseases. Clin Exp Allergy 35:1511–1520
108. Noverr MC, Noggle RM, Toews GB et al (2004) Role of antibiotics and fungal microbiota in driving pulmonary allergic responses. Infect Immun 72:4996–5003
109. Mizgerd JP (2006) Lung infection – a public health priority. PLoS Med 3(2):e76
110. Cho I, Blaser MJ (2012) The human microbiome: at the interface of health and disease. Nat Rev Genet 13(4):260–270
111. Gill SR, Pop M, Deboy RT et al (2006) Metagenomic analysis of the human distal gut microbiome. Science (New York, NY) 312(5778):1355–1359
112. Dominguez-Bello MG, Costello EK, Contreras M et al (2010) Delivery mode shapes the acquisition and structure of the initial microbiota across multiple body habitats in newborns. Proc Natl Acad Sci U S A 107(26):11971–11975
113. Capone KA, Dowd SE, Stamatas GN, Nikolovski J (2011) Diversity of the human skin microbiome early in life. J Invest Dermatol 131(10):2026–2032
114. Madan JC, Koestler DC, Stanton BA et al (2012) Serial analysis of the gut and respiratory microbiome in cystic fibrosis in infancy: interaction between intestinal and respiratory tracts and impact of nutritional exposures. Am Soc Microbiol 3(4):e00251–e00212
115. Charlson ES, Chen J, Custers-Allen R et al (2010) Disordered microbial communities in the upper respiratory tract of cigarette smokers. PLoS One 5(12):e15216
116. Pragman AA, Kim HB, Reilly CS et al (2012) The lung microbiome in moderate and severe chronic obstructive pulmonary disease. PLoS One 7(10):e47305
117. Stressmann FA, Rogers GB, Klem ER et al (2011) Analysis of the bacterial communities present in lungs of patients with cystic fibrosis from American and British centers. J Clin Microbiol 49(1):281–291
118. Andersen DH (1938) Cystic fibrosis of the pancreas and its relation to celiac disease clinical and pathologic study. Arch Pediatr Adolesc Med 56(2):344–399
119. Han MLK, Huang YJ, Lipuma JJ et al (2012) Significance of the microbiome in obstructive lung disease. Thorax 67(5):456–463
120. Rogers G, Carroll M, Serisier D et al (2004) Characterization of bacterial community diversity in cystic fibrosis lung infections by use of 16s ribosomal DNA terminal restriction fragment length polymorphism profiling. J Clin Microbiol 42(11):5176–5183
121. Rogers GB, Carroll MP, Serisier DJ et al (2005) Bacterial activity in cystic fibrosis lung infections. Respir Res 6(1):49
122. Maughan H, Cunningham KS, Wang PW et al (2012) Pulmonary bacterial communities in surgically resected noncystic fibrosis bronchiectasis lungs are similar to those in cystic fibrosis. Pulm Med 2012:746358
123. Tunney MM, Field TR, Moriarty TF et al (2008) Detection of anaerobic bacteria in high numbers in sputum from patients with cystic fibrosis. Am J Respir Crit Care Med 177 (9):995–1001
124. Klepac-Ceraj V, Lemon KP, Martin TR et al (2010) Relationship between cystic fibrosis respiratory tract bacterial communities and age, genotype, antibiotics and Pseudomonas aeruginosa. Environ Microbiol 12(5):1293–1303
125. Cox MJ, Allgaier M, Taylor B et al (2010) Airway microbiota and pathogen abundance in age-stratified cystic fibrosis patients. PLoS One 5(6):e11044

126. Wills-Karp M, Santeliz J, Karp CL (2001) The germless theory of allergic disease: revisiting the hygiene hypothesis. Nat Rev Immunol 1(1):69–75
127. Hilty M, Burke C, Pedro H et al (2010) Disordered microbial communities in asthmatic airways. PLoS One 5(1):e8578
128. Huang YJ, Nelson CE, Brodie EL et al (2011) Airway microbiota and bronchial hyperresponsiveness in patients with suboptimally controlled asthma. J Allergy Clin Immunol 127(2):372–381. e371 373
129. Alm JS, Swartz J, Lilja G et al (1999) Atopy in children of families with an anthroposophic lifestyle. Lancet 353(9163):1485–1488
130. Bjorksten B, Naaber P, Sepp E et al (1999) The intestinal microflora in allergic Estonian and Swedish 2-year-old children. Clin Exp Allergy 29(3):342–346
131. Noverr MC, Falkowski NR, Mcdonald RA et al (2005) Development of allergic airway disease in mice following antibiotic therapy and fungal microbiota increase: role of host genetics, antigen, and interleukin-13. Infect Immun 73(1):30–38
132. Stuart-Harris CH, Pownall M, Scothorne CM et al (1953) The factor of infection in chronic bronchitis. Q J Med 22(86):121–132
133. Tager I, Speizer FE (1975) Role of infection in chronic bronchitis. N Engl J Med 292 (11):563–571
134. Albert RK, Connett J, Bailey WC et al (2011) Azithromycin for prevention of exacerbations of COPD. N Engl J Med 365(8):689–698
135. Erb-Downward JR, Thompson DL, Han MK et al (2011) Analysis of the lung microbiome in the "healthy" smoker and in COPD. PLoS One 6(2):e16384
136. Sze MA, Dimitriu PA, Hayashi S et al (2012) The lung tissue microbiome in chronic obstructive pulmonary disease. Am J Respir Crit Care Med 185(10):1073–1080
137. Huang YJ, Kim E, Cox MJ et al (2010) A persistent and diverse airway microbiota present during chronic obstructive pulmonary disease exacerbations. OMICS 14(1):9–59
138. Singh S, Amin AV, Loke YK (2009) Long-term use of inhaled corticosteroids and the risk of pneumonia in chronic obstructive pulmonary disease: a meta-analysis. Arch Intern Med 169 (3):219
139. Botha P, Archer L, Anderson RL et al (2008) Pseudomonas aeruginosa colonization of the allograft after lung transplantation and the risk of bronchiolitis obliterans syndrome. Transplantation 85(5):771–774
140. Borewicz K, Pragman AA, Kim HB et al (2012) Longitudinal analysis of the lung microbiome in lung transplantation. FEMS Microbiol Lett 339:57–65
141. Charlson ES, Diamond JM, Bittinger K et al (2012) Lung-enriched organisms and aberrant bacterial and fungal respiratory microbiota after lung transplant. Am J Respir Crit Care Med 186(6):536–545
142. Friaza V, La Horra C, Rodriguez-Dominguez MJ et al (2010) Metagenomic analysis of bronchoalveolar lavage samples from patients with idiopathic interstitial pneumonia and its antagonic relation with *Pneumocystis jirovecii* colonization. J Microbiol Methods 82 (1):98–101

# 5 Microbiome in Asthma

Khalid Saad Alharbi, Sattam Khulaif Alenezi, and Sulaiman Mohammed Alnasser

**Abstract**

Asthma is a prevalent persistent pulmonary illness that affects people all over the world. It affects people of all ages, although it is more common in children. Individual predisposition, virus infections, allergens contact, cigarette smoke exposure, and exposure to air pollution may all have a role in the onset and aggravation of asthma. Many studies have shown that the microbiome has a role in the regulation of immune function and the development of atopy and asthma. The lack of early life experience to the varied ambient microbiome required for colonisation of the pulmonary tracts and/or gastrointestinal appears to be the foundation of many clinical disorders. The establishment of a balanced, tolerogenic immune response requires symbiotic microorganisms. The discovery of symbiotic microorganisms in the pulmonary tracts and gastroenterological might be a game-changing and crucial development. The importance of microbiome in a healthy immune reaction is widely recognised, and gut dysbacteriosis has been linked to severe inflammatory respiratory diseases, notably asthma. To further understand the role of the microbiota in inflammatory process and its influence on important asthma risk variables including cigarette smoke and host genetic features, more study is needed.

**Keywords**

Microbiome · Asthma · Smoking · Interleukins · Cytokines

K. S. Alharbi (✉)
Department of Pharmacology, College of Pharmacy, Jouf University, Sakaka, Al-Jouf, Saudi Arabia
e-mail: Kssalharbi@ju.edu.sa

S. K. Alenezi · S. M. Alnasser
Department of Pharmacology and Toxicology, Unaizah College of Pharmacy, Qassim University, Unaizah City, Qassim, Saudi Arabia





G. Gupta et al. (eds.), *Microbiome in Inflammatory Lung Diseases*,
https://doi.org/10.1007/978-981-16-8957-4_5

## 5.1 Introduction

Asthma is a chronic illness that affects about 300 million individuals worldwide and is expected to reach 400 million by 2025. Every year, over 250,000 asthma-related fatalities are documented, most of which might be avoided. It is a chronic inflammatory condition of the lower respiratory tract marked by wheeze, breathlessness, chest tightening, and coughing, all of which can vary in incidence, recurrence, and intensity over time. Varying exhalation airflow problems, such as difficulties in breathing due to bronchospasm (constriction of the airways), stiffness of the airway wall, and enhanced mucus secretion resulting to extended exhalation time, are linked to these indications. Asthma is a multi-phenotypic illness with a wide range of clinical and pathophysiology features. Individual predisposition, virus infections, exposure to air pollution, tobacco smoke, and allergen may all have contribution in the establishment and aggravation of asthma. Ambient allergen exposure has a significant influence in all allergic inflammation causes [1, 2]. Mould fungus, animal epithelia, tree pollen, grass, and dust mites are the most frequent allergens implicated in asthma onset and aggravation. The frequency of allergic illnesses has risen rapidly during the last five decades, with significant differences in asthma prevalence between nations. Despite the fact that asthma complaints are particularly frequent in certain high-income nations, asthma is also prevalent in numerous middle- and low-income regions. Asthma is highly chronic in children in middle- and low-income nations than it is in high-income nations. Following that, TH2 cells play a key role in allergic disease development, causing B cells to flip isotypes and produce IgE antibodies that are unique to the allergen in question. TH2 cells recruit more mast cells and eosinophils, cause goblet cell proliferation, and increase bronchial over reactivity [3, 4]. The airway blockage that characterises asthma's diagnosis occurs in phases of symptom-free periods preceded by variable periods of aggravation, with a trigger or triggers in most cases. Exacerbations cause symptoms and lung function to deteriorate, and while they are usually recoverable and preceded by a recovery to normal pulmonary function, in some individuals, recurring exacerbations can result in a new, impaired baseline. Medications, pollution, aeroallergens, infection, and physical stimulation are examples of aggravating triggers. Enhanced inhalation treatment, systemic corticosteroid, and antibiotics are frequently used to treat these exacerbations [5].

The decrease in infection rates in Western nations, and most subsequently in developing nations, appears to be at the root of the rise in autoimmune and allergy illness cases. Several regulatory T cell subgroups and toll-like receptors are involved in the basic processes of this process, which are many and complicated (TLRs). Modifications in the microbiota induced by modifications in lifestyle may contribute to some of these processes [6, 7]. The first to indicate a relationship among microorganisms and allergies was the "hygiene theory". The hygiene hypothesis has now been broadened to include increased antibiotic usage and vaccines, as well as other lifestyle modification that have lowered adolescence illnesses and changed the microflora. Caesarean delivery and milk formula breastfeeding are also major prenatal and early postnatal variables. Another important concern is the modern diet's shift to a high-fat, low-fibre diet, which has substantial implications for the composition of the gut microbiome [8, 9].

The human microbiome is no longer regarded inactive, given the large number of diverse microbes that populate the human body, their close interaction with human cells, and their numerous impacts on the host. Human health is dependent on microbial populations in the digestive and pulmonary systems. The varied, sturdy, robust, and resilient adult microbiome, which comprises fungi, viruses, bacteria, and even archaea, is rich in many distinct species and can usually recover to its pre-perturbation condition. Many variables linked to lifestyle and ecological exposures can contribute to microbiota dysbiosis in the respiratory and gut in asthmatics [10, 11]. In the gut and lungs, an imbalance among symbiotic and pathogenic bacterial strains can promote immunological dysfunction and improper inflammatory reactions, but it is uncertain whether the imbalance is induced or result of disease. The function of the pulmonary and intestinal microflora in the establishment and improvement of respiratory health, as well as its abnormalities in various subgroups of asthma, will be the focus of this chapter.

## 5.2 Microbiome and Atopy

When exposed to an ecological antigen, particularly one inhaled or swallowed, atopy is characterised as a hereditary propensity to have an allergic response and create high amounts of IgE. Along with their genetic history, vulnerable people have an increased Th2-type immune reaction versus prevalent, non-pathogenic milieu antigen, which causes persistent inflammation in atopic illness. Th cells are key regulators of the immunological responses: Th17and Th2cells, in especially, direct immunological responses and integrate other cells (that is neutrophils, mast cells, eosinophils, or B cells,). Interleukin (IL)-13, IL-10, IL-5, IL-9, and IL-4 are among the cytokines produced by Th2 cells. IL-4 also activates B cells to generate IgE antibodies and eosinophils, which trigger mast cells to secrete leukotrienes, serotonin, and histamine, causing bronchoconstriction and leading to allergic reactions. The transition of naive T cells into IL-4-secreting T cells is among the hallmarks of allergies. The finding of germ-free animals, that are born and reared in a sterile milieu, suggests that the GIT microbiome may have a function in atopy [12, 13]. When compared to germ-free mice, they are more vulnerable to anaphylactic caused by oral antigens, indicating how challenging it is to develop oral resistance in animals with a different microbiome. The involvement of regulating T cells (Tregs) in inducing and maintaining oral tolerance is critical. Tregs are a kind of T cell that regulates immune functions, maintains self-antigen resistance, and protects from initiation of autoimmune diseases. In the past few years, it also became evident that symbiotic bacteria impact Treg induction, suggesting a connection among our ecology and allergy sensitivity. In the last 10 years, many investigations have proven the microbiota's significance in immunological regulation. Bacteroides fragilis, for example, regulates the Th1/2 equilibrium, and branched filamentous bacteria drive Th17 cell development directly, while Clostridium spp. increase Treg formation. In addition, the microbiota generates a variety of mediators, including short-chain fatty acids (SCFAs), peptidoglycans, lipopolysaccharides (LPS), and compounds having gaseous nature, all have impact on host biology based on tissue type, developmental time period, and dosage [14, 15]. In addition, giving LPS to germ-free mice were sufficient to reestablish oral resistance. Clostridium spp., for example, generate propionic acid after fermenting complex carbohydrate fibres, which is involved in cell signalling, immunological function (the formation of Tregs), and neurotransmitters production and secretion. Tryptophan is another essential molecule that is dependent on microbial metabolism. It can control serotonin synthesis and have a variety of major impacts on brain function, resulting to mental disorders. The innate immune reaction is made up of a multitude of defence mechanisms that respond to external stimuli in a nonspecific way. Both cellular and soluble components are involved in such reactions. The airway epithelium and mucosal layer offer innate immunity through immune cells such as innate lymphoid cells (ILCs), dendritic cells, and leukocytes such as macrophages, eosinophils, and neutrophils, in order to serving as a mechanical barricade. Nucleotide-binding oligomerization domain-like receptors, TLRs, and receptors like retinoic acid inducible gene I-like receptors detect various substrate patterns. These receptors perform a crucial function in enabling prompt reactions that lead to adaptive immunity. Leukocyte protease inhibitor, cathelicidin, interferons, lactoferrin, and defensins are among the noncellular released antimicrobial components in the host defence [16, 17].

### 5.2.1 Immune Mechanisms Influenced by the Microbiome

Bacteria can trigger regulatory reactions or decrease chronic inflammation via a variety of pathways, which have been identified. Inside the mucosa, both metabolites of microbiota and elements of bacterial cell wall have been linked to immunomodulatory actions. In mice, particular symbiotic bacteria including *Clostridium, Lactobacillus*, and *Bifidobacterium* species have been found to boost the percentage of T regulatory cells. Clostridium also causes ILC3s to produce IL-22, which helps to strengthen the epithelial barrier and decreases intestinal permeability to dietary peptides. Moreover, *Lactobacilli* and *Bifidobacteria* can enhance T regulatory cell activation by stimulating metabolic functions in dendritic cells including heme oxygenase-1, tryptophan metabolism, and vitamin A degradation. *Bacteroides fragilis* capsular polysaccharide A has been found to react effectively with mice plasmacytoid dendritic cells, promoting IL-10 production from $CD4^+$ T cells [18, 19]. Furthermore, a *Bifidobacterium longum* exopolysaccharide has recently been demonstrated to inhibit Th17 responses in the gastrointestinal and lungs. Bacterial-derived metabolites, in addition to bacterial-associated substances, have substantial impact on immunomodulatory mechanisms. The gut microbiome produces small-chain fatty acids like butyrate, propionate, and acetate, which have been found to impact T cell and dendritic cell reactions by interacting to GPCRs and inhibiting histone deacetylases, encouraging chromatin modifications. Biogenic

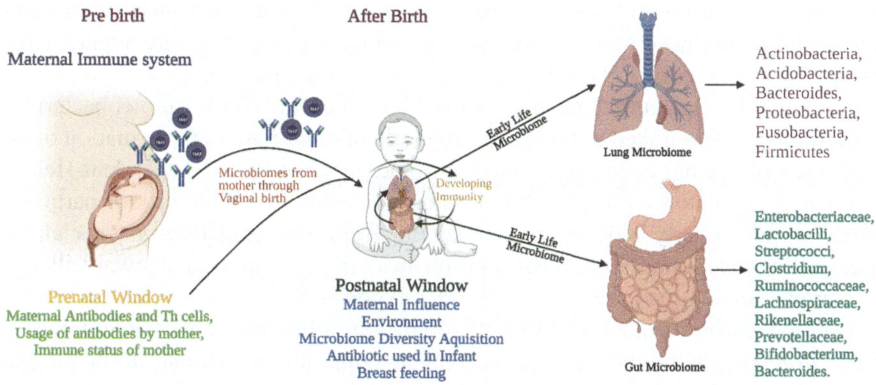


**Fig. 5.1** Influence of Maternal Immunity and Microbiome development in Neonate

amines are produced by microbes in the human GIT, and they can impact immunological and inflammatory activities. Microbiota-derived spermine, histamine, and taurine have been demonstrated to affect host–microbiome interaction in mice models by regulating downstream antimicrobial protein production, epithelial IL-18 production, and NLRP6 inflammasome signalling [20, 21] (Fig. 5.1).

### 5.2.2 Microbiome in Preclinical Asthma Models

A lot of preclinical investigations back up the idea that the microbiota has a contribution in the establishment of airway disorders. GF animals, who have never been exposed to non-pathogenic or pathogenic microbes, have given significant insights into the molecular involvement of the microbiome in the establishment of allergic airway inflammation. The scientists discovered that OVA causes considerably hypersensitivity and more type 2 airway inflammation in GF mice than in mice colonised with symbiotic microorganisms in a specific pathogen-free environment (SPF). Furthermore, for 3 weeks when GF animals were placed in same compartment for with SPF animals, the increased allergic inflammatory process in GF mice was decreased to the identical extent as in SPF mice, showing that GIT and lung recolonisation with symbiotic microorganisms had protective benefits. Furthermore, colonisation of GF mice early in life reduced the intensity of allergic airway reactions by preventing unique natural-killer-T-cells development in the lungs and GIT lamina propria [22–24]. Colonisation later in life showed no impact on unique natural killer T cell proliferation, regulatory T cell proliferation, or disease phenotypes. In addition, antimicrobial therapy of newborn mice led to a stronger Th2 cells and fewer regulatory T cells activity, which could be avoided by restoring a symbiotic gut microbiome. The researchers looked studied the vulnerability of mouse of various ages to allergic airway inflammation induced by HDM-house dust mite, replicating the circumstances of progressive colonisation of the human baby airways. In comparison to older mice, neonatal mice had increased airway eosinophilia, produced more type 2 cytokines, and had greater airway hyperreactivity after being exposed to HDM. The colonisation of the mice lungs with increasing quantities of bacteria and a transition from Firmicutes and Gammaproteobacteria to Bacteroidetes were linked to this preventive role in older mice. The formation of the lung microbiome has been connected to the emergence of PDL-1-dependent-Helios-negative-T-regulatory-cells [25, 26]. This research shows that the lack of particular bacterial species early in development may impact proper regulatory systems later in life, shifting the immunological equilibrium away from tolerance and toward allergy.

Various findings have shown that directly exposing the murine respiratory system to microorganism metabolites like CpG-containing oligonucleotides, endotoxin, or other TLR ligands might suppress the typical symptoms of asthma. In the allergic airway inflammation induced by the OVA model, for ex. intranasal administration to the bacteria E. coli proved preventive. New data linking the preventive impact of the agricultural atmosphere with the endotoxin and microbiome concentrations in home dirt have recently been added to these investigations. Furthermore, via Myd88 and Trif-dependent pathways, intranasal implantation of Amish households dust, but not Hutterite homes dust, decreased allergic airway inflammatory response induced by the OVA in mice [27, 28]. When compared to Hutterite household dirt, Amish household dust showed distinct microbial communities (particularly greater in Bartonellaceae) and greater endotoxin concentrations [29].

## 5.2.3 The Respiratory Microbiome's Role in Asthma

Because normal human pulmonary tissue was considered to be sterile at the period, the-Human-Microbiome-Project, which began in 2007, did not involve respiratory track tissue specimens. Nevertheless, a number of ground-breaking papers in this subject came immediately after, and numerous study associations and separate group began extensive investigations to define and explain the makeup and functioning of the pulmonary microbiome in wellness and illness. Presently, it is recognised that niche-specific microbial populations live in the healthy pulmonary mucosa. The upper respiratory tract has the densest bacterial populations, with around 103 live bacteria/nasal sample from nasopharynx and the nasal passage, and around 106/ml live cells from oropharyngeal lavages [30–32]. The predicted amount of bacteria in the trachea and lungs is smaller, with bronchoalveolar lavages from normal lungs containing about 102/ml bacterial cells. Actinobacteria, Acidobacteria, Bacteroides, Proteobacteria, Fusobacteria, and Firmicutes, are the six most common phyla reported in the lungs. Microbiome evaluations of brushings of bronchia, oropharynx, the nose, and BAL specimens from the lower respiratory track demonstrated that the phylum of *Proteobacteria*, particularly Haemophilus species, are frequently observed in the lower and upper airways of COPD and asthmatic patients of any age in the original evidence-based research. The research found that asthma patients which are suboptimally regulated, characterised as chronic signs on the Asthma-Control-Questionnaire-after-4-weeks of standardised therapy with aerosolized fluticasone, had a more diverse airway microflora than control participants, which was linked to bronchial hyper reactivity. In asthma patients, the phylum Proteobacteria, which includes the Pseudomonadaceae, Oxalobacteraceae, Nitrosomonadaceae, Sphingomonadaceae, Comamonadaceae, and families, increased [33, 34]. The patients who mostly benefitted from clarithromycin therapy, as measured by a decrease in airway hypersensitivity to methacholine, had considerably more microbial communities before to the treatment. Proteobacteria were found in greater numbers in asthmatic airways, according to subsequent investigations. Furthermore, when compared to mild-to-moderate asthma patients and normal individuals, species of Klebsiella were abundant in severe asthmatic patients. Additionally, Proteobacteria was linked to a TH17-related gene signature in the airway epithelium, which was linked to deteriorating asthma management and total leucocytes in the sputum in chronic asthma patients, whereas Firmicutes/Bacteroides was more prevalent in chronic asthmatic obese patients. The existence of Actinobacteria, on the other hand, was linked to improved and/or no change in asthma control [35, 36]. The existence of Mycoplasma pneumoniae and Chlamydophila pneumoniae has long been linked to chronic asthma, prompting numerous therapeutic studies with macrolide antibiotics in patients group. Nonetheless, in light of the conflicting research findings and the risk that essential bacterial species may be harmed, additional clinical studies with thorough microbiota investigations are required. The makeup of the airway microbiota grows dramatically early in life, and the lifestyle, health condition, and age can all impact it later in life. Birth method (vaginal or by caesarean surgery), early contacts, and the atmosphere of the first 3–4 months of life have all been demonstrated to have a critical role in the establishment of a healthy pulmonary and GIT microbiome, which is essential for pulmonary health later in life. The dust particulates that are inhaled can transmit a mixture of microorganisms and bacterial variables that impact asthma risk through their impact on innate and adaptive immune mechanisms, according to both animal and human research. The researcher examined the nasopharynx microbiota in a randomised cohort of children aged 2 to 2 years and found a link between the existence of particular microbial groups and acute pulmonary illnesses. Up to 2 months of age, healthy babies in this cohort were infected with Corynebacterium or Staphylococcus species, with subsequent sustained colonisation by Moraxella or Alloiococcus [37, 38]. In 1st 60 weeks after birth, however, colonisation with Streptococcus, Haemophilus, or Moraxella was linked to virus-associated acute pulmonary illnesses. Earlier asymptomatic Streptococcus colonisation, which is uncommon in children from pet-owning homes, elevated the incidence of asthma at the age of five. Initial colonisation of the upper pulmonary tract with, Mycobacterium catarrhalis, Haemophilus influenzae, and/or Streptococcus pneumoniae in 4-week-old infants from different birth groups was similarly linked to an elevated incidence of pneumonia, bronchiolitis, and asthma by the age of 5 [39, 40].

### 5.2.4 The Gut Microbiome's Function in Asthma

The human GIT microbiota is the body's biggest assemblage of bacteria, with 500–1000 various microorganisms and about eight million genes that might influence the immune function of the host. Verrucomicrobia, Proteobacteria, Actinobacteria, Firmicutes, and Bacteroidetes are the most common bacteria in the GIT microbiome of European adults. Aerobic bacteria such as Enterobacteriaceae, Lactobacilli, and Streptococci colonise the proximal small intestine, duodenum, and stomach, whereas anaerobic bacteria like Clostridium, Ruminococcaceae, Lachnospiraceae, Rikenellaceae, Prevotellaceae, Bifidobacterium, and Bacteroides species colonise the colon and distal small intestine. Multiple pathways exist for the gut microbiome to impact immunological activity in distant locations (such as the lungs) [41, 42]. It was also observed, for example, that the bacteria competent of secreting histamine from the stomach of adult asthma sufferers is higher than that of healthy subjects. Nevertheless, because histamine may generate defensive actions in the lungs through $H_2$ receptors as well as harmful ones through$H_1$ and $H_4$ receptors, it is unclear whether enhanced histamine production by GIT microorganisms can have an overall deleterious or preventive impact. By the age of three, the GIT microbiota is considered to have reached adult-like variety. Many environmental variables impact the establishment of the early life GIT microbiota, like living in microbially abundance surroundings (e.g., on a farmland or with regular exposure with animals and pets) or eating a varied diet, all of which have been linked to a lower risk of childhood asthma. It is considered that early life contact to and colonisation by specific microorganisms is crucial for pathogen tolerance, immune cell development, and gut development, all can safeguard versus the progression of asthma [43, 44]. The method of distribution has a big impact on colonisation. Infant born through caesarean delivery had higher levels of Acinetobacter, Firmicutes, Corynebacterineae, Propionobacterineae, Bacillales, and Staphylococcus species, with lower levels of Bacteroidetes and Actinobacteria, whereas vaginal birth has been related to greater Clostridia colonisation. Clostridia convert fibres to SCFAs, which have anti-inflammatory properties throughout the body. In addition to birth method and nutrition, antibiotic therapy in early infancy or maternal antibiotic usage during pregnancy has been linked to long-term consequences such as enhanced Bacteroidetes and Proteobacteria and reduced Actinobacteria. Numerous researches have connected early life GIT microbiome dysbiosis to a higher incidence of asthma later in life. The colonisation of C. difficile at 30 days of age was linked to wheezing for 1st 6–7 years after birth and asthma for the 12–14 years age of life [45, 46]. When compared to non-asthmatic infants, children who acquired asthma at school age had a reduced GIT microbiota diversification at age of 1 week or 30 days, but not at age of 1 year. The comparative richness of the bacterial genera Rothia, Faecalibacterium, Veillonella, and Lachnospira, in kids at danger of asthma was dramatically reduced in another research. Decreased faecal acetate levels and dysregulation of enterohepatic metabolites were also associated with this dysbiosis. Furthermore, newborns with the fewest comparative number of Faecalibacterium, Akkermansia, and Bifidobacteria, and the greatest comparative number of certain fungi

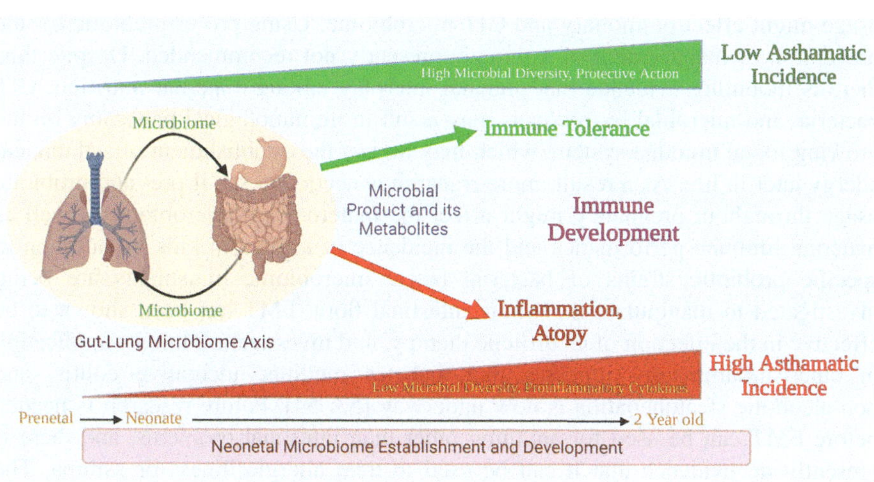


**Fig. 5.2** Gut–Lung Axis, Microbiome and Progression of Asthma

(Rhodotorula and Candida) had the greatest risk of developing asthma and atopy [47, 48] (Fig. 5.2).

## 5.3 Asthma Control, Therapy, and Management Using Microbiome Strategies

One possible approach presently being evaluated is the purposeful restoration of GIT and lung microbiome by the administration of synbiotics, probiotics, or prebiotics. Research in prebiotics and probiotics for their possible advantages in preventing airway inflammation is growing, especially as multiple lines of evidence show that a "healthy" bacterial population aids in the establishment of immunological tolerance. In vitro as well as animal experiments have consistently demonstrated that specific probiotic strains can reduce lung inflammatory responses, however not all probiotics have the same impact. Many variables confound the study of dysbiosis in asthma patients, which might explain why the results of intervention and preventive research in people are contradictory [49, 50]. Because of the significant variability in the prebiotics and/or probiotics utilised, research design, age of the sample participant, geographical area, sample size, and living variables, comparing human studies is challenging. In allergic asthmatic patients, one early investigation found that symbiotic (pro and prebiotic) usage increased maximum exhalation flow and decreased systemic synthesis of Th2 cytokines. Another current research found that combining nutritional interventions (extracts of vegetable and fish oils) with a probiotic resulted in considerable improvements in lung function variables and a lower need for short-acting corticosteroids and bronchodilators in asthma children, implying that combining various strategies may yield the best results [51, 52]. These outcomes are encouraging, but additional research is necessary to establish if pro and/or prebiotic usage might affect pulmonary and GIT microbiome. Using pro or prebiotic for the prevention or management of asthma is presently not recommended. Despite this, there is mounting evidence that prenatal interplay among maternal nutrition, GIT bacteria, and microbial by-products may result in immunological patterning on the growing foetal immune system, which may impact the establishment of asthma and allergy later in life. As a result, more research is needed to see if pre- and probiotic usage throughout pregnancy might affect the maternal gut microbiota, as well as maternal immune performance and the incidence of asthma in kids. In addition to specific probiotic strains of bacteria, faecal microbiome transplants are being investigated to manipulate the whole intestinal flora. FMT has been shown to be effective in the infection of C. difficile therapy, and investigation into its application in other inflammatory illnesses like diabetes mellitus, ulcerative colitis, and non-alcoholic steatohepatitis is now underway [53, 54]. Future research is needed before FMT can be used for anything other than intestinal problems, and there is presently no evidence that it can be used to treat allergic illness or asthma. The oncology area has done the most research on the impact of the microbiome in affecting precision medical strategies to patient treatment. According to mounting data, the microbiota impacts the intensity of treatment-related adverse actions in carcinoma patients as well as has a significant impact on effectiveness of therapy through pharmacodynamics and immune processes. A melanoma mice model, for example, demonstrated symbiotic microbe-derived anticancer immunity as shown by increased intratumoral CD8+ T cell proliferation. Bifidobacteria were shown to have the largest link to anticancer T cell immunity and the capacity to increase the efficiency of carcinoma immunotherapy anti-PD-L1-specific-antibody-therapy in this microbiome. Dendritic cell activity was improved by bifidobacteria, resulting in increased CD8+ T cell priming and concentration inside the tumour [55, 56]. For discovering novel microbiota-related disease processes, additional describing and characterisation of the microbiota linked with various asthma phenotypes are required. Furthermore, recognising these important bacterial species and their functional impacts would aid in the more accurate characterisation of asthma phenotypes, as well as the discovery of more appropriate “phenotype-specific” therapy options [57, 58].

## 5.4 Conclusion

In the last several years, significant research has emerged that link alterations in gut or lung microbial communities to asthma. Moreover, more research into the mechanisms by which components of the microbiome cause or modify inflammatory activities in asthmatic individuals is required. We expect that the ongoing discovery of novel bacterial species, their constituents, and metabolic products that affect mucosal immunomodulatory mechanisms will open up noval avenues for a bug-to-drug strategy in the management of asthmatic patients and the avoidance of lung disorders.

## References


1. Abdel-Aziz MI, Vijverberg SJH, Neerincx AH, Kraneveld AD (2019) Maitland-van der Zee AH. The crosstalk between microbiome and asthma: Exploring associations and challenges. Clin Exp Allergy 49(8):1067–1086
2. Adami AJ, Bracken SJ (2016) Breathing better through bugs: asthma and the microbiome. Yale J Biol Med 89(3):309–324
3. Adelman MW, Woodworth MH, Langelier C, Busch LM, Kempker JA, Kraft CS et al (2020) The gut microbiome's role in the development, maintenance, and outcomes of sepsis. Crit Care 24(1):278
4. Banskar S, Detzner AA, Juarez-Rodriguez MD, Hozo I, Gupta D, Dziarski R (2019) The Pglyrp1-regulated microbiome enhances experimental allergic asthma. J Immunol. 203(12): 3113–3125
5. Barcik W, Boutin RCT, Sokolowska M, Finlay BB (2020) The role of lung and gut microbiota in the pathology of asthma. Immunity 52(2):241–255
6. Barko PC, McMichael MA, Swanson KS, Williams DA (2018) The gastrointestinal microbiome: a review. J Vet Intern Med 32(1):9–25
7. Barnig C, Martin C (2018) [Asthma and the microbiome]. Rev Mal Respir 35(2):103–115
8. Bonamichi-Santos R, Aun MV, Agondi RC, Kalil J, Giavina-Bianchi P (2015) Microbiome and asthma: what have experimental models already taught us? J Immunol Res 2015:614758
9. Borbet TC, Zhang X, Müller A, Blaser MJ (2019) The role of the changing human microbiome in the asthma pandemic. J Allergy Clin Immunol 144(6):1457–1466
10. Budden KF, Shukla SD, Rehman SF, Bowerman KL, Keely S, Hugenholtz P et al (2019) Functional effects of the microbiota in chronic respiratory disease. Lancet Respir Med 7(10): 907–920
11. Carr TF, Alkatib R, Kraft M (2019) Microbiome in mechanisms of asthma. Clin Chest Med 40(1):87–96
12. Chen R, Wang L, Koch T, Curtis V, Yin-DeClue H, Handley SA et al (2020) Sex effects in the association between airway microbiome and asthma. Ann Allergy Asthma Immunol. 125(6): 652–7.e3
13. Cho Y, Shore SA (2016) Obesity, asthma, and the microbiome. Physiology (Bethesda). 31(2): 108–116
14. Das B, Nair GB (2019) Homeostasis and dysbiosis of the gut microbiome in health and disease. J Biosci. 44(5):117
15. Depner M, Taft DH, Kirjavainen PV, Kalanetra KM, Karvonen AM, Peschel S et al (2020) Maturation of the gut microbiome during the first year of life contributes to the protective farm effect on childhood asthma. Nat Med 26(11):1766–1775
16. Di Cicco M, Pistello M, Jacinto T, Ragazzo V, Piras M, Freer G et al (2018) Does lung microbiome play a causal or casual role in asthma? Pediatr Pulmonol 53(10):1340–1345
17. Dick S, Turner S (2020) The airway microbiome and childhood asthma - what is the link? Acta Med Acad 49(2):156–163
18. Dickson RP (2016) The microbiome and critical illness. Lancet Respir Med 4(1):59–72
19. Durack J, Lynch SV, Nariya S, Bhakta NR, Beigelman A, Castro M et al (2017) Features of the bronchial bacterial microbiome associated with atopy, asthma, and responsiveness to inhaled corticosteroid treatment. J Allergy Clin Immunol 140(1):63–75
20. Ege MJ (2017) The hygiene hypothesis in the age of the microbiome. Ann Am Thorac Soc. 14 (Supplement_5):S348–Ss53
21. Fazlollahi M, Lee TD, Andrade J, Oguntuyo K, Chun Y, Grishina G et al (2018) The nasal microbiome in asthma. J Allergy Clin Immunol. 142(3):834–43.e2
22. Frati F, Salvatori C, Incorvaia C, Bellucci A, Di Cara G, Marcucci F et al (2018) The role of the microbiome in asthma: the gut–lung axis. Int J Mol Sci. 20(1):123
23. Fujimura KE, Lynch SV (2015) Microbiota in allergy and asthma and the emerging relationship with the gut microbiome. Cell Host Microbe 17(5):592–602

24. García-Rivero JL (2020) The microbiome and asthma. Arch Bronconeumol 56(1):1–2
25. Guilleminault L, Williams EJ, Scott HA, Berthon BS, Jensen M, Wood LG (2017) Diet and asthma: is it time to adapt our message? Nutrients. 9(11):1227
26. Huang C, Shi G (2019) Smoking and microbiome in oral, airway, gut and some systemic diseases. J Transl Med 17(1):225
27. Huang YJ (2015) The respiratory microbiome and innate immunity in asthma. Curr Opin Pulm Med 21(1):27–32
28. Huang YJ, Boushey HA (2015) The microbiome in asthma. J Allergy Clin Immunol 135(1): 25–30
29. Huang YJ, Marsland BJ, Bunyavanich S, O'Mahony L, Leung DY, Muraro A et al (2017) The microbiome in allergic disease: Current understanding and future opportunities-2017 PRACTALL document of the American Academy of Allergy, Asthma & Immunology and the European Academy of Allergy and Clinical Immunology. J Allergy Clin Immunol 139(4): 1099–1110
30. Hufnagl K, Pali-Schöll I, Roth-Walter F, Jensen-Jarolim E (2020) Dysbiosis of the gut and lung microbiome has a role in asthma. Semin Immunopathol 42(1):75–93
31. Jartti T, Bønnelykke K, Elenius V, Feleszko W (2020) Role of viruses in asthma. Semin Immunopathol 42(1):61–74
32. Jartti T, Gern JE (2017) Role of viral infections in the development and exacerbation of asthma in children. J Allergy Clin Immunol 140(4):895–906
33. Kozik AJ, Huang YJ (2019) The microbiome in asthma: Role in pathogenesis, phenotype, and response to treatment. Ann Allergy Asthma Immunol 122(3):270–275
34. Langan SM, Irvine AD, Weidinger S (2020) Atopic dermatitis. Lancet. 396(10247):345–360
35. Mammen MJ, Scannapieco FA, Sethi S (2020) Oral-lung microbiome interactions in lung diseases. Periodontology 2000. 83(1):234–241
36. McAleer JP, Kolls JK (2018) Contributions of the intestinal microbiome in lung immunity. Eur J Immunol 48(1):39–49
37. McKenzie C, Tan J, Macia L, Mackay CR (2017) The nutrition-gut microbiome-physiology axis and allergic diseases. Immunol Rev 278(1):277–295
38. Michalovich D, Rodriguez-Perez N, Smolinska S, Pirozynski M, Mayhew D, Uddin S et al (2019) Obesity and disease severity magnify disturbed microbiome-immune interactions in asthma patients. Nat Commun 10(1):5711
39. Milani C, Duranti S, Bottacini F, Casey E, Turroni F, Mahony J et al (2017) The first microbial colonizers of the human gut: composition, activities, and health implications of the infant gut microbiota. Microbiol Mol Biol Rev. 81(4):e00036–e00017
40. Milliken S, Allen RM, Lamont RF (2019) The role of antimicrobial treatment during pregnancy on the neonatal gut microbiome and the development of atopy, asthma, allergy and obesity in childhood. Expert Opin Drug Saf 18(3):173–185
41. Moffatt MF, Cookson WO (2017) The lung microbiome in health and disease. Clin Med (Lond). 17(6):525–529
42. Mohan A, Grace J, Wang BR, Lugogo N (2019) The effects of obesity in asthma. Curr Allergy Asthma Rep 19(10):49
43. Moraes TJ, Sears MR, Subbarao P (2018) Epidemiology of asthma and influence of ethnicity. Semin Respir Crit Care Med 39(1):3–11
44. Noval Rivas M, Crother TR, Arditi M (2016) The microbiome in asthma. Curr Opin Pediatr 28(6):764–771
45. Ozturk AB, Turturice BA, Perkins DL, Finn PW (2017) The potential for emerging microbiome-mediated therapeutics in asthma. Curr Allergy Asthma Rep 17(9):62
46. Perdijk O, Marsland BJ (2019) The microbiome: toward preventing allergies and asthma by nutritional intervention. Curr Opin Immunol 60:10–18
47. Peroni DG, Nuzzi G, Trambusti I, Di Cicco ME, Comberiati P (2020) Microbiome composition and its impact on the development of allergic diseases. Front Immunol 11:700

48. Peters U, Dixon AE, Forno E (2018) Obesity and asthma. J Allergy Clin Immunol 141(4): 1169–1179
49. Pettersen VK, Arrieta MC (2020) Host-microbiome intestinal interactions during early life: considerations for atopy and asthma development. Curr Opin Allergy Clin Immunol 20(2): 138–148
50. Pisi G, Fainardi V, Aiello M, Bertorelli G, Crisafulli E, Chetta A (2017) The role of the microbiome in childhood asthma. Immunotherapy 9(15):1295–1304
51. Prince BT, Mandel MJ, Nadeau K, Singh AM (2015) Gut microbiome and the development of food allergy and allergic disease. Pediatr Clin N Am 62(6):1479–1492
52. Rick EM, Woolnough KF, Seear PJ, Fairs A, Satchwell J, Richardson M et al (2020) The airway fungal microbiome in asthma. Clin Exp Allergy 50(12):1325–1341
53. Scherzer R, Grayson MH (2018) Heterogeneity and the origins of asthma. Ann Allergy Asthma Immunol 121(4):400–405
54. Shore SA, Cho Y (2016) Obesity and asthma: microbiome-metabolome interactions. Am J Respir Cell Mol Biol 54(5):609–617
55. Shukla SD, Shastri MD, Chong WC, Dua K, Budden KF, Mahmood MQ et al (2019) Microbiome-focused asthma management strategies. Curr Opin Pharmacol 46:143–149
56. Sozańska B (2019) Microbiome in the primary prevention of allergic diseases and bronchial asthma. Allergol Immunopathol 47(1):79–84
57. Stiemsma LT, Michels KB (2018) The role of the microbiome in the developmental origins of health and disease. Pediatrics. 141(4):e20172437
58. Stokholm J, Blaser MJ, Thorsen J, Rasmussen MA, Waage J, Vinding RK et al (2018) Maturation of the gut microbiome and risk of asthma in childhood. Nat Commun 9(1):141

# Microbiome in Chronic Obstructive Pulmonary Disease (COPD)

6

C. Sarath Chandran, Anitha Jose Subin, Alan Raj, K. K. Swathy, and Indu Raghunath

**Abstract**

Chronic obstructive pulmonary disease (COPD) is a progressive irreversible lung disease, distinguished from other lung diseases by obstructed airflow from the lungs, manifested as breathlessness and predisposes to exacerbations and serious illness. Microbiome denotes the collective genome of all microbes comprising bacteria, bacteriophage, fungi, protozoa and viruses along with the host environment that they inhabit. The respiratory tract of healthful individuals accommodates an extensive microbiota that diminished biomass from the upper to the lower respiratory tract. In healthy persons, the respiratory microbiome has a balance by migration and eradication, whereas, in chronic diseases like COPD, reproduction of the resident bacteria occurs. Tobacco smoke and other chemicals can damage or weaken the lung's immune defences, thereby allowing pathogenic microbes to interact with epithelial and immune cells in the airways, which will trigger inflammation. Such inflammation can lead to acute changes in the microbiome of airways leading to extensive inflammatory responses, presented as exacerbations of COPD. Thus, the presence of a pathogenic microbiome in the airways and lungs can act as a marker for the progression of the disease and can act as a target for the therapy of the disease.

C. Sarath Chandran (✉) · K. K. Swathy
College of Pharmaceutical Sciences, Govt. Medical College Kannur, Pariyaram, India

A. J. Subin
Learning and Development, Life Health Group, Dubai, UAE

A. Raj
Manipal College of Pharmaceutical Sciences, Manipal, India

I. Raghunath
NSGM Institute of Pharmaceutical Sciences, Mangalore, India


G. Gupta et al. (eds.), *Microbiome in Inflammatory Lung Diseases*, https://doi.org/10.1007/978-981-16-8957-4_6

**Keywords**

COPD · Microbiome · Inflammation · Biomarker · Microbiota · Therapeutic target

## 6.1 Introduction

Chronic obstructive pulmonary disease (COPD) is a heterogeneous, advanced and irreversible lethal lung disorder. COPD is characterized by obstructed airflow from the lungs, manifested as breathlessness and predisposes to exacerbations as well as serious illness, in which frequent contagious diseases of bacterial origin are the most important etiological agent for the disease [1, 2]. COPD is considered a multidimensional disease condition with extremely complicated pathogenesis. The genetic risk factors to lung microbiomes are associated with the pathogenesis of COPD. This disease is associated with increased morbidity and mortality that leads to immense healthcare costs for those who are affected [3]. As per the Global Burden of Disease (GBD) study 2000, COPD was accountable for a projected 2.75 million mortalities globally [4]. Currently, COPD is the third foremost reason of death in the USA [5]. COPD marks 15.4 million doctor appointments, 1.5 million emergency department (ED) calls or drop-ins and 726,000 hospitalizations annually in the USA as well as is accountable for high healthcare costs ("Morbidity & Mortality: 2012 Chart Book on Cardiovascular, Lung, and Blood Diseases", [6]).

The microbiome refers to the "ecological community of commensal, symbiotic and pathogenic organisms that share our body space" which includes bacteria, bacteriophage, fungi, protozoa and viruses [7]. To be more specific, microbiome, microbiota and metagenome are related terms that should be defined separately. Microbiome refers to the microorganisms and their genes and how they confer to the wellbeing (or worsening of disease) of the human body, whereas microbiota refers only to various microbial taxa associated with humans. Metagenome simply refers to the genes in the host environment [7]. Microbiota that inhabits various parts of the human body like the gut, skin and lungs are important in maintaining tissue, organ as well as immune homeostasis [8]. Similarly, the respiratory microbiome in COPD gives a clear idea of disease genesis, progression, exacerbation and severity [9]. Microbiomes existing in as well as on the human body are not always invaders causing diseases, but beneficial colonizers. But the dysfunction of homeostasis (dysbiosis) causes the accumulation of disease-causing microbes, which changes the gene activity, metabolic responses or lead to abnormal immune response (Fig. 6.1) ("The Human Microbiome: What It Is, Why It Is Important and Opportunities for Microbiome-Based Therapeutics | American Pharmaceutical Review - The Review of American Pharmaceutical Business & Technology", [10]). The recent research suggested the significant role of the respiratory microbiome on airway illnesses like asthma as well as COPD. The respiratory microbiome identified in patients suffering from COPD was different compared to those not having the disease [3, 11]. Pathogenesis and manifestation of COPD or

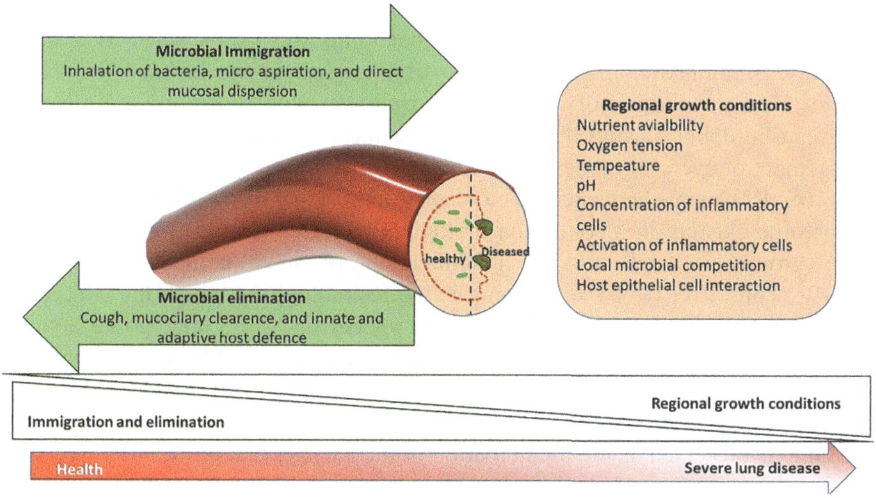


**Fig. 6.1** Determinants of the lung microbiome in robust and diseased lungs

other airway diseases with respect to the microbiome present at the specific site were little-noticed for a long time.

For a long time, the lung was considered a sterile body part inside the human body. So, the presence of lung microbiota went unnoticed for a quite long time. Healthy lungs have a mucous layer lining that can inhibit growth as well as the reproduction of microbes and acts as a low nutrient medium for the microbiota. In other words, bacterial growth or replication is not favoured by the environment of normal or healthy lungs. But it is proved that the respiratory tract of a healthy individual hosts an extensive microbiota which declines in biomass from the upper to the lower respiratory way. The upper respiratory tract and oropharynx which is abundant in microbiota are in direct communication from lungs, which in turn leads to the existence of microbiota in lungs too [12]. *Haemophilus, Fusobacterium, Neisseria, Streptococcus, Veillonella and Prevotella* are the utmost plentiful types in the lungs of healthy individuals, belonging to various phyla such as *Bacteroidetes, Firmicutes, Proteobacteria and Actinobacteria* [13].

The Human Microbiome Project (HMP) by the National Institute of Health (NIH) was originated in 2007 with the fundamental goal of identifying the major microbiome sites throughout the human body [14]. The lungs and airways are mislaid within the series of structures to be considered at first because of the false belief that a healthy lung is sterile as well as the difficulties to access the lower airways without invasive methods such as bronchoscopy [12, 15]. The first major tool HMP used for studying the microbiome in lungs was 16S ribosomal RNA (16S rRNA) sequencing, which is a culture-independent technique [16]. Cultures that are dependent upon the reproduction of microorganisms in conventional growth conditions fail to recognize much human-linked microbiota which became suggestive of many culture-independent analyses [17]. Culture-independent analysis of respiratory samples confirmed that lungs and airways act as hosts for large microbiota. Investigations based on gene encoding 16S rRNA acted as a strong ground for finding out the composition of microbes in the isolated respiratory samples. The distribution of microbial environment both in healthy and diseased lungs can be defined using the 16S rRNA sequencing technique that permits the arrangement of every microbe identified from the respiratory sample taxonomically [17]. 16S rRNA gene sequencing is a readily appropriate method that describes the microbiome existing in a composite biological mixture, permitting examination of entire groups as well as the identification of their component members [18, 19]. Thus, 16S rRNA sequencing is the technology by which the concept of "sterile lung" became outdated. Respiratory specimens while undergoing a single sequencing run may produce thousands of DNA sequences, whereas 16S ribosomal genes contain many "hypervariable regions" and may accurately give sequences that are highly specific among certain taxa of microbes [20].

COPD is presented as the inflammation of the airways that primes to irreversible deterioration of functioning of lungs and its capacity. There are numerous inflammatory cells as well as intermediaries related to the pathophysiology of COPD, out of which inflammation of the airways is of primary importance. The presence of the respiratory microbiome enhances airway inflammation, because of the interaction between pathogen and inflammatory response of the host [21–23] (Fig. 6.2).

### 6.1.1 Lung Microbiome

The microbiome in the lungs changes depending on the severity of the disease and its exacerbation [24]. Three factors influence the lung microbiome's configuration [25]. The first is the immigration of the microbiome either via direct mucosal dispersion, inhalation of microbes or microaspiration. The second factor is continuous and steady-state elimination by making use of the host's inborn as well as adaptive immune response, cough or mucociliary transport. The third factor is various growth conditions such as availability of nutrients, pH, the concentration of various immune cells, temperature, microbial competition, activation of inflammatory cells to carry out the host immune defence, and sometimes oxygen stress or oxygen tension [26]. In many cases, healthy individuals have unreceptive growing circumstances for the microbiota, whereas in the case of diseases such as COPD, conditions become advantageous for microbial growth. The use of corticosteroids and certain antibiotics is also a factor that may promote the growth of the lung microbiome [27].

The lung microbiome is transient as well as dynamic with no physical blockage in the airway that helps the migration of bacteria in a bidirectional way [28]. The composition of the microbiome in the bronchial tree is analogous to the one in the upper airways although the biomass is lower in the bronchial tree compared to the upper airways, implying that the microbiome migrates by aspiration [29, 30]. As per the results of research conducted by Charlson et al., the similarity pattern of the microbiome in the bronchial tree, as well as the oropharynx, indicated

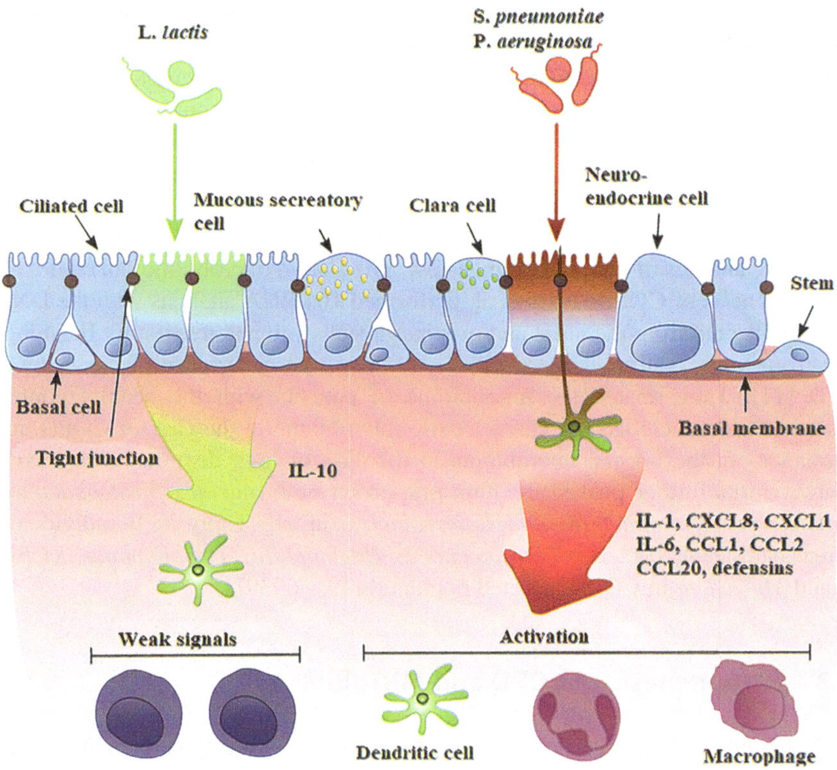


**Fig. 6.2** Interaction between microbiome and host inflammatory system

the absence of a unique microbiome in the lungs [29]. But the abundance of the microbiome in the lungs is lesser compared to that in the upper respiratory airways [31]. Also, the presence of more dead bacteria in the lung parenchyma of a healthy individual is observed compared to a mixture of dead and live bacteria in the upper airways.

One of the major etiological factors of COPD is found to be smoke exposure to tobacco products, and it was found that smoking changes the microbiome pattern in the upper respiratory system. Though there will not be significant changes in the microbiome among the non-smokers, smokers with insignificant COPD symptoms, and healthy smokers. But it can tremendously vary in the case of individuals with severe COPD or recurrent exacerbations [32].

### 6.1.2 Lung Microbiome in COPD

COPD is a progressive lung disorder that includes emphysema and chronic bronchitis that contributes to a blockage of the airways resulting in a decrease in lung function. Lung irritants say, for example, cigarette smoke, chemical fumes, dust as well as air pollution can adversely contribute to the disease along with the genetic factors [33]. The heterogeneity in pathogenesis, severity, and symptoms developing raise lots of challenges in the management of COPD [34]. Acute exacerbation is another difficulty associated with the disease, with a temporary span of augmented COPD manifestations which invites remedial treatment as well as even admission to hospital [35].

The pathogenesis, as well as progression of COPD, is related to the changed lung microbiome and it is of high importance now. The altered microbiome can confer for the pathogenesis of COPD. Hilty et al. performed 16s RNA analysis with the DNA samples collected from the nasal and oropharyngeal swab and respiratory brushings taken from lungs of five diseased people, to prove the occurrence of dysbiosis in COPD [11]. They related the microbiome of patients with the healthy control population and concluded the characteristic microbiome in the case of COPD and disturbances in the normal microbiome in the case of lung disorders. Successive research works further proved the high proportion of *Firmicutes, Proteobacteria, Fusobacteria, Actinobacteria and Bacteroidetes*, in physically well individuals, whereas pathogenic bacteria of *Streptococcus, Haemophilus, Pseudomonas, Klebsiella* and *Moraxella* in COPD affected population [9, 36–39].

### 6.1.3 Pathogenesis of COPD and Microbiome

It is clear from the earlier studies that respiratory pathogens are existing in the human body affected by COPD [40]. The colonization of microbes in the lungs may be correlated with inflammatory reactions, radiological and pathological changes in airway obstruction, increased daily symptoms of COPD, etc. [41, 42]. This finding is supporting the hypothesis of a vicious cycle. This theory suggests changes in inborn lung protection persuaded by inhalational toxins like tobacco smoke or chemical smoke exposure permit definite disease-causing microbes to persist and proliferate which arrive in the lower respiratory tract by microaspiration as well as inhalation. The epithelial, as well as immune cells in the respiratory airways, has specific receptors through which the microbes can signal the inflammation process, where the lung attempts to clear the contagion. The defence mechanism which is inborn in the lung is damaged by the inflammation-induced already, which allows proliferation as well as the persistence of the bacteria again which will lead to chronic inflammation of airways leading to a vicious cycle (Fig. 6.3). If the chronic infection and inflammation cycle repeatedly occurs in airways, that may lead to more airflow obstruction and lung damage that is persistent and is a characteristic feature of COPD. These changes in the airways may be irreversible even after removing the primary insult such as cigarette smoke [41].

Bacteria and viruses are reasons for a considerable share of exacerbations of COPD. Chronic infections by such microbes may lead to inflammatory reactions of the airways, causing more adverse consequences on the airways which helps in the pathogenesis of COPD [43].

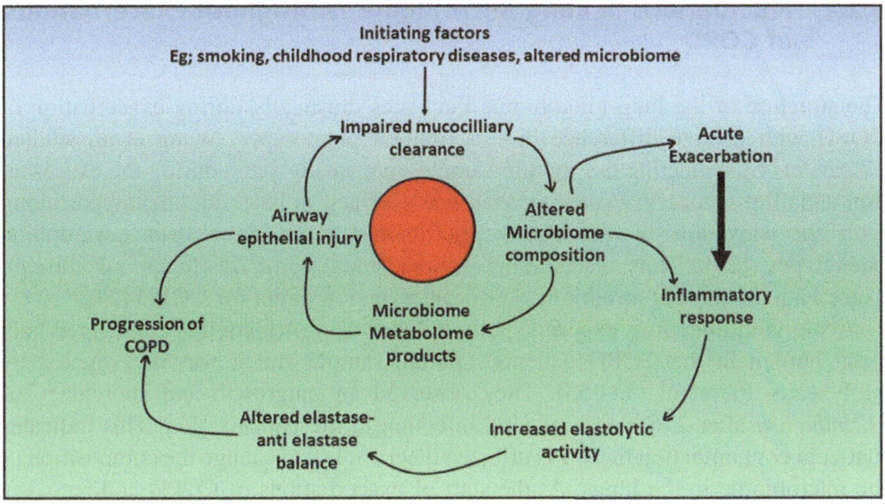


**Fig. 6.3** Vicious Circle

## 6.2 Acute Exacerbation of COPD (AECOPD)

Various infectious, as well as non-infectious factors, may contribute to AECOPD. Viral and bacterial respiratory tract infections lead to the occurrence of AECOPD [44, 45]. Studies conducted in the Asia-Pacific region showed a prevalence rate of 33% in viral infections associated with COPD. Various viruses can contribute to the AECOPD, where influenza virus is prevalent in Asia, whereas picornavirus is prevalent in Australia, America and Europe [46]. Other causative viruses include adenovirus, coronavirus, parainfluenza virus, respiratory syncytial virus, etc. More severe clinical infection (with increased hospital stay, hypoxemia and lung function abnormalities) is observed in diseased people having viral infections than in non-infectious cases [46, 47].

Bacterial infections contribute to AECOPD with a prevalence rate of 26–81% [48]. *Pseudomonas aeruginosa, Klebsiella pneumoniae, Moraxella catarrhalis, Haemophilus influenzae, Streptococcus pneumoniae* are some of the causative organisms for AECOPD. Concurrent infection with bacteria and viruses is an etiological factor contributing to AECOPD, which leads to lower forced expiratory volume resulting in possible hospitalization or frequent readmissions to the hospitals [45, 48, 49]. Non-infectious causes of AECOPD include air pollution (exposure to sulphur dioxide, carbon monoxide, nitrogen dioxide, ozone and other small particulate matter) and comorbidities such as heart failure and pulmonary embolism [47, 50–52]. But AECOPD, without unknown aetiology contributes 30% of total occurrence.

### 6.2.1 Fluctuations in Lung Microbiome Throughout Exacerbations of COPD

The structure of the lung microbiome fluctuates drastically during exacerbation of COPD with a huge difference in exacerbation phenotypes. Wang et al. studied 87 patients by collecting the sputum samples at a steady-state, during the exacerbation and after recovery. *Proteobacteria* was elevated in bacterial phenotypes along with the elevation in *Firmicutes* microbiome was reported in eosinophilic phenotypes [22]. Thus fluctuations in the microbiome of lungs all through exacerbations may act as a biomarker or intervention target for COPD [24].

Acute viral infections experimentally induced using Rhinovirus augmented bacterial burden in the COPD patients' sputum sample and it was investigated by Molyneaux PL et al., in [53]. They observed an outgrowth and supremacy of *H. Influenza* after 2–6 weeks of viral infection in the patients [53]. This indicates that viral contamination in COPD affected that people can change the composition of the microbiome in the lungs. At the start of exacerbations of COPD and recovery over 3 months cases, a more stable lung microbiome was observed [54]. But in hospitalized patients with severe COPD exacerbations, the structure, as well as variety of fungal as well as bacterial genera, change rapidly by each day [55].

During the exacerbation of COPD in most cases, the etiological factor is considered as increased abundance of specific types of microbes as per the recent microbiome investigations, whereas some microbial flora remains unchanged during the exacerbations [54]. However, extensive investigations are suggestive to understand the role of restoration of altered microbiome via treatment, thereby it acts as a biomarker for recovery from COPD exacerbations [56].

Thus, respiratory dysbiosis is considered a common explanation for the exacerbation of COPD (Fig. 6.4). Any inflammatory trigger such as viral infection, air pollution, allergic exposure or toxic inhalation can initiate the inflammatory response in the host that can significantly change the growth conditions of microbes in the airways. Mucus production and vascular permeability change in such a way that all these promote bacterial growth in the respiratory way [57]. The proliferation of explicit bacterial classes such as *Pseudomonas aeruginosa, Streptococcus aureus, Streptococcus pneumoniae and Burkholderia cepacia* is enhanced by various inflammatory cytokines as well as catecholamines [58–62]. Inflammatory cell activation leads to clearing bacteria by killing it or reducing the variable effectiveness of bacteria [63]. Decreased oxygen tension and elevated temperature are created by airway mucus, thereby favouring the growth of pathogenic microbes [64, 65]. Thus, the dynamic homeostasis of the microbiome in the airway is dysregulated resulting in respiratory dysbiosis. The new and more virulent as well as high immunogenicity microbiome can further progress the airway inflammation, resulting in tissue injury. Thus, the airway inflammation in acute exacerbations persists for a longer period, even long after exposure to the precipitating factors [53, 66].

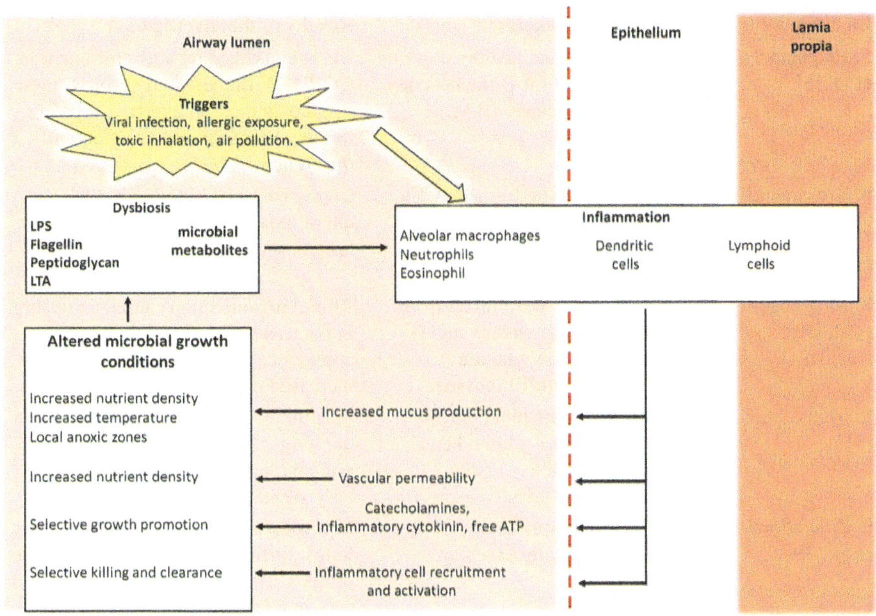


**Fig. 6.4** Dysbiosis – Inflammation cycle

Various studies are done on the microbiome in COPD patients and smokers.

| Investigation | Populace who undergone study | Results of the investigation |
|---|---|---|
| 1. Huang et al. [67] | COPD patients with assisted ventilation from whom the aspirates are taken from inside of trachea. | Observed various bacterial communities in respiratory ways of patients who are experiencing severe exacerbations of COPD frequently. |
| 2. Hilty et al. [11] | Samples are taken by bronchial brushing from 5 COPD, 11 asthma, 8 healthy individuals for investigation. | The presence of various microbiota was observed in the airways of patients among which Proteobacteria was most prominent in both Asthma and COPD cases. |
| 3. Erb-Downward et al. [36] | Bronchoalveolar lavage (BAL) was taken from smokers, non-smokers and COPD patients. In patients with severe COPD, the sample was tissue explants from the lungs. | The lung microbiome was seen to be different from the microbiome of the oral mucosa. As the severity of COPD progresses, the diversity of microbiota reduces. Pseudomonas was found to be the predominant group. |
| 4. Cabrera-Rubio et al. [68] | Samples are taken from moderately severe COPD patients. Samples were sputum, BAL, bronchial wash and tissue biopsy. | Increased microbial variety was found out. Microbiota observed in different kinds of samples were different. Sputum as well as bronchial washings obtained similar microbiota which is entirely different from one observed in biopsy and BAL. |

(continued)

| Investigation | Populace who undergone study | Results of the investigation |
|---|---|---|
| 5. Pragman et al. [3] | COPD patients and healthy controls were considered for Bronchoscopy with Alveolar Lavage (BAL). | Microbial diversity was prominent in COPD. With increased severity, there was no change in diversity. Inhaler therapy was found to bring about differences. |
| 6. Sze et al. [26] | Patients with cystic fibrosis (CF), COPD, smokers in addition to non-smokers were included in the study | Lesser bacterial biomass as well as alterations in bacterial inhabitants in lung tissue affected with severe COPD. |
| 7. Molyneaux et al. [53] | Sputum samples were taken from COPD patients as well as the control population who are contaminated with Rhinovirus. | Rhinovirus infection caused an upsurge in bacterial load especially the abundance of *Haemophilus influenza* increased drastically. |
| 8. Morris et al. [31] | Smokers and non-smokers from whom the samples were taken as an oral wash as well as bronchoalveolar lavage. | Specific bacteria were predominant in the lungs. Smoker, as well as non-smoker lung microbiota, did not vary significantly. |
| 9. Zakharkina et al. [39] | 9 COPD, 9 controls—Bronchoscopy with Alveolar Lavage | In healthy lungs and COPD patient's lungs, different kinds of bacteria were present. There were different bacterial taxa in COPD. |
| 10. Galiana et al. [69] | Sputum sample of 9 mild/moderate COPD and 10 severe COPD. | In severe COPD patients, bacterial load was found to be very high. |
| 11. Huang et al. [54] | Sputum trials are taken from COPD patients of varying severity. | Microbial communities are found to be altered in case of acute exacerbations of COPD as well as therapy using antibiotics and steroids or both. |
| 12. Millares et al. [70] | Sputum sample of 11 severe COPD cases, 5 COPD colonized by *Pseudomonas aeruginosa* (PA) | PA-infected sputum showed an upsurge in the diversity of microbes throughout the exacerbation of the disease. |

### 6.2.2 Microbiome as a Biomarker in COPD

The use of the microbiome as a biological marker of COPD is still a concept that is in its inception stage. Biomarkers usually need large cohorts for different phases such as discovery, validation and forthcoming clinical trials. Besides, the development of biomarkers should meet the standards of clinical validity, analytical validity and clinical utility. The lower biomass of the microbiome in the lower respiratory tract led to analytical validity problems for the researchers. Apart from the other challenges, there was always a possible threat of reagent contamination in lower respiratory airway samples. Even though the sampling technique involved was simple as well as non-invasive sputum collection, but limited numbers of airway microbiome study centres was another reason behind insufficient reports generated in this area. Moreover, the researchers need effective policies to confirm whether the clinical outcomes can be affected using microbiome screening data. As an initial step in biomarker discovery, a study was conducted to find out the connection among specific taxa characteristics of the sputum microbiome as well as the augmented death rate due to COPD [71]. The profiling of the microbiome is done using 16S rRNA genomic sequencing method. To specify variations in the abundance of the different microbiome, alpha diversity (diversity of microbiome within a single sample) and beta diversity (diversity between the groups, or similarity between the samples from different individuals) are calculated among survivors and non-survivors of COPD. Interestingly lower alpha diversity was observed with non-survivor groups. The survivor's microbiota checked in sputum was rich in *Fusobacterium, Rothia, Prevotella, Veillonella* and *Actinomyces,* whereas the non-survivors microbiota investigated using sputum sample was rich in *Escherichia-Shigella* as well as *Staphylococcus. Veillonella* was found to be a useful bacterium and its absence in the sputum indicated increased mortality around 13.5 folds. *Staphylococcus* in the sputum sample is associated with increased hospital stay as well as increased mortality by 7.3 folds. But a detailed validation in different cohorts may be needed in this investigation and the confounders such as antibiotic usage before sampling could change the results of these assessments. Moreover, a sputum microbiota may reflect the oral microbiome in a better way than the lung microbiome [72].

Another investigation among a group of 55 COPD affected people was conducted to find out the constitution of microbiome throughout worsening of COPD as well as post-stabilization phase [73]. The study determined the presence of viruses along with bacteria. The patients who are stabilized after the acute attacks of COPD are observed with stable microbiome, whereas patients with exacerbation phenotype showed dysbiosis phenomena in their microbiome composition. Proteobacteria was found to be predominant in exacerbation phenotypes. As per the study, New Generation Sequencing (NGS) can be utilized to stratify patients suffering from COPD exacerbations, by identifying the causative microbe of exacerbation. This, in turn, can lead to the characterization of major microbiome biomarkers and their utilization in the management of COPD patients and reduction of healthcare costs.

The lung microbiome varies during COPD and the variation is related to both clinical and biochemical features during the disease. This can provide clarity in the association between lung microbiome, inflammatory responses in the host as well as the pathogenesis of the disease. The normal lung microbiome composition in the phyla level is *Firmicutes* (51.4%), *Proteobacteria* (35.9%), *Actinobacteria* (6.5%) or *Bacteroidetes* (4.6%). At the genera level of microbes, the utmost plentiful were *Streptococcus* (41.1%), *Haemophilus* (18.9%), *Moraxella* (5.6%) and *Pseudomonas* (4.4%), which are characteristic components of the lung microbiota [74]. During exacerbations, there is a shift in the microbiome with a total diminution in alpha diversity with a rise in the comparative richness of *Proteobacteria* with a decrease in *Firmicutes. Moraxella* also exhibits greater relative abundance by 5% [22].

COPD exacerbation phenotypes are classified as bacterial, eosinophilic, viral, bacterial-eosinophilic blend, bacterial-viral combination and pauci-inflammatory

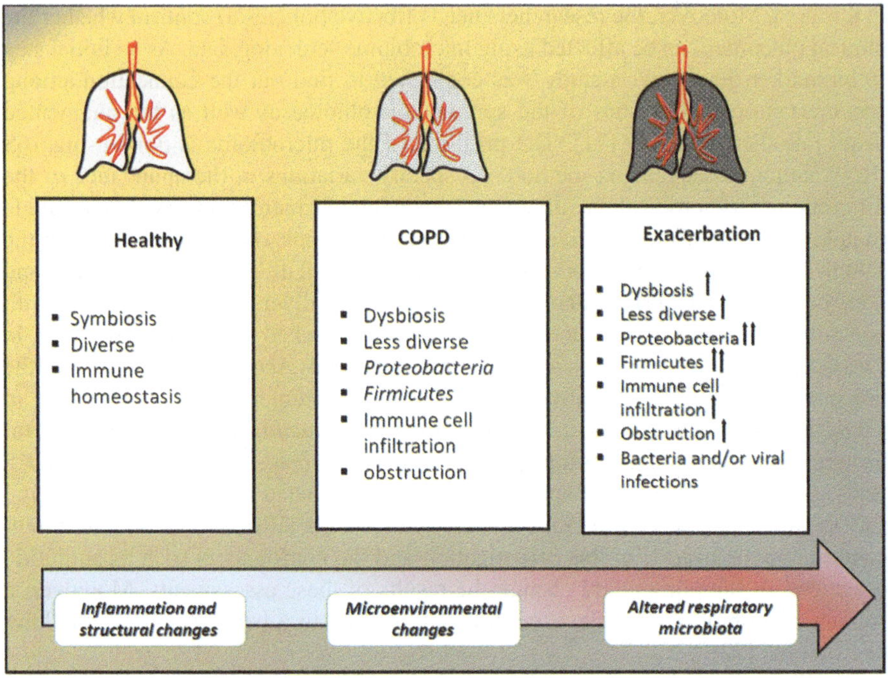


**Fig. 6.5** Common lung microbial features in healthy, COPD and exacerbation cases

[75]. During investigations, different microbiome profile was observed among different phenotypes of COPD exacerbations at both phylum and genus level. The difference is more prominent among bacterial and eosinophilic phenotypes. Bacterial subgroup compared to eosinophilic phenotype has a typical decrease in alpha diversity and *Firmicutes,* in addition to an upsurge in *Proteobacteria* (Fig. 6.5). At the genus level, a decline in *Streptococcus* in addition to an increase in *Hemophilus* is detected in bacterial phenotype. *Proteobacteria: Firmicutes* proportion is diminished in eosinophilic phenotypes throughout worsening of disease compared to other exacerbation phenotypes [22].

The relative abundance of various operational taxonomic units (OTU) of various microbial drivers is studied for the severity of symptoms. In a study group of populations with severe symptoms, the OTU, *Granulicatella species* and *Neisseria subflava* were reduced in diseased people presenting with serious signs of COPD [76].

While considering the fungal microbiome, *Ascomycota* is the predominant fungal microbiome over *Basidiomycota* in COPD patients. *Meyerozyma*, *Aspergillus, Candida* and *Schizophyllum* were utmost plentiful at the genus level. Bacterial microbial diversity is inversely correlated with fungal microbial diversity. Alteration to both mycobiome and bacterial microbiome occurs in contrary directions in the case of patients of COPD and healthy individuals. The same may happen with frequent and non-frequent exacerbators. Fungal OTUs in *Candida palmioleophila*,

*Sordariomycetes* and *Aspergillus* mutually occur with other mycological taxonomic units in the frequent exacerbators than in non-frequent exacerbators. Similarly, the common co-occurrence amongst bacterial OTUs in *Rothia mucilaginosa, Prevotella,* Streptococcus and *Veillonella* are seen in non-frequent exacerbators but not in the frequent exacerbators. Thus, the airway mycobiome can also act as an excellent biomarker in the distinction between exacerbator and non-exacerbator phenotypes ("Airway bacterial and fungal microbiome in chronic obstructive pulmonary disease", n.d.).

### 6.2.3 Microbiome During Drug Therapy of COPD

Early and accurate diagnosis, as well as timely treatment, is essential for reducing the harmful impacts of COPD in individuals. Inhibition of worsening of disease is important in the treatment of COPD. Smoking cessation is one of the major interventions as cigarette smoke is a very important etiological factor [77]. Non-pharmacological methods such as increased physical activity, education of the affected for the self-management of acute conditions, pulmonary rehabilitation and vaccination with pneumococcal vaccines (to reduce infection-related exacerbations) are recommended for effective management of COPD [78–80]. Various maintenance pharmacotherapies are used for reducing the symptoms, frequency and severity of the disease. Common medications used for the management of COPD are bronchodilators, anticholinergics, anti-inflammatory drugs (corticosteroids that are either inhalational or oral as well as antimicrobial medicines), mucolytic medicines and phosphodiesterase-4 inhibitors [66]. Presently the Global Initiative for Chronic Obstructive Lung Disease (GOLD) strategy file endorses treatment of mild exacerbations using bronchodilators. Antibiotics with or without corticosteroids can be used in moderate or serious exacerbations. Furthermore, to the previously cited treatment using various medicines, respiratory aids like oxygen therapy as well as (non-) invasive ventilation can be used for management of exacerbations in a hospital setting.

The effects of medication used in COPD on the microbiome are now getting more investigated. By regulating the lung microbiota, the efficiency of medications can be extended or side effects can be prevented [81]. Also, targeting the specific microbe or pathogen may bring about more therapeutic potential.

#### 6.2.3.1 Antibiotics

Antibiotic therapy especially using Macrolide antibiotics is used for the long-term management of COPD [82]. The severity and duration of Chronic Respiratory Diseases (CRDs) induced by *Haemophilus, mycoplasma* and *Chlamydia* are reduced by the usage of long-term antibiotics [83]. Long-term antibiotic therapy has an irresistible risk of antimicrobial resistance [84]. Brill et al. reported at least one macrolide-resistant gene in whole-genome sequenced from COPD patients on long-term macrolide antibiotic treatment [85]. In asthma sufferers, azithromycin decreased lung Prevotella, Haemophilus and Staphylococcus [86]. Long-term treatment with erythromycin in case of bronchiectasis increases *H parainfluenzae* load and decreases *Actinomyces odontolyticus* and *Streptococcus pseudopneumoniae* abundance [87]. In emphysema affected people who are smokers, the bacterial load is not found to be reduced by azithromycin. Instead, it can reduce alpha diversity as well as proinflammatory cytokines. Azithromycin is found to be associated with increased anti-inflammatory bacterial metabolites. Therefore it is proved that even though some antibiotics cannot beneficially reduce the microbial load they can favourably act on bacterial metabolism that brings about anti-inflammatory effects [88].

#### 6.2.3.2 Corticosteroids

In COPD patients, corticosteroids are related to expanded microbial extravagance and diversity. Management throughout worsening of the disease improved Bacteroidetes, Proteobacteria, as well as Firmicutes in the lung [70]. Corticosteroids easily hinder inborn immunity mediated by type I interferon as well as adaptive immunity mediated by T cell antiviral reactions, prompting deferred viral clearance as well as expanded lung bacterial burdens. Steroid-resistant air route swelling and irritation and airway hyper-responsiveness can be mediated by a respiratory syncytial virus, Chlamydia, H influenzae and influenza A virus [89]. Patients treated with concurrent administration of Antibiotics and Corticosteroids showed a considerable upsurge in Proteobacteria richness.

In short, COPD patients treated with corticosteroids only lead to a decrease in alpha diversity where *Proteobacteria* increases in abundance over *Firmicutes*. This can cause a reduction in *Streptococcus* and arise in the abundance of *Moraxella* in addition to *Haemophilus*. In the case of antibiotic therapy with or without corticosteroids, the reverse process happens in terms of microbial composition changes as well as alpha diversity. The changes that happen in the microbiome as a result of therapy with either antibiotics or corticosteroids is found to be maintained for the long term [22, 54].

## 6.3 Dietary Fibre: A Beneficial treatment in COPD

In COPD patients, dysbiosis of the gut microbiome occurs because of various known factors (tobacco smoke, gender, age, diet and BMI, therapeutic agents such as steroids and antibiotics) causing COPD [90, 91]. A healthy gut microbiome is rich in microbial diversity. Diet is an important factor that can have an impact on microbiome diversity. It is proved that the western diet mainly based on animal fat as well as protein can cause the increased colonization of *Bacteroides* species leading to decreased microbial diversity, whereas a prudent diet with legumes—fruits and vegetables enhance the diversity of microbiota—by utilizing indigestible fibre from the prudent diet as their basis of energy [92, 93].

Airway inflammation is one among the chief factors in the pathogenesis of COPD and noxious inhalants such as cigarette smoke are an inducer of the inflammation [21]. Smoke exposure has an effect on not only the airway inflammation but also increases systemic inflammation and disturbs the gut microbiota [21]. The most effective intervention for COPD was found to be smoking cessation. Even though smoking cessation is effectively done, the clinical condition of many patients continues to weaken. Cigarette smoke is likely to be the reason for reduced diversity in the gut microbiome. Lack of fermentable fibre in the diet will lead to poor nutrition of the microbiota, causing dysbiosis in gut microbiota and an increase in the local or systemic chronic inflammation. In short, cigarette smoke promotes to the pathogenesis of the disease by enhancing the inflammation via dysbiosis of the gut microbiota [94]. Therefore, dietary fibres also show a significant part in the management of COPD via gut-based intervention.

Breakdown of dietary fibre through the intestine microbiome can increase the production of certain short-chain fatty acids (SCFA) by the commensal microorganisms and SCFA are anti-inflammatory in their function, which can protect the lungs against inflammation [24]. Acetate, butyrate, and propionate are the organic components of SCFA and these components play a significant part in the metabolism, cell multiplication and inflammation processes [95]. Thus, the risk of COPD is reduced by a high fibre diet. This is by the anti-inflammatory actions of the SCFAs produced from the dietary fibre.

## 6.4 β Agonists

Changes in lung microbiome upon the administration of β-agonists in chronic respiratory diseases including COPD are not clearly defined. Experimental data shows that Salmeterol can decrease the clinging of microorganisms to the airway mucosa. It also reduces epithelial damage of airway mucosa which can be induced by microbes like *H influenzae* and *P aeruginosa* [96, 97]. An opposite effect has been observed in experimental studies where the inhalational β agonists weaken the clearance of microbes such as *H influenzae* [98].

## 6.5 Prebiotics and Probiotics

Prebiotics are non-digestible carbohydrates, which can be metabolized by bacteria in the gut and can stimulate the growth as well as actions of beneficial colonic bacteria. Probiotics are microbes that can maintain a balance of bacteria in the intestine. Both prebiotics and probiotics can vary the equilibrium of gut microbiota. They act together with inborn as well as adaptive immunity to encourage the discharge of metabolites which are anti-inflammatory in nature as well as secretory products, both could result in health benefits in long-lasting airway ailments such as COPD. *Bifidobacterium bifidum* and *Lactobacillus acidophilus* administered in children with atopic asthma considerably enhanced lung function and resulted in less frequent exacerbations [99]. Also, probiotics can restore the altered microbiome because of antibiotic usage. This will lead to gut motility, increased immune functions of lungs and gut and inhibition of pathogenic bacteria. Probiotics are proved to be capable of inhibiting of exacerbation of pulmonary diseases also [100].

## 6.6 Pathogen Targeting

Treatment strategies that focus on the various pathogen of bacterial origin are mandatory to reduce injury as well as death due to long-lasting airway diseases such as COPD. This can be achieved by making use of various vaccines against pathogenic bacteria in COPD patients, which may result in decreased pathogen load as well as inflammation associated with it [101]. Chronic infection with *H. Influenzae* enhances the inflammation of airways resulting in COPD, which is insensitive to steroids. Preventing bacterial accumulation is one of the useful adjunct therapies for COPD as well as asthma [101]. Streptococcus pneumoniae being an important pathogen in a susceptible population should be prevented from colonizing in the airway as well as lungs by making use of the vaccination against it [102]. Similarly, secondary infection of airways with *Streptococcus pyogenes* can be prevented by using the influenza vaccine in susceptible individuals [103]. The common commensal bacteria of the lungs can have a certain level of protective effects against chronic respiratory diseases which can be harnessed in the therapy of such diseases. The growth of pathogenic bacteria *H influenzae* can be inhibited by the common lung commensal bacteria, *Haemophilus haemolyticus* [104]. Therefore, various approaches that can enhance the growth of commensal respiratory microbes can bring about positive changes in the treatment of COPD.

## 6.7 Future Aspects

The chapter provides a brief understanding of the variations in the lung microbiome in COPD patients, the potential influence of the microbiome in the pathophysiology as well as exacerbation of the disease, the characteristic of the microbiome as a potential biomarker for COPD and finally as a target for respiratory therapeutics in future. The detection and management of respiratory diseases such as COPD have improved a lot with the next-generation sequencing procedures. But the application of such information to patient care, monitoring of patients and treatment of the disease may still be a challenging task, which needs more investigation.

## References

1. Ball P (1995) Epidemiology and treatment of chronic bronchitis and its exacerbations. Chest 108:43S–52S. https://doi.org/10.1378/chest.108.2_Supplement.43S
2. Wang Z, Maschera B, Lea S, Kolsum U, Michalovich D, Van Horn S, Traini C, Brown JR, Hessel EM, Singh D (2019) Airway host-microbiome interactions in chronic obstructive pulmonary disease. Respir Res 20:113. https://doi.org/10.1186/s12931-019-1085-z
3. Pragman AA, Knutson KA, Gould TJ, Hodgson SW, Isaacson RE, Reilly CS, Wendt CH (2019) Chronic obstructive pulmonary disease upper airway microbiome is associated with select clinical characteristics. PLoS One 14:e0219962. https://doi.org/10.1371/journal.pone.0219962

4. Lopez AD (2006) Chronic obstructive pulmonary disease: current burden and future projections. Eur Respir J 27:397–412. https://doi.org/10.1183/09031936.06.00025805
5. Mannino DM, Kiri VA (2006) Changing the burden of COPD mortality. Int J COPD 1:219–233. https://doi.org/10.2147/copd.2006.1.3.219
6. National Heart, Lung, and Blood Institute (NHLBI) (2012) Morbidity & Mortality: 2012 Chart book on cardiovascular. Lung Blood Dis, 2012. 117
7. Lederberg J, McCray AT (2001) 'Ome Sweet' Omics – a genealogical treasury of words 2. Scientist 15(7):8
8. O'Dwyer DN, Dickson RP, Moore BB (2016) The lung microbiome, immunity, and the pathogenesis of chronic lung disease. J Immunol. 196:4839–4847. https://doi.org/10.4049/jimmunol.1600279
9. Martinez FJ, Erb-Downward JR, Huffnagle GB (2013) Significance of the microbiome in chronic obstructive pulmonary disease. Ann Am Thorac Soc 10:S170–S179. https://doi.org/10.1513/AnnalsATS.201306-204AW
10. The Human Microbiome: What It Is, Why It Is Important and Opportunities for Microbiome-Based Therapeutics | American Pharmaceutical Review - The Review of American Pharmaceutical Business & Technology [WWW Document] (n.d.). URL https://www.americanpharmaceuticalreview.com/Featured-Articles/562548-The-Human-Microbiome-What-It-Is-Why-It-Is-Important-and-Opportunities-for-Microbiome-Based-Therapeutics/. Accessed 06 Sep 2021)
11. Hilty M, Burke C, Pedro H, Cardenas P, Bush A, Bossley C, Davies J, Ervine A, Poulter L, Pachter L, Moffatt MF, Cookson WOC (2010) Disordered microbial communities in asthmatic airways. PLoS One 5:e8578. https://doi.org/10.1371/journal.pone.0008578
12. Dickson RP, Erb-Downward JR, Huffnagle GB (2013) The role of the bacterial microbiome in lung disease. Expert Rev Respir Med 7:245–257. https://doi.org/10.1586/ers.13.24
13. Sommariva M, Le Noci V, Bianchi F, Camelliti S, Balsari A, Tagliabue E, Sfondrini L (2020) The lung microbiota: role in maintaining pulmonary immune homeostasis and its implications in cancer development and therapy. Cell Mol Life Sci 77:2739–2749. https://doi.org/10.1007/s00018-020-03452-8
14. Turnbaugh PJ, Ley RE, Hamady M, Fraser-Liggett CM, Knight R, Gordon JI (2007) The human microbiome project. Nature 449:804–810. https://doi.org/10.1038/nature06244
15. Moffatt MF, Cookson WO (2017) The lung microbiome in health and disease. Clin Med 17:525–529. https://doi.org/10.7861/clinmedicine.17-6-525
16. Fox GE, Magrum LJ, Balch WE, Wolfe RS, Woese CR (1977) Classification of methanogenic bacteria by 16S ribosomal RNA characterization. Proc Natl Acad Sci 74:4537–4541. https://doi.org/10.1073/pnas.74.10.4537
17. Suau A, Bonnet R, Sutren M, Godon J-J, Gibson GR, Collins MD, Doré J (1999) Direct analysis of genes encoding 16S rRNA from complex communities reveals many novel molecular species within the human gut. Appl Environ Microbiol 65:4799–4807. https://doi.org/10.1128/AEM.65.11.4799-4807.1999
18. Hamady M, Knight R (2009) Microbial community profiling for human microbiome projects: tools, techniques, and challenges. Genome Res 19:1141–1152. https://doi.org/10.1101/gr.085464.108
19. Woese CR, Kandler O, Wheelis ML (1990) Towards a natural system of organisms: proposal for the domains Archaea, Bacteria, and Eucarya. Proc Natl Acad Sci U S A 87(12):4576–4579. https://doi.org/10.1073/pnas.87.12.4576
20. Dickson RP, Erb-Downward JR, Freeman CM, Walker N, Scales BS, Beck JM, Martinez FJ, Curtis JL, Lama VN, Huffnagle GB (2014) Changes in the lung microbiome following lung transplantation include the emergence of two distinct pseudomonas species with distinct clinical associations. PLoS One 9:e97214. https://doi.org/10.1371/journal.pone.0097214
21. Shaw JG, Vaughan A, Dent AG, O'Hare PE, Goh F, Fong KM, Yang IA (2014) Biomarkers of progression of chronic obstructive pulmonary disease (COPD). J Thorac Dis 6:16

22. Wang Z, Bafadhel M, Haldar K, Spivak A, Mayhew D, Miller BE, Tal-Singer R, Johnston SL, Ramsheh MY, Barer MR, Brightling CE, Brown JR (2016) Lung microbiome dynamics in COPD exacerbations. Eur Respir J 47:1082–1092. https://doi.org/10.1183/13993003.01406-2015
23. Wilkinson TMA, Patel IS, Wilks M, Donaldson GC, Wedzicha JA (2003) Airway bacterial load and FEV 1 decline in patients with chronic obstructive pulmonary disease. Am J Respir Crit Care Med 167:1090–1095. https://doi.org/10.1164/rccm.200210-1179OC
24. Wang L, Hao K, Yang T, Wang C (2017) Role of the lung microbiome in the pathogenesis of chronic obstructive pulmonary disease. Chin Med J 130:2107–2111. https://doi.org/10.4103/0366-6999.211452
25. Dickson RP, Huffnagle GB (2015) The lung microbiome: new principles for respiratory bacteriology in health and disease. PLoS Pathog 11:e1004923. https://doi.org/10.1371/journal.ppat.1004923
26. Sze MA, Dimitriu PA, Hayashi S, Elliott WM, McDonough JE, Gosselink JV, Cooper J, Sin DD, Mohn WW, Hogg JC (2012) The lung tissue microbiome in chronic obstructive pulmonary disease. Am J Respir Crit Care Med 185:1073–1080. https://doi.org/10.1164/rccm.201111-2075OC
27. Barfod KK, Vrankx K, Mirsepasi-Lauridsen HC, Hansen JS, Hougaard KS, Larsen ST, Ouwenhand AC, Krogfelt KA (2015) The murine lung microbiome changes during lung inflammation and intranasal vancomycin treatment. TOMICROJ 9:167–179. https://doi.org/10.2174/1874285801509010167
28. Saeedi P, Salimian J, Ahmadi A, Imani Fooladi AA (2015) The transient but not resident (TBNR) microbiome: a Yin Yang model for lung immune system. Inhal Toxicol 27:451–461. https://doi.org/10.3109/08958378.2015.1070220
29. Charlson ES, Bittinger K, Haas AR, Fitzgerald AS, Frank I, Yadav A, Bushman FD, Collman RG (2011) Topographical continuity of bacterial populations in the healthy human respiratory tract. Am J Respir Crit Care Med 184:957–963. https://doi.org/10.1164/rccm.201104-0655OC
30. Gleeson K, Maxwell SL, Eggli DF (1997) Quantitative aspiration during sleep in normal subjects. Chest 111:1266–1272. https://doi.org/10.1378/chest.111.5.1266
31. Morris A, Beck JM, Schloss PD, Campbell TB, Crothers K, Curtis JL, Flores SC, Fontenot AP, Ghedin E, Huang L, Jablonski K, Kleerup E, Lynch SV, Sodergren E, Twigg H, Young VB, Bassis CM, Venkataraman A, Schmidt TM, Weinstock GM (2013) Comparison of the respiratory microbiome in healthy nonsmokers and smokers. Am J Respir Crit Care Med 187: 1067–1075. https://doi.org/10.1164/rccm.201210-1913OC
32. Charlson ES, Chen J, Custers-Allen R, Bittinger K, Li H, Sinha R, Hwang J, Bushman FD, Collman RG (2010) Disordered microbial communities in the upper respiratory tract of cigarette smokers. PLoS One 5:e15216. https://doi.org/10.1371/journal.pone.0015216
33. Yang IV, Schwartz DA (2011) Epigenetic control of gene expression in the lung. Am J Respir Crit Care Med 183:1295–1301. https://doi.org/10.1164/rccm.201010-1579PP
34. Roca J, Vargas C, Cano I, Selivanov V, Barreiro E, Maier D, Falciani F, Wagner P, Cascante M, Garcia-Aymerich J, Kalko S, De Mas I, Tegnér J, Escarrabill J, Agustí A, Gomez-Cabrero D, The Synergy-COPD Consortium (2014) Chronic Obstructive Pulmonary Disease heterogeneity: challenges for health risk assessment, stratification and management. J Transl Med 12:S3. https://doi.org/10.1186/1479-5876-12-S2-S3
35. Mayhew D, Devos N, Lambert C, Brown JR, Clarke SC, Kim VL, Magid-Slav M, Miller BE, Ostridge KK, Patel R, Sathe G, Simola DF, Staples KJ, Sung R, Tal-Singer R, Tuck AC, Van Horn S, Weynants V, Williams NP, Devaster J-M, Wilkinson TMA (2018) Longitudinal profiling of the lung microbiome in the AERIS study demonstrates repeatability of bacterial and eosinophilic COPD exacerbations. Thorax 73:422–430. https://doi.org/10.1136/thoraxjnl-2017-210408
36. Erb-Downward JR, Thompson DL, Han MK, Freeman CM, McCloskey L, Schmidt LA, Young VB, Toews GB, Curtis JL, Sundaram B, Martinez FJ, Huffnagle GB (2011) Analysis

of the lung microbiome in the "healthy" smoker and in COPD. PLoS One 6:e16384. https://doi.org/10.1371/journal.pone.0016384
37. Murphy TF, Brauer AL, Grant BJB, Sethi S (2005) Moraxella catarrhalis in chronic obstructive pulmonary disease: burden of disease and immune response. Am J Respir Crit Care Med 172:195–199. https://doi.org/10.1164/rccm.200412-1747OC
38. Wu D, Hou C, Li Y, Zhao Z, Liu J, Lu X, Shang X, Xin Y (2014) Analysis of the bacterial community in chronic obstructive pulmonary disease sputum samples by denaturing gradient gel electrophoresis and real-time PCR. BMC Pulm Med 14:179. https://doi.org/10.1186/1471-2466-14-179
39. Zakharkina T, Heinzel E, Koczulla RA, Greulich T, Rentz K, Pauling JK, Baumbach J, Herrmann M, Grünewald C, Dienemann H, von Müller L, Bals R (2013) Analysis of the airway microbiota of healthy individuals and patients with chronic obstructive pulmonary disease by T-RFLP and Clone sequencing. PLoS One 8:e68302. https://doi.org/10.1371/journal.pone.0068302
40. Sethi S, Maloney J, Grove L, Wrona C, Berenson CS (2006) Airway inflammation and bronchial bacterial colonization in chronic obstructive pulmonary disease. Am J Respir Crit Care Med 173:991–998. https://doi.org/10.1164/rccm.200509-1525OC
41. Sethi S, Murphy TF (2008) Infection in the pathogenesis and course of chronic obstructive pulmonary disease. N Engl J Med 359:2355–2365. https://doi.org/10.1056/NEJMra0800353
42. Soler N, Ewig S, Torres A, Filella X, Gonzalez J, Zaubet A (1999) Airway inflammation and bronchial microbial patterns in patients with stable chronic obstructive pulmonary disease. Eur Respir J 14:1015–1022. https://doi.org/10.1183/09031936.99.14510159
43. Sethi S, Mallia P, Johnston SL (2009) New paradigms in the pathogenesis of chronic obstructive pulmonary disease II. Proc Am Thorac Soc 6:532–534. https://doi.org/10.1513/pats.200905-025DS
44. Choi K-J, Cha S-I, Shin K-M, Lee J, Hwangbo Y, Yoo S-S, Lee J, Lee S-Y, Kim C-H, Park J-Y, Jung T-H (2013) Prevalence and predictors of pulmonary embolism in korean patients with exacerbation of chronic obstructive pulmonary disease. Respiration 85:203–209. https://doi.org/10.1159/000335904
45. Wark PAB, Tooze M, Powell H, Parsons K (2013) Viral and bacterial infection in acute asthma and chronic obstructive pulmonary disease increases the risk of readmission: asthma & COPD viral/bacterial infection. Respirology 18:996–1002. https://doi.org/10.1111/resp.12099
46. Mohan A, Chandra S, Agarwal D, Guleria R, Broor S, Gaur B, Pandey RM (2010) Prevalence of viral infection detected by PCR and RT-PCR in patients with acute exacerbation of COPD: a systematic review. Respirology 15:536–542. https://doi.org/10.1111/j.1440-1843.2010.01722.x
47. Ko FWS, Ip M, Chan PKS, Fok JPC, Chan MCH, Ngai JC, Chan DPS, Hui DSC (2007) A 1-year prospective study of the infectious etiology in patients hospitalized with acute exacerbations of COPD. Chest 131:44–52. https://doi.org/10.1378/chest.06-1355
48. Chang CL, Sullivan GD, Karalus NC, Mills GD, Mclachlan JD, Hancox RJ (2011) Predicting early mortality in acute exacerbation of chronic obstructive pulmonary disease using the CURB65 score: CURB65 in acute exacerbation of COPD. Respirology 16:146–151. https://doi.org/10.1111/j.1440-1843.2010.01866.x
49. Sapey E (2006) COPD exacerbations {middle dot} 2: aetiology. Thorax 61:250–258. https://doi.org/10.1136/thx.2005.041822
50. Lee I-M, Tsai S-S, Chang C-C, Ho C-K, Yang C-Y (2007) Air pollution and hospital admissions for chronic obstructive pulmonary disease in a tropical city: Kaohsiung, Taiwan. Inhal Toxicol 19:393–398. https://doi.org/10.1080/08958370601174818
51. Qiu H, Yu IT, Tian L, Wang X, Tse LA, Tam W, Wong TW (2012) Effects of coarse particulate matter on emergency hospital admissions for respiratory diseases: a time-series analysis in Hong Kong. Environ Health Perspect 120:572–576. https://doi.org/10.1289/ehp.1104002
52. Tian L, Ho K-F, Wang T, Qiu H, Pun VC, Chan CS, Louie PKK, Yu ITS (2014) Ambient carbon monoxide and the risk of hospitalization due to chronic obstructive pulmonary disease. Am J Epidemiol 180:1159–1167. https://doi.org/10.1093/aje/kwu248

53. Molyneaux PL, Mallia P, Cox MJ, Footitt J, Willis-Owen SAG, Homola D, Trujillo-Torralbo M-B, Elkin S, Kon OM, Cookson WOC, Moffatt MF, Johnston SL (2013) Outgrowth of the bacterial airway microbiome after rhinovirus exacerbation of chronic obstructive pulmonary disease. Am J Respir Crit Care Med 188:1224–1231. https://doi.org/10.1164/rccm.201302-0341OC
54. Huang YJ, Sethi S, Murphy T, Nariya S, Boushey HA, Lynch SV (2014) Airway microbiome dynamics in exacerbations of chronic obstructive pulmonary disease. J Clin Microbiol 52: 2813–2823. https://doi.org/10.1128/JCM.00035-14
55. Su J, Liu H, Tan X, Ji Y, Jiang Y, Prabhakar M, Rong Z, Zhou H, Zhang G (2015) Sputum bacterial and fungal dynamics during exacerbations of severe COPD. PLoS One 10:e0130736. https://doi.org/10.1371/journal.pone.0130736
56. Ko FW, Chan KP, Hui DS, Goddard JR, Shaw JG, Reid DW, Yang IA (2016) Acute exacerbation of COPD: hot topics on acute exacerbation of COPD. Respirology 21:1152–1165. https://doi.org/10.1111/resp.12780
57. Van De Graaf EA, Out TA, Roos CM, Jansen HM (1991) Respiratory membrane permeability and bronchial hyperreactivity in patients with stable asthma: effects of therapy with inhaled steroids. Am Rev Respir Dis 143:362–368. https://doi.org/10.1164/ajrccm/143.2.362
58. Freestone PP, Hirst RA, Sandrini SM, Sharaff F, Fry H, Hyman S, O'Callaghan C (2012) Pseudomonas aeruginosa -Catecholamine Inotrope Interactions. Chest 142:1200–1210. https://doi.org/10.1378/chest.11-2614
59. Kanangat S, Meduri GU, Tolley EA, Patterson DR, Meduri CU, Pak C, Griffin JP, Bronze MS, Schaberg DR (1999) Effects of Cytokines and Endotoxin on the Intracellular Growth of Bacteria. Infect Immun 67:2834–2840. https://doi.org/10.1128/IAI.67.6.2834-2840.1999
60. Kaza SK, McClean S, Callaghan M (2011) IL-8 released from human lung epithelial cells induced by cystic fibrosis pathogens Burkholderia cepacia complex affects the growth and intracellular survival of bacteria. Int J Med Microbiol 301:26–33. https://doi.org/10.1016/j.ijmm.2010.06.005
61. Marks LR, Davidson BA, Knight PR, Hakansson AP (2013) Interkingdom signaling induces streptococcus pneumoniae biofilm dispersion and transition from asymptomatic colonization to disease. mBio 4. https://doi.org/10.1128/mBio.00438-13
62. Porat R, Clark B, Wolff S, Dinarello C (1991) Enhancement of growth of virulent strains of Escherichia coli by interleukin-1. Science 254:430–432. https://doi.org/10.1126/science.1833820
63. Finlay BB, McFadden G (2006) Anti-immunology: evasion of the host immune system by bacterial and viral pathogens. Cell 124:767–782. https://doi.org/10.1016/j.cell.2006.01.034
64. Schmidt A, Belaaouaj A, Bissinger R, Koller G, Malleret L, D'Orazio C, Facchinelli M, Schulte-Hubbert B, Molinaro A, Holst O, Hammermann J, Schniederjans M, Meyer KC, Damkiaer S, Piacentini G, Assael B, Bruce K, Häußler S, LiPuma JJ, Seelig J, Worlitzsch D, Döring G (2014) Neutrophil elastase-mediated increase in airway temperature during inflammation. J Cyst Fibros 13:623–631. https://doi.org/10.1016/j.jcf.2014.03.004
65. Worlitzsch D, Tarran R, Ulrich M, Schwab U, Cekici A, Meyer KC, Birrer P, Bellon G, Berger J, Weiss T, Botzenhart K, Yankaskas JR, Randell S, Boucher RC, Döring G (2002) Effects of reduced mucus oxygen concentration in airway Pseudomonas infections of cystic fibrosis patients. J Clin Invest 109:317–325. https://doi.org/10.1172/JCI0213870
66. Wedzicha JA, Seemungal TA (2007) COPD exacerbations: defining their cause and prevention. Lancet 370:786–796. https://doi.org/10.1016/S0140-6736(07)61382-8
67. Huang YJ, Kim E, Cox MJ, Brodie EL, Brown R, Wiener-Kronish JP, Lynch SV (2010) A persistent and diverse airway microbiota present during chronic obstructive pulmonary disease exacerbations. OMICS 14:9–59. https://doi.org/10.1089/omi.2009.0100
68. Cabrera-Rubio R, Garcia-Núñez M, Setó L, Antó JM, Moya A, Monsó E, Mira A (2012) Microbiome diversity in the bronchial tracts of patients with chronic obstructive pulmonary disease. J Clin Microbiol 50:3562–3568. https://doi.org/10.1128/JCM.00767-12
69. Galiana A, Aguirre E, Rodriguez JC, Mira A, Santibanez M, Candela I, Llavero J, Garcinuno P, Lopez F, Ruiz M, Garcia-Pachon E, Royo G (2014) Sputum microbiota in moderate versus severe patients with COPD. Eur Respir J 43:1787–1790. https://doi.org/10.1183/09031936.00191513

70. Millares L, Ferrari R, Gallego M, Garcia-Nuñez M, Pérez-Brocal V, Espasa M, Pomares X, Monton C, Moya A, Monsó E (2014) Bronchial microbiome of severe COPD patients colonised by Pseudomonas aeruginosa. Eur J Clin Microbiol Infect Dis 33:1101–1111. https://doi.org/10.1007/s10096-013-2044-0
71. Leitao Filho FS, Alotaibi NM, Ngan D, Tam S, Yang J, Hollander Z, Chen V, FitzGerald JM, Nislow C, Leung JM, Man SFP, Sin DD (2019) Sputum microbiome is associated with 1-year mortality after chronic obstructive pulmonary disease hospitalizations. Am J Respir Crit Care Med 199:1205–1213. https://doi.org/10.1164/rccm.201806-1135OC
72. Sulaiman I, Wu BG, Li Y, Scott AS, Malecha P, Scaglione B, Wang J, Basavaraj A, Chung S, Bantis K, Carpenito J, Clemente JC, Shen N, Bessich J, Rafeq S, Michaud G, Donington J, Naidoo C, Theron G, Schattner G, Garofano S, Condos R, Kamelhar D, Addrizzo-Harris D, Segal LN (2018) Evaluation of the airway microbiome in nontuberculous mycobacteria disease. Eur Respir J 52:1800810. https://doi.org/10.1183/13993003.00810-2018
73. López Caro JC, Santibáñez M, García Rivero JL, Villanueva M, Sainz J, González Astorqui P, Hierro M, Rodríguez Porres M, Paras Bravo P, Mira A, Rodriguez JC, Galiana A, on behalf of the ACINAR-microbiome study group (2019) Sputum microbiome dynamics in chronic obstructive pulmonary disease patients during an exacerbation event and post-stabilization. Respiration 98:447–454. https://doi.org/10.1159/000501988
74. Bathoorn E (2008) Airways inflammation and treatment during acute exacerbations of COPD. Int J Chron Obstruct Pulmon Dis 3:217–229. https://doi.org/10.2147/COPD.S1210
75. Maurer JR (2012) Acute exacerbations of chronic obstructive pulmonary disease: identification of biologic clusters and their biomarkers. Yearbook of Pulmonary Disease 2012: 42–43. https://doi.org/10.1016/j.ypdi.2012.01.014
76. Diao W, Shen N, Du Y, Erb-Downward J, Sun X, Guo C, Ke Q, Huffnagle G, Gyetko M, He B (2018) Symptom-related sputum microbiota in stable chronic obstructive pulmonary disease. Int J Chron Obstruct Pulmon Dis 13:2289–2299. https://doi.org/10.2147/COPD.S167618
77. Au DH, Bryson CL, Chien JW, Sun H, Udris EM, Evans LE, Bradley KA (2009) The effects of smoking cessation on the risk of chronic obstructive pulmonary disease exacerbations. J Gen Intern Med 24:457–463. https://doi.org/10.1007/s11606-009-0907-y
78. Lenferink A, Brusse-Keizer M, van der Valk PD, Frith PA, Zwerink M, Monninkhof EM, van der Palen J, Effing TW (2017) Self-management interventions including action plans for exacerbations versus usual care in patients with chronic obstructive pulmonary disease. Cochrane Database Syst Rev. https://doi.org/10.1002/14651858.CD011682.pub2
79. Moore E, Newson R, Joshi M, Palmer T, Rothnie KJ, Singh S, Majeed A, Soljak M, Quint JK (2017) Effects of pulmonary rehabilitation on exacerbation number and severity in people with copd. Chest 152:1188–1202. https://doi.org/10.1016/j.chest.2017.05.006
80. Poole P, Chacko EE, Wood-Baker R, Cates CJ (2006) Influenza vaccine for patients with chronic obstructive pulmonary disease. Cochrane Database Syst Rev. https://doi.org/10.1002/14651858.CD002733.pub2
81. Budden KF, Shukla SD, Rehman SF, Bowerman KL, Keely S, Hugenholtz P, Armstrong-James DPH, Adcock IM, Chotirmall SH, Chung KF, Hansbro PM (2019) Functional effects of the microbiota in chronic respiratory disease. Lancet Respir Med 7:907–920. https://doi.org/10.1016/S2213-2600(18)30510-1
82. Suresh Babu K, Kastelik J, Morjaria JB (2013) Role of long term antibiotics in chronic respiratory diseases. Respir Med 107:800–815. https://doi.org/10.1016/j.rmed.2013.02.009
83. Essilfie A-T, Horvat JC, Kim RY, Mayall JR, Pinkerton JW, Beckett EL, Starkey MR, Simpson JL, Foster PS, Gibson PG, Hansbro PM (2015) Macrolide therapy suppresses key features of experimental steroid-sensitive and steroid-insensitive asthma. Thorax 70:458–467. https://doi.org/10.1136/thoraxjnl-2014-206067
84. Brill SE, Law M, El-Emir E, Allinson JP, James P, Maddox V, Donaldson GC, McHugh TD, Cookson WO, Moffatt MF, Nazareth I, Hurst JR, Calverley PMA, Sweeting MJ, Wedzicha JA (2015) Effects of different antibiotic classes on airway bacteria in stable COPD using culture

and molecular techniques: a randomised controlled trial. Thorax 70:930–938. https://doi.org/10.1136/thoraxjnl-2015-207194
85. Brill SE, James PL, Cuthbertson L, Zhu A, Lawley T, Cookson WO, Cox MJ, Wedzicha JA, Moffatt MF (2018) Haemophilus, antibiotic therapy and the airway microbiome in chronic obstructive pulmonarydisease (preprint). Microbiology. https://doi.org/10.1101/419127
86. Slater M, Rivett DW, Williams L, Martin M, Harrison T, Sayers I, Bruce KD, Shaw D (2014) The impact of azithromycin therapy on the airway microbiota in asthma. Thorax 69:673–674. https://doi.org/10.1136/thoraxjnl-2013-204517
87. Choo JM, Abell GCJ, Thomson R, Morgan L, Waterer G, Gordon DL, Taylor SL, Leong LEX, Wesselingh SL, Burr LD, Rogers GB (2018) Impact of long-term erythromycin therapy on the oropharyngeal microbiome and resistance gene reservoir in non-cystic fibrosis bronchiectasis. mSphere:3. https://doi.org/10.1128/mSphere.00103-18
88. Segal LN, Clemente JC, Wu BG, Wikoff WR, Gao Z, Li Y, Ko JP, Rom WN, Blaser MJ, Weiden MD (2017) Randomised, double-blind, placebo-controlled trial with azithromycin selects for anti-inflammatory microbial metabolites in the emphysematous lung. Thorax 72: 13–22. https://doi.org/10.1136/thoraxjnl-2016-208599
89. Singanayagam A, Glanville N, Girkin JL, Ching YM, Marcellini A, Porter JD, Toussaint M, Walton RP, Finney LJ, Aniscenko J, Zhu J, Trujillo-Torralbo M-B, Calderazzo MA, Grainge C, Loo S-L, Veerati PC, Pathinayake PS, Nichol KS, Reid AT, James PL, Solari R, Wark PAB, Knight DA, Moffatt MF, Cookson WO, Edwards MR, Mallia P, Bartlett NW, Johnston SL (2018) Corticosteroid suppression of antiviral immunity increases bacterial loads and mucus production in COPD exacerbations. Nat Commun 9:2229. https://doi.org/10.1038/s41467-018-04574-1
90. Kimokoti RW, Millen BE (2016) Nutrition for the prevention of chronic diseases. Med Clin N Am 100:1185–1198. https://doi.org/10.1016/j.mcna.2016.06.003
91. Shaheen SO, Jameson KA, Syddall HE, Aihie Sayer A, Dennison EM, Cooper C, Robinson SM, The Hertfordshire Cohort Study Group (2010) The relationship of dietary patterns with adult lung function and COPD. Eur Respir J 36:277–284. https://doi.org/10.1183/09031936.00114709
92. Kaluza J, Larsson SC, Orsini N, Linden A, Wolk A (2017) Fruit and vegetable consumption and risk of COPD: a prospective cohort study of men. Thorax 72:500–509. https://doi.org/10.1136/thoraxjnl-2015-207851
93. Senghor B, Sokhna C, Ruimy R, Lagier J-C (2018) Gut microbiota diversity according to dietary habits and geographical provenance. Hum Microbiome J 7–8:1–9. https://doi.org/10.1016/j.humic.2018.01.001
94. Chassaing B, Vijay-Kumar M, Gewirtz AT (2017) How diet can impact gut microbiota to promote or endanger health. Curr Opin Gastroenterol 33:417–421. https://doi.org/10.1097/MOG.0000000000000401
95. Vaughan A, Frazer ZA, Hansbro PM, Yang IA (2019) COPD and the gut-lung axis: the therapeutic potential of fibre. J Thorac Dis 11:S2173–S2180. https://doi.org/10.21037/jtd.2019.10.40
96. Dowling RB, Johnson M, Cole PJ, Wilson R (1998) Effect of salmeterol on Haemophilus influenzae infection of respiratory mucosa in vitro. Eur Respir J 11:86–90. https://doi.org/10.1183/09031936.98.11010086
97. Dowling RB, Rayner CF, Rutman A, Jackson AD, Kanthakumar K, Dewar A, Taylor GW, Cole PJ, Johnson M, Wilson R (1997) Effect of salmeterol on Pseudomonas aeruginosa infection of respiratory mucosa. Am J Respir Crit Care Med 155:327–336. https://doi.org/10.1164/ajrccm.155.1.9001332
98. Maris NA, Florquin S, van't Veer C, de Vos AF, Buurman W, Jansen HM, van der Poll T (2006) Inhalation of β2 agonists impairs the clearance of nontypable Haemophilus influenzae from the murine respiratory tract. Respir Res 7:57. https://doi.org/10.1186/1465-9921-7-57

99. Chung KF (2017) Airway microbial dysbiosis in asthmatic patients: a target for prevention and treatment? J Allergy Clin Immunol 139:1071–1081. https://doi.org/10.1016/j.jaci.2017.02.004
100. Anderson JL, Miles C, Tierney AC (2017) Effect of probiotics on respiratory, gastrointestinal and nutritional outcomes in patients with cystic fibrosis: a systematic review. J Cyst Fibros 16: 186–197. https://doi.org/10.1016/j.jcf.2016.09.004
101. Essilfie A-T, Simpson JL, Dunkley ML, Morgan LC, Oliver BG, Gibson PG, Foster PS, Hansbro PM (2012) Combined Haemophilus influenzae respiratory infection and allergic airways disease drives chronic infection and features of neutrophilic asthma. Thorax 67: 588–599. https://doi.org/10.1136/thoraxjnl-2011-200160
102. Jain S, Williams DJ, Arnold SR, Ampofo K, Bramley AM, Reed C, Stockmann C, Anderson EJ, Grijalva CG, Self WH, Zhu Y, Patel A, Hymas W, Chappell JD, Kaufman RA, Kan JH, Dansie D, Lenny N, Hillyard DR, Haynes LM, Levine M, Lindstrom S, Winchell JM, Katz JM, Erdman D, Schneider E, Hicks LA, Wunderink RG, Edwards KM, Pavia AT, McCullers JA, Finelli L (2015) Community-acquired pneumonia requiring hospitalization among U.-S. children. N Engl J Med 372:835–845. https://doi.org/10.1056/NEJMoa1405870
103. Lee S, Eick A, Bloom MS, Brundage JF (2008) Influenza immunization and subsequent diagnoses of group A streptococcus-illnesses among U.S. Army trainees, 2002–2006. Vaccine 26:3383–3386. https://doi.org/10.1016/j.vaccine.2008.04.041
104. Latham RD, Gell DA, Fairbairn RL, Lyons AB, Shukla SD, Cho KY, Jones DA, Harkness NM, Tristram SG (2017) An isolate of Haemophilus haemolyticus produces a bacteriocin-like substance that inhibits the growth of nontypeable Haemophilus influenzae. Int J Antimicrob Agents 49:503–506. https://doi.org/10.1016/j.ijantimicag.2016.12.010

# Microbiome in Asthma-COPD Overlap (ACO) 7

Shibi Muralidar, Gayathri Gopal, and Senthil Visaga Ambi

**Abstract**

Microbiome plays a pivotal role in maintaining host physiological homeostasis through degenerating toxicants, resisting pathogens, absorbing nutrients, and immune system regulation. Each component of the human body has a unique set of microbiome compositions made for specific roles; the respiratory tree also has its own microbiome commonly referred to as the lung microbiome. Crucial composition of the lung microbiome is considered as a decisive influencer in both health and disease, including obstructive lung diseases. Asthma and chronic obstructive pulmonary disease (COPD) are some of the common obstructive lung diseases that are prevalently found in the world population. Patients with overlapping spirometry data and inflammatory markers that feature a mix of both asthma and COPD are referred to as patients with Asthma-COPD Overlap (ACO). With developing evidence on the crucial role of the lung microbiome in several respiratory disorders, the underlying mechanistic link between lung

Shibi Muralidar and Gayathri Gopal contributed equally with all other contributors.

S. Muralidar · G. Gopal
Biopharmaceutical Research Lab, Anusandhan Kendra-1, SASTRA Deemed-to-be-University, Thanjavur, India

School of Chemical and Biotechnology, SASTRA Deemed-to-be-University, Thanjavur, India

S. V. Ambi (✉)
Biopharmaceutical Research Lab, Anusandhan Kendra-1, SASTRA Deemed-to-be-University, Thanjavur, India

School of Chemical and Biotechnology, SASTRA Deemed-to-be-University, Thanjavur, India

Department of Bioengineering, School of Chemical and Biotechnology, SASTRA Deemed-to-be-University, Thanjavur, India
e-mail: senthilvisagaambi@scbt.sastra.edu



G. Gupta et al. (eds.), *Microbiome in Inflammatory Lung Diseases*,
https://doi.org/10.1007/978-981-16-8957-4_7

microbiome and ACO still needs to be deciphered. This chapter will focus on the pathogenesis and severity of ACO, the role of the lung microbiome, and its impact on ACO. Further, the chapter would also provide insights to the readers on various therapeutic strategies targeting the lung microbiome and ACO.

**Keywords**

Asthma · COPD · Asthma-COPD overlap · Exacerbations · Inflammation · Lung microbiome

## 7.1 Introduction

Human bodies are known to be a host for many microbes which are generally found to be present in all mucosal sites. A healthy individual's lungs were previously considered sterile, however recent research showed the presence of a large population of microbial communities that are collectively referred to as microbiome [1, 2]. Being coexisted with the human body for millions of years, this microbiome plays a pivotal role in maintaining host physiological homeostasis through degenerating toxicants, resisting pathogens, absorbing nutrients, and immune system regulation [2, 3]. Each component of the human body has a unique set of microbiome compositions made for specific roles; the respiratory tree also has its own microbiome commonly referred to as the lung microbiome. With ambient temperature, mucus, moisture, and large surface area with periodic contact to the external environment, lungs are a prominent site for rich microbiome composition. Bidirectional movement of mucus and air makes the lung microbiome more transient and dynamic than the gastrointestinal tract. Further, this crucial composition of the lung microbiome is considered as a decisive influencer in both health and disease, including obstructive lung diseases [2, 4–6].

Asthma and chronic obstructive pulmonary disease (COPD) are some of the common obstructive lung diseases that are prevalently found in the world population. Allergic asthma, chronic inflammatory lung disease is characterized by reversible airflow obstruction, abnormal airway mucosa, airway hyperresponsiveness, wheezing, chest tightness, and breathlessness. In contrast, COPD is considered as a progressive functional deterioration of the pulmonary network, characterized by increased inflammation of small airways, persistent airflow limitation that is primarily mediated by cigarette smoking and tobacco [7–10]. In spite of varying etiology and pathophysiology, diagnosing and differentiating asthma and COPD are still a great challenge. Especially in the case of patients with overlapping syndrome, the spirometry data and inflammatory markers feature a mix of both asthma and COPD. These groups of individuals with overlapping clinical features were latterly referred to as patients with Asthma-COPD Overlap (ACO). In 2014, The Global Initiative for Asthma (GINA) and Global Initiative for Chronic Obstructive Lung Disease

(GOLD) reported a generalized definition for ACO as a unique entity with chronic airflow limitation with overlapping clinical features that are consistent with both asthma and COPD [11–14]. With developing evidence on the crucial role of the lung microbiome in several respiratory disorders, the underlying mechanistic link between lung microbiome and ACO still needs to be deciphered. This chapter will focus on the pathogenesis and severity of ACO, the role of the lung microbiome, and its impact on ACO. Further, the chapter would also provide insights to the readers on various therapeutic strategies targeting the lung microbiome and ACO.

## 7.2 Asthma-COPD Overlap

### 7.2.1 Pathogenesis of ACO

Asthma is distinguished by a reversible inflammatory process mediated by $T_h2$ cytokines, $CD^{4+}$ lymphocytes, or eosinophils and generally responds to inhaled corticosteroids (ICS), whereas COPD inflammation is dominated by $T_h1$ cytokines, $CD^{8+}$ lymphocytes, or neutrophils and is known for its progressive airway obstruction. In 2007, the Canadian Thoracic Society brought up the term ACOS to categorize the patients who were observed with signs and symptoms of both diseases [15]. Several minor and major criteria were developed for the diagnosis and categorize the patients with ACO. A very positive bronchodilator test (increased Forced Expiratory Volume in one second ($FEV_1$) $\geq$15% and $\geq$400 ml over baseline), personal history of asthma with eosinophilia in sputum are considered as some major criteria, whereas high total Immunoglobulin E (IgE), personal history of atopy, and positive bronchodilator test (increase in $FEV_1$ $\geq$12% and 200 ml over baseline) on two or more occasions are considered as minor criteria. ACO diagnosis necessitates two major and two minor criteria, or one major and two minor criteria. Airway inflammation, bronchial hyperresponsiveness, and airway obstruction are crucial components that are common to obstructive pulmonary diseases. Exposure of lung airways to gases and noxious particles like smoke and tobacco can potentially result in the dysfunction of smooth muscle and small airway inflammation which ultimately leads to exacerbations in ACO patients. While most of the asthmatic patients are observed with $CD^{4+}$-eosinophilia mediated inflammation, COPD patients are observed with $CD^{8+}$-neutrophilia mediated inflammation. Interestingly, patients with ACO have both eosinophilia and neutrophilia mediated inflammation [15–18].

A population-based cohort study conducted by de Marco et al. observed that patients diagnosed with ACO have comparatively worse basal pulmonary functions than both asthma and COPD patients. ACO patients exhibit a similar decline in $FEV_1$ and forced vital capacity (FVC) to asthma patients is slightly less than COPD patients. However, their prevalence of emergency admissions in hospitals due to respiratory problems was found twice than that of asthma and COPD patients [19]. Individuals having early-onset allergic asthma with smoking habits in later life could potentially result in fixed airflow limitation and COPD. At the same time, patients with a prolonged smoking history that subsequently leads to the development of COPD might show late-onset/adult-onset eosinophilic asthma or COPD driven by eosinophilic inflammation. Smoking asthmatics have higher neutrophils in airways leading to comparatively more resistance to steroids than non-smoking eosinophilic asthmatic patients and COPD patients with eosinophilic inflammation [20–22].

Several structural changes in small airways can pathologically contribute to the phenotypic overlap of COPD and asthma. Patients with ACO are found to have elevated airway wall thickness as a result of remodeling comprising of inflammation, hypersecretion, mucus plugs hypertrophy, mucosal edema, and hyperplasia of airway smooth muscles. These pathological changes are also observed in both asthma and COPD which ultimately leads to airway obstruction. Even though the structures that are remodeled in ACO, asthma, and COPD are found to be the same, their respective degree of remodeling varies from one another [23–25]. When compared to COPD patients, patients with ACO had thicker airway walls and greater gas trapping on inspiratory and expiratory CT scans, respectively. Clinically, asthma and COPD share a crucial risk factor with ACO. Despite aging and exposure to noxious agents, bronchial hyper-reactivity, a characteristic feature of the asthmatic population has also been considered as an important risk factor in COPD. In case of both asthma and chronic exposure to noxious agents or biomass smoke, patients may end up with elevated levels of airway obstruction which is considered as an indispensable factor for ACO development [24].

### 7.2.2 Severity of ACOS

An epidemiological study carried out by Brzostek et al., assessed the severity of ACOS based on a one-year follow-up on the number of exacerbations in outpatients and hospitalization of patients during their lifetime. Out of the total group of observed patients, 68.6% (with an exacerbation mean of $2.11 \pm 1.76$) of the study patients were found to have exacerbations and the other 31.4% of study patients had a stable course of disease without any clinically diagnosed exacerbations. Additionally, the study also pointed out the mean of nearly 4 hospitalizations ($3.82 \pm 3.67$ hospital stays) during their lifetime. Further, patients with ACO were noted to have three times the severity and rate of exacerbation than that of patients with only asthma or COPD. These prevalent exacerbations are majorly caused due to viral or bacterial infections and can often result in augmented loss of lung function [26–29].

With greater severity and elevated symptoms, patients with ACO experience frequent exacerbations that further worsen their general conditions with swifter deterioration. Additionally, the cost of medical care (consultations, medications, and hospitalizations) required to treat ACO patients are twice high as compared to the cost required to treat COPD alone. The average annual medical cost was found to be way higher for ACO ($14,914) when related to asthma ($2307) or COPD ($4879) alone. ACO is found prevalently among elderly patients which may get associated with multiple clinical conditions leading to an adverse impact on the patient's health.

Apart from frequent exacerbation, patients with ACO have more frequent symptoms of wheezing, dyspnoea, lower respiratory-specific quality, and decreased level of physical activity than patients with COPD alone. Further, COPD patients accompanied with atopy (a feature of ACO) has been found to have a high prevalence of sputum production and chronic cough than those without atopy [10, 24, 26, 30–32].

An 18-year follow-up study observed the risk of death in ACO, asthma, and COPD using a hazard ratio (HR). The group reported the highest risk of death in patients with ACO (with HR of 1.83, 95% CI: 1.34–2.49), followed by patients with COPD (with HR of 1.44, 95% CI: 1.28–1.62), and patients with asthma (with HR of 1.16, 95% CI: 0.94–1.42). Even after adjustments in baseline lung function, patients with ACO still had the highest risk of death when compared to patients with COPS and asthma alone [24, 33]. These findings clearly show that the patients with ACO are at higher risk in comparison to the patients with only asthma and COPD.

## 7.3 Lung Microbiome

### 7.3.1 Roles of the Lung Microbiome

Being persistently exposed to a broad variety of external environments, lungs are always placed at the frontline of our immunity. The respiratory pathway acts as a crucial component in several pulmonary infections, where the pathogenic microorganisms access and adhere to the epithelial cells of the respiratory tract via the respiratory route. The microbiome is known to be an indispensable mediator in influencing pulmonary immunity and a healthy lung is characterized by a vast collection of microbiota. This broad set of microbes boosts both innate and adaptive immunity (site-specific in lungs, and systemic) to release multiple factors for assisting and preventing respiratory functions and infections, respectively. The normal resident microbiota hampers the growth of several harmful lung pathogens which makes their way through the respiratory route. This growth restriction and prevention of respiratory infection by the microbiome may occur through several mechanisms including restricting nutrient access and growth inhibitors secretion against the invading pathogen [6, 9]. Further, researchers have also hypothesized that the establishment of initial microbiota at birth may contribute to pulmonary immune system development. Dysbiosis at these crucial initial stages can potentially lay a foundation for subsequent respiratory diseases. Thus, the development and establishment of proper healthy lung microbiota play a crucial role in protecting the lungs from future pathogenic attacks by impelling the development of local immune response [9, 34, 35].

γδ T cells, one of the crucial regulator and effector cells of the lungs rely on pathogen invasion. Exposure to peculiar bacterial strains can potentially protect neonates from extreme airway inflammation via immune cells modification. Further evidence shows that microbial exposure in children is an indispensable part of developing innate immunity. Children pre-exposed to the microbial surplus environment are at a lower risk rate of allergic asthma, sensitization, and are found to have a stronger immunity when compared to children who were not exposed to microbial-rich environments. Some of the products of lung's normal flora have also shown allegro-protective effects in animal models with airway inflammation [6]. Lung immune cells are responsible for several important roles like patrolling the airway to react against incoming pathogens and restricting the pathogen spread. Apart from this crucial function, lung immune cells have an indispensable duty to avoid unwanted and exaggerated inflammatory responses to harmless environmental particles or stimuli. The subpopulation of dendritic cells (DCs) and alveolar macrophages (AMs) primarily maintains the high immune tolerance nature of the lung microenvironment. Both AMs and DCs exhibit their immunoregulatory properties by influencing the production of regulatory T cells ($T_{reg}$) followed by the release of interleukin-10 (IL-10), prostaglandin E2 ($PGE_2$), and tumor growth factor-beta (TGF-β). Recent increasing evidence suggests that the lung microbiota acting on resident lung immune cells plays a chief role in stimulating immune tolerance in the lungs [36–38].

A certain population of the lung microbiota is often termed as keystone species which may potentially show high beneficial effects on the function, health, and microbial ecosystem balance. Some of the potential keystone species found in the upper respiratory tract (URT) microbiota are *Corynebacterium* spp. and *Dolosigranulum* spp. These notable species are observed to be strongly linked to respiratory health and hampering potential pathogens, especially *Streptococcus pneumoniae* [39–41]. The primary role of a microbial ecosystem is to stimulate a state of symbiosis in order to deliver colonization resistance against the incoming pathogens. This colonization resistance mechanism is dependent on the presence of a local diverse microbiome which can potentially consume all of the nutrients available and thereby averting the incoming pathogens from utilizing the necessary source of nutrients for colonization. Even though there is no direct evidence to demonstrate elevated microbial diversity's role in protecting the respiratory tract against incoming pathogens, specific members of microbiota were found to actively eliminate pathogens from the nasopharyngeal niche. *Staphylococcus epidermidis*, a specific member of microbiota was shown to effectively exclude and destroy *Staphylococcus aureus* and other pre-existing biofilms via the secretion of serine proteases. Moreover, the microbiome's interaction with the host immune system is found to be a potential enhancing mechanism for the process of colonization resistance. For instance, priming with *Haemophilus influenzae* potentially increases the ability of neutrophils to kill *S. pneumoniae*. With these indispensable beneficiary roles, establishing and retaining a balanced microbiota are vital to maintain respiratory health due to their resilience towards detrimental pathogenic expansion [41–44].

### 7.3.2 Composition of Lung Microbiome in the Healthy Lung

Even though it is clear that healthy lungs are not sterile and harbor a phylogenetically diverse microbial community, the study of the normal human lung microbiome needs more attention which is still in its infancy. The lungs are colonized by a distinct set of microbial populations when compared to that of the gut. Although microbiota found in both the human gut and lungs are similar at the phylum structure level, they differ from each other in terms of bacterial species composition. The gut microbiome is the most studied microbiome compared to the less known lung microbiome. The human newborn is found to be deprived of bacteria prior to birth, and the establishment of standard microflora is a constant continuous process which gets initiated during the delivery of the newborn [1, 45, 46]. Normally, bacteria of maternal origin colonize in the newborn oral cavity in the course of vaginal delivery. It was observed that the lower respiratory tract of healthy people contains bacterial DNA of numerous oral bacterial species such as *Veillonella* and *Prevotella*. Moreover, microbial colonization and its species variation are found to be influenced by the exposure configuration during the neonatal period [46].

*Prevotella, Veillonella, Fusobacteria, Pseudomonas,* and *Streptococcus* are some of the dominant genus found in the lungs, whereas *Neisseria* and *Haemophilus* are some of the rare genus found in the lungs. These genera are found to be easily colonizing in the oxygen-rich environment, larynx, damp ciliated mucosa, and the tracheobronchial tree [46–48]. *Bacteroidetes* and *Firmicutes* are some of the species that are predominantly found in bronchoalveolar lavage (BAL) samples of healthy volunteers (HV). However, extensive study reports have also identified the presence of other dominant phyla such as *Fusobacterium, Actinobacteria,* and *Proteobacteria* in several lung-tissue samples. Interestingly, the bacterial communities that are prevalently found in the lungs are usually reflected in the oral cavity but not in the case of the nose. Similarly, few species that show a high prevalent presence in the mouth are found to be less abundant in the lower lungs. These variations in abundance of certain species in the mouth and lower lungs led to a belief where the oral microbes that migrate from the oral cavity to the lungs can be selectively eradicated in order to prevent low-level inflammatory processes [9, 49].

### 7.3.3 Composition of the Lung Microbiome in Lung Diseases

Acute and chronic lung diseases are potential influencers in altering the ecological factors of lung microbiome—microbial immigration, elimination, and regional growth condition leading to noticeably different microbial population. Dysbiosis is a negative impact that arises due to this irregular distribution of microbial communities. In several chronic disease states of the respiratory tract, the commonly observed characteristics are a higher abundance of selective species, species variation, and a shift in microbial populations. *Proteobacteria* including the genus of *Neisseria, Haemophilus, Rickettsia, Pseudomonas,* and *Moraxella* were found to be associated with both controlled and uncontrolled asthma. Similarly, *Firmicutes* with the genus *Lactobacillus* and genus *Clostridium* were isolated from a large set of asthmatic patients and children with airway allergies, respectively [46, 50–52]. One of the critical factors of the microbiome in individuals with diseased lower airways is the shift in microbial populations away from the phylum *Bacteroidetes*, which dominates in the microbiome composition of healthy individuals. These changes in lung microbiota are known to be linked with crucial clinical features and disease prognosis including frequency of exacerbations in bronchiectasis, the mortality rate in idiopathic pulmonary fibrosis, and asthmatic patient's responsiveness towards corticosteroids and antibiotics [51].

*Pseudomonas*, *Rothia*, and *Corynebacterium* were found abundant; *Prevotella* and *Streptococcus* were found less abundant in the intubated patients with pneumonia when compared to patients without pneumonia. Similarly, the *Malassezia* genus, *Cladosporium*, and *Aspergillus penicillium* were some of the over-presented, abundantly found microbial populations in the asthmatic patients. In the case of COPD, a decrease in diversity of lung microbiota and shifts in microbial profiles are often observed characteristics that drive disease prognosis. *Haemophilus influenzae* and *Pseudomonas aeruginosa* are some of the microbial species found during COPD exacerbations of intubated patients. Similar to COPD, patients with cystic fibrosis have also been found to have decreased diversity of lung microbiota. Further, the genus *Prevotella*, *Streptococcus*, *Rothia*, *Actinomyces*, *Veillonella*, *Neisseria*, *Gemellaare*, and *Haemophilus* were isolated during exacerbations in pediatric patients with cystic fibrosis [46, 53–55].

## 7.4 Impact of Lung Microbiome in ACOS

### 7.4.1 Microbiome Mediated Inflammation and Immune Response

The role of the airway microbiome in the lungs can directly impact either lung's immunity or disease. Chronic inflammation and repeated exacerbations that characterize COPD impair the innate immune defense of the lungs leading to alternations in the lung microbiome. In return, these changes in the lung microbiome (diversity, composition, and abundance) affect the host defense. The gram-negative *Prevotella* spp. which are prevalently found in healthy individuals exhibit variations in lipopolysaccharide (LPS) structure when compared to gram-negative *Gammaproteobacteria*. Further bacterial load in the airways is found to be associated with elevated levels of interleukin (IL)-10, IL-1β, and tumor necrosis factor (TNF)-α in the sputum. Additionally, the degree of airway inflammation in stable COPD is related to the composition of the microbiota [56–58]. Moreover, the lung microbiota may also play a crucial role in immune tolerance, recruitment, and activation of antigen-presenting cells (APC) and $T_{reg}$. Multiple lines of evidence have supported the need of specific bacterial species subsequent to birth for the establishment of $T_{reg}$. After the first 2 weeks of birth, the bacterial population in the lungs gets elevated with a parallel progressive shift of bacterial phyla from *Firmicutes* and *Gammaproteobacteria* to *Bacteroidetes*. These crucial modifications of microbiome composition are some of the decisive determinants for reduced responsiveness towards aeroallergens [37].

Recent studies have reported that specific pathogen-free (SPF) mice are at high risk of acute inflammation-induced death to subsequent influenza virus challenges than mice existing in a natural environment. The same group has also demonstrated that the existence of commensal *Staphylococcus aureus* (commonly colonizes in the URT) is an indispensable factor to resist a lethal inflammatory response. This protective effect of *S. aureus* is mediated by the recruitment of $CCR2^{+}CD11b^{+}$ monocytes from the bloodstream into the alveoli, which is followed by polarization and maturation of M2 alveolar macrophages (AMs) in a toll-like receptor 2 (TLR2) pathway-dependent manner. Further, M2 AMs release anti-inflammatory molecules and express immunomodulatory ligands to suppress the lethal inflammation facilitated by influenza virus infection [37, 59].

### 7.4.2 Microbiome Mediated Exacerbations

The overlap syndrome is also characterized by punctuated frequent exacerbations as such asthma and COPD. But, the severity and frequency of exacerbations of ACO are three times than that of both asthma and COPD. Acute exacerbations contribute to 50–75% of COPD healthcare costs in the USA and this exacerbation is responsible for increased morbidity, mortality, and economic burden of the disease. While severe to very severe COPD patients and asthma patients experience 2 or more exacerbations annually, the overlap patient population suffers from a considerably higher rate of exacerbations (up to 2 or 2.5 times) [10, 60, 61]. One of the most common factors triggering exacerbation in asthmatic patients is viral respiratory tract infections (especially rhinoviruses). In contrast, bacterial lung infections are mostly not found to be a part of exacerbation triggering pathogens in asthma. However, in some minority cases, an atypical bacterial infection caused by *Chlamydophila pneumoniae* and *Mycoplasma pneumoniae* has been significantly detected in asthmatic attacks in both adults and children. Similarly, potentially pathogenic *Streptococcus pneumoniae*, *Moraxella catarrhalis,* and Proteobacteria *Haemophilus influenzae* are also detected (culture-based and molecular techniques) in both upper and lower respiratory tract samples in exacerbating asthmatic children [62].

Remarkably, in the course of exacerbations, bacterial taxa that are narrowly related to the specific phylogenetic tree were found to be enriched, whereas phylogenetically distant taxa were found to decline. These observations suggest that taxa which are closely related to potential pathogens could possibly play an inducing role in the exacerbation procedure. Collectively, these reported data insist on the crucial dynamic interlink between microbiota composition and exacerbation mechanisms [54, 56].

## 7.5 Treatment Strategies that Target Lung Microbiome and ACOS

The treatment of ACOS depends on the guidelines of asthma and COPD. The aim of the treatment is to reduce the symptoms and ultimately improve lung functions. Treatment strategies are expected to prevent the progression of the disease and overcome airway remodeling and exacerbations. No single medicine can improve the symptoms of the disease, but combination therapies can help treat the pathologies [63]. Long-term management of ACOS and asthma/COPD alone is needed for the improvement of patient complete recovery. In addition, the treatment should address both COPD and asthma. Similarly, ACOS requires intensive treatment as the disease tends to be more severe when compared to either COPD or asthma alone. Low-dose inhaled corticosteroids (ICS) could be of choice which is a long-term treatment that treats airway inflammation. The dose of corticosteroid is prescribed based on the level of symptoms that the patients exhibit. Long-acting beta-agonist (LABA) and long-acting muscarinic antagonist (LAMA) are the bronchodilators prescribed along with ICS [64].

It is always advised not to treat patients with LABA alone as it may worsen the treatment regimen. This combinatorial therapy of ICS along with bronchodilators was found beneficial to patients to some extent. Rather, there are no definite comparative studies on ICS-bronchodilators therapy stating the better preventive measures and treatment benefits of ACOS. Also, the doses to be recommended for ACOS treatment need to be established. Mostly, the chain smokers with ACOS do not respond to low and mid doses of ICS and even combinatorial therapy. According to the present treatment scenario and patient response to treatment, a combination of ACOS and long-acting bronchodilators is essential from the beginning of the treatment schedule with the doses determined based on the disease severity [13]. Another possible treatment choice would be macrolides which can decrease ACOS exacerbations. Due to their anti-viral property, macrolides are capable of suppressing the activation of neutrophils and mucus hypersecretion. Macrolides were also identified to reduce the frequency of exacerbations by acting on the neutrophil-mediated inflammation caused by lung microbes. Eosinophil inflammation can be inhibited by anti-IgE and anti-IL-5 antibodies which could treat asthmatic components in the pathogenesis of ACOS [65]. The treatment response in patients is also assessed using different parameters including forced expiratory volume per second, maximum expiratory flow using spirometry, exacerbation frequency, sputum eosinophil ratio, peripheral blood eosinophil, and neutrophil ratio, and SpO2 analysis. The above-mentioned approaches are expected to improve the pathology of ACOS and help in the management of ACOS with prescribed clinical benefits [66].

## 7.6 Conclusion

The study of the lung microbiome and its role in asthma and COPD has paved the way to a new question of the role of the lung microbiome in ACO. While asthma and COPD are relatively studied well, ACO is still in its preliminary infancy stage and needs more extensive research. Even though the microbiome, its composition, and its dynamic role in lung airways are reported in several ongoing researches, still the strategic role of the microbiome in the pathogenesis and prognosis of several lung diseases is unexplored. Especially, in terms of the relation between the microbiome and ACO, detailed mechanistic studies deciphering the role of the microbiome in

ACO pathogenesis are still an important question to be addressed. ACO patients with a comparatively higher risk of exacerbations and other detrimental effects, it is an indispensable need to study the patient's microbiome composition in the disease prognosis and exacerbations. With very limited recommendations and guidelines, management of the ACO is still a hard task for most clinicians. Well-conducted clinical trials targeting detrimental alterations of the microbiome in ACO and specific inhibitors targeting disease prognosis of ACO are need of the hour. As a conclusive point, dynamic interplay between the lung microbiome and ACO pathogenesis is an essential area of research that needs to be well explored in the upcoming future.

**Acknowledgments** The authors express their gratitude to SASTRA-Deemed-to-be-University, Tamil Nadu, India for infrastructure and financial support. The authors also extend their appreciation for the contribution of Biopharmaceutical research lab members, SASTRA-Deemed-to-be-University.

**Author Contribution** **Shibi Muralidar**: Writing - Original Draft, Collection of data, Writing - Review & Editing; **Gayathri Gopal**: Writing - Original Draft, Collection of data, Writing - Review & Editing; **Senthil Visaga Ambi**: Conceptualization, Visualization, Supervision, and Writing - Review & Editing.

**Conflict of Interest** The authors declare no conflict of interest.

## References

1. Dumas A, Bernard L, Poquet Y et al (2018) The role of the lung microbiota and the gut–lung axis in respiratory infectious diseases. Cell Microbiol 20. https://doi.org/10.1111/cmi.12966
2. Ruta V, Alexescu T, Ungur A et al (2020) Composition and modification of the lung microbiome in patients with idiopathic pulmonary fibrosis. J Mind Med Sci 7:162–167. https://doi.org/10.22543/7674.72.P162167
3. Li KJ, Chen ZL, Huang Y et al (2019) Dysbiosis of lower respiratory tract microbiome are associated with inflammation and microbial function variety. Respir Res 20:1–16. https://doi.org/10.1186/s12931-019-1246-0
4. Gill SR, Pop M, RT DB et al (2006) Metagenomic analysis of the human distal gut microbiome. Science. 312:1355–1359. https://doi.org/10.1126/science.1124234
5. Han MLK, Huang YJ, LiPuma JJ et al (2012) Significance of the microbiome in obstructive lung disease. Thorax 67:456–463. https://doi.org/10.1136/thoraxjnl-2011-201183
6. Khatiwada S, Subedi A (2020) Lung microbiome and coronavirus disease 2019 (COVID-19): possible link and implications. Hum Microbiome J 17. https://doi.org/10.1016/j.humic.2020.100073
7. Chunxi L, Haiyue L, Yanxia L et al (2020) The gut microbiota and respiratory diseases: new evidence. J Immunol Res 2020. https://doi.org/10.1155/2020/2340670
8. Kusumoto M, Mathis BJ (2021) Biologic treatments for asthma and chronic obstructive pulmonary disorder. Allergie 1:92–107. https://doi.org/10.3390/allergies1020007
9. Marimón JM (2018) The lung microbiome in health and respiratory diseases. Clin Pulm Med 25:131–137. https://doi.org/10.1097/CPM.0000000000000268
10. Papaiwannou A, Zarogoulidis P, Porpodis K et al (2014) Asthma-chronic obstructive pulmonary disease overlap syndrome (ACOS): current literature review. J Thorac Dis 6:3–8. https://doi.org/10.3978/j.issn.2072-1439.2014.03.04

11. GINA (2021) Global Initiative for Asthma. Global Strategy for Asthma Management and Prevention, 2021
12. Hikichi M, Hashimoto S, Gon Y (2018) Asthma and COPD overlap pathophysiology of ACO. Allergol Int 67:179–186. https://doi.org/10.1016/j.alit.2018.01.001
13. Tu X, Donovan C, Kim RY et al (2021) Asthma-COPD overlap: current understanding and the utility of experimental models. Eur Respir Rev 30. https://doi.org/10.1183/16000617.0185-2019
14. Zhou XL, Zhao LY (2021) Comparison of clinical features and outcomes for asthma-COPD overlap syndrome vs. COPD patients: a systematic review and meta-analysis. Eur Rev Med Pharmacol Sci 25:1495–1510. https://doi.org/10.26355/eurrev_202102_24857
15. Kocakaya D (2016) Asthma-COPD overlap syndrome (ACOS). Marmara Med J 29:3–9. https://doi.org/10.5472/mmjsi.2902.02
16. Akwe J (2016) Asthma chronic obstructive pulmonary disease overlap syndrome (ACOS): where we stand. Int J Clin Exp Med Sci 2:59. https://doi.org/10.11648/j.ijcems.20160204.12
17. Chanez P, Vignola AM, O'Shaugnessy T et al (1997) Corticosteroid reversibility in COPD is related to feature of asthma. Pneumologie 51:923
18. Soler-Cataluña JJ, Cosío B, Izquierdo JL et al (2012) Consensus document on the overlap phenotype COPD-asthma in COPD. Arch Bronconeumol 48:331–337. https://doi.org/10.1016/j.arbr.2012.06.017
19. De Marco R, Marcon A, Rossi A et al (2015) Asthma, COPD and overlap syndrome: a longitudinal study in young European adults. Eur Respir J 46:671–679. https://doi.org/10.1183/09031936.00008615
20. Bobolea I, de Llano LAP (2016) Asthma-COPD overlap syndrome (ACOS): current understanding and future perspectives. Asthma - from child asthma to ACOS phenotypes. IntechOpen, London. https://doi.org/10.5772/62412
21. Cosío BG, Dacal D, Pérez de Llano L (2018) Asthma–COPD overlap: identification and optimal treatment. Ther Adv Respir Dis 12:1–11. https://doi.org/10.1177/1753466618805662
22. Shaw DE, Berry MA, Hargadon B et al (2007) Association between neutrophilic airway inflammation and airflow limitation in adults with asthma. Chest 132:1871–1875. https://doi.org/10.1378/chest.07-1047
23. Bumbacea D, Campbell D, Nguyen L et al (2004) Parameters associated with persistent airflow obstruction in chronic severe asthma. Eur Respir J 24:122–128. https://doi.org/10.1183/09031936.04.00077803
24. Van Tho N, Park HY, Nakano Y (2016) Asthma–COPD overlap syndrome (ACOS) A diagnostic challenge. Respirology. 21:410–418
25. Vonk JM, Jongepier H, Panhuysen CIM et al (2003) Risk factors associated with the presence of irreversible airflow limitation and reduced transfer coefficient in patients with asthma after 26 years of follow up. Thorax 58:322–327. https://doi.org/10.1136/thorax.58.4.322
26. Brzostek D, Kokot M (2014) Asthma-chronic obstructive pulmonary disease overlap syndrome in Poland. Findings of an epidemiological study. Postep Dermatol Alergol 31:372–379. https://doi.org/10.5114/pdia.2014.47120
27. Menezes AMB, De Oca MM, Pérez-Padilla R et al (2014) Increased risk of exacerbation and hospitalization in subjects with an overlap phenotype: COPD-asthma. Chest 145:297–304. https://doi.org/10.1378/chest.13-0622
28. Lange P et al (1998) A 15-year follow-up study of ventilatory function in adults with asthma. N Engl J Med. 339:1194–1200
29. Zeki AA, Schivo M, Chan A et al (2011) The asthma-COPD overlap syndrome: a common clinical problem in the elderly. J Allergy 2011:1–10. https://doi.org/10.1155/2011/861926
30. McDonald VM, Simpson JL, Higgins I, Gibson PG (2011) Multidimensional assessment of older people with asthma and COPD: clinical management and health status. Age Ageing 40: 42–49. https://doi.org/10.1093/ageing/afq134

31. Rhee CK, Yoon HK, Yoo KH et al (2014) Medical utilization and cost in patients with overlap syndrome of chronic obstructive pulmonary disease and asthma. COPD 11:163–170. https://doi.org/10.3109/15412555.2013.831061
32. Shaya FT, Dongyi D, Akazawa MO et al (2008) Burden of concomitant asthma and COPD in a Medicaid population. Chest 134:14–19. https://doi.org/10.1378/chest.07-2317
33. Enrique DG, Khosravi M, Mannino DM (2011) Asthma, chronic obstructive pulmonary disease, and mortality in the U.S. population. COPD 8:400–407. https://doi.org/10.3109/15412555.2011.611200
34. Hall H (2017) Maturation of the infant respiratory microbiota, environmental drivers and health consequences: a prospective cohort study. Am J Respir Crit Care Med 196(12):1582–1590. https://doi.org/10.1164/rccm.201703-0554OC
35. Lal CV, Travers C, Aghai ZH et al (2016) The airway microbiome at birth. Sci Rep 6:1–13. https://doi.org/10.1038/srep31023
36. Hussell T, Bell TJ (2014) Alveolar macrophages: plasticity in a tissue-specific context. Nat Rev Immunol 14:81–93. https://doi.org/10.1038/nri3600
37. Sommariva M, Le Noci V, Bianchi F et al (2020) The lung microbiota: role in maintaining pulmonary immune homeostasis and its implications in cancer development and therapy. Cell Mol Life Sci 77:2739–2749. https://doi.org/10.1007/s00018-020-03452-8
38. Soroosh P, Doherty TA, Duan W et al (2013) Lung-resident tissue macrophages generate Foxp3+ regulatory T cells and promote airway tolerance. J Exp Med 210:775–788. https://doi.org/10.1084/jem.20121849
39. Goodrich JK, Waters JL, Poole AC et al (2014) Human genetics shape the gut microbiome. Cell 159:789–799. https://doi.org/10.1016/j.cell.2014.09.053
40. Laufer AS, Metlay JP, Gent JF et al (2011) Microbial communities of the upper respiratory tract and otitis media in children. MBio 2:1–6. https://doi.org/10.1128/mBio.00245-10
41. Man WH, De Steenhuijsen Piters WAA, Bogaert D (2017) The microbiota of the respiratory tract: gatekeeper to respiratory health. Nat Rev Microbiol 15:259–270. https://doi.org/10.1038/nrmicro.2017.14
42. Baümler AJ, Sperandio V (2016) Interactions between the microbiota and pathogenic bacteria in the gut. Nature 535:85–93. https://doi.org/10.1038/nature18849
43. Iwase T, Uehara Y, Shinji H et al (2010) Staphylococcus epidermidis Esp inhibits Staphylococcus aureus biofilm formation and nasal colonization. Nature 465:346–349. https://doi.org/10.1038/nature09074
44. Kamada N, Chen GY, Inohara N, Núñez G (2013) Control of pathogens and pathobionts by the gut microbiota. Nat Immunol 14:685–690. https://doi.org/10.1038/ni.2608
45. Faner R, Sibila O, Agustí A et al (2017) The microbiome in respiratory medicine: current challenges and future perspectives. Eur Respir J 49:1–12. https://doi.org/10.1183/13993003.02086-2016
46. Stavropoulou E, Kantartzi K, Tsigalou C et al (2021) Unraveling the interconnection patterns across lung microbiome, respiratory diseases, and COVID-19. Front Cell Infect Microbiol 10:1–13. https://doi.org/10.3389/fcimb.2020.619075
47. Beck JM, Young VB, Huffnagle GB (2012) The microbiome of the lung. Transl Res 160:258–266. https://doi.org/10.1016/j.trsl.2012.02.005
48. Dickson RP, Erb-Downward JR, Martinez FJ, Huffnagle GB (2016) The microbiome and the respiratory tract. Annu Rev Physiol 78:481–504. https://doi.org/10.1146/annurev-physiol-021115-105238
49. O'Carroll O, Peart J, Mullen E, Burke C (2018) The respiratory microbiome in COPD. In: COPD - An Update in Pathogenesis and Clinical Management. IntechOpen, London. https://doi.org/10.5772/intechopen.70776
50. Chiu CY, Chan YL, Tsai MH et al (2019) Gut microbial dysbiosis is associated with allergen-specific IgE responses in young children with airway allergies. World Allergy Organ J 12:100021. https://doi.org/10.1016/j.waojou.2019.100021

51. Huffnagle GB, Dickson RP, Lukacs NW (2017) The respiratory tract microbiome and lung inflammation: a two-way street. Mucosal Immunol 10:299–306. https://doi.org/10.1038/mi.2016.108
52. Park HK, Shin JW, Park SG, Kim W (2014) Microbial communities in the upper respiratory tract of patients with asthma and chronic obstructive pulmonary disease. PLoS One 9. https://doi.org/10.1371/journal.pone.0109710
53. Cox MJ, Allgaier M, Taylor B et al (2010) Airway microbiota and pathogen abundance in age-stratified cystic fibrosis patients. PLoS One:5. https://doi.org/10.1371/journal.pone.0011044
54. Huang YJ, Kim E, Cox MJ et al (2010) A persistent and diverse airway microbiota present during chronic obstructive pulmonary disease exacerbations. OMICS 14:9–59. https://doi.org/10.1089/omi.2009.0100
55. Worlitzsch D, Rintelen C, Böhm K et al (2009) Antibiotic-resistant obligate anaerobes during exacerbations of cystic fibrosis patients. Clin Microbiol Infect 15:454–460. https://doi.org/10.1111/j.1469-0691.2008.02659.x
56. Dima E, Kyriakoudi A, Kaponi M et al (2019) The lung microbiome dynamics between stability and exacerbation in chronic obstructive pulmonary disease (COPD): current perspectives. Respir Med 157:1–6. https://doi.org/10.1016/j.rmed.2019.08.012
57. Larsen JM, Musavian HS, Butt TM et al (2015) Chronic obstructive pulmonary disease and asthma-associated Proteobacteria, but not commensal Prevotella spp., promote Toll-like receptor 2-independent lung inflammation and pathology. Immunology 144:333–342. https://doi.org/10.1111/imm.12376
58. Larsen JM, Steen-Jensen DB, Laursen JM et al (2012) Divergent pro-inflammatory profile of human dendritic cells in response to commensal and pathogenic bacteria associated with the airway microbiota. PLoS One:7. https://doi.org/10.1371/journal.pone.0031976
59. Wang J, Li F, Sun R et al (2013) Bacterial colonization dampens influenza-mediated acute lung injury via induction of M2 alveolar macrophages. Nat Commun 4:1–10. https://doi.org/10.1038/ncomms3106
60. Hardin M, Silverman EK, Barr RG et al (2011) The clinical features of the overlap between COPD and asthma. Respir Res 12:127. https://doi.org/10.1186/1465-9921-12-127
61. Soriano JB, Visick GT, Muellerova H et al (2005) Patterns of comorbidities in newly diagnosed COPD and asthma in primary care. Chest 128:2099–2107. https://doi.org/10.1378/chest.128.4.2099
62. Loverdos K, Bellos G, Kokolatou L et al (2019) Lung microbiome in asthma: current perspectives. J Clin Med 8:1967. https://doi.org/10.3390/jcm8111967
63. Kondo M, Tamaoki J (2018) Therapeutic approaches of asthma and COPD overlap. Allergol Int 67:187–190. https://doi.org/10.1016/j.alit.2017.09.002
64. Papi A (2020) Asthma COPD overlap PRO-CON debate. ACO: the mistaken term. COPD 17: 474–476. https://doi.org/10.1080/15412555.2020.1817882
65. Alshabanat A, Zafari Z, Albanyan O et al (2015) Asthma and COPD overlap syndrome (ACOS): a systematic review and meta analysis. PLoS One 10:1–15. https://doi.org/10.1371/journal.pone.0136065
66. Baarnes C, Kjeldgaard P, Nielsen M et al (2017) Identifying possible asthma–COPD overlap syndrome in patients with a new diagnosis of COPD in primary care. NPJ Prim Care Resp Med 27:16084. https://doi.org/10.1038/npjpcrm.2016.84

# Microbiome in Acute Respiratory Distress Syndrome (ARDS)

# 8

Gayathri Gopal, Shibi Muralidar, Abishek Kamalakkannan, and Senthil Visaga Ambi

**Abstract**

Acute respiratory distress syndrome (ARDS) is a sign of severe injury to the epithelial cells of the alveoli. It is hallmarked by pulmonary oedema and inflammation leading to respiratory failure. The incidence of ARDS is reported to be 78 in 1 lakh people every year. The disease is caused by micro-aspiration (direct) or by indirect lesions including sepsis. A high mortality rate of up to 40% is observed in lung fibrosis. In patients with ARDS, the lung microbiome is flourished with gut bacteria. The gut-associated microbiota serves as the critical marker in the pathogenesis of ARDS. The alteration of microbial population in the lungs tends to induce alveolar inflammation and lung injury. The imbalance in microbial pathogens in terms of migration, elimination, and reproduction contributes to the pathophysiological changes in the lungs. The translocation of the enteric bacteria into the lungs is the key feature in ARDS development. The predominant gut bacteria rich in lung microbiota of ARDS are *Bacteroidetes* and

Gayathri Gopal and Shibi Muralidar contributed equally with all other contributors.

G. Gopal · S. Muralidar · A. Kamalakkannan
Biopharmaceutical Research Lab, Anusandhan Kendra-1, SASTRA Deemed-to-be-University, Thanjavur, India

School of Chemical and Biotechnology, SASTRA Deemed-to-be-University, Thanjavur, India

S. V. Ambi (✉)
Biopharmaceutical Research Lab, Anusandhan Kendra-1, SASTRA Deemed-to-be-University, Thanjavur, India

School of Chemical and Biotechnology, SASTRA Deemed-to-be-University, Thanjavur, India

Department of Bioengineering, School of Chemical and Biotechnology, SASTRA Deemed-to-be-University, Thanjavur, India
e-mail: senthilvisagaambi@scbt.sastra.edu



G. Gupta et al. (eds.), *Microbiome in Inflammatory Lung Diseases*,
https://doi.org/10.1007/978-981-16-8957-4_8

*Enterobacteriaceae* commonly referred to as 'more gut in the lung'. The predominance of a particular bacterial population in the entire microbiota in critically ill patients in comparison to healthy individuals is referred to as the modification of the lung microbiome. Advances in genome sequencing have led to the detection of unique microbial communities in patients with ARDS. BALF specimens are suitable for the detection of the bacterial burden by 16S rRNA sequencing. Meta transcriptome analysis and imaging techniques including CT scan, X-ray, and histopathological analysis could be of choice for the identification of lung microbial population. The lung microbiome can be therapeutically modulated by antibiotic treatments and the pathologies may be solved by improving the oxygen supplementation. The oxygen deprivation in ARDS can be supported by mechanical ventilation or intubation and ECMO in certain severe ARDS conditions. Ultimately, understanding the lung microbiome in ARDS and the influence of modified microbiome in the outcomes of the disease may help in arriving at therapeutic interventions in preventing and treating the disease. This chapter will summarize the pathogenesis of ARDS, modification of lung microbiome involved in the pathophysiology of ARDS, analysis of microbial population with the focus on its detection and treatment strategies for the management of ARDS.

**Keywords**

Alveolar inflammation · Hypoxemia · Microbiota · Sepsis · Gut–lung axis · Mechanical ventilation

## 8.1 Introduction to ARDS

Acute Respiratory Distress Syndrome (ARDS) is a pathetic lung condition that lowers blood oxygen (hypoxemia). ARDS is associated with severe injury in the alveolar epithelial cells with subsequent respiratory failure leading to mortality rates of 40% approximately. The incidence of ARDS is reported to be 78 in 1 lakh people every year [1]. ARDS is often characterized by inflammation in the lungs [2]. In ARDS, hypoxemia is caused by the accumulation of fluid in the distal air spaces of the lung, thereby interrupting the blood gas exchanges. Fluid from the blood vessels drains into the damaged alveolar walls, limiting the exchange of oxygen and carbon-dioxide. As a result, the lung surfactant breaks down preventing the lung from properly occupied with air causing stiffness of the lung tissues. Damage of the air sacs is due to lung infection or continuous inhalation of smoke which triggers inflammation in the air sacs. This, in turn, causes difficulty in breathing and most cases require endotracheal intubation and ventilation [3]. The development of ARDS is associated with several clinical factors, most importantly the pulmonary and non-pulmonary infections. The majority of the patients develop ARDS due to pneumonia caused by bacteria or viral infections and also due to sepsis-associated with pulmonary infections, more recently COVID-19. Other common causes of the syndrome include transfusion-associated lung injury, acute pancreatitis, adverse drug reactions, aspiration of gastric contents, and patient lifestyle [4]. There is also the risk of multiple organ failure including cardiovascular failure requiring vasopressor support, renal failure, haematological impairments such as anaemia and thrombocytopenia, and abnormal liver function [5].

The symptoms of ARDS include shortness of breath, low blood oxygen levels, bubbling or rattling of the lungs while breathing. These symptoms may develop over time or occur quickly depending on the age and medical history. In some patients, fast breathing, coughing with phlegm, chest pain, and fatigue worsen the syndrome. Other medical problems associated with ARDS pulmonary hypertension and long-period treatment in hospitals. ARDS is usually diagnosed with a physical examination or medical history. At times, it becomes difficult for the physicians to diagnose as the medical signs coincide with other lung disorders [2]. It was reported that 10% of the patients under intensive care unit (ICU) develop ARDS while the patient is being treated for any infections or trauma. Patients with ARDS require ventilator support or even extracorporeal membrane oxygenation (ECMO). Unfortunately, these life support treatments tend to develop new problems including lung collapse, infections from the catheter, and blood clots as the patients are lying still for long periods [2, 6].

Among the causes of ARDS, the most unfavourable cause is sepsis, accounting for high mortality. The pathophysiological changes occurring in the lungs contribute to the clinical course of ARDS. The entry, migration, and elimination of pathogens in the respiratory tract are the preliminary factors to be considered for understanding the pathophysiology of ARDS [7]. Recently, Dickson and co-workers have reported that lung microbiota in ARDS is enriched with bacteria found in the gastrointestinal tract. This was identified through 16S rRNA sequencing of the bacteria isolated from the mouse models of ARDS, which could not be possible with conventional culturing methods. It was also found that these gut bacteria are responsible for the severity of the disease, indicating that the translocation of the bacteria from the gut to the lung might be the renowned mechanism for the development of ARDS or sepsis [8].

## 8.2 Pathogenesis of ARDS

ARDS is characterized by several noticeable conditions that lead to a common pathophysiological pathway. The subsequent events thereafter are classified into two classes namely:

1. Direct 'pulmonary'
2. Indirect 'extrapulmonary' conditions.

The explicit causes include distinct conditions that damage the lung parenchyma along with cases of pneumonia, pulmonary contusion due to trauma, aspiration, and inhalation or consumption of toxic agents. One of the recurrent indirect causes is sepsis syndrome which is a familiar and fatal cause of ARDS. The indirect causes also include acute pancreatitis, the overdose of certain drugs such as opiates of thiazides, propagative intravascular coagulation, and multiple blood transfusions. Amidst various triggers, the culminated ARDS in its subsequent later stages displays a systematic clinical and pathological pattern, irrespective of its pathophysiological pathways and the symptoms may differ according to the event that injures the lungs [9].

Lung injury is commenced by a particular cause that can be aggravated by poor ventilation. Concisely, alveolar over-distension can produce pro-inflammatory feedback that is intensified by repeated opening and closing of alveoli employing improper low levels of positive end-expiratory pressure [2, 5]. Undoubtedly, the recurring opening and closing of alveoli can stimulate systemic injury to the lungs. The adverse effects resulting in high levels of oxygen during the disease are highly unsettled, especially in humans. Nevertheless, protracted subjection to 100% oxygen is lethal in most animal models where the neutrophil influx and alveolar oedema can be restrained in using anti-inflammatory actions by inhalation of low dose carbon monoxide [10].

There are widely numerous pathophysiological features that are prime factors involved in the incidence of ARDS. Irregular inflammatory events and increased permeability of lung epithelium are the primary reasons for pathogenesis. In the early stages, as detected, acute lung injury is navigated by dysregulated inflammation. The metabolic products of the microbes or cell-injury-related endogenous molecules which cohere to toll-like receptors (TLR) on the lung epithelium and alveolar macrophages stimulate the innate immune responses. Certain events of the innate immune defense system include neutrophil extracellular traps formation and histone releases, which is advantageous in capturing pathogens but might intensify the alveolar injury. The immune system producing ROS (reactive oxygen species), leukocyte proteases, chemokines, and cytokines that help in counterbalancing the pathogens could result in worse outcomes. [11].

Besides elevated inflammation, another notable pathophysiological feature is the interference of the lung microvascular barrier which is due to increased permeability of endothelium and epithelium. In the lungs of healthy individuals, vascular endothelial cadherin (VE-cadherin) mediates endothelial stabilization. VE-cadherin is an endothelial-specific adherent junction protein, required for the endothelial barrier coherence within the lung micro-vessels. Unfortunate lung damage, the proliferation of thrombin, tumour necrosis factor-α (TNF-α), vascular endothelial growth factor (VEGF), and leukocyte signals in the lungs to restrict the VE-cadherin bonds, has been established in the murine models. Precisely, alveolar fluid compiles up in lipopolysaccharide (LPS) induced mice models—promoted lung injury, in according to that VE-cadherin bonds are sustained by a genetic modification that prevents mishap or by intercepting VE phosphodiesterase, with the reduction in edema formation. Thereby, the inflammatory-promoted injury to lung endothelium results in an upsurge of capillary permeability, promoting pulmonary edema formation [11].

The pathogenesis of ARDS is a highly complex mechanism involving various factors and the roles they play in the processes. However, recent advancements in pathogenesis study involve various chemokines that influence various developments in endothelial and epithelial functioning. The key factors of pathogenesis include VEGF, interleukin (IL-8), and transforming growth factor (TGF-β) which helps understand the process.

**Vascular Endothelial Growth Factor (VEGF):** VEGF is a glycoprotein that is isolated as a permeability factor. Normally, vascular endothelial cells, lung epithelium, platelets, and leukocytes are involved in the production of the release of VEGF. It is widely helpful to amplify angiogenesis and microvascular permeability by receptor binding. Generally, the concentration of VEGF in the normal alveolar space is low. An upsurge in its concentration has led to the attention that VEGF is firmly associated with the improper function of the ACM in various respiratory disorders, including ARDS. In the genome analysis-based studies, a polymorphism linked to decreased plasma concentration of VEGF was observed to be elevated in ARDS patients, and with sufficient modifications within the lungs during the early, exudative phase of lung injury [12].

Researchers hypothesize that VEGF would be solely responsible for pulmonary haemorrhage, endothelial destruction, and alveolar remodeling in an emphysema-like phenotype. However, the production of surfactant protein-B was not distressed. In models of ARDS, the high tidal and detrimental volume of ventilation support escalated the lung VEGF-R2 (the primary VEGF bioactivity signaling receptor) protein concentrations. But protective ventilatory strategies did not bring down the concentration of VEGF-R2, proposing that the response of response might be secondary to more critical events [13].

VEGF interactions are monitored at various levels; alternate transcript splicing of exons (6–8 leads) and production of several isoforms, with diversified properties. Additionally, receptors VEGFR-1 and VEGFR-2 are subdued to their expression in various tissues, and alternative splicing of VEGFR-1 results in its soluble form (soluble-FMS-like tyrosine kinase, s-FLT), which is suppressive. Expression of s-FLT has been observed to be altering in the pathogenesis of ARDS, with an upsurge of plasma levels in all the stages of the disease. During the proliferative and fibro-proliferative stages of ARDS, both VEGFR-1/2 were found to be unregulated. VEGF-2 (soluble) is also detected in traceable amounts in the BAL fluid of patients with ARDS. The varying expression of VEGF with changes in its isoforms suggests that regulation of VEGF bioactivity is highly essential in determining the pathogenesis and disease progression of ARDS [12].

**IL-8 (Interleukin-8):** IL-8 (CXCL8) is a chemokine with the characteristic of captivation and stimulation and neutrophils. It has a definite role in ARDS, where they exist at higher concentrations in the bronchoalveolar lavage fluids (BALF) from ARDS patients in contrast to controls. A higher concentration of IL-8 is correlated to a high mortality rate [12]. In 1992, Miller and his colleagues found that IL-8 was abundant in BALF collected from patients with ARDS patients as compared to controls [14].

The presence of IL-8 was exclusive during the early and exudative phase of ARDS, where IL-8 was found to be moderate in cellular infiltration [12]. In subsequent research work carried by Kurdowska and his colleagues, they noticed that a significant amount of IL-8 in BALF isolated from ARDS patients is correlated with anti-IL-8 autoantibody, which attaches with large rapport to IL-8. Anti-IL-8 autoantibody consists of the following:

- 1 immunoglobulin G (IgG) molecule where a larger amount is IgG3 and IgG4 subclasses
- 1 IL-8 molecule

Anti-IL-8 autoantibodies hindering the synergy of IL-8 with certain receptors on neutrophils indicating their role in balancing the IL-8 activity in ARDS [14].

IL-8's role in ARDS is highly complex; certain activity seems to be balanced by an anti-IL-8 autoantibody. This phenomenon was directed to the common hypothesis that vast infection or other affronts exceeds the capacity of the removal mechanisms. The enduring amounts of anti-IL-8: IL-8 complexes in the lung may be indicative of ARDS pathologies. These antibody-IL-8 complexes control the capacity to activate the signaling of neutrophil and respiratory burst/degranulation; consequently, they are still capable of triggering other inflammatory responses [12].

**Transforming growth factor (TGF-β):** One of the prominent arbitrators in ARDS is TGF-β with an influence on epithelial and endothelial permeability through protease-activated receptor-1 to develop alveolar flooding. Compared to other mediators involved in ARDS developments, TGF-β in BAL fluid was identified at increased levels. They are observed in lower levels in patients who are free from ventilator support and intensive care unit (ICU). These observations suggest that TGF-β plays a primary role in the pathology of the syndrome [12].

TGF-β regulating mechanisms help to devise some of the evident dysregulations of the disease. TGF-β is produced as an inactive complex, attached to the latency-associated peptide, which inhibits it from binding to other specific receptors. The inactive complex is activated by pH, heat, proteases, thrombospondin-1, reactive oxygen species (ROS), and integrins. This complex might have a prolonged half-life than TGF-β 1and might bind to its receptor to strive anti-inflammatory effects through the Foxp3-dependent mechanism. The latency-associated peptide also binds to TGF-β 1 as a formidable inhibitor monitoring its bioactivity [5].

**Inflammation:** The beginning of inflammation is seen when leucocyte production and engagement to the inflamed site upsurge. Stimulation of mediator cascades includes the synthesis of cytokines, chemokines, acute phase proteins, free radicals, complement, coagulation pathway components, and focal upregulation of adhesion molecule expression. The 'anti-inflammatory' response includes the

- glucocorticoids,
- cytokines and other mechanisms,
- shedding of adhesion molecules.

From the above mentioned, there is a widely known cytokine called TNF-α which plays a prominent role in inflammation, thus serving as inflammatory mediators in the process [10].

They are synthesized by inflammatory cells and can enhance neutrophil-endothelial adhesion, microvascular leakage, magnify other pro-inflammatory responses. Despite their definitive profile in the septic response, the importance of these cytokines like TNF- α in the pathogenesis of ARDS is unclear. Studies of the levels of TNF-α are not consistently increasing in cases of reported lung injury and anti-TNF-α therapies have been upsetting. An upsurge of TNF-α level occurs very early in the clinical course and may be missed by the time of presentation although anti-TNF-α therapies can be helpful in some cases of sepsis. The huge redundancy in the pro-inflammatory mediator systems suggests that the search for a 'common pathway' susceptible to inhibition may be too simplistic [10].

Citing Animal references, free radicals are vital factors involved in tissue damage from pro-inflammatory catalysts and antioxidants including glutathione and superoxide dismutase. In humans, oxidative stress is upsurged as plasma antioxidant ranges are significantly decreased in patients with ARDS. Nitric oxide is instrumental in septic lung injury as nitro- tyrosine, a derivative from peroxynitrite is predominantly found in patients with ARDS [10].

## 8.3 Pulmonary Microbiome in Healthy and Critically Ill Patients

The researchers were hypothesizing the sterility of the lungs and focus on the qualitative and quantitative analysis of the upper and lower respiratory tracts. The presence of bacterial 16S rRNA sequences in the LRT confirms the microbiota in the healthy lungs. The normal microbiota in healthy individuals consists of microbial communities which may be symbiotic and pathogenic. The microbiome consists of the genetic material of the microbiota. More than 100 billion microorganisms colonize the human body that is discovered by NGS and metagenome analysis. The imbalance in the microbial community (dysbiosis) is associated with the disease or any organ failure [15]. The microbial population throughout the body is crucial for human physiology. The lung microbiota contains a low density of bacteria and is essential for good health. Microbes of the lungs enter the body through the oral cavity, pass through the saliva as microparticles, and reach the lungs. Normally the lung microbiota disperses from the mouth or nose and a balance is maintained between immigration and elimination. The host immune response is responsible for eliminating the microbes through mucociliary movements, sneezing, and coughing [16, 17]. The microbial community keeps renewing and is often being replaced. In healthy individuals, the most predominant phyla are Bacteroidetes and Firmicutes, analysed by whole-genome sequencing. The lung microbiota is unique for every individual, constituting the homeostasis between the microbes and the host tissues. In respiratory diseases, particularly in ARDS, an imbalance or dysbiosis is observed due to a shift in the normal microbiota. This change can be the predominance of particular bacteria or modification in the entire microbial community. Though microbial immigration and reproduction decide the composition of healthy lung microbiota, certain physiological factors of the respiratory tract also influence the incidence of pulmonary disorders [18].

The infiltration of bacteria into the lungs and the imbalance of the bacterial flora will result in the modification of pulmonary microbiome composition. In critically ill patients, pulmonary dysbiosis can be due to several factors including antibiotic treatment, ICU observations, or recurrence of infections. *Streptococcus pneumonia* infection was found to be higher in patients with lung inflammation and consistent antibiotic treatment ([19, 20]. In addition, ventilation of patients with high tidal volume increases the incidence of lung microbiome modification. ARDS is characterized by pulmonary edema wherein the cytokine signaling could alter the metabolic and physiochemical status of the lungs. The presence of *Bacteroidetes* and *Enterobacteriaceae* in patients with ARDS increased pro-inflammatory markers contributing to the pathogenesis of the disease. In particular, Enterobacteriaceae induce the production of IL-22, TLR-2, and TLR-4 leading to severe lung injury. The bronchoalveolar lavage (BAL) analysis in adults with ARDS has provided insight on the modification of lung microbiome and difference in the inflammatory response. The activity of the anti-inflammatory markers is affected by the modification of the lung microbiome, resulting in the severity of the disease. The close relation of lung microbiome between COVID-19 and ARDS was identified by the representation of similar bacterial phylum and genera [19, 21].

Patients with ARDS were observed to exhibit a reduction in microbial diversity and overgrowth of one particular flora. The disease pathology aggravates the modification of the pulmonary microbiome and favours the significant reduction of bacterial diversity. At the onset of the disease, the dominance of a particular microbial population hampers the function of alveolar capillaries through the migration of macrophages and granulocytes. As a result, the epithelial damage would hinder the normal epithelial barrier against the modified microbiome leading to further lung injury. Further, the mechanically ventilated patients undergoing sedation fail in mucociliary clearance which may interfere in the oxygen concentration and increase the chances of specific pathogenic infection [2, 7, 15].

## 8.4 Lung Microbiota in ARDS

The lung microbiota of patients with lung disorders is critically important concerning alveolar and systemic inflammation. In ARDS, the lung microbiome is enriched with gut-associated bacteria, serving as the primary biomarker in the development of the disease [8]. The altered microbiome in critically ill patients intensifies the alveolar inflammation leading to severe clinical outcomes. The transmission of gut microbiota from the GI tract to the lungs is identified to be the fundamental cause of sepsis and ARDS. However, the role of gut microbes in lung pathology remains unidentified. Dickson and his team have reported that the lung microbiota diagnosed in ARDS is enriched with bacteria which are usually found in

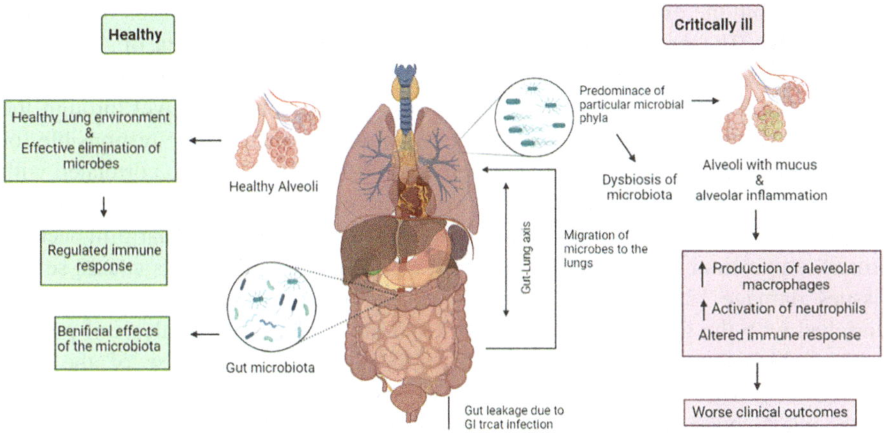


**Fig. 8.1** Role of lung microbiota in healthy individuals and critically ill patients

the lower GI tract. The team also found that the presence of gut microbes in ARDS patient samples was associated with the severity of the disease [22] (Fig. 8.1).

Few pre-clinical studies have demonstrated the specific effect of the microbiome in asthma and cystic fibrosis. But there is a gap in bridging the fate of lung and gut microbiome in ARDS. A recent study with a murine model of sepsis has identified the enrichment of gut bacteria including *Enterococcus* and *Lachnospiracea*e species and Bacteroidetes order. It was reported that the microbiome was altering in the initial days of infection and normalizes by 2–3 weeks. Also, critically ill bedridden mechanically ventilated patients are more susceptible to the disease due to the evolution of airway microbiota during the prolonged treatment [23]. Two interesting facts have been observed in the recent clinical study by Kyo and co-workers: (1) the composition of lung microbiome alters in patients with injuries due to smoking, (2) the bacterial load and composition significantly differ over the time of patient observation. These findings could critically help in developing a therapeutic intervention concerning the mechanically ventilated patients and patients with smoking habits, to avoid the subsequent development of ARDS [6]. Yet, the actual role and significance of gut bacteria in the lung microbiota are still not solved. It was reported that patients with both acute and severe pulmonary infections exhibit gut-associated complications, indicating that there is a gut-lung communication [24].

### 8.4.1 Microbial Burden

ARDS is generally characterized by a hyper-inflammatory response of the immune system and the microbial population is involved in regulating the immune response. The microbiome plays a significant role in the development of ARDS. The bacteria and fungi have a substantial role in uplifting the inflammation leading to frequent pulmonary exacerbations. The mechanism of micro-aspiration (immigration of microbes from the oropharynx to the lungs) determines the composition of the lung microbiome. The predominant gut bacteria rich in lung microbiota of ARDS are Bacteroidetes and *Enterobacteriaceae* commonly referred to as 'more gut in the lung' [24]. The permeability of the gut allows the microbes to travel through the colon and reach the lungs, regulating the inflammation and become responsible for acute pulmonary damage. This hyper-permeability of the gut is also correlated with the alveolus-capillary permeability, in which the mucus containing Gram-negative pathogens triggers the pro-inflammatory environment. The travel of harmful bacteria from the intestinal lumen through the mesenteric lymph to the lungs causes severe tissues damage [25]. Numerous researches supported the fact that the lung is enriched with the gut-associated microbes and the experimental analysis of pulmonary dysbiosis revealed that the gut would likely be the source of the lung microbiome. This hypothesis was supported by the findings that specific bacteria of the intestine are abundant in patients with ARDS [6, 24, 26, 27].

The pro-inflammatory markers including IL-6 and IL-8 tend to increase in the microbial environment. This is primarily due to the prevalence of *Enterococcus* spp, *Clostridium difficile*, *Pseudomonas* spp., *Salmonella* spp., in the lungs transported from the intestine. The microbial colonization can be due to multiple factors such as nutritional diet, intestinal ischemia, antibiotic treatment, vasoactive drugs, intestinal ulcers, which weaken lung function. *S. pneumonia* and *K. pneumonia* alter the regulation of IL-7 and GM-CSF which are ultimately required for the lung defense mechanism. When the lung dysbiosis remains intact in the patients, it can facilitate multiple organ failure with high chances of mortality. The presence of two common groups of bacteria *Enterobacteriaceae* and *Lachnospiraceae* was responsible for frequent bacterial load in the lungs, resulting in the severity of the disease with worse outcomes [3, 28]. The researchers also hypothesize that the fungal population could be dominating bacterial load in the lungs and this would help understand some people with lung disorders develop ARDS and some do not. In a recent study clinical carried out with COVID19, most of the patients exhibited the presence of *Candida spp.* dominating in their lung microbiome. In critically ill patients with mechanical ventilation, 100 different types of fungi were diagnosed but the diversity of the samples was much lower when compared to healthy humans. This decreased diversity is associated with the protein pentraxin-3, a critical inflammatory marker that is responsible for the severity of the disease [25].

### 8.4.2 Enrichment of Lungs with Microbiota

Micro-aspiration or gastro-oesophageal reflux allows the transmission of microbes from the nasopharyngeal cavity to the alveoli. The low density of the lung microbiome was observed at the range of $10^3$–$10^5$ CFU/g of lung tissue. A healthy lung microbiome consists of the bacterial communities in the upper respiratory tract that entered the lungs through micro-aspiration. The lung microbiome is formed by the immigration, elimination, and reproduction of the microorganisms in the airways. The protective factor is the balance of the microbiome in the lungs, helping in the production of antibacterial peptides in the epithelial mucus which is responsible for the inhibition of bacterial multiplication [17]. Gut dysbiosis has been the reason for the development of infection and inflammatory responses in the respiratory system. The leak of gut microbes alters the lung normal microbiome. The metabolic products of the bacteria in the gut stimulate the immune responses in the lungs thereby sensitizing the travel of microbes from the gut to the lungs. The microbial population in the upper and lower respiratory tract are distinct with more Firmicutes and actinobacteria in the nostril and proteobacteria and Bacteroidetes in the oropharynx. The lung microbiome is endowed with the prominent genera of *Streptococcus* and *Veillonella* [16, 26]. The colonization of the microbes in the epithelial environment of the lung and gut are similar indicating the crosstalk between them (gut–lung axis). Studies have identified that even transient shifts of the gut microbes to the lung can alter the lung microbiome. The alteration of the microbiome in the lungs is correlated to alveolar inflammation. Proteobacteria enrich the lung microbiome during lung disorders and this elevates the concentration of TNF-α which is a sign of severe pulmonary infection. As a result, ARDS is developed and increases the risk of mortality. In healthy individuals, the enrichment of the lung microbiome with Bacteroidetes exhibit lower concentrations of TNF-α [17].

## 8.5 Analysis of Respiratory Microbiota Associated with ARDS

The current diagnosis of ARDS requires culture-independent techniques to analyse the respiratory microbiota. The conventional culture-dependent methods could not describe the predominance of a bacterial population. The respiratory microbiota of ARDS is rich in GI tract bacteria that usually colonize in the lower respiratory tract due to the interaction between the GI tract and lower respiratory tract. Next-generation sequencing is found to be the most promising method of estimating the bacterial load in BALF, patient serum, and cytokine levels. The advancements in sequencing techniques of the whole genome have critically helped in overcoming the challenges in the culture-independent diagnosis of the microbial genome in the lung microbiota [29, 30].

### 8.5.1 Bacterial Identification and Characterization

Kyo et al. have attempted to measure the copy numbers of 16S rRNA in the patient BALF sample and the cytokines released as a result of microbial infection. It was found that the diversity of the microbes was decreased, with the dominance of a particular bacterial phylum. The cytokine levels differ among the patients but IL-6 and IL-8 were constantly high in all the patients [31]. The decrease of proteobacteria and abundance of Enterobacteriaceae in ARDS patients were analysed by the copy numbers of 16S rRNA. This represents the unique pattern of the microbial community associated with relevant cytokine levels, in ARDS patients. The 16S rRNA gene amplicon sequencing using NGS is being the choice of diagnosis to preferable analyse the microbial population in BALF and sputum samples of ARDS patients. Culture-based methods have also been used to diagnose bacterial infection. The BALF and other clinical specimens were cultured on a specific growth medium and the colonization of the particular bacteria was identified. In addition, colorimetric methods were used to analyse the bacterial load in the patient sample [32, 33]. On advancements of diagnostic strategies, meta-transcriptomic sequencing has found a place in analysing the total RNA sequences of the microbes. Normally, the initial phase of ARDS would be identified with laboratory diagnosis of bacterial and fungal infections wherein the sputum and nasal secretions would be cultured according to standard procedures [33].

### 8.5.2 Imaging Techniques

The Chest X-ray and Computed Tomography (CT) are the two primary techniques used in the detection of ARDS development. X-ray is useful in detecting an unpredictable condition in the lungs and can show the accumulation of fluid due to bacterial colonization. The pattern of the X-ray would be cloudy indicating the restriction of gaseous exchange. The loss of aeration can be analysed by CT. The obstruction of the aeration due to imbalanced airway pressure and the improved re-aeration of the lung tissues upon treatment can be visualized by CT [34, 35]. Furthermore, histopathological analysis of the lung biopsies can reveal the colonization of microbial communities concerning the healthy lung microbiota [36].

## 8.6 Patient Care in ARDS

Patients with ARDS have to be managed with the maintenance of the airway, adequate oxygenation, and perfusion in the alveolar capillaries. Care must be taken to maximize the perfusion in the pulmonary capillary system, thereby increasing the oxygen transport. The blood pressure and perfusion can be increased by vasopressors. The positioning of the patients with ARDS also influences the recovery of patients. The prone positing can be a benefit in curing severe respiratory complications. During the initial stages of ARDS, mechanical ventilation could help the patients to get rid of the alveolar obstruction. The goal of the treatment is to support the patient by providing an adequate supply of oxygen to prevent the damage of lungs, recover from the injury caused by ARDS [37, 38].

### 8.6.1 Therapeutic Interventions

The optimal care of ARDS has improved over the years where the ultimate goal is to support the gaseous exchange and minimize the risk of microbiome modification in the lungs. Several therapeutic interventions targeting the lung microbiome or other pathophysiology mechanisms of ARDS are under pre-clinical and clinical trials that would positively result in promising findings. The early phase of ARDS is associated with pro-inflammatory responses and the accumulation of fluid in the alveolar spaces [29]. Therapies targeting the immune responses could reduce pulmonary edema. Gluco-corticosteroids are found efficient in overcoming the inflammatory responses and bring down the risk of infection. Treatment of patients with methylprednisolone can help in limiting the mortality rates. Still, clear clinical data on gluco-corticosteroids is necessary as strong evidence for their benefit [39]. Salbutamol was also prescribed for the reduction of pulmonary edema but found harmful. Numerous surfactants were of treatment choice to improve the gaseous exchanges between the alveolar and pulmonary capillaries. Oxidative pulmonary damage is one of the important physiologies of ARDS and could be minimized by N-acetylcysteine, a good antioxidant. But the administration of N-acetylcysteine does not reflect in the mortality rates of the patients. ARDS development is correlated with the activation of neutrophils during the lung microbiome modification and neutrophil esterase plays a crucial role in it. Sivelestat, a neutrophil esterase inhibitor, is found useful in restricting the neutrophil activation, thereby reducing lung injury [40]. Granulocyte-Macrophage Colony Stimulating Factor (GMSF) mechanism in ARDS is inactive due to the colonization of microbes in the lungs. GMSF was administered externally to revert the lung injury. Most of the drugs fail to bring down the mortality rates rather helps in keeping the patient alive. Poor clinical outcomes of most of the tested drugs would result in unresponsive of the patients towards the treatment. Several clinical trials also suggest that drugs that may be beneficial in a particular patient population may not work with patients with a different biological response. Antibiotic and pro-biotic treatment to rectify the dominating microbial load in the lungs was found disturbing the normal flora of the lungs and did not improve the pathology caused by microbiota [29, 41].

### 8.6.2 Critical Care

As of now, there is no positive treatment for ARDS. Ultimately the treatment of ARDS focuses on supporting the patients with oxygen to be delivered into the capillaries so that the body recovers from the damage due to ARDS [37]. Mechanical ventilation support will give extra oxygen support, opening the airspaces and assist in breathing. Patients who develop ARDS typically lie on their back in the bed. When ventilation support could not deliver enough oxygen, patients are advised to lie on their stomach to have adequate oxygen flow into the blood, called proning. There are cases where patients are unable to perform the task [42]. To get rid of shortness of breath and avoid movement during the oxygen supply, patients require sedation. The adjustment of the patients for the mechanical ventilation, they are sometimes treated with paralytic medications. However, this medication has side effects and is of high risk, requiring continuous monitoring. Excess fluid build-up in the lungs can be eliminated with diuretic medications to elevate the frequency of urination with the aim of removing excess fluid from the body. Extracorporeal

**Table 8.1** Management of ARDS with treatment strategies

| Treatment strategies | Management of the pathology |
|---|---|
| *Non-pharmacological* | |
| Oxygen supplementation | Intubation or support by mechanical ventilation for both mild and severe ARDS patients |
| Prone positioning | Increase the delivery of oxygen and decrease the risk of microbial infection by mechanical ventilation |
| Sedation and analgesia | Adjustment of patients during oxygen supply and restrict movement |
| ECMO | The external supply of oxygen to the blood through a membrane helping in a consistent oxygen supply |
| *Pharmacological* | |
| Corticosteroids | Accelerate ARDS resolution and manage the pulmonary inflammation |
| Diuretics | Removes excess fluid from the body and improves the clinical outcomes |
| Antibiotics and probiotics | Reduces the predominant bacterial load in the lungs |
| Antioxidants | Restricts the oxidative pulmonary damage |
| Neutrophil blockade | Inhibition of neutrophil activation, thereby limiting the lung microbiome modification |
| Inhaled vasodilators and vasopressors | Management of blood pressures throughout the treatment of ARDS |

membrane oxygenation (ECMO) is the least choice of treatment in ARDS, which derives blood from the body, allows it to pump through a membrane, removes $CO_2$, and adds oxygen to pump the blood back into the body. This treatment is very complicated and does not apply to patients with high co-morbidities. ARDS patients often require a minimum treatment period of 14 days. Any severe cases require tracheostomy along with ventilation support. Most of the patients somehow drive hard to survive the disease and could preferably regain lung function. Only patients at high risk due to age factor, 70% of lung infection, and co-morbidities suffer long term hospitalization and are susceptible to death [29, 38, 43] (Table 8.1).

## 8.7 Conclusion and Future Directions

ARDS is the serious syndrome of acute respiratory failure which is a resultant of pulmonary edema and acute inflammation. The development of ARDS is correlated to the aspiration of microbes or gastric contents and pulmonary injuries such as sepsis and pancreatitis as indirect causes and other lung disorders including pneumonia and pulmonary contusion as a direct cause. There have been consistent researches on ARDS in terms of pathogenesis and lung microbiome in ARDS. Still, a better understanding of the microbiota responsible for ARDS and the interaction of gut microbiome and lung microbiome is required for investigating the therapeutic strategies against the microbiota. The incidence of gut microbial transfer into the lung, disturbing the normal microbiome of the lungs is considered the primary reason for the modification of lung microbiota. Several studies have been published in investigating the prevention and treatment of ARDS. Yet, ARDS remains a syndrome that is difficult to diagnose and failure of treatments, resulting in high mortality and morbidity rates. Clinical data on the pathogenesis focussing on lung microbiome are scarce which may hinder the treatment response against disease outcomes. The conventional culture techniques in the detection of lung microbiota have not been useful anymore due to the modification of the microbes. Culture-independent genome sequencing could potentially help in a precise diagnosis of the microbes. Future research directions may focus on ARDS susceptibility, mechanism of primary microbiome modification, and treatment strategies at the early phase and targeted therapies.

**Acknowledgement** The authors express their gratitude to SASTRA-Deemed-to-be-University, Tamil Nadu, India for infrastructure and financial support. Authors also extend their appreciation for the contribution of Biopharmaceutical research lab members, SASTRA-Deemed-to-be-University.

**Author Contribution** **Gayathri Gopal:** Writing - Original Draft, Writing - Review & Editing Collection of data; **Shibi Muralidar:** Writing - Original Draft, Writing - Review & Editing, Collection of data; **Abishek Kamalakkannan:** Writing - Original Draft, Collection of data; **Senthil Visaga Ambi:** Conceptualization, Visualization, Supervision, and Writing - Review & Editing.

**Conflict of Interest** The authors declare no conflict of interest.

## References

1. Matthay MA, Zemans RL, Zimmerman GA, Arabi YM, Beitler JR, Mercat A, Herridge M, Randolph AG, Calfee CS (2018) Acute respiratory distress syndrome. Nat Rev Dis Primers 5(1). https://doi.org/10.1038/s41572-019-0069-0
2. Mason C, Dooley N, Griffiths M (2016) Acute respiratory distress syndrome. Clin Med (Lond) 16:s66–s70. https://doi.org/10.7861/clinmedicine.16-6-s66
3. Lall R, Hamilton P, Young D, Hulme C, Hall P, Shah S, MacKenzie I, Tunnicliffe W, Rowan K, Cuthbertson B, McCabe C, Lamb S, OSCAR collaborators (2015) A randomised controlled trial and cost-effectiveness analysis of high-frequency oscillatory ventilation against conventional artificial ventilation for adults with acute respiratory distress syndrome. The OSCAR (OSCillation in ARDS) study. Health Technol Assess (Winchester, England) 19(23):1–vii. https://doi.org/10.3310/hta19230
4. Matthay MA, Zemans RL (2011) The acute respiratory distress syndrome: pathogenesis and treatment. Annu Rev Pathol 6:147–163. https://doi.org/10.1146/annurev-pathol-011110-130158
5. Fan E, Brodie D, Slutsky AS (2018) Acute respiratory distress syndrome advances in diagnosis and treatment. JAMA 319(7):698–710. https://doi.org/10.1001/jama.2017.21907
6. Kyo M, Nishioka K, Nakaya T, Kida Y, Tanabe Y, Ohshimo S, Shime N (2019) Unique patterns of lower respiratory tract microbiota are associated with inflammation and hospital mortality in acute respiratory distress syndrome. Respir Res 20(1):1–2. https://doi.org/10.1186/s12931-019-1203-y
7. Schmitt FCF, Lipinski A, Hofer S, Uhle F, Nusshag C, Hackert T, Dalpke AH, Weigand MA, Brenner T, Boutin S (2020) Pulmonary microbiome patterns correlate with the course of disease in patients with sepsis-induced ARDS following major abdominal surgery. J Hosp Infect 105(3):438–446. https://doi.org/10.1016/j.jhin.2020.04.028
8. Dickson RP, Singer BH, Newstead MW, Falkowski NR, Erb-Downward JR, Standiford TJ, Huffnagle GB (2016) Enrichment of the lung microbiome with gut bacteria in sepsis and the

acute respiratory distress syndrome. Nat Microbiol 1(10):1–9. https://doi.org/10.1038/nmicrobiol.2016.113
9. Pitoyo CW (2008) Acute respiratory distress syndrome. Acta Med Indones 40(1):48–52
10. Goldstone J (2002) The pulmonary physician in critical care • 10: Difficult weaning. Thorax 57(11):986–991. https://doi.org/10.1136/thorax.57.11.986
11. Torres Acosta MA, Singer BD (2020) Pathogenesis of COVID-19-induced ARDS: implications for an ageing population. Eur Respir J 56(3):2002049. https://doi.org/10.1183/13993003.02049-2020
12. Sharp C, Millar AB, Medford ARL (2015) Advances in understanding of the pathogenesis of acute respiratory distress syndrome. Respiration 89(5):420–434. https://doi.org/10.1159/000381102
13. Medford ARL, Millar AB (2006) Vascular endothelial growth factor (VEGF) in acute lung injury (ALI) and acute respiratory distress syndrome (ARDS): Paradox or paradigm? Thorax 61(7):621–626. https://doi.org/10.1136/thx.2005.040204
14. Allen TC, Kurdowska A (2014) Interleukin 8 and acute lung injury. Arch Pathol Lab Med 138(2):266–269. https://doi.org/10.5858/arpa.2013-0182-RA
15. Martin-Loeches I, Dickson R, Torres A, Hanberger H, Lipman J, Antonelli M, De Pascale G, Bozza F, Vincent JL, Murthy S, Bauer M, Marshall J, Cilloniz C, Bos LD (2020) The importance of airway and lung microbiome in the critically ill. Crit Care 24(1):1–9. https://doi.org/10.1186/s13054-020-03219-4
16. Dickson RP (2018) The lung microbiome and ARDS it is time to broaden the model. Am J Respir Crit Care Med 197(5):549–551. https://doi.org/10.1164/rccm.201710-2096ED
17. Siwicka-Gieroba D, Czarko-Wicha K (2021) Lung microbiome - a modern knowledge. Centr Eur J Immunol 45(3):342–345. https://doi.org/10.5114/CEJI.2020.101266
18. Gaibani P, Viciani E, Bartoletti M, Lewis RE, Tonetti T, Lombardo D, Castagnetti A, Bovo F, Horna CS, Ranieri M, Viale P, Re MC, Ambretti S (2021) The lower respiratory tract microbiome of critically ill patients with COVID-19. Sci Rep 11(1):1–11. https://doi.org/10.1038/s41598-021-89516-6
19. Evsyutina Y, Komkova I, Zolnikova O, Tkachenko P, Ivashkin V (2017) Lung microbiome in healthy and diseased individuals. World J Respirol 7(2):39. https://doi.org/10.5320/wjr.v7.i2.39
20. Huang YJ, Charlson ES, Collman RG, Colombini-Hatch S, Martinez FD, Senior RM (2013) The role of the lung microbiome in health and disease: a national heart, lung, and blood institute workshop report. Am J Respir Crit Care Med 187(12):1382–1387. https://doi.org/10.1164/rccm.201303-0488WS
21. Weeks H (1978) H.E.W. data plan seen threat to hospitals. Hosp Peer Rev 3(9):114–116. https://doi.org/10.1016/S0140-6736(14)61136-3.The
22. Dickson RP, Schultz MJ, Van Der Poll T, Schouten LR, Falkowski NR, Luth JE, Sjoding MW, Brown CA, Chanderraj R, Huffnagle GB, Bos LDJ (2020) Lung microbiota predict clinical outcomes in critically ill patients. Am J Respir Crit Care Med 201(5):555–563. https://doi.org/10.1164/rccm.201907-1487OC
23. Fromentin M, Ricard JD, Roux D (2021) Respiratory microbiome in mechanically ventilated patients: a narrative review. Intensive Care Med 47(3):292–306. https://doi.org/10.1007/s00134-020-06338-2
24. Fanos V, Pintus MC, Pintus R, Marcialis MA (2020) Lung microbiota in the acute respiratory disease: from coronavirus to metabolomics. J Pediatr Neonatal Individ Med 9(1):1–10. https://doi.org/10.7363/090139
25. Khatiwada S, Subedi A (2020) Lung microbiome and coronavirus disease 2019 (COVID-19): Possible link and implications. Hum Microbiome J 17:100073. https://doi.org/10.1016/j.humic.2020.100073
26. Chunxi, L., Haiyue, L., Yanxia, L., Jianbing, P., & Jin, S. (2020). The gut microbiota and respiratory diseases: new evidence. J Immunol Res, 2020. https://doi.org/10.1155/2020/2340670
27. Wu C, Chen X, Cai Y, Xia J, Zhou X, Xu S, Huang H, Zhang L, Zhou X, Du C, Zhang Y, Song J, Wang S, Chao Y, Yang Z, Xu J, Zhou X, Chen D, Xiong W, Xu L, Zhou F, Jiang J,

Bai C, Zheng J, Song Y (2020) Risk factors associated with acute respiratory distress syndrome and death in patients with coronavirus disease 2019 Pneumonia in Wuhan, China. JAMA Intern Med 180(7):934–943. https://doi.org/10.1001/jamainternmed.2020.0994
28. Madan JC, Koestle DC, Stanton BA, Davidson L, Moulton LA, Housman ML, Moore JH, Guill MF, Morrison HG, Sogin ML, Hampton TH, Karagas MR, Palumbo PE, Foster JA, Hibberd PL, O'Toole GA (2012) Serial analysis of the gut and respiratory microbiome in cystic fibrosis in infancy: Interaction between intestinal and respiratory tracts and impact of nutritional exposures. MBio 3(4). https://doi.org/10.1128/mBio.00251-12
29. Griffiths MJD, McAuley DF, Perkins GD, Barrett N, Blackwood B, Boyle A, Chee N, Connolly B, Dark P, Finney S, Salam A, Silversides J, Tarmey N, Wise MP, Baudouin SV (2019) Guidelines on the management of acute respiratory distress syndrome. BMJ Open Respir Res 6(1). https://doi.org/10.1136/bmjresp-2019-000420
30. Zhong H, Wang Y, Shi Z, Zhang L, Ren H, He W, Zhang Z, Zhu A, Zhao J, Xiao F, Yang F, Liang T, Ye F, Zhong B, Ruan S, Gan M, Zhu J, Li F, Li F, Wang D, Li J, Ren P, Zhu S, Yang H, Wang J, Kristiansen K, Tun HM, Chen W, Zhong N, Xu X, Li Y-m, Li J, Zhao J (2020) Characterization of microbial co-infections in the respiratory tract of hospitalized COVID-19 patients. MedRxiv. https://doi.org/10.1101/2020.07.02.20143032
31. Chen C, Shen T, Tian F, Lin P, Li Q, Cui Z, Zhang Y, Xue M, Ye J, Guo X, Zhou Y (2013) New microbiota found in sputum from patients with community-acquired pneumonia. Acta Biochim Biophys Sin 45(12):1039–1048. https://doi.org/10.1093/abbs/gmt116
32. Emonet S, Lazarevic V, Leemann Refondini C, Gaïa N, Leo S, Girard M, Nocquet Boyer V, Wozniak H, Després L, Renzi G, Mostaguir K, Dupuis Lozeron E, Schrenzel J, Pugin J (2019) Identification of respiratory microbiota markers in ventilator-associated pneumonia. Intensive Care Med 45(8):1082–1092. https://doi.org/10.1007/s00134-019-05660-8
33. Langelier C, Kalantar KL, Moazed F, Wilson MR, Crawford ED, Deiss T, Belzer A, Bolourchi S, Caldera S, Fung M, Jauregui A, Malcolm K, Lyden A, Khan L, Vessel K, Quan J, Zinter M, Chiu CY, Chow ED, Wilson J, Miller S, Matthay MA, Pollard KS, Christenson S, Calfee CS, DeRisi JL (2018) Integrating host response and unbiased microbe detection for lower respiratory tract infection diagnosis in critically ill adults. Proc Natl Acad Sci U S A 115(52):E12353–E12362. https://doi.org/10.1073/pnas.1809700115
34. Ball L, Vercesi V, Costantino F, Chandrapatham K, Pelosi P (2017) Lung imaging: how to get a better look inside the lung. Ann Transl Med 5(14):1–11. https://doi.org/10.21037/atm.2017.07.20
35. Man WH, De Steenhuijsen Piters WAA, Bogaert D (2017) The microbiota of the respiratory tract: gatekeeper to respiratory health. Nat Rev Microbiol 15(5):259–270. https://doi.org/10.1038/nrmicro.2017.14
36. Hernández-Beeftink T, Guillen-Guio B, Villar J, Flores C (2019) Genomics and the acute respiratory distress syndrome: current and future directions. Int J Mol Sci 20(16):1–19. https://doi.org/10.3390/ijms20164004
37. Bein T, Grasso S, Moerer O, Quintel M, Guerin C, Deja M, Brondani A, Mehta S (2016) The standard of care of patients with ARDS: ventilatory settings and rescue therapies for refractory hypoxemia. Intensive Care Med 42(5):699–711. https://doi.org/10.1007/s00134-016-4325-4
38. Powers J (2007) The five P's spell positive outcomes for ARDS patients. Am Nurse Today:1–6. https://www.americannursetoday.com/the-five-ps-spell-positive-outcomes-for-ards-patients/
39. Bein T, Briegel J, Annane D (2016) Steroids are part of rescue therapy in ARDS patients with refractory hypoxemia: yes. Intensive Care Med 42(5):918–920. https://doi.org/10.1007/s00134-015-4162-x

40. Zhang LN, Sun JP, xue, X. Y., & Wang, J. X. (2013) Exogenous pulmonary surfactant for acute respiratory distress syndrome in adults: a systematic review and meta-analysis. Exp Ther Med 5(1):237–242. https://doi.org/10.3892/etm.2012.746
41. Albert RK, Keniston A, Baboi L, Ayzac L, Guérin C (2014) Prone position–induced improvement in gas exchange does not predict improved survival in the acute respiratory distress syndrome. Am J Respir Crit Care Med 189(4):494–496. https://www.atsjournals.org/doi/full/10.1164/rccm.201311-2056LE. *(P/F) ratio and Pa*
42. Chiumello D, Brochard L, Marini JJ, Slutsky AS, Mancebo J, Ranieri VM, Thompson BT, Papazian L, Schultz MJ, Amato M, Gattinoni L, Mercat A, Pesenti A, Talmor D, Vincent JL (2017) Respiratory support in patients with acute respiratory distress syndrome: an expert opinion. Crit Care 21(1):1–8. https://doi.org/10.1186/s13054-017-1820-0
43. Restrepo RD, Walsh BK (2012) Humidification during invasive and noninvasive mechanical ventilation: 2012. Respir Care 57(5):782–788. https://doi.org/10.4187/respcare.01766

# Role of Brain–Gut–Microbiome Axis in Depression Comorbid with Asthma

9

Shvetank Bhatt, K. Sreedhara R. Pai, C. R. Patil, S. N. Manjula, and S. Mohana Lakshmi

**Abstract**

Asthma (ATA) is a long term inflammatory condition of the respiratory tract (RT) where stress and psychological factors play a significant role. A high rate of comorbidity of ATA and depression or major depressive disorder (MDD) is observed in many patients. The proposed correlations between ATA and depression include a vulnerability (trait) and state connection. Vulnerability for both ATA and depression may utilize genetic and early development causes. In addition, some other factors are common in both the conditions, such as obstructive factors, factors associated with inflammation, insomnia, psychological reactions to long term medical illness. The recent research advocates the participation of the central nervous system (CNS) in ATA. Recently, the role of the brain–gut–microbiome (BGM) and gut–lung–microbiome (GLM) axis is studied, and both the pathways have exhibited strong interconnection with each other. Commensal microbes are crucial for the formation of a proper immune system.

S. Bhatt (✉) · S. Mohana Lakshmi
Amity Institute of Pharmacy, Amity University Madhya Pradesh (AUMP), Gwalior, India
e-mail: sbhatt@gwa.amity.edu

K. S. R. Pai
Manipal College of Pharmaceutical Sciences (MCOPS), Manipal Academy of Higher Education (MAHE), Manipal, India

C. R. Patil
Department of Pharmacology, Delhi Pharmaceutical Sciences and Research University, New Delhi, India

S. N. Manjula
Department of Pharmacology, JSS College of Pharmacy, Mysuru, JSS Academy of Higher Education & Research [JSSAHER], Mysuru, India



G. Gupta et al. (eds.), *Microbiome in Inflammatory Lung Diseases*,
https://doi.org/10.1007/978-981-16-8957-4_9

The role of commensal bacteria in both the respiratory and gastrointestinal tracts can be a crucial factor in treating ATA.

Similarly, human gut microbiota (GM) exhibits a marked role in the pathophysiology of depression. Recent studies suggest correlations between the altered GM and major depressive disorders (MDD). Further characterization of clinical, psychological, cellular, and molecular associations between ATA and depression is required to evaluate and treat these patients in a better way. The present book chapter mainly focuses on the influence of the brain–gut–microbiome axis with the involvement of lungs in the pathophysiology and treatment of depression comorbid with ATA.

**Keywords**

Asthma · Depression · Microbes · Inflammation

## 9.1 Introduction

ATA is one of the chronic inflammatory airway diseases that affects 300 million of worldwide populations and is predicted to be nearly 400 million in the coming 5 years. High incidence of ATA accounts for 1 out of 250 deaths worldwide [1]. ATA is a chronic allergic condition that exhibits the hyper-responsiveness of bronchi towards some allergens. The immune-histopathologic features of ATA involve infiltration of cells involved in inflammation, such as neutrophils, eosinophils, lymphocytes, activation of mast cells, and injury of epithelial cells [2]. The improper treatment of ATA leads to chronic inflammation and further damage of airways, including hypertrophy of smooth muscle cells, hyperplasia of epithelial cells, and airway connective tissue [3]. The airway obstruction leads to increased resistance of airways, decrease in maximum flow of expiration, trapping of air, increased pressure of airways, reduced $O_2$ and increased $CO_2$ levels, pulsus paradoxus, and fatigue and failure respiration [4]. In addition, atopy, the genetic predisposition is responsible for the progression of an immunoglobulin E antibodies (IgEA) driven response to normal aeroallergens, is the common distinguishable predisposing factor responsible for the development of ATA. Moreover, infections caused due to virus also play a predominant role in ATA progression [5].

The comorbidity of depression or MDD is very high with ATA [6]. Depression is a mental disorder characterized by a prolonged sad mood [7]. The emotional factors of depression and stress further exacerbate ATA. Moreover, increased levels of various inflammatory markers and amplified oxidative stress are also common in the pathophysiology of depression and ATA [8, 9]. Depression is a very serious and recurring neuropsychiatric disorder affecting more than 264 million people worldwide. It is the fourth leading cause of ill health, improper quality of life and economic burden [10]. The most accepted theory for depression progression is monoamine theory. According to this theory, decreased levels of neurotransmitters (NTs), such as noradrenaline (NA), dopamine (DA), and 5-hydroxytryptamine (serotonin, 5-HT) were observed in depression [11]. The abnormality in the negative feedback mechanism of the hypothalamic–pituitary–adrenal (HPA) axis is also found in MDD patients compared to normal subjects [12]. In addition, genetic participation in depression development has also been identified. Genetic modifications are also exhibited a significant role in the abnormally high progression of depression [13]. (discussed in the later section).

Moreover, imbalance in the antioxidant and oxidant mechanisms also has a predominant role in depression progression. Various research groups have indicated that depression exhibits increased levels of different peripheral inflammatory biomarkers compared with non-depressed individuals. Increased C-reactive protein levels, TNF- α, Interferon-α, have been observed in depressed patients [14]. Recently, the BGM axis with the involvement of lungs has emerged as a new treatment modality for MDDs [15, 16]. In the coming sections of this book chapter, we will discuss the significance of the BGM axis and lungs in the progression of MDD comorbid with ATA, with the main emphasis on the modulations of the GM as a mode of treating MDD [17].

## 9.2 Pathophysiology of ATA

The exacerbation of ATA can be divided into two phases, namely early and late. The early phase contains sensitization, stimulation, and finally secretion of IgEA by plasma cells. These antibodies are stimulated due to the response to certain environmental factors and triggers known as an allergen [18]. Afterwards, the IgEA bind to high-affinity cells such as mast cells and basophils [19]. The inhalation of pollutants or specific allergens leads to the release of cytokines and mast cells degranulation. The release of histamine, prostaglandins, and leukotrienes from the mast cells results in the contraction of smooth muscle cells and causes airway tightening. The T-helper lymphocytes (Th2) plays a vital role in the release of interleukins (IL-4, IL-5, IL-13) and granulocyte-macrophage- colony-stimulating factor (GM-CSF), which take part in signalling and networking with adjacent cells and withstand inflammation [20]. IL-3 and IL-5 assist survival of eosinophils and basophils. Moreover, IL-13 contributes to remodelling, scarring of tissues, and hyperplasia. The hyper-responsiveness of the airways and inflammation are crucial factors for the progression of the disease [21].

Moreover, in the late phase of ATA, the remodelling transition of the epithelial cells to mesenchymal cells increases the smooth muscle contents [22]. Furthermore, eosinophils can also amplify remodelling of the airway due to their release of transforming growth factor (TGF-β) and cytokines by communications of mast cells. The ATA condition may worsen over time if the mechanism of airway remodelling and inflammation is not corrected properly [23]. The role of inflammatory mediators in the progression of ATA has been shown in Fig. 9.1.

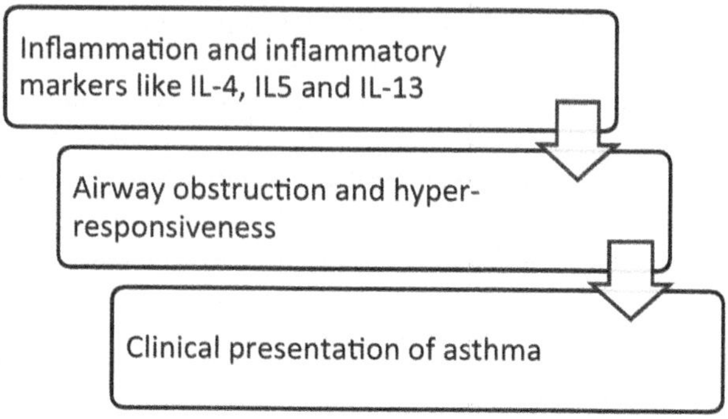


**Fig. 9.1** The airway inflammation in asthma leads to pathological progression of asthma

### 9.2.1 Pathophysiology of Depression

Depression is the condition of prolonging a sad mood. The DSM-IV has described nine symptoms for the assessment of MDD [7]. If 5 out of the 9 symptoms are present for more than 2 weeks, the patient is said to be affected by depression; however, further confirmation is warranted. The decreased level of NTs, namely 5-HT, NE, and DA, is observed in depression, as discussed earlier. The NTs NE is mainly involved in the fight, flight and fear response. DA is involved in the pleasure activity, and serotonin is involved in the obsessions and compulsions and the regulation of sexual behaviour [24, 25]. In addition to NTs, increased oxidative stress and inflammation, HPA axis dysregulation, increased cortisol levels are observed in depression. Various studies indicated that MDD demonstrates increased levels of various peripheral inflammatory biomarkers when compared with non-depressed individuals. Increased C-reactive protein levels, TNF- α, Interferon-α, have been observed in depressed patients [26].

### 9.2.2 Association of Monoamines and Other Neurotransmitters with Depression

The three major NTs involved in depression are NE, 5-HT, and DA. As discussed earlier, the levels of all these NTS are reduced in MDD [25]. Most of the serotonergic, noradrenergic, and dopaminergic neurons are positioned in the nuclei of the midbrain and brainstem as well as projected to the large areas of the complete brain [25, 27]. This suggests the importance of the monoaminergic system in various physiological functions, namely mood, attention, processing of reward, sleep, appetite, and cognition [28]. The compounds that inhibit the uptake of monoamines and increase their synapses levels act as an antidepressant. In addition, inhibition of enzyme monoamine oxidase also leads to the antidepressant effect. The above statements further potentiate the monoamine hypothesis. In addition, γ-aminobutyric acid (GABA) and glutamate also have an essential role in the pathogenesis of MDD. Moreover, recent studies also relate depression with modifications in the physiology of the brain, neuronal plasticity, and reduction in the volume of the frontal cortex and hippocampus region [29].

### 9.2.3 Genetic Influence on the Progression of Depression

The genetic studies with family, twins, and adoption have provided very important evidence about the influence of genetic makeup on the progression of depression. Depression is a familial disorder. The genetic factors influence around 30–40% of cases of depression, while the remaining 60–70% cases have found the presence of non-genetic factors. However, genome-wide association studies have exhibited that various genes with minor effects are linked with complicated diseases, increasing the difficulty in identifying such types of genes. Genes like SLC6A4 (previously known as SERT), DRDR4, SLC6A4 or 5-HTT and TPH2 influence the pathological progression of depression [13]. Now clinicians also understood that family history would be considered the most concrete data source to study the genetic risk of depression.

### 9.2.4 Stress Hormone in the Progression of Depression

The level of the stress hormone cortisol is found consistently more in the blood of a depressed person. The release of cortisol is taken place under the influence of corticotropin-releasing hormone (CRH). The CRH released from the hypothalamus leads to the release of adrenocorticotropic hormone (ACTH) from the anterior-pituitary gland (APG). This ACTH hormone travels via blood and acts on the cortex part of the adrenal gland and causes the release of the glucocorticoid, i.e., cortisol. The involvement of the hypothalamus-pituitary and adrenal gland makes an axis known as the HPA-axis [30]. When this cortisol level is more in the blood, it sends the signal to the hypothalamus and APG to reduce its hormone/factor release. This is called hormonal or negative feedback mechanism (NFM). The NFM works properly in a normal individual, and on the other hand, it disrupts in the patient of MDD and this results in high levels of cortisol in the blood. The level of CRH in cerebrospinal fluid (CSF) is increased in some MDD patients. The disruption of HPA axis signalling is considered as one of the major causes of depression or MDD as cortisol is involved in chronic stress [31].

### 9.2.5 Inflammatory Mediators in the Progression of Depression

The 'sickness behaviour' occurs as the result of the upregulation of the different inflammatory responses. These have many common symptoms like depression, namely fatigue, loss of pleasure, delay in psychomotor activity, and cognitive impairment. The proinflammatory cytokines are involved in depression are IL-1α, tumour necrosis factor-α (TNFα), and IL-6, which influence the HPA axis and impair the central serotonin system [32]. The blocking of these anti-inflammatory signalling in animal models resulted in an antidepressant-like effect [33]. Human trial data suggest that cytokines may have a crucial role in the pathophysiology of a subgroup of MDD subjects, specifically those with different comorbid physical conditions [34]. Moreover, animal studies using rodents have shown that administration of celecoxib was linked with a reduction in the PGE2 levels and a reversal of stress-induced MDD behaviours [35]. In addition, the COX-2 inhibitors are involved in reduction of activity of indoleamine 2, 3-dioxygenase (IDO) and subsequently responsible for decrease in glutamine-mediated excitotoxicity [36].

### 9.2.6 Oxidative Stress (OS) in Depression

The reactive oxygen species (ROS) have a crucial role in normal brain physiology as well as in the pathological progression of MDD. Amplified OS is demonstrated as an inequity between production of ROS and the antioxidant ability of the cell [14]. Oxidative phosphorylation is the main source of ATP that occurs in cell mitochondria. During this process, ROS, reactive nitrogen species (RNS), and carbon- and sulphur-centred radicals have been formed as a by-product [37]. The brain has high metabolic demand hence more susceptible to OS. The difference between oxidative and antioxidative mechanisms leads to pathogenic brain physiology and deformities in the signalling of nerve cells. Increased peroxidation of lipids has an important role in the progression of MDD. The antioxidant enzymes, namely catalase (CAT) and superoxide dismutase (SOD), along with reduced glutathione (GSH), signify the involvement of OS in MDD. GSH is the most predominant non-enzymatic endogenous antioxidant [14, 38] (Fig. 9.2).

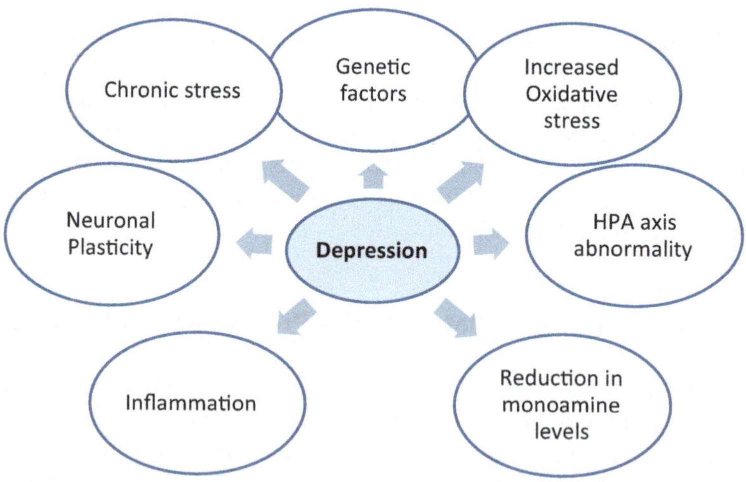


**Fig. 9.2** Various factors involved in the pathophysiological progression of depression

### 9.2.7 Comorbid Relation of ATA with Depression

The functional relation between asthma and MDD has been known for a long time. Depression is commonly present in patients with asthma and vice versa. The association of MDD and asthma is well studied and reported [39]. However, it is still unclear whether asthma precedes or presents concurrently with MDD. The increased prevalence of these comorbid disorders further enhances the scientific community's interest in studying the association of their comorbid mechanisms. More are, female sex and low socio-demographic status increase the comorbidity of these disorders to a great extent. Patients with asthma have approximately double the chances to affect by depression as compared to non-asthmatics [6].

The affected individuals have shown poor patient compliance. They cannot adhere to specific treatments and avoid the allergen, impair functioning, and increase utilization of health care, resulting in poor control of ATA. According to Jiang et al., 2014 the inflammatory pathway is common between ATA and depression [39]. However, according to Valença et al., 2006 high morbidity rate of MDD and anxiety in asthmatic patients is not depend on the severity of ATA [40]. Severe asthma leads to a decrease in oxygen level or saturation and fatigue, which are symptoms of depression [41]. In addition, some studies also have suggested the association between ATA and affective traits, including depression and neuroticism [42]. This suggests the significant involvement of shared cellular pathways underlying both ATA and depression. As inflammation is a crucial player in both ATA and depression, the involvement of proinflammatory pathways controlled by genes may be plausible. Several human trials have been conducted in patients with ATA who reported high rates of depression and anxiety symptoms compared to normal healthy subjects. Moreover, some studies also reveal that ATA is linked with increased suicidal ideation due to comorbidity [43, 44].

### 9.2.8 The Human Lung Environment: Sterile or Not?

The lungs were thought to be sterile for a long time, but various researchers have demonstrated that it harbours its microbiota [45]. The first application of next-generation sequencing (NGS) on samples of lower airway compared the microbiome levels in healthy and ATAadults with ATA and children. According to this research, bacteria from the phylum Bacteroidetes and species Prevotella were evident more in healthy person as compared to volunteers of ATA. These bacteria are coming under the gram-negative category and cannot be cultivated easily. Further, some research studies exhibited that additionally phyla Firmicutes, Proteobacteria, and Actinobacteria are more common in the lungs of healthy volunteers [46].

The upper portion of the RT is covered by cylindrical respiratory epithelium and mucus membrane [47]. The constant turbulence of mucus fluid and airflow determines the bacterial load in the upper and lower RT film [47]. The mucociliary clearance and host immune mechanism assist in the elimination of microorganisms. Overall the shifts from the upper to the lower RT, the gradient pressure and change of temperature favour the growth of bacteria. Some bacteria adapt to stay in the anaerobic zone in the lower RT [45, 47]. The alveoli or air vacuoles of lungs consist of pneumocytes (type 1), a thin layer of squamous epithelial cell, and type II pneumocytes producing a lung surfactant [48]. Surfactants are composed of phospholipids (90%) and proteins, i.e., surfactant proteins A to D, with a marked innate role in the clearance virus and bacteria. The concept of connection of gut-lung is influenced by the opinion that different lung disorders can be exaggerated or influenced by the gut's microenvironment [49]. Change in the microenvironment of the gut leads to modification in the other disease. The microbiome is a critical factor responsible for linking gut and lung in ATA [50]. The GM in the patient with bacterial pneumonia, cystic fibrosis, and influenza varied with respect to healthy control subjects [51]. Several studies demonstrated that initial life is the vital period in which the GM's dysbiosis may be the prime cause of numerous respiratory diseases, as GM has a predominant impact on the maturation of immune cells and resistance to pathogens [52].

### 9.2.9 Gut and Lung Microbes in ATA

Dysbiosis or imbalance of the GLM axis has been highly linked with the development of allergic diseases such as ATA. External factors have a high impact on the composition of lung microbiota [46, 53]. Its growth is positively correlated with the farming environment and negatively associated with allergens and air pollutants. The use of antibiotics, antiulcer medications, and other drugs harshly dysregulates the lung and GM [46]. The overall result is hypersensitivity and hyperactivity to respiratory and food allergens. Reconstitution of microbiota with probiotics, prebiotics, or any other approach will be helpful for maintaining the proper immune response and prevention of ATA. The dysbiosis of the gut and lung seems to be critical and worsens ATA's criticalness [54]. The commensal microorganisms are important for the induction of a proper, tolerogenic immune system development. Further studies are needed to improve the understanding of the role of microbes in immune response and its role on principal risk factors for ATA, including tobacco smoke and genetic features of the host [55, 56].

### 9.2.10 Influence of Immune Mechanism by the Microbiome

The bacteria or other microbes can influence the individual's immune response. Both components of bacteria's cell wall and their metabolites have been associated with mucosal immunoregulatory effects [57]. Commensal bacteria like Bifidobacterium, Lactobacillus, and Clostridium strains are linked with an increase in the proportion of T regulatory cells (Tregs) in mice [58]. Clostridia sp. is also responsible for reducing the intestinal permeability towards dietary proteins [59]. Moreover, Bifidobacteria and Lactobacilli can amplify the metabolic processes in dendritic cells, such as biotransformation of vitamin A and tryptophan metabolism and heme oxygenase-1, which has an influence on Tregs [60]. The consumption of Bifidobacterium longum by healthy subjects enhanced the levels of Foxp3+ Tregs in peripheral blood. At the same time, insertion or administration of bacterial strain in chronic fatigue, psoriasis, and ulcerative colitis patients leads to reduced levels of inflammatory markers in serum like CRP. Increased numbers of Tregs mediate the mechanism that may be involved [61, 62].

### 9.2.11 Communication Between Gut and Lung

The mechanism or link of communication between gut and lung is still not known. Still, it is well demonstrated that the signals from the endothelium are absorbed by the epithelial other structural, and immune cells to produce a cytokine controlled local microenvironment, which causes the alterations in immune responses at remote sites. The naïve cells on the starting phase are triggered in the gut and then move via lymph and blood vessels to the RT, i.e., lungs. In the lungs, these cells have effector functions. The immune reaction that is produced locally in gut-associated lymphoid tissue (GALT) and Inducible bronchus-associated lymphoid tissue (iBALT) can alter the systemic immune responses; at the same time, mucosal immune system can also act separately at the systemic site [63, 64]. The flow of communication between the gut and the lung axis happens in a bidirectional way. For example, when a mouse treated with lipopolysaccharide (LPS) leads to a significant rise in the bacterial gut count, and it is also observed that pneumonia stimulates intestinal damage and reduces the proliferation of gut epithelium [65, 66].

### 9.2.12 Involvement of Gut and Lung Microbiome (GLM) Axis in ATA

Pathophysiology of ATA involves both innate and adaptive components [67]. The prolonged exposure of the respiratory system to agents present in the air in patients with a genetic predisposition offers repetitive opportunities for immune response development [68]. The epithelial mucosa and the dendritic cells are in constant contact with the airway lumen, and immune cells' antimicrobial peptides play an important role in response to environmental agents [52]. The human gut is abundantly colonized with bacteria, and these bacteria also correlate with ATA [69]. For instance, Lachnospira, Veillonella, Faecalibacterium, and Rothia genera are markedly decreased in significant quantity in faecal samples of 3-month-old Canadian children with high susceptibility to ATA. The reduction of the levels of these bacteria leads to a reduction in the acetate of faecal matter and impaired regulation of enterohepatic metabolites [70]. In addition, analysis of bacterial community function shows depletion in the synthesis of LPS in the microbiota of children susceptible to ATA [70].

Moreover, reports suggested that decreased airway inflammation has been observed when the faecal slurry from an asthmatic infant is transplanted into germ-free mice with species of lachnospira, villanelle, faecal bacterium, and Rothia in ovalbumin-induced airway inflammation model [71]. Metabolites of bacteria partially mediate the effect of GM in ATA. These bacteria influence the immune response in the distal part of the body. The most common metabolite in this regard is short-chain fatty acids (SCFAs). These short-chain fatty acids (SCFAs) have a protective effect on human airway inflammation [72]. Children with a significant quantity of SCFAs at 1 year have low atopic sensitization predominantly and are less likely to get ATA at 3–6 years. Moreover, supplementation with soluble fibre has been reported to decrease sputum eosinophilia and the expression of sputum histone deacetylase 9 in ATA patients. It is also observed that SCFAs also decrease the inflammation in airway inflammation preclinical models induced by OVA and house dust mite (HDM [73, 74]. However, some bacteria in the human gut are responsible for producing biogenic amines like histamine, which are well known to produce pro- and anti-inflammatory effects. The number of bacteria that leads to histamine secretion is more in the case of faecal samples of asthmatic patients than non-asthmatics [75].

In the lungs, Proteobacteria is more abundant than non-asthmatic volunteers, as seen across various clinical trials. The use of corticosteroids reduces the levels of certain bacteria [69, 70]. These bacteria can be used as a marker for diagnostic purposes. On the other hand, Klebsiella, one of the genera of bacteria belonging to Proteobacteria phylum is abundantly present, specifically in patients with severe ATA. In contrast, the phylum Actinobacteria is linked with enhanced or no change in ATA control [69, 76]. Moreover, *Akkermansia muciniphila* was reported to be very less in the faecal matter of both obese and non-obese ATA patients with severe disease. *Streptococcus* is one more genus seen in the lungs of patients with severe ATA. It may have a causal relationship as demonstrated in acute and chronic airway inflammation induced murine models [70]. It is not clear of the mechanism by which the bacteria or other microbes are involved in the pathophysiological modifications of ATA. While most of the microbes have good effects on ATA, some of them worsen the symptoms of the disease. More studies are required to find out the exact mechanism by which the gut and lungs microbes are involved in ATA.

### 9.2.13 GM in the Development of the Brain

The GM plays a unique role in the development of the brain. The gut microbes have a predominant effect on the normal functions of the host and early life programming of the circuits of the brain with an impact on the regulation of the responses to stress, memory functions, anxiety-like behaviour, and neuromuscular activity [77]. The various preclinical reports tried to establish the logical relation between the development of the neonatal brain and GM. During birth and infancy, trillions of microorganisms present in the gut are liable for the development of the immune system, proper physiology of the epithelial barrier, and homeostasis of the gut. In addition, the GM has a significant role in the development of the HPA axis. The elevation of stress hormone is seen in germ-free mice, which indicated the role of microbes in the development of the HPA axis [78, 79].

In addition, the evidence suggests the importance of GM in the development of the blood–brain barrier (BBB). The permeability of BBB is more in germ-free (GF) mice as compared to specific pathogen-free (SPF) mice with normal gut flora from early faetal brain development to adulthood. In addition, GF mice with a lack of microbial flora demonstrate defects in learning, recognition, and memory [80].

### 9.2.14 Brain–Gut–Microbiome (BGM) Axis in Depression

The BGM axis is important in the pathophysiological modifications of depression. Now according to some classifications, the human gut is also considered as a separate nervous system, well known as the 'enteric nervous system' [81]. The gut of humans contains high levels of various microorganisms, namely archaea, fungi, viruses, protozoa, and bacteria. The effect of microbiota on the physiology of the brain is predominant, and the diet with proper nutritious value has an effect on the growth of the microbes [82]. The use of antibiotics has a detrimental impact on the GM number and may lead to the progression of various CNS disorders, including depression. In addition, stress also predominantly modifies the number of microbes in the gut [83].

The patients affected with MDD have dysfunctions of the gut–brain axis. The various factors such as the immune system, HPA axis and stress hormone, oxidative stress, abnormal neurotransmitter levels are responsible for the progression of depression, as discussed above [14, 84]. In addition, the role of microbes in the progression of depression is getting important. The long term stress has a significant reduction in the GM community. The above statement further validates the evidence that high correlation and comorbidity are found between patients of irritable bowel syndrome (IBS) and depression and anxiety [85]. More than 50% of IBS patients have comorbid depression or anxiety. In an experiment conducted by Liu et al., transplantation of faecal microbiota of MDD patients to the rat leads to depression-like symptoms in the animal [86]. The rats that received faecal matter from patients exhibited depressive symptoms, like diminished pleasure activity, the elevation of symptoms of anxiety, and abnormal tryptophan metabolism, similar to those of their microbiota donors. Animal studies also exhibited the differences in the microbiota of depressive and control animals [87]. A number of depression models have shown the events, including the olfactory bulbectomy [88], maternal separation, [89] social disruption [90], chronic unpredictable stress [91], and chronic restraint stress model [92].

The evidence suggested that some bacteria are involved in the secretion of biologically active compounds in nature. The biologically active compounds can affect the various functional activities of the body, such as sleep, appetite, mood, and cognition. These bioactive compounds can communicate between the gut and brain. Moreover, these compounds can cross the BBB [93]. For example, *Lactobacillus* can secrete acetylcholine; *E.coli*, *Streptococcus*, *Enterococcus*, *Candida* can secrete serotonin; *Bacilli* and *Serratia* can secrete DA. Several bacteria have also been shown to produce DA and NE [94].

Moreover, In vitro, E. coli, Proteus Vulgaris, Serratia marcescens, Bacillus subtilis, and Bacillus mycoides were found to release relatively more levels (0.45–2.13 mM) of NE in their biomass [95]. Notably, while various strains of bacteria are revealed to produce 5-HT, the capabilities of producing 5-HT have not been found in the GM. However, further studies are required to prove the functional effectiveness of these neurotransmitters/bioactive compounds. The levels of brain-derived neurotrophic factor (BDNF) in the hippocampus are also influenced by GM. The administration of prebiotics is also leading to enhance the BDNF levels [96].

### 9.2.15 Role of the Microbiota in Depression Comorbid with ATA

GM has emerged as an important pathophysiological marker that has a role in ATA and depression. The main approaches used to the restoration of microbes are the use of probiotics, prebiotics, and proper diet. The live microorganisms such as bacteria and yeasts generally ingested in adequate amounts to maintain the normal gut physiology are known as probiotics. Probiotics are frequently used in various intestinal disorders such as Crohn's and inflammatory bowel disease (IBD) [97]. There are various preclinical evidence available that have shown the importance of probiotics and their potential to change behaviour, improve the mood, anxiety and cognition of rodents by regulating the levels of neurotransmitters [98]. The use of probiotics modifies the levels of various inflammatory markers like cytokines IL-1β and interleukin-6 (IL-6), as well as TNFα and microglial activation markers [99]. The above mechanism may be a helpful approach for the treatment of ATA and depression. In addition, some studies suggest that consumption of probiotics is associated with a reduction in the metabolism of serotonin and leads to modification of hyperactivity of the HPA axis [83]. Preclinical studies demonstrated that probiotics could effectively suppress IgEA production and in turn inhibits the accumulation of eosinophils.

Prebiotics are the food substances that induce the growth or activity of beneficial microbiota, such as bacteria and fungi. Prebiotics and regulating GM also improve behaviour and cognition and act as psychobiotics [100]. The probiotics exert their effects through the functional augmentation of the BGM axis. Use of prebiotics helps to the reduction in physical activity-induced ATA. Probiotics and prebiotics can improve the host immune system through the gut ecosystem and may be helpful/useful/valuable in treating allergic diseases like ATA. Probiotics and prebiotics also showed/exhibited/demonstrated the preventive effect of allergic diseases [101]. According to a study, prebiotics and synbiotics (mixture of probiotics and prebiotics) reduce airway inflammation in murine ATA models, whereas recurrent wheeze in infants is not improved [102]. Since the BGM and GLM axis is involved in the modification of immune system. The use of probiotics, prebiotics, and synbiotics has also shown beneficial effects in ATA and depression. The treatment approach has the potential to treat the comorbidity of depression with ATA.

However, further trials are needed to establish the efficacy of prebiotics and synbiotics on ATA outcomes in adults.

## 9.3 Conclusion

ATA and depression are highly comorbid with each other. ATA and depression both have common components of inflammation, immune system, and stress. The levels of inflammatory markers are more in depression comorbid with ATA. Certain microbes are involved in the synthesis of neurotransmitters. Neurotransmitters like NE have a bronchodilator effect and are helpful to cope up with depression. However, these neurotransmitters' functional effectiveness is still unclear, and further, some preclinical and clinical studies are warranted to validate the efficacy of these neurotransmitters. Intranasal exposure to certain bacteria also reduces allergic inflammation. In addition, fermentation of dietary fibre by the mouse intestinal microbiota protects against inflammatory diseases, including ATA. The positive overall effects have been seen through modification of the BGM and GLM axis by using probiotics, prebiotics, synbiotics, and proper diet.

## References


1. Dharmage SC et al (2019) Epidemiology of asthma in children and adults. Front Pediatr 7:246
2. Hall S, Agrawal DK (2014) Key mediators in the immunopathogenesis of allergic asthma. Int Immunopharmacol 23(1):316–329
3. Kudo M et al (2013) Pathology of asthma. Front Microbiol 4:263
4. Papiris S et al (2002) Clinical review: severe asthma. Crit Care 6(1):30–44
5. Ober C, Yao TC (2011) The genetics of asthma and allergic disease: a 21st century perspective. Immunol Rev 242(1):10–30
6. Kewalramani A et al (2008) Asthma and Mood Disorders. Int J Child Health Hum Dev 1(2): 115–123
7. Center for Substance Abuse Treatment. Managing Depressive Symptoms in Substance Abuse Clients During Early Recovery. Rockville (MD): Substance Abuse and Mental Health Services Administration (US); 2008. (Treatment Improvement Protocol (TIP) Series, No. 48.) Appendix D—DSM-IV-TR Mood Disorders. Available from: https://www.ncbi.nlm.nih.gov/books/NBK64063/
8. Lee CH, Giuliani F (2019) The role of inflammation in depression and fatigue. Front Immunol 10:1696
9. Dhar AK et al (2014) Design, synthesis, and pharmacological evaluation of novel 2-(4-substituted piperazin-1-yl)1, 8 naphthyridine 3-carboxylic acids as 5-HT3 receptor antagonists for the management of depression. Chem Biol Drug Des 84(6):721–731
10. https://www.who.int/news-room/fact-sheets/detail/depression
11. Brigitta B (2002) Pathophysiology of depression and mechanisms of treatment. Dialogues Clin Neurosci 4(1):7–20
12. Varghese FP, Brown ES (2001) The hypothalamic-pituitary-adrenal axis in major depressive disorder: a brief primer for primary care physicians. Prim Care Companion J Clin Psychiatry 3(4):151–155
13. Shadrina M et al (2018) Genetics factors in major depression disease. Front Psych 9:334

14. Bhatt S et al (2020) Role of oxidative stress in depression. Drug Discov Today 25(7): 1270–1276
15. Dash S et al (2015) The gut microbiome and diet in psychiatry: focus on depression. Curr Opin Psychiatry 28(1):1–6
16. Schnorr SL, Bachner HA (2016) Integrative therapies in anxiety treatment with special emphasis on the gut microbiome. Yale J Biol Med 89(3):397–422
17. Enaud R et al (2020) The gut-lung axis in health and respiratory diseases: a place for inter-organ and inter-kingdom crosstalks. Front Cell Infect Microbiol 10:9
18. Matucci A et al (2018) Is IgE or eosinophils the key player in allergic asthma pathogenesis? Are we asking the right question? Respir Res 19:113
19. Stone KD et al (2010) IgE, mast cells, basophils, and eosinophils. J Allergy Clin Immunol 125(2 Suppl 2):S73–S80
20. Krystel-Whittemore M et al (2016) Mast cell: a multi-functional master cell. Front Immunol 6: 620
21. McBrien CN, Menzies-Gow A (2017) The biology of eosinophils and their role in asthma. Front Med (Lausanne) 4:93
22. Lamouille S et al (2014) Molecular mechanisms of epithelial-mesenchymal transition. Nat Rev Mol Cell Biol 15(3):178–196
23. Venge P (2010) The eosinophil and airway remodelling in asthma. Clin Respir J. 4 Suppl 1: 15–19
24. Jacobsen JP et al (2012) The 5-HT deficiency theory of depression: perspectives from a naturalistic 5-HT deficiency model, the tryptophan hydroxylase 2Arg439His knockin mouse. Philos Trans R Soc Lond Ser B Biol Sci 367(1601):2444–2459
25. Nutt DJ (2008) Relationship of neurotransmitters to the symptoms of major depressive disorder. J Clin Psychiatry. 69 Suppl E1:4–7
26. Osimo EF et al (2020) Inflammatory markers in depression: a meta-analysis of mean differences and variability in 5,166 patients and 5,083 controls. Brain Behav Immun 87: 901–909
27. Ranjbar-Slamloo Y, Fazlali Z (2020) Dopamine and noradrenaline in the brain; overlapping or dissociate functions? Front Mol Neurosci 12:334
28. Drevets WC et al (2008) Brain structural and functional abnormalities in mood disorders: implications for neurocircuitry models of depression. Brain Struct Funct 213(1–2):93–118
29. Lener MS et al (2017) Glutamate and gamma-aminobutyric acid systems in the pathophysiology of major depression and antidepressant response to ketamine. Biol Psychiatry 81(10): 886–897
30. Stephens MA, Wand G (2012) Stress and the HPA axis: role of glucocorticoids in alcohol dependence. Alcohol Res 34(4):468–483
31. Mello AF et al (2003) Update on stress and depression: the role of the hypothalamic-pituitary-adrenal (HPA) axis. Braz J Psychiatry 25(4):231–238
32. Farooq RK et al (2017) Role of inflammatory cytokines in depression: focus on interleukin-1β. Biomed Rep 6(1):15–20
33. Di Paolo NC, Shayakhmetov DM (2016) Interleukin 1α and the inflammatory process. Nat Immunol 17(8):906–913
34. Hasler G (2010) Pathophysiology of depression: do we have any solid evidence of interest to clinicians? World Psychiatry 9(3):155–161
35. Chen Q et al (2017) Cyclooxygenase-2 signalling pathway in the cortex is involved in the pathophysiological mechanisms in the rat model of depression. Sci Rep 7:488
36. Cesario A et al (2011) The interplay between indoleamine 2,3-dioxygenase 1 (IDO1) and cyclooxygenase (COX)-2 in chronic inflammation and cancer. Curr Med Chem 18(15): 2263–2271
37. Phaniendra A et al (2015) Free radicals: properties, sources, targets, and their implication in various diseases. Indian J Clin Biochem 30(1):11–26

38. Maes M et al (2011) A review on the oxidative and nitrosative stress (O&NS) pathways in major depression and their possible contribution to the (neuro)degenerative processes in that illness. Prog Neuro-Psychopharmacol Biol Psychiatry 35(3):676–692
39. Jiang M et al (2014) Comorbidity between depression and asthma via immune-inflammatory pathways: a meta-analysis. J Affect Disord 166:22–29
40. Valença AM et al (2006) The relationship between the severity of asthma and comorbidites with anxiety and depressive disorders. Braz J Psychiatry 28(3):206–208
41. Van Lieshout RJ (2008) Psychological factors in asthma. Allergy Asthma Clin Immunol 4(1): 12–28
42. Lehto K (2019) Asthma and affective traits in adults: a genetically informative study. Eur Respir J 53(5):1802142
43. Slavich GM, Irwin MR (2014) From stress to inflammation and major depressive disorder: a social signal transduction theory of depression. Psychol Bull 140(3):774–815
44. Barker E et al (2015) The relationship between asthma and suicidal behaviours: a systematic literature review. Eur Respir J 46(1):96–106
45. Dickson RP (2016) The microbiome and the respiratory tract. Annu Rev Physiol 78:481–504
46. Hufnagl K (2020) Dysbiosis of the gut and lung microbiome has a role in asthma. Semin Immunopathol 42(1):75–93
47. Bustamante-Marin XM, Ostrowski LE (2017) Cilia and mucociliary clearance. Cold Spring Harb Perspect Biol 9(4):a028241
48. Ward HE, Nicholas TE (1984) Alveolar type I and type II cells. Aust NZ J Med 14(5 Suppl 3): 731–734
49. Nayak A (2012) An insight into the diverse roles of surfactant proteins, SP-A and SP-D in innate and adaptive immunity. Front Immunol 3:131
50. Kho ZY, Lal SK (2018) The human gut microbiome - a potential controller of wellness and disease. Front Microbiol 9:1835
51. Hanada S (2018) Respiratory viral infection-induced microbiome alterations and secondary bacterial pneumonia. Front Immunol 9:2640
52. Belkaid Y, Hand TW (2014) Role of the microbiota in immunity and inflammation. Cell 157(1):121–141
53. Dickson RP et al (2013) The role of the bacterial microbiome in lung disease. Expert Rev Respir Med 7(3):245–257
54. Pickard JM (2017) Gut microbiota: role in pathogen colonization, immune responses, and inflammatory disease. Immunol Rev 279(1):70–89
55. Frati F et al (2018) The role of the microbiome in asthma: the gut–lung axis. Int J Mol Sci 20(1):123
56. Subbarao P et al (2009) Asthma: epidemiology, etiology and risk factors. CMAJ 181(9):E181–E190
57. Zheng D et al (2020) Interaction between microbiota and immunity in health and disease. Cell Res 30:492–506
58. Atarashi K et al (2011) Induction of colonic regulatory T cells by indigenous Clostridium species. Science 331(6015):337–341
59. Guo P et al (2020) Clostridium species as probiotics: potentials and challenges. J Anim Sci Biotechnol 11:24
60. Sokolowska M (2018) Microbiome and asthma. Asthma Res Pract 4:1
61. Jakubczyk D et al (2020) The effectiveness of probiotics in the treatment of inflammatory bowel disease (IBD)-a critical review. Nutrients 12(7):1973
62. Furrie E et al (2005) Synbiotic therapy (Bifidobacterium longum/Synergy 1) initiates resolution of inflammation in patients with active ulcerative colitis: a randomised controlled pilot trial. Gut 54(2):242–249
63. Bienenstock J, Befus D (1984) Gut- and bronchus-associated lymphoid tissue. Am J Anat 170(3):437–445

64. Silva-Sanchez A, Randall TD (2020) Anatomical uniqueness of the mucosal immune system (GALT, NALT, iBALT) for the induction and regulation of mucosal immunity and tolerance. Mucosal Vaccines:21–54. https://doi.org/10.1016/B978-0-12-811924-2.00002-X
65. Budden K et al (2017) Emerging pathogenic links between microbiota and the gut–lung axis. Nat Rev Microbiol 15:55–63
66. Zhang D (2020) The cross-talk between gut microbiota and lungs in common lung diseases. Front Microbiol 11:301
67. Holtzman MJ (2012) Asthma as a chronic disease of the innate and adaptive immune systems responding to viruses and allergens. J Clin Invest 122(8):2741–2748
68. Nicholson LB (2016) The immune system. Essays Biochem 60(3):275–301
69. Loverdos K et al (2019) Lung microbiome in asthma: current perspectives. J Clin Med 8(11):1967
70. Barcik W (2020) The role of lung and gut microbiota in the pathology of asthma. Immunity 52(2):241–255
71. Arrieta MC (2015) Early infancy microbial and metabolic alterations affect risk of childhood asthma. Sci Transl Med 7(307):307ra152
72. He J et al (2020) Short-chain fatty acids and their association with signalling pathways in inflammation, glucose and lipid metabolism. Int J Mol Sci 21(17):6356
73. Trompette A et al (2014) Gut microbiota metabolism of dietary fiber influences allergic airway disease and hematopoiesis. Nat Med 20(2):159–166
74. Cait A (2018) Microbiome-driven allergic lung inflammation is ameliorated by short-chain fatty acids. Mucosal Immunol 11(3):785–795
75. Pugin B et al (2017) A wide diversity of bacteria from the human gut produces and degrades biogenic amines. Microb Ecol Health Dis 28(1):1353881
76. Podschun R, Ullmann U (1998) Klebsiella spp. as nosocomial pathogens: epidemiology, taxonomy, typing methods, and pathogenicity factors. Clin Microbiol Rev 11(4):589–603
77. Sudo N (2019) Role of gut microbiota in brain function and stress-related pathology. Biosci Microbiota Food Health 38(3):75–80
78. Farzi A (2018) Gut microbiota and the neuroendocrine system. Neurotherapeutics 15(1):5–22
79. Bhatt S et al (2021) Role of reactive oxygen species in the progression of Alzheimer's disease. Drug Discov Today 26(3):794–803
80. Braniste V (2014) The gut microbiota influences blood-brain barrier permeability in mice. Sci Transl Med 6(263):263ra158
81. Nezami BG, Srinivasan S (2010) Enteric nervous system in the small intestine: pathophysiology and clinical implications. Curr Gastroenterol Rep 12(5):358–365
82. Thursby E, Juge N (2017) Introduction to the human gut microbiota. Biochem J 474(11):1823–1836
83. Carabotti M (2015) The gut-brain axis: interactions between enteric microbiota, central and enteric nervous systems. Ann Gastroenterol 28(2):203–209
84. Gupta D (2013) Role of hypothalamic-pituitary-adrenal-axis in affective disorders: anti-depressant and anxiolytic activity of partial 5-HT1A agonist in adrenalectomised rats. Indian J Psychol Med 35(3):290–298
85. Cheung SG (2019) Systematic review of gut microbiota and major depression. Front Psych 10:34
86. Liu S (2020) Gut microbiota regulates depression-like behavior in rats through the neuroendocrine-immune-mitochondrial pathway. Neuropsychiatr Dis Treat 16:859–869
87. Bosi A et al (2020) Tryptophan metabolites along the microbiota-gut-brain axis: an interkingdom communication system influencing the gut in health and disease. Int J Tryptophan Res 13:1178646920928984
88. Park AJ (2013) Altered colonic function and microbiota profile in a mouse model of chronic depression. Neurogastroenterol Motil 25:733–e575

89. O'mahony SM (2009) Early life stress alters behavior, immunity, and microbiota in rats: implications for irritable bowel syndrome and psychiatric illnesses. Biol Psychiatry 65:263–267
90. Bharwani A (2017) Oral treatment with lactobacillus rhamnosus attenuates behavioural deficits and immune changes in chronic social stress. BMC Med 15:7
91. Yu M (2017) Variations in gut microbiota and fecal metabolic phenotype associated with depression by 16s rRNA gene sequencing and Lc/Ms-based metabolomics. J Pharm Biomed Anal 138:231–239
92. Liang S (2015) Administration of lactobacillus helveticus Ns8 improves behavioral, cognitive, and biochemical aberrations caused by chronic restraint stress. Neuroscience 310:561–577
93. Banks WA (2008) The blood-brain barrier: connecting the gut and the brain. Regul Pept 149(1–3):11–14
94. Liu L, Zhu G (2018) Gut-brain axis and mood disorder. Front Psych 9:223
95. Tsavkelova EA (2000) Detection of neurotransmitter amines in microorganisms with the use of high-performance liquid chromatography. Dokl Biochem 372(1–6):115–117
96. Savignac HM et al (2013) Prebiotic feeding elevates central brain derived neurotrophic factor, N-methyl-D-aspartate receptor subunits and D-serine. Neurochem Int 63(8):756–764
97. Jonkers D, Stockbrügger R (2003) Probiotics and inflammatory bowel disease. J R Soc Med 96(4):167–171
98. Wallace CJK, Milev R (2017) The effects of probiotics on depressive symptoms in humans: a systematic review [published correction appears in]. Ann Gen Psychiatry 16:14
99. Wang IK et al (2015) The effect of probiotics on serum levels of cytokine and endotoxin in peritoneal dialysis patients: a randomised, double-blind, placebo-controlled trial. Benef Microbes 6(4):423–430
100. Sarkar A (2016) Psychobiotics and the manipulation of bacteria-gut-brain signals. Trends Neurosci 39(11):763–781
101. Sestito S (2020) The role of prebiotics and probiotics in prevention of allergic diseases in infants. Front Pediatr 8:583946
102. Clapp M et al (2017) Gut microbiota's effect on mental health: the gut-brain axis. Clin Pract 7(4):987

# Understanding the Impact of the Microbiome on Lung Cancer

10

Anindita Goswami, Sanchita Chandra, Sarmistha Adhikari, and Paramita Mandal

**Abstract**

Higher incidence of prevalence and mortality is recorded with lung cancer among all the cancers worldwide. Human microbiomes are considered as one of the major hallmarks of cancer. Microbial dysbiosis is a multifactorial event that leads to tumor initiation, progression, and development of cancer. Metagenomics is used to analyze the genomes of microbial community in the sample and able to explore the identity of microbes including anaerobic bacteria also. In cancer research, this approach also creates a revolution to unravel the impact of microbiome in cancer. In case of carcinogenesis, lung microbiota plays a major role by regulating the specific oncogenic pathways and bacterial metabolites were reported to modulate various host metabolic pathways. Healthy microbiome diversity in lung can be disrupted through several environmental factors and induce the chances of cancer. These alterations are caused due to several biochemical alterations associated with reactive oxygen species, nitric oxide, interferons, and interleukins which induce genomic instability. Gut microbiota and their association with lung play an essential role for the cure in case of lung cancer. *Lactobacillus, Bifidobacterium,* and *Bacteroides* are well known probiotics residing in gut and in case of cancer progression that was observed to decrease in number. But, when these probiotics were used along with conventional chemotherapy, it exhibited a remarkable result in lung cancer by decreasing the frequency of gastrointestinal microflora that was responsible for cancer progression.

A. Goswami · S. Chandra · S. Adhikari · P. Mandal (✉)
Biomedical Genetics Laboratory, Department of Zoology, The University of Burdwan, Burdwan, India


G. Gupta et al. (eds.), *Microbiome in Inflammatory Lung Diseases*,
https://doi.org/10.1007/978-981-16-8957-4_10

**Keywords**

Lung cancer · Microbiome · Metagenomics · Risk factors · Probiotics

## 10.1 Introduction

In comparison to other complex diseases, cancer is the second most prevalent disorder related death worldwide and its susceptibility and progression are primarily influenced by gene environmental interaction. Worldwide higher prevalence and mortality (11.6% of all cancers) were recorded in lung cancer among all the cancer types. Similar to other cancers, heterogeneity based on pathological and clinical features is observed in lung cancer. Globally, human health is greatly affected by lung cancer. As majority of the cases are detected at very advanced stage, therefore higher mortality rate is associated with this disorder. The lung is an organ of the body which is continuously exposed to the environmental factors. These environmental risk factors including microbes regulate host immune responses. Microbiota might be considered as crucial environmental factors that we are continuously exposed to and they can influence carcinogenesis. Recent studies identified the impact of specific microbiota association in our system as key determinants of major health diseases, including cancer [1]. Recent studies established that lung has distinct microbiome and that may influence the development of lung cancer [2, 3]. These studies also suggested that global change of microbiome diversity contributes to such disease pathogenesis. Surprisingly, total metagenomic content in our system is about 100 fold to our genome and plays a key role in metabolism and inflammation [4].

The microbiome is considered as the ecological community of microorganisms that colonizes in our system and forms a micro-ecological system [5]. Diverse variety of microorganisms including fungi, bacteria, and viruses develop their symbiotic or pathogenic relationship with the host during long term evolution [6]. A stable beneficial microbial association within human body is termed as "eubiosis." In addition to that, prepotency of pathogenic bacteria is entitled as "dysbiosis." These pathogenic relationships are assumed to influence carcinogenesis. Pathogenic microbial organisms are known as the primary causal factors of many infectious diseases and they altered metabolites, dysregulate hormonal balance, modulate immune functions, DNA damage, and mutagenesis [7].

Currently, human microbiomes are considered as one of the major hallmarks of cancer. Several epidemiological studies are indicated that many association with several microorganisms can influence the process of carcinogenesis. Several studies identified the association of various bacteria and with chronic disorders including cancer [3]. It was also reported that infectious agents contribute 20% cancer burden globally. Chronic infections of many pathogens, particularly viruses and bacteria promote the development of cancer. From previous studies, it is clear that microbial dysbiosis is a multifactorial event that leads to tumor initiation, progression, and development of cancer [8]. Several oncogenic viruses that may potentially contribute their role in the development of carcinogenesis, such as human papilloma viruses (HPV 16 and 18), hepatitis B virus (HBV), hepatitis C virus (HCV), human herpes viruses (HHV 4 and 8), and human T-lymphotropic virus-I (HTLV-I). Among bacteria, *Salmonella typhi, Helicobacter pylori, Mycobacterium tuberculosis* primarily associated with gastric cancer, colon cancer, and lung cancer. The protozoal parasites also play a role in carcinogenesis such as *Schistosoma haematobium* and *Opisthorchis viverrini* which are associated with urinary bladder and gallbladder cancer [3].

In this communication we present the putative role of the microbiome in lung cancer progression. Finally, we describe the potential usage of therapy based on microbiota and its applications.

## 10.2 Association of Microbiota with Lung Cancer

### 10.2.1 Determination of Microbial Diversity in Lung Cancer

The term metagenome refers to the genetic content of any group of microorganisms. Previously we have already mentioned that 20% of cancer cases are associated with microbes but due to vast majority of microorganisms and limitation of culture-based method most of them were not identified. Metagenomics can solve this problem and opened up a horizon for identification of many novel microorganisms. Metagenomics is used to analyze the genomes of microbial community in the sample and able to explore the identity of microbes including anaerobic bacteria also. In cancer research, this approach also creates a revolution to identify the influence of diverse microorganisms on carcinogenesis. Metagenomics solved the dilemma to identify the microbial diversity in various types of cancer including lung cancer [9, 10].

Some studies based on 16S rRNA sequencing have identified a set of genera with maximum or minimum abundances in lung cancer tissue as well as normal lung tissue[11–15]. Previous studies reported that phylum Proteobacteria and Firmicutes and another three most frequent bacterial species such as *Achromobacter xylosoxidans*, *Schizothorax sinensis,* and *Staphylococcus sciuri* were detected in lung cancer patient based on metagenomic assessment of bronchoalveolar lavage fluid (BALF) sample [16]. Another study revealed the presence of *Streptococcus* in cancerous tissue of lung [14, 15]. Other studies identified total of diverse phyla and genera by metagenomics based approach. Most predominant phyla were Proteobacteria, Actinobacteria, Firmicutes, and Bacteroidetes and among genera *Haemophilus* (Proteobacteria); Corynebacterium, and Actinomyces (Actinobacteria); *Streptococcus* and *Veillonella*, (Firmicutes) and *Prevotella* (Bacteroidetes) were the most prevailing type based on pool sequencing approach (16S rRNA V3-V6) using BALF sample [17]. Another study based on 16SrRNA based metagenomic sequencing revealed that *Streptococcus* was mostly found in lung cancer patient and two bacteria such as *Corynebacterium tuberculostearicum* and *Keratinibaculum paraultunense* were limited in lung cancer group patients.

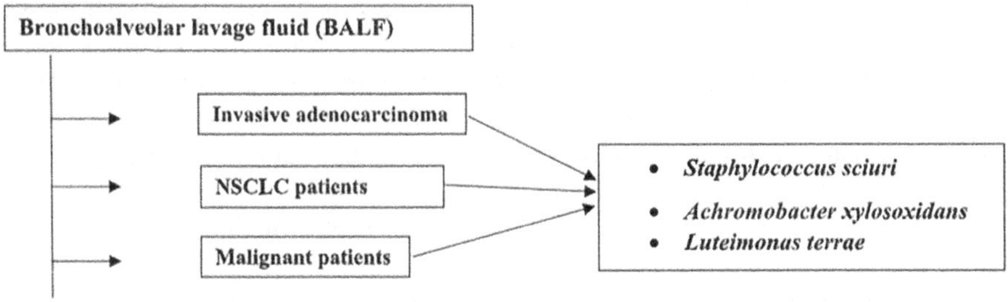


**Fig. 10.1** Common three microbiota identified in three different stages of lung cancer [16]

*Corynebacterium tuberculostearicum* has was previously detected in small cell lung cancer (SCLC) [14–16, 18]. Another interesting study found that consumption of alcohol for prolonged period alters the microbial diversity in lung. Samples (BALF) taken from invasive adenocarcinoma, non-small cell lung cancer (NSCLC), and invasive malignant patient showed only three similar species common to all after metagenomics analysis such as *Staphylococcus sciuri, Luteimonas terrae, and Achromobacter xylosoxidans.* Results showed that the invasive adenocarcinoma sample revealed maximum diversity including 67 bacterial species. In case of NSCLC, *Nocardia pneumoniae, Actinomyces israelii, Fusobacterium periodonticum, Legionella hackeliae, and Treponema socranskii* were observed solely. Some of species were found in both advanced stage of adenocarcinoma and NSCLC patients' sample but not present in cancer patient such as *Nesterenkonia lacusekhoensis, Corynebacterium jeikeium, Brevundimonas albigilva, Propionibacterium acnes, and Comamonas denitrificans* [16] (Fig. 10.1).

Using metagenomics of sputum sample, seven types of pathogens including *Granulicatella adiacens* were detected in lung cancer patients. It was observed that G. *adiacens* had significant co-relationship with other six bacterial sp. such as *Enterococcus* sp. 130, *Streptococcus intermedius*, *Escherichia coli*, *Streptococcus viridans*, *Acinetobacter junii*, and *Streptococcus* sp. 6 only in case of lung cancer positive patient. Some interesting studies also reported that specially genus *Granulicatella* was observed in non-smoker lung cancer patients, which reflect an ideal lung cancer scenario without any exposure to tobacco [19–21]. It was also identified by another study that bacteria within cancerous cells can inhibit the efficacy of chemotherapeutic drug [22]. Other study reported that the application of metagenomic based next generation sequencing (mNGS) identified the pulmonary invasive fungal infections, fastidious, and anaerobic pathogen which were previously not found by culture method in laboratory. Some novel microorganisms such as Rhizopus and Mucor were exclusively identified by metagenomic analysis which were usually not detected by culture-based methods [23]. Parallelly, the presence of *Capnocytophaga* and *Veillonella* like oral microorganisms in the saliva of lung cancer patients also pointed the use of biomarkers for detection of early stages of lung cancer [20, 24]. Previous study demonstrated that squamous cell lung carcinoma often correlated with Acidovorax and *Acidovorax* genus was especially

**Table 10.1** Microbial diversity in different stages of lung cancer

| Sp. name | Invasive adenocarcinoma | NSCLC | SCLC | Malignant case |
|---|---|---|---|---|
| *Staphylococcus sciuri* | + | + | – | + |
| *Achromobacter xylosoxidans* | + | + | – | + |
| *Luteimonas terrae* | + | + | – | + |
| *Nocardia pneumoniae* | – | + | – | – |
| *Actinomyces israelii* | – | + | – | – |
| *Fusobacterium periodonticum* | – | + | – | – |
| *Legionella hackeliae* | – | + | – | – |
| *Treponema socranskii* | – | + | – | – |
| *Corynebacterium jeikeium* | + | + | – | – |
| *Nesterenkonia lacusekhoensis* | + | + | – | – |
| *Propionibacterium acnes* | + | + | – | – |
| *Brevundimonas albigilva* | + | + | – | – |
| *Comamonas denitrificans* | + | + | – | – |
| *Streptococcus intermedius* | – | – | – | + |
| *Enterococcus* sp. 130 | – | – | – | + |
| *Acinetobacter junii* | – | – | – | + |
| *Escherichia coli* | – | – | – | + |
| *Streptococcus viridans* | – | – | – | + |
| *Streptococcus* sp. 6 | – | – | – | + |
| *Granulicatella adiacens* | – | – | – | + |
| *Corynebacterium tuberculostearicum* | – | – | + | – |
| *Veillonella* sp. | + | – | + | – |
| *Capnocytophaga* sp. | + | – | + | – |
| *Acidovorax* sp. | – | – | + | – |

References: [14–16, 18–21]

abundant in patient with TP53 mutation [11, 20]. Table 10.1 represented the diversity of microbiotas detected in different stages of lung cancer by metagenomics.

### 10.2.2 Role of Microbiota in Lung Cancer Progression

It was already reported that among wide diversity of airway microbiota, metagenomic analysis identified Firmicutes, Proteobacteria, Bacteroidetes, and Actinobacteria as a major phylum of lung cancer microbiota and genera included *Veillonella, Prevotella, Neisseria, Streptococcus, Haemophilus, Sphingomonas, Fusobacterium, Pseudomonas, Megasphaera, Acinetobacter, Staphylococcus,* and *Corynebacterium*. Several epidemiological data identified some mutual association of bacteria with lung carcinoma. Pulmonary infections are reported to be associated with majority of lung cancer cases and post-obstructive pneumonia was frequently reported which negatively impacted upon the survival of lung cancer individuals

[20, 25]. In case of carcinogenesis, lung microbiota plays a major role by regulating the specific oncogenic pathways and bacterial metabolites can modulate various metabolic pathways. Elevation of PI3K and ERK1/2 signaling pathways was observed when the human adenocarcinoma cell line was treated with bacterial products of lung cancer patients. *Veillonella parvula* was mostly abundant microorganism which was found to upregulate PI3K [20, 26]. Upregulation of PI3K pathway was considered as the key step in the lung carcinogenesis, which suggests the influence of microbiota on lung carcinogenesis [20, 27]. To promote the tumor growth, microbiota influence lung microenvironment and induce lung inflammation. Thus, lung microbiota contribute to tumor progression by modulation of tumor cells or by altering immune response.

## 10.3 Risk Factors of Microbiome Diversity in Lung Cancer

Healthy microbiome diversity in lung can be disrupted through several mechanisms and induce the chances of cancer. These alterations are caused due to several biochemical alterations associated with reactive oxygen species, interleukins, and interferons and induce genomic instability [28]. There are several associated factors for disrupting the microbiome balance which are discussed below:

### 10.3.1 Upregulation of ERK & PI3K Signaling Pathways

Extracellular signal regulated kinases (ERK) and Phosphoinositide 3- kinase (PI3K) signaling pathways were frequently upregulated in lung cancer by enhancing the number of oral bacteria like *Streptococcus* and *Veillonella* in the lower respiratory tract of lung [26, 29]. So, the upregulation of both pathways indicates that transcriptome signature is associated with lung cancer etiology and promotes the enrichment of oral bacteria (*Veillonella, Streptococcus, Prevotella*) in the lower respiratory tract. Not only in lower airway, *Streptococcus* was also found in airway brushings in case of lung cancer patients [14, 15]. Additionally, *Haemophilus influenza* was also responsible to promote lung carcinogenesis with the help of cigarette smoking. This smoking phenomenon drives the metastatic growth influenced by neutrophil infiltration and IL-17C resulted in tumor formation [30, 31].

### 10.3.2 Altered Expression of CD36

CD 36 is a key factor for inflammation which interacts with pathogen borne ligands or toxins [32]. In case of lung microbiome scenario, CD36 influences the initialization of generation of *Cyanobacteria* generated microcystin residues in lung alveoli and thereby upregulate PARP1 [33]. Studies also demonstrated that there is a connection between lung microbiota and lung carcinogenesis that is closely associated with modulation of CD36 which can affect the poly(ADP-ribose) polymerase 1 (PARP1) signaling pathway and these changes increased the number of Bacteroidetes and Probacteria, mainly *Cyanobacteria* by 53% [34, 35].

### 10.3.3 TP53 Mutation with Tobacco Smoking

Tobacco smoking induced TP53 mutations which help to invade some oncogenic bacteria in lung epithelial cell and these bacterial species take advantages to act as promoters for lung tumorigenesis and disrupted the healthy lung microbiota [11].

### 10.3.4 COPD and Microbial Dysbiosis

Microbial dysbiosis is referred to a state of microbial imbalance and elevates a pathway of colonization and chronic inflammation which have a direct effect on leaking of serum proteins and loss of epithelial integrity that leads to lung carcinogenesis [5]. These phenomena are observed mostly in chronic obstructive pulmonary disease (COPD), Studies also suggested that the imbalance in microbiota contributes to damage of lung parenchymal cells, degradation of alveolar attachment, and chronic inflammation by several bacterial products and formyl peptides which migrate from alveoli with the help of strong chemo alterations [3, 5].

### 10.3.5 Cytokines and NF-κβ1 Pathway

*H. pylori* is reported in chronic bronchitis and lung cancer. Proinflammatory cytokines like IL-6 along with tumors necrosis factors are associated with development of lung cancer scenario [36]. Some in vitro studies suggested that IL-6 was involved in activating NF-κβ1 pathway which directly involved in tumorigenesis by stimulating cellular proliferation, migration, invasion, and cancer progression. Induction of IL-8 in lung epithelial cells increases the mucus secretion and affected the epithelial layer integrity with the association of elastase, cathepsin G, and proteinase 3 [37]. Therefore, the loss of integrity in lung epithelial cells associated with some cytokines which is responsible for lung tumorigenesis in some affected patients.

### 10.3.6 Mycobacterium Tuberculosis

Prolonged tuberculosis infection can induce TNF-α which is an important mediator of chronic inflammation and carcinogenesis [38]. Usually, the lung immune system defends itself against pathogen invasion by increasing the alveolar surfactant, blocking the pathogen translocation by epithelial cells with the help of TLRs and prevents the metabolite overload of pathogens [39]. Apart from this, pulmonary

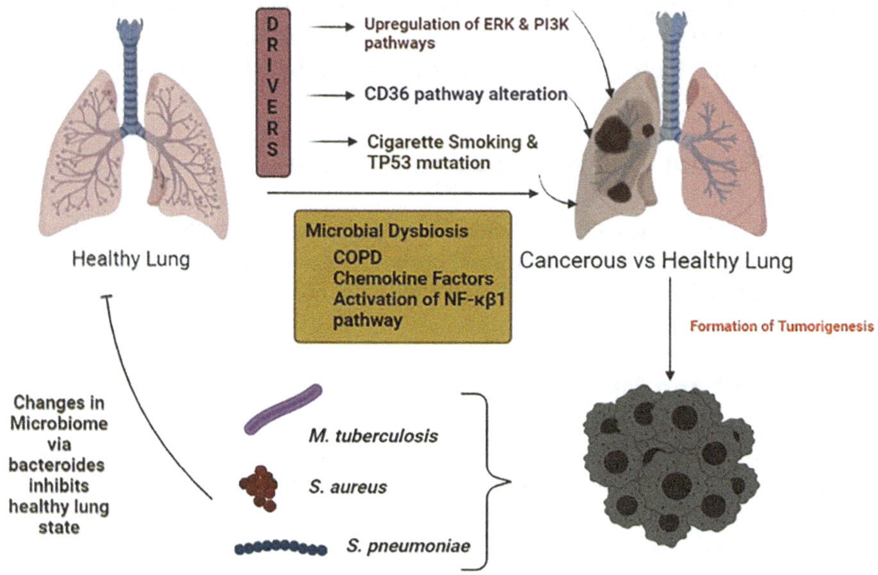


**Fig. 10.2** Risk factors of lung cancer and microbial dysbiosis

fibrosis was also mediated by tuberculosis which also responsible for development of lung cancer.

### 10.3.7 Bacteroides Fragilis and Activation of STAT3

Activation of STAT3 is triggered by *Bacteroides fragilis* with the help of T-helper type 17 (Th17) and induces the carcinogenesis via Th-17 dependent pathway via supporting cancer cell proliferation and angiogenesis [38, 40]. Furthermore, *Bacteroides fragilis* produces the chemicals like superoxide dismutase which are associated with genomic instability and responsible for the disruption in DNA repair system [41].

Therefore several risk factors may modulate microbiome diversity which play role in lung cancer pathogenesis (Fig. 10.2). So, in future, along with the conventional therapeutic techniques microbiome diversity can be used to detect or serve as potential biomarker to state the health of lung and disease progression scenario.

## 10.4 Use of Probiotics for Lung Cancer Prevention and Therapy

With all the therapeutic advancements in lung cancer scenario, still the survival rate is very much limited among male and females worldwide (American Cancer Society, 2017). So, search for new and novel therapeutic strategy to prevent or treat lung cancer is essential. Healthy microbiome and their association are an essential factor for lung immunity. With all the check-point inhibitors or immunotherapy to block the disease progression, probiotics are also a good option to maintain healthy microbiome diversity in lung cancer cases. Though, several limitations of probiotics are there, but after understanding the state of health, right probiotics combined with conventional or chemotherapy, could be a better option to restore the healthy microbiome flora and maintain the broad spectrum of microbiome diversity. Generally, probiotics are useful in eliminating anti-microbial and anti-tumor and improving immunity [42]. They are also reported to decrease the side effects of chemotherapy [14, 15].

Gut microbiota and their association with lung play an essential role for the cure in case of lung cancer. *Lactobacillus, Bifidobacterium, and Bacteroides* are well-known probiotics resides in gut and in case of cancer progression that was observed to decrease in number. But, when these probiotics were used along with conventional chemotherapy, it exhibited a remarkable result in lung cancer by decreasing the frequency of gastrointestinal microflora that was responsible for cancer progression [43]. The following probiotics can be used for prevention and therapy of lung cancer.

### 10.4.1 Usage of Different Species of Lactobacillus

- ***Lactobacillus acidophilus:*** Cisplatin/Lactobacillus acidophilus mixture showed some positive results in decreasing the size of tumor and improved survival rates in case of Lewis lung cancer bearing mouse rather than the other group treated with Cisplatin/ABX (only antibiotic mixture). In this study, probiotic treated group showed downregulation of vega and Ras oncogenes and upregulation of tumor suppressor genes like cdkn1b and Bax gene. Upregulation of IFN-λ, Gzmb and Prf1 mRNA expression were also observed and play a protective role in tumorigenesis [44].
- ***Lactobacillus casei:*** Introduction of *L. casei* (YIT 9018) can reduce tumor on Lewis Lung carcinoma (3LL) mice was observed and it suppressed pulmonary and regional metastasis in mice and guinea pigs [45].

### 10.4.2 Probiotics Mixture of Enterococcus Faecium & Saccharomyces Cerevisiae

These two probiotic mixture and their metabolite products were used against Lewis Lung carcinoma (3LL) as combined vaccine and had synergistic effect to inhibit metastasis by 2.5 fold compared to other therapeutic options [46]

### 10.4.3 Bifidobacterium Cocktail

*Bifidobacterium infantis* is a recombinant therapeutic agent used as potential probiotic in lung cancer patients. *Bifidobacterium* cocktail (*B. bifidum, B. lactis, B. breve*) manage to improve immunity by cytokine–cytokine interaction, (CD8+) T cell activation, co-stimulation, and chemokine regulation [47]. *Bifidobacterium infantis* mediated gene transfer also reported to inhibit tumor growth and prolong the survival rates [48].

#### 10.4.3.1 Bacteroides (B. thetaiotaomicron & B. Fragilis)

*Bacteroides* mixture influences immunostimulatory effects of CTLA4 blockade. It increased the cross-reactivity of tumor and bacterial epitopes by restoring therapeutic response to CTLA4 antibody treatment [49].

All those data suggested that cancer therapy with microbiota or their products has opportunity to cure tumorigenesis and microbial agents possibly inhibit the cancer progression when used with combined therapy [50, 51].

## 10.5 Conclusion

The human and microbiomes communities play an important role in regulating host functions. Various in vitro and in vivo studies identified the association of microbiota in cancer development in response to the host's ever-changing internal and environmental factors. Revolution of metagenomics has enhanced our ability to study the lung cancer microbiome diversity which play a crucial role in cancer progression. Lastly, biotherapeutics along with probiotics or engineered recombinant probiotic agents have enormous possibility to treat or inhibit the carcinogenesis or cancer progression in upcoming days. Proper use of probiotics and probiotics mixture along with conventional therapy showed promising results in various experiments and also be a good option to combat the severity of lung cancer in suffered patients.

**Acknowledgement** The authors acknowledge Department of Zoology, The University of Burdwan.

**Conflicts of Interest Statement** There are none.

## References

1. Vivarelli S, Salemi R, Candido S, Falzone L, Santagati M et al (2019) Gut microbiota and cancer: from pathogenesis to therapy. Cancers 11(1):38
2. Halley A, Leonetti A, Gregori A, Tiseo M, Deng DM et al (2020) The role of the microbiome in cancer and therapy efficacy: focus on lung cancer. Anticancer Res 40(9):4807–4818
3. Martins D, Mendes F, Schmitt F (2021) Microbiome: a supportive or a leading actor in lung cancer? Pathobiology 88(2):198–207

4. Schwabe RF, Jobin C (2013) The microbiome and cancer. Nat Rev Cancer 13(11):800–812
5. Maddi A, Sabharwal A, Violante T, Manuballa S, Genco R et al (2019) The microbiome and lung cancer. J Thorac Dis 11(1):280–291
6. Liu N-N, Ma Q, Ge Y, Yi C-X, Wei L-Q et al (2020) Microbiome dysbiosis in lung cancer: from composition to therapy. NPJ Precis Oncol 4(1):33
7. Parida S, Sharma D (2021) The microbiome and cancer: creating friendly neighborhoods and removing the foes within. Cancer Res 81(4):790–800
8. Bhatt AP, Redinbo MR, Bultman SJ (2017) The role of the microbiome in cancer development and therapy. CA Cancer J Clin 67(4):326–344
9. Banerjee J, Mishra N, Dhas Y (2015) Metagenomics: a new horizon in cancer research. Meta Gene 5:84–89
10. Handelsman J, Tiedje J, Alvarez-Cohen L, Ashburner M, Cann IKO, Delong EF, Schmidt TM (2007) The new science of metagenomics: revealing the secrets of our microbial planet, vol 13. National Academies Press, Washington, DC. Nat Res CouncRepub
11. Greathouse KL, White JR, Vargas AJ, Bliskovsky VV, Beck JA, von Muhlinen N, Polley EC, Bowman ED, Khan MA, Robles AI et al (2018) Interaction between the microbiome and TP53 in human lung cancer. Genome Biol 19:123
12. Jin J, Gan Y, Liu H, Wang Z, Yuan J, Deng T, Zhou Y, Zhu Y, Zhu H, Yang S et al (2019) Diminishing microbiome richness and distinction in the lower respiratory tract of lung cancer patients: a multiple comparative study design with independent validation. Lung Cancer 136: 129–135
13. Leng Q, Holden VK, Deepak J, Todd NW, Jiang F (2021) microbiota biomarkers of lung cancer. Diagnostics 11(407):3–15
14. Liu HX, Tao LL, Zhang J, Zhu YG, Zheng Y, Liu D et al (2018) Difference of lower airway microbiome in bilateral protected specimen brush between lung cancer patients with unilateral lobar masses and control subjects. Int J Cancer 142(4):769–778
15. Liu Y, O'Brien JL, Ajami NJ, Scheurer ME, Amirian ES, Armstrong G, Tsavachidis S, Thrift AP, Jiao L, Wong MC et al (2018) Lung tissue microbial profile in lung cancer is distinct from emphysema. Am J Cancer Res 8:1775–1178
16. Ekanayake A, Madegedara D, Chandrasekharan V, Arachchi DM (2019) Respiratory bacterial microbiota and individual bacterial variability in lung cancer and bronchiectasis patients. Indian J Microbiol. https://doi.org/10.1007/s12088-019-00850-w
17. Gomes S, Cavadas B, Ferreira JC, Marques PI, Monteiro C, Sucena M, Sousa C, Rodrigues LV, Teixeira PP, Abreu TTD, Barbara C, Semedo J, Mota L, Carvalho AS, Matthiesen R, Pereira L, Seixas S (2019) Profiling of lung microbiota discloses differences in adenocarcinoma and squamous cell carcinoma. Sci Rep 9:12838
18. Hinic V, Lang C, Weisser M et al (2012) Corynebacterium tuberculostearicum: a potentially misidentified and multiresistant Corynebacterium species isolated from clinical specimens. J Clin Microbiol 50:2561–2567
19. Cameron SJS, Lewis KE, Huws SA, Hegarty MJ, Lewis PD, Pachebat JA et al (2017) A pilot study using metagenomic sequencing of the sputum microbiome suggests potential bacterial biomarkers for lung cancer. PloS One 12(5):e0177062
20. Dong Q, Chen ES, Zhao C, Jin C (2021) Host-microbiome interaction in lung cancer. Front Immunol 12(679829):1–9
21. Hosgood HD, Sapkota AR, Rothman N, Rohan T, Hu W, Xu J et al (2014) The potential role of lung microbiota in lung cancer attributed to household coal burning exposures. Environ Mol Mutagen 55(8):643–651
22. Geller LT, Barzily-Rokni M, Danino T et al (2017) Potential role of intratumor bacteria in mediating tumor resistance to the chemotherapeutic drug gemcitabine. Science 357:1156–1160
23. Li H, Gao H, Meng H, Wang Q, Li S, Chen H, Li Y, Wang H (2018) Detection of pulmonary infectious pathogens from lung biopsy tissues by metagenomic next-generation sequencing. Front Cell Infect Microbiol 8(205):1–11

24. Yan X, Yang M, Liu J, Gao R, Hu J, Li J et al (2015) Discovery and validation of potential bacterial biomarkers for lung cancer. Am J Cancer Res 5(10):3111–3122
25. Qiao D, Wang Z, Lu Y, Wen X, Li H, Zhao H (2015) A retrospective study of risk and prognostic factors in relation to lower respiratory tract infection in elderly lung cancer patients. Am J Cancer Res 5(1):423–432
26. Tsay JCJ, Wu BG, Badri MH, Clemente JC, Shen N, Meyn P, Li Y, Yie TA, Lhakhang T, Olsen E et al (2018) Airway microbiota is associated with upregulation of the PI3K pathway in lung cancer. Am J Respir Crit Care Med 198:1188–1198
27. Gustafson AM, Soldi R, Anderlind C, Scholand MB, Qian J, Zhang X et al (2010) Airway pi3k pathway activation is an early and reversible event in lung cancer development. Sci Trans Med 2(26):26ra5–26ra5
28. Chumduri C, Gurumurthy RK, Zietlow R, Meyer TF (2016) Subversion of host genome integrity by bacterial pathogens. Nat Rev Mol Cell Biol 17(10):659–673
29. De Marco C, Laudanna C, Rinaldo N, Oliveira DM, Ravo M, Weisz A et al (2017) Specific gene expression signatures induced by the multiple oncogenic alterations that occur within the PTEN/PI3K/AKT pathway in lung cancer. PLoS One 12(6):e0178865
30. Goto T (2020) Airway microbiota as a modulator of lung cancer. Int J Mol Sci 21(9):3044
31. Jungnickel C, Schmidt LH, Bittigkoffer L, Wolf L, Wolf A, Ritzmann F et al (2017) IL-17C mediates the recruitment of tumor-associated neutrophils and lung tumor growth. Oncogene 36(29):4182–4190
32. Koch M, Hussein F, Woest A, Grundker C, Frontzek K, Emons G, Hawighorst T (2011) CD36-mediated activation of endothelial cell apoptosis by an n-terminal recombinant fragment of thrombospondin-2 inhibits breast cancer growth and metastasis in vivo. Breast Cancer Res Treat 128:337–346
33. Apopa PL, Alley L, Penney RB, Arnaoutakis K, Steliga MA, Jeffus S, Bircan E, Gopalan B, Jin J, Patumcharoenpol P et al (2018) Parp1 is up-regulated in non-small cell lung cancer tissues in the presence of the cyanobacterial toxin microcystin. Front Microbiol 9:1757
34. Bai L, Zhao J, Yu H, Zhao N, Liu D, Zhong W, Zhao Y (2013) The cd36 dynamic change after radiation therapy in lung cancer patients and its correlation with symptomatic radiation pneumonitis. Radiother Oncol 107:389–391
35. Mehan MR, Ayers D, Thirstrup D, Xiong W, Ostroff RM, Brody EN, Walker JJ, Gold L, Jarvis TC, Janjic N et al (2012) Protein signature of lung cancer tissues. PLoS one 7:e35157
36. Koshiol J, Flores R, Lam TK, Taylor PR, Weinstein SJ, Virtamo J et al (2012) Helicobacter pylori seropositivity and risk of lung cancer. PLoS One 7(2):e32106
37. Bingula R, Filaire M, Radosevic-Robin N, Bey M, Berthon JY, Bernalier-Donadille A et al (2017) Desired turbulence? Gut-lung axis, immunity, and lung cancer. J Oncol 2017:5035371
38. Mao Q, Jiang F, Yin R, Wang J, Xia W, Dong G et al (2018) Interplay between the lung microbiome and lung cancer. Cancer Lett 415:40–48
39. Uehara A, Fujimoto Y, Fukase K, Takada H (2007) Various human epithelial cells express functional Toll-like receptors, NOD1 and NOD2 to produce anti-microbial peptides, but not proinflammatory cytokines. Mol Immunol 44(12):3100–3111
40. Garcia-Castillo V, Sanhueza E, McNerney E, Onate SA, Garcia A (2016) Microbiota dysbiosis: a new piece in the understanding of the carcinogenesis puzzle. J Med Microbiol 65(12): 1347–1362
41. Carbonero F, Benefiel AC, Alizadeh-Ghamsari AH, Gaskins HR (2012) Microbial pathways in colonic sulfur metabolism and links with health and disease. Front Physiol 3(November):448
42. Kumar M, Kumar A, Nagpal R, Mohania D, Behare P, Verma V, Yadav H (2010) Cancer preventing attributes of probiotics: an update. Int J Food Sci Nutr 61(5):473–496
43. Mlu S, Urtenova MA, Tkachenko EL, Avalueva EB, Orlov SV, Ivanov SV, Skazyyaeya EV (2013) On the possibilities of correction of changes of the gastrointestinal tract microbiota in patients with lung cancer treated receiving chemotherapy. Exp Clin Gastroenterol 11:15–20

44. Gui QF, Lu HF, Zhang CX, Xu ZR, Yang YM (2015) Well-balanced commensal microbiota contributes to anti-cancer response in a lung cancer mouse model. Genet Mol Res 14(2): 5642–5651
45. Matsuzaki T, Yokokura T, Azuma I (1985) Anti-tumour activity of Lactobacillus casei on Lewis lung carcinoma and line-10 hepatoma in syngeneic mice and guinea pigs. Cancer Immunol Immunother 20(1):18–22
46. Tanasienko OA, Cheremshenko NL, Titova GP, Potebnya MG, Gavrilenko MM, Nagorna SS, Kovalenko NK (2005) Elevation of the efficacy of antitumor vaccine prepared on the base of lectines from B. subtilis B-7025 upon its combined application with probiotics in vivo. Exp Oncol 27(4):336–338
47. Sivan A, Corrales L, Hubert N (2015) Commensal Bifidobacterium promotes antitumor immunity and facilitates anti PD-L1 efficacy. Science 350:1084–1089
48. Zhu H, Li Z, Mao S, Ma B, Zhou S, Deng L, Huang Y (2011) Antitumor effect of sFlt-1 gene therapy system mediated by Bifidobacterium Infantis on Lewis lung cancer in mice. Cancer Gene Ther 18:884–896
49. Vétizou M, Pitt JM, Daillère R, Lepage P, Waldschmitt N, Flament C, Zitvogel L (2015) Anticancer immunotherapy by CTLA-4 blockade relies on the gut microbiota. Science 350(6264):1079–1084
50. Sharma A, Viswanath B, Park YS (2018) Role of probiotics in the management of lung cancer and related diseases: an update. J Funct Foods 40:625–633
51. Zitvogel L, Daillère R, Roberti MP, Routy B, Kroemer G (2017) Anticancer effects of the microbiome and its products. Nat Rev Microbiol 15(8):465–478

# Microbiome in Pulmonary Tuberculosis 11

Arnab Rakshit, Aarti Verma, Saloni Verma, Gurjit Kaur Bhatti, Amit Khurana, Jasvinder Singh Bhatti, Snehal Sainath Jawalekar, and Umashanker Navik

**Abstract**

Tuberculosis (TB) is among the global dominant fatal infection caused by a single organism, and it is still holding its position in spite of the golden age of the antibiotics. The recent studies are mostly focused on finding the prevention of TB rather than curing it because the antimycobacterial chemotherapy is failing constantly due to emerging multidrug resistance (MDR). Further, the intestinal microbiota is the central command for maintaining the homeostasis of the microbial profile of different organs. The change in the intestinal microbiota effects homeostasis by impacting the immune response to the microbial profile of various organs. Thus, it also affects the chance of contracting the infections. Here in this

---

Arnab Rakshit, Aarti Verma and Saloni Verma contributed equally with all other contributors.

A. Rakshit · A. Verma · S. Verma · U. Navik (✉)
Department of Pharmacology, Central University of Punjab, Bathinda, India

G. K. Bhatti
Department of Medical Lab Technology, University Institute of Applied Health Sciences, Chandigarh University, Mohali, India

A. Khurana
Department of Veterinary Pharmacology and Toxicology, College of Veterinary Science (CVSc), Hyderabad, India

Centre for Biomedical Engineering (CBME), Indian Institute of Technology (IIT) Delhi, New Delhi, India

J. S. Bhatti
Department of Human Genetics and Molecular Medicine, School of Health Sciences, Central University of Punjab, Bathinda, India

S. S. Jawalekar
Department of Biotechnology, National Institute of Pharmaceutical Education and Research, S.A.S. Nagar, India



G. Gupta et al. (eds.), *Microbiome in Inflammatory Lung Diseases*,
https://doi.org/10.1007/978-981-16-8957-4_11

chapter, it is mostly focused on the reason behind the TB getting chance to infect the healthy lung tissue. It is also found that dysbiosis in gut microbiota, which directly affects the lung, plays a key role in giving TB a chance to hold its ground. It also highlights the new curative method which we can apply by correcting the gut microbial profile, which in turn corrects the lung microbial profile and rest of the function will done by body's own immune system. It is thus found that proper restoration of the microbial profile enhances the immune response and could restore the homeostasis.

**Keywords**

Tuberculosis · Microbiome · Gut–lung axis · Inflammatory cascade · Probiotics

## 11.1 Introduction

The appellation tuberculosis is associated with a disease that is a leading cause of mortality due to one microorganism caused by most commonly *Mycobacterium tuberculosis* (less commonly *Mycobacterium bovis*) [1, 2]. In history, it is found that in the Neolithic and Pre-Columbian periods, tuberculosis (TB) was prevalent; however, in the modern world after the Industrial Revolution, the crowing of people increased, favouring the spread of TB in society [3]. During the seventeenth and eighteenth centuries, one-fourth of adult death in Europe was caused by TB. Before the antimicrobial age, the only treatment for TB was to rest in an open area. The modern treatment of TB began in 1946 with streptomycin demonstrated efficacy against TB. Later in 1952, with the discovery of isoniazid and in 1970 with rifampin, TB becomes curable to most of the patients [3]. After 1992 the first problem of drug resistance was found in the case of TB, which cause the failure in the treatment. The other cause of the failure was nonadherence and nosocomial transmission. The incidence of infection due to *Mycobacterium tuberculosis* was approx. 10.4 million people, with a mortality of 1.7 million; among this, there are about 490,000 new multidrug-resistant cases of infection found in 2016 [4]. In 2019, according to the WHO, approx. 10 million new cases of TB were reported out of which 1.4 million patients died due to the infection. The main two reasons for spreading this disease are the low living conditions in developing countries and the spread of other diseases that cause the decrease of immunity (HIV) [4]. At the epilogue of the nineteenth century, the declining cases gave hope that the disease would be eradicated in the future. Still, the neoteric unfolding of multidrug resistance TB (MDR-TB) destroys that. The drug resistance can be either primary ,i.e., already resistant before the initiation of the treatment, or secondary ,i.e., emergence of resistance after initiating antituberculosis chemotherapy [5]. Infection wherein the strains resistant to rifampicin and isoniazid are called Multi-Drug Resistant (MDR)-TB and the infection wherein strains are resistant to isoniazid, rifampicin, fluoroquinolone, and second-line injectable drugs are called Extensively Drug-Resistant (XDR)-TB. Globally, the incidence of new cases is 3.3% and 20% of the cases of treatment-experienced TB cases are Multi-Drug Resistant [6]. Thus, the search of a new way is to prevent the disease in underway. One of the recent advancements was the establishment of a link between the gut flora with the probability of getting mycobacterium infection. Hence, the authors of present chapter aim to discuss how gut as well as lung microbiota affects the progression of TB, as well as their association with host vulnerability to *M. tuberculosis* infection.

## 11.2 Tuberculosis and Its Pathophysiology

Pulmonary TB is airborne as well as contagious caused by *Mycobacterium tuberculosis* and pulmonary tuberculosis patients are the most common infection source. Generally, infection is caused by inhaling airborne droplet nuclei, 1–5 μm in dia. Patients having *M. tuberculosis* through cough exhale some fine particles that are suspended in the form of aerosols for a prolonged time (minutes to hours) because of the small size of the droplet nuclei. It can transmit from one person to another through inhaling the droplet nuclei that contains *M. tuberculosis* and crosses the nasal passages, respiratory tract, bronchioles and reaches the lungs alveoli. The chances of infection are governed by several criterion, including the virulence of the source case, the proximity, the bacillary load inhaled, and the immunity of the potential host [7–9].

### 11.2.1 Microbiology

Globally, TB illness is caused by the ongoing transmission of *M. tuberculosis* infection (Firdessa et al.) and latent tuberculosis infection (LTBI) reactivation [10]. The causative organism for most TB infections is *M. tuberculosis, M. africanum,* or (sensu stricto), a closely linked organism; a small number of people are caused by zoonotic components of the *M. tuberculosis* complex *M. caprae* or *bovis* [11]. An ecological reservoir for *M. tuberculosis* is not known exist except humans [12]. Hence, *M. tuberculosis* is a pathogen and a symbiont, that has ramifications for our knowledge of the relationship between the host and the pathogen.

### 11.2.2 Interaction of Host–pathogen

Genomics has revealed significant variable genotypes among isolates worldwide , i.e., hundreds of single-nucleotide polymorphisms across a genome of 4.4 million bp, that could reflect clustered genetic drift linked to human migration patterns or variable pathogenicity of different lineages [13]. Based on epidemiological investigations, the existence of hypervirulent strains was postulated. If this is the case, genomic analysis of these strains could reveal lineage-specific virulence factors [14], which are perhaps used to prioritize patient care and decisions for infection control. There exists a complicated communication between *M. tuberculosis* and the host. In absence of the host determining susceptibility, investigating *M. tuberculosis* virulence factors can mask cooperative interactions. For example, a particular host–pathogen interaction could describe why East-Asian lineage strains are highly morbific in Asian people [15] but have a standard clinical and epidemiological demonstration in Canada [16] and Switzerland [17]. In the right epidemiological and social context, strains that are not remarkable in genetic and laboratory characterization can be linked to outbreaks [18].

### 11.2.3 Virulence

The chances of the succession of active tuberculosis from latent tuberculosis are much higher than the live vaccination strains: *M. Bovis*, Bacillus Calmette–Guérin (BCG), it follows that chromosomal variations are the basis of reduced virulence that can be found using *M. Tuberculosis* and BCG [19]. Indeed, genetic comparisons revealed various changes, the most notable of which is the region of difference 1 (RD1) which may explain the reason for the vaccine being given to neonates each year with a very low risk of progression of illness (Behr et al. Lewis et al. Mahairas et al.). Categorization of virulence determinants is based on molecular characteristics and cellular localization. The catagories are (a) metabolism of lipids FAs, comprising of cholesterol catabolism, (b) proteins of cell envelope: containing lipoproteins, secretion systems, and cell-wall proteins, (c) proteins that suppress macrophage antimicrobial effects, such as those involved in oxidative and nitrosative stress responses, phagosome arrest, and apoptosis inhibition by protein kinases, (d) proteases, with metalloproteases, (e) metal-transporter proteins, partition into importer & exporters, (f) 2-component systems, sigma factors, as well as other transcriptional regulators are among the gene expression regulators, (g) PE and PGRS glycine rich families of proteins with unclear functions [20].

When *M. tuberculosis* is exposed, one of two things happens: the pathogen is either eliminated or persists. In first situation, the pathogen is either annihilated by innate immune responses in which tuberculin skin tests (TSTs), interferon-γ release assays (IGRAs) may be −ve, whereas in the adaptive immune response, TSTs as well as IGRAs may show +ve or −ve, which depends on whether the priming of memory T lymphocyte expression was completed [21, 22]. Here, the patients will not benefit from LTBI therapy, regardless of how the infection is eradicated. It has long been known that over half of those who are exposed to TB patients' close household connections have negative results [23]. One possible theory which explains for why innately resistant to tuberculosis peoples are found is discovering that there is a genetic predisposition to remain TST −ve in spite of having an extensive exposure [24].

### 11.2.4 Immunology and Granulomas Formation

There is limited evidence of the premature *M. tuberculosis* infection in humans, and research in small animals (like guinea pigs and mice) and other primates which have greatly aided in identifying the significance of early issues in the course of primary infection [25]. *M. tuberculosis* enters the body through the upper respiratory tract; after inhalation, droplets of M. tuberculosis travel through the lower respiratory tract, infecting alveolar macrophages, the most common cell type. The bacteria are internalized by these cells through receptor-mediated phagocytosis, which involves a variety of receptors. This process had been investigated for a long time without considering into account the microenvironment in the alveolus. Surfactants, plentiful in the fluid that coats the epithelium, may have a vital role to play in the early interaction between host and the pathogen [26]. For instance, surfactant protein D can inhibit *M. tuberculosis* from being phagocytosed by alveolar macrophages [27]

Infected and stimulated macrophages transformed into multinucleated cells, epithelioid cells as well as foamy macrophages loaded with droplets of lipids, and neutrophils fabricate the human TB granulomas (Fig. 11.1) [28]. The internal accretion of cells gets enclosed by primarily cluster of differentiation cells-4 T lymphocytes but also Cluster of Differentiation-8 T lymphocytes. B-cells, lymphocytes, fibroblasts surround the inner concentration of cells, forming a peripheral fibrotic capsule [29]. Still, in the granuloma, T cells have limited cell function [30]. In addition to adhesion molecules, other proinflammatory and inhibitory cytokines and chemokines play essential roles in forming granulomas [29]. Further, Ulrichs et al. collected the lungs samples from the MDR-TB patients and observed that the granulomas formation with central necrosis required a minimum size of 0.1 $mm^3$ and demonstrated occurrence of lymphoid follicle-like structures in the peripheral space of the granulomas. The lymphoid follicle like structure is primarily formed from the B-cells, a smatter of CD4 and CD8 T lymphocytes and contiguous

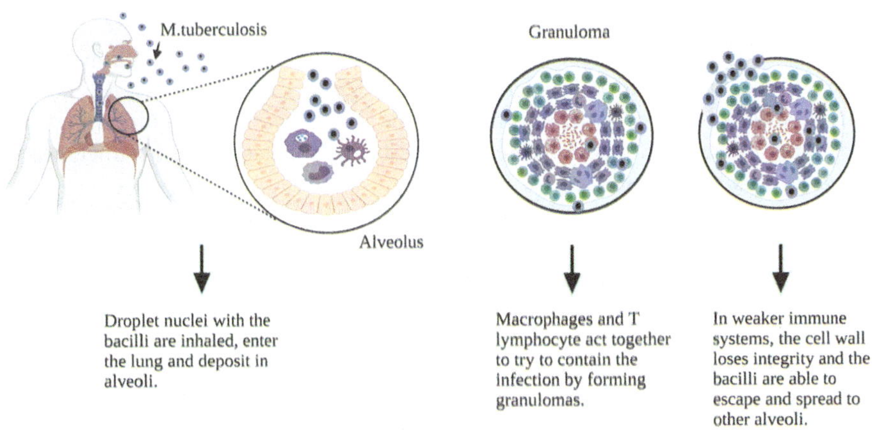


**Fig. 11.1** Pathophysiology of tuberculosis and formation of granuloma

infected macrophages [31]. Mycobacteria can survive within granulomas, in the periphery of the granulomas and even further afield in apparently normal healthy parenchymal tissues [31, 32].

### 11.2.5 Apoptosis of Macrophage as a Defence Mechanism for *M. tuberculosis* Infection

Necrotic death might induce by the infection of macrophages with *M. Tuberculosis*, which is described by cell lysis, allowing the bacilli to depart macrophages and spread from cell to cell. Infection can also cause death of macrophages by apoptosis via an intact plasma membrane [33], which is linked to decreased pathogen viability and increased immunity [33]. Apoptosis was initially identified as an antimycobacterial action in alveolar macrophages, in which attenuated mycobacterial strains, including *M. tuberculosis* H37Ra, showed decreased viability when their host macrophages undergo apoptosis [34]. Further study shows bystander macrophages' contact-dependent death after infection with *M. tuberculosis* H37Ra, implying yet another mechanism through which the host limits bacterial propagation [35]. Virulent *M. tuberculosis* strains cause low macrophage apoptosis and proliferate intracellularly and gradually in these cells [35]. The fact that *M. tuberculosis* prevents apoptosis which is the mechanism of virulence was already demonstrated in vivo by the diminution of proapoptotic *M. tuberculosis* secA2 and nuoG deletion mutants on the infection [33, 36, 37].

SecA2 system in *M. tuberculosis* plays an important role in pathogenesis by adjusting host innate immune defences. SecA2 infected macrophages generate more significant conc. of the pro-inflammatory cytokines-TNF-α and IL-6 as well as high levels of RNI. These immunomodulatory components are activated by *M. tuberculosis* and have a role in regulating *M. tuberculosis* during host infection [38–40]. Apoptosis is also increased in macrophages infected with the secA2 mutant, attributable to the mutant's deficient SodA secretion [36]. Reducing the production of mycobacterial superoxide dismutase, inactivation of the secA2 gene, which encrypts a constituent of a virulence-associated protein secretion system, increased death in infected macrophages. In vivo, deletion of secA2 boosted antigen-specific CD8+ T cell priming, while immunization of guinea pigs and mice with a secA2 mutant increased CD4+ cell reactivity and tolerance to *M. tuberculosis* infection [36].

## 11.3 Microbiome and Lung Health

In acknowledgment of the fundamental role, it plays an important part in human health and disease. The term 'microbiome' now covers every creature (not just bacteria) and every method of accessing its DNA (metagenomics), metabolites, RNA species, and proteins [41]. The US NIH launched the HMP in 2007, resulting in a surge in human microbiome research [42]. The gut microbiota includes the

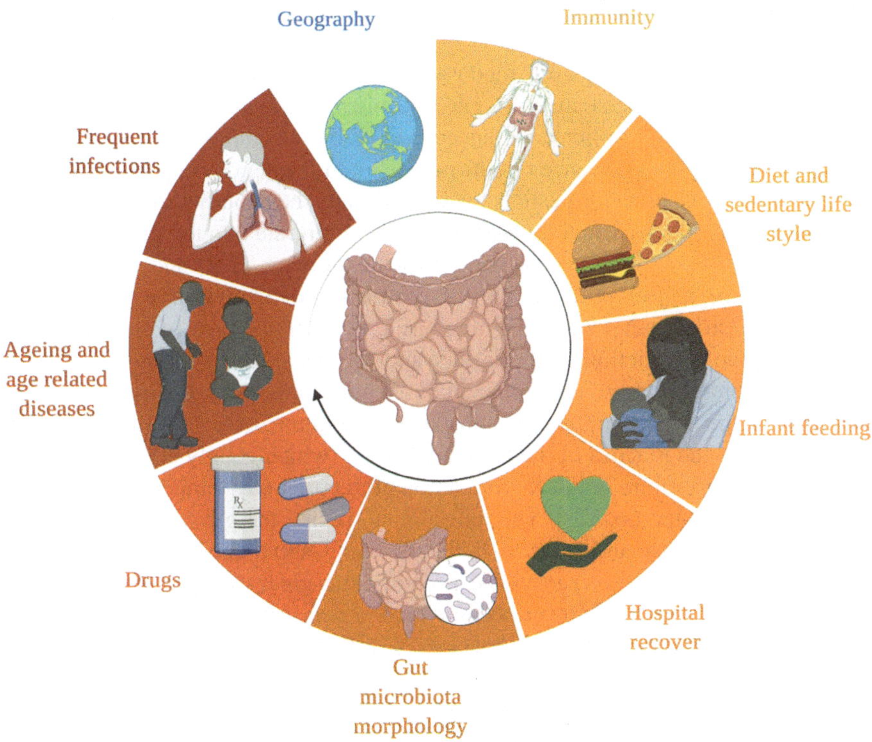


**Fig. 11.2** Major factors influencing gut microbiome development: Factors like age, diet, and antibiotics treatment influence the gut microbial profile. A high fibre diet causes an increase in the versatility and richness of the gut microbiome. On the other hand, broad-spectrum antibiotics kill the commensal bacteria, thus disrupting the species' balance present in the gut microbiome

various microbial communities that reside in the host's GIT [43]. The gut microbiome comprises up to 100 trillion bacteria that live in a nutrient-rich environment in the GI tract [44]. The gut microbiota substantially affects human physiology and nutrition, which is necessary for human survival [45]. Eating habits, age, medications like antibiotics, lifestyle choices, and the delivery method at birth are all significant factors of the gut microbiome's composition (Fig. 11.2) [46]. 16S rRNA gene and shotgun metagenomic analysis are two techniques for identifying microbial sequences that have provided new-fangled perception into the range of microorganisms found in infected and healthy guts [47]. The gut microbiome is consistent within individuals and is mostly shared among healthy peoples [48, 49]. Ailment and abnormalities, including interstitial lung diseases (ILDs), are linked to microbial imbalance or dysbacteriosis in the gut flora [50, 51].

### 11.3.1 Mucosal Biology

Although the gut and lungs are mucosa-lined luminal organs having a common embryological origin, their macroscopic and micro-anatomical features are highly different, resulting in significant variances in microbiota composition and population dynamics. Through the absence of anti-peristaltic movement or oesophageal reflux, microbe migration in the digestive tract is through the mouth to the anus and is serially blocked by a variety of impediments. To immigrate into the cecum, an orally introduced microbe must survive the pH of the stomach (2.0) and the pH of the duodenum (8.0), as well as a struggle for resources with a highly populated resident microbiome. On the other hand, the lung has a bidirectional flow of air, mucus, and germs, with no obstruction in the larynx and the distal alveolus. As a result, the lungs' microbiome is more influential and transitory to the microbiome of the lower GIT. While the GIT maintains a constant temperature of 37 °C throughout its 9-metre length, the mucosa of the respiratory tract (which is only a ½ metre length) represents a temperature gradient from ambient to core body temperature inside the alveoli during inhalation. [52]. The lung milieu, unlike the gut, is much oxygenic. However, the trachea and bronchi both, the same about the gut, are lined with produced mucus' highly glycosylated proteins, the great most of the lung's total surface area is covered with a surfactant which is rich in lipid, having bacteriostatic effects against a different strain of bacteria [52]. Because bacterial concentration in the airways is modest in contrast to that of the duodenum [53]. Finally, the nature of host–bacterial interactions vary in the gut and lungs. Luminal IgA levels are much greater in the gut, and bacterial–host leukocyte interactions outside the blood stream are much more common in the lungs (alveolar macrophages). As a result of these vastly different environmental conditions, microbial populations are also vastly different.

### 11.3.2 Lung Microbiome Determination

The balance of three parameters determines the makeup of the lung microbiome [54]: (1) microbial relocation into the airways, (2) microbe removal from the airways, (3) the proportional reproduction frequencies of its members of the community, as dictated by regional growth nutrient availability. Alteration to the microbiome in a particular person or over-patient population must be related to a change in these variables. Inhalation of air which includes 104 to 106 bacteria/mm$^3$ even before contacting the bacterial-compacted upper airways [55], subclinical micro aspiration of upper respiratory tract contents, and direct dissemination along airway mucosa are also drivers of microbial immigration. Mucociliary clearance, cough which is common even among healthy people and host immunological systems all contribute to microbial elimination (both innate and adaptive). The copiousness and activation state of host inflammatory cells are among the environmental factors that affect regional growth circumstances in the lungs and those that are universal to all environmental niches [56],

### 11.3.3 The Primary Source of Bacterial Microbiota in the Lungs: The Oral Microbiome

Since then, various culture-independent investigations have established that the lungs' microbiome matches to the oropharynx more consistently than competing source communities such as breathed air, the nasopharynx, or the lower GIT through hematogenous transmit [57–59]. According to direct research in people and a heavily populated-based model, the intranasal microbiome adds small to lung colonies in health [60]. The nasal microbiome is more similar to the skin than to the lungs. This resemblance amid lung and oral microbiota is visible even though the lung is tested by a nasally injected bronchoscope, suggesting that upper respiratory tract infection has a minor impact on biomicroscopically obtained specimens [54, 61]. The oropharynx generates two litres of saliva daily, significantly more than the nasal mucosa releases in normal circumstances.

### 11.3.4 Alteration in the Lung Microbiome During Disease

Throughout the lung diseases, the ecological factors of the lung microbiome—immigration, removal, and regional growing environments alter drastically [54]. As a result, in disease conditions, the lung microbiome's communal membership changes. Almost all of the hundreds of investigations comparing the microbiota of sick lungs to that of healthy people reported significant changes in community composition [62]. The significance of interference in the gut microbiome and the lungs, known as the gut–lung axis, is highlighted by epidemiological and experimental findings. According to the gut–lung axis, variations in gut microbial populations can significantly impact respiratory illness. Changes in immunological responses, airway homeostasis, and inflammatory situations in the GI tract are all associated with dysbiosis in the gut microbe population [63]. The incidence of recurrent exacerbations in bronchiectasis [64], fatality in idiopathic pulmonary fibrosis [65], and response to corticosteroids and antibacterial drugs in asthma [66] have all been linked to changes in lung microbiota at the start of the disease.

### 11.3.5 Pulmonary Health Depends on the Gut Microbiome

In formative years, the respiratory microbiome builds in tandem with the gut microbiome [67]. As defined by observational and clinical data, the gut–lung axis is a critical communication amid the gut microbiota and lungs. Current findings have connected dysbiosis in the gut microbiota to variations in the immune system and pulmonary homeostasis response in various lung diseases, including asthma, ILD, and pneumonia. Explains the ways through which the gut has been displayed to influences of lung morphology.

#### 11.3.5.1 Gut–Lung Axis

Our body consists of symbiotic and commensal microbiota. Traditionally it was not known the exact role of this microbial colony in the homeostasis, but recent studies and complex sequencing analysis led us to the more complex profiling of symbiotic microbiota diversity and its diverse prevalence (i.e., skin, mouth, respiratory, intestinal, and urinary tract) in our body. At first, it was known that the gut microbiota helps in digestion and metabolism and they provide some essential biomolecules that contribute to homeostasis. Later improved understanding unfolded the interconnection between host and microbiota in different diseases [68, 69]. The majority of the microbial abundance is in the gut. The gut microbiota, through the intercommunicated axis, influences the microbial diversity and prevalence of other organ systems of the body. The gut microbiota acts as the central command for microbial symbiotic maintenance [70].

Here we have discussed the overall dynamic mechanism of diet and species of microbiome available in the gut and how it influences pulmonary microbiota development, which affects the lung immunity and development of respiratory infectious diseases. The potential of therapeutics for diseases like the infection of *M. tuberculosis* with the help of the human microbiome is another vital point to be discussed. The dysbiosis of gut microbiota due to different factors like diet, lifestyle causes the change of the human immune response as the immune response perfectly orchestrates with the microbiome in gut and also causes the alteration in microbial profile in different organ which also, in turn, causes the immune response in the respective organ. Thus, the gut microbial profile, the gut–lung axis, and lung microbiota have significant role in lung immune response and development of tuberculosis [71].

### 11.3.6 Dynamics of Diet, Gut, and Lung Microbiota

Dynamics of gut and lung microbiota can be briefly explained into five key concepts, Firstly, the microbial profile that is in a symbiotic relationship with the human body has co-evolved and drawn from a relatively restricted level of phyla [72–74]. There is an astounding similarity between the bacterial species at different compartments of the body. The higher level of taxonomic classification describes the different body site-specific complexity of microbial profile [75, 76]. Secondly, the complete mapping of the microbiological profile is yet to be discovered both at organism and the genetic level [77]. Third, the body sites such as healthy lungs are previously considered sterile, but it unfolded its variable but non-transient microbial profile [78], which helps in maintaining the immunologic as well as tissue homeostasis and also has a defensive role against the pathogens [79]. Fourth, with the advancement in the technology and also in the medical sciences how the food technology derived food, packed food, frozen food and also, with the use of different antibiotics against the pathogenic organisms sometimes unnecessary; creates a selective pressure on the commensal microorganisms thus changing their diversity, richness, and constituent profile. This diminished diversity directly links with the increasing development of the chronic and susceptibility of the pathogenic infection [80, 81]. Although it is still unclear that the exact mechanism by which the symbionts exert their affect that dysbiosis is the hallmark of many diseases [82, 83]. Lastly, there is an orchestral system between the gut microbiota, lung microbiota, and local and systemic immunity. Prime example of this is the interconnected immune response in lung pathogenesis with the change in the gut microbial profile [84]. A well-balanced diet that includes dietary fibres is linked to a more diverse gut microbiota. [85]. The metabolic product of these microbiome profiles influences the host's health [86, 87]. The research by different scientist [86, 88, 89] shows the link between the gut dysbiosis and malnutrition which further prove the theory. Due to dysbiosis on the gut microbiota apart from the intestinal disorders like IBD, DM, Ca colon and diseases of lungs like Asthma, COPD, TB are also impacted [90].

The role of diet in the gut microbial profile is studied at the various stages of life [91]. The initial colonization that took place at the time of birth depends on the mode of delivery, antibiotic exposure, and surrounding environment. It also keeps changing in different stages of life [92]. The taxonomic studies show that in breast feed babies, the profile consists of higher Bifidobacteria, Lactobacilli, Staphylococci, and Streptococci, which considerably differs from the formulae fed infants (higher Bacteroides, Clostridia, and Proteobacteria) [93]. The infant microbial profile stabilizes after introducing the solid food; after that, it changes dynamically with the dietary intake [93].

The younger adults show a marked difference in their microbial profile depending on their food habits and lifestyle [94]. The one with low fat and high plant product diet (Fig. 11.3) has a significantly higher quantity of Bacteroidetes than one with a western diet (High fat and sugar) [95]. Children from rural Africa have a diet of plant polysaccharide/protein and fibre and less animal protein than European children with a diet of protein, sugar, fat and low fibre (Figs. 11.4 and 11.5). The African children have a higher level of Actinobacteria and Bacteroidetes, whereas the European children have higher constituents of Firmicutes and Proteobacteria [95].

The short-chain fatty acids (SCFA) produced by the gut microbiota promote recruitment and mature the immune cells responsible for the anti-inflammatory mechanism [95]. The produced SCFA gets distributed systematically and generally utilized for energy production or signaling molecule (Figs. 11.3 and 11.4). Increased SCFA production reduces gut pH, in turn reducing pathogenic species like *Escherichia coli* and Enterobacteriaceae [96]. The African population shows a higher prevalence of bacterial species like Xylanibacter, Prevotella, Butyrivibrio, and Treponema that can digest plant products to yield SCFA [95]. In the later stages of life, other factors like immune strength, gut morphology, hospitalization, and medication affect the gut microbial profile [97]. The age and diversity of gut microbiota are inversely related though it varies with environmental and geographical factors. However, it is still not known that these changes with age are either causes of ageing or effect of ageing. Apart from the change in diversity, the metabolic strength of the microbiome and rate and amount of SCFA production also decrease [98, 99].

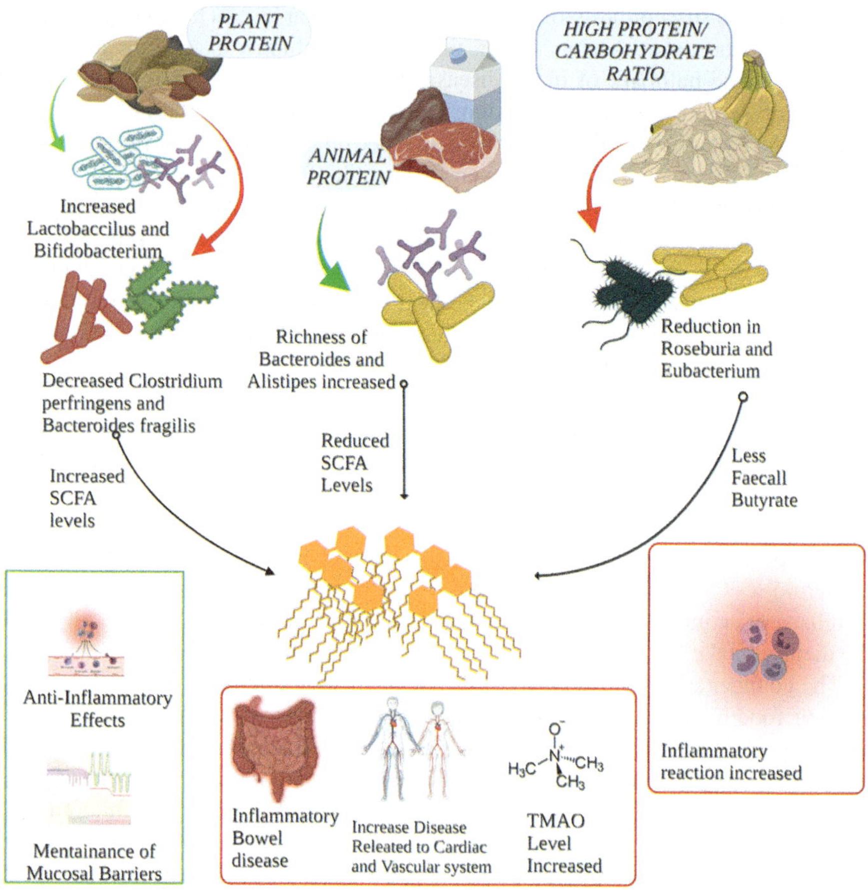


**Fig. 11.3** Influence of Dietary Component on Gut Microbiota. A person with a diet rich in plant protein shows an increase in Lactobacillus and Bifidobacterium levels and also shows a decrease in *Clostridium perfringens* and *Bacteroides fragilis*. Further, a diet rich in animal protein can increase the abundance of Bacteroides and Alistipesis. The diet with a high protein and carbohydrate ratio reduced Roseburia and Eubacterium. All three types of diet cause the increase in SCFA level (Plant protein), decrease in SCFA Level (animal protein), and Less Faecal Butyrate (high protein:carbohydrate ratio). The increased SCFA, in turn, shows the anti-inflammatory effect and helps in the maintenance of the mucosal barrier. The decrease in SCFA and faecal butyrate levels may cause the increase in the occurrence of the inflammatory bowel disease and diseases related to the cardiac and vascular system, and the body level of Trimethylamine N-oxide (TMAO) is increased.

Further, it is reported that fungi are also a part of the commensal microbial profile [100, 101]. The fungi, 100 times bigger than the bacteria, although the fungal sequence is 100 or 1000 times less frequent than the bacterial sequence; still the role of fungi cannot be ignored. The fungal genera which already been found are far less than the bacterial genera—Saccharomyces cerevisiae, *Malassezia restricta*, and *Candida albicans* [102]. The gut acts as an ecosystem where the inter-kingdom

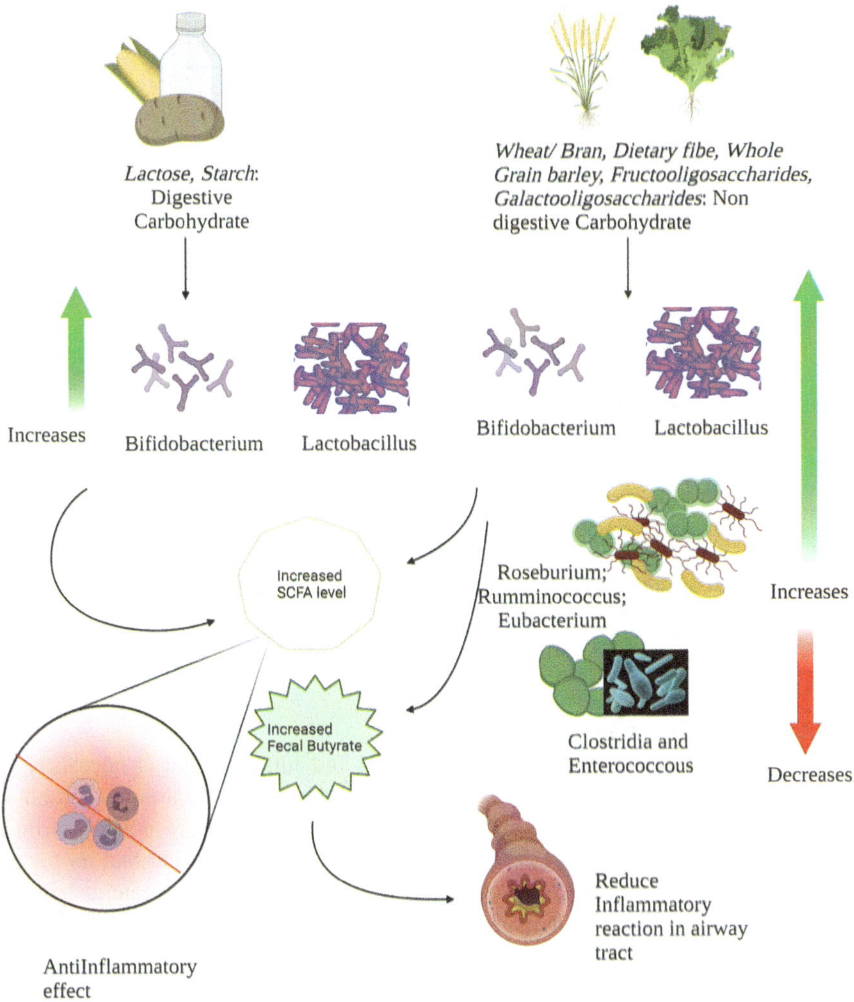


**Fig. 11.4** Effect of dietary carbohydrate on gut microbiota: Carbohydrates like lactose, starch, digestive carbohydrate, high fibre diet cause an increase in abundance of the Bifidobacterium, Lactobacillus which in turn increased the level of SCFA, thus demonstrating the anti-inflammatory effect

interaction occurs [103, 104]. Yeast Saccharomyces boulardii and *C. albicans* and some other fungal wall component β-glucans are inhibitors of different pathogenic microbial development [105, 106]. *S. boulardii* also produces proteases and phosphates, which acts as an inhibitor of toxin produced by *Clostridium difficile* and *Escherichia coli* [107, 108]. In addition, in the case of gut dysbiosis, like after antibiotic treatment, the fungal species may take over the role of bacteria in immunomodulation, thus prevents the damage to the mucosal tissue [109].

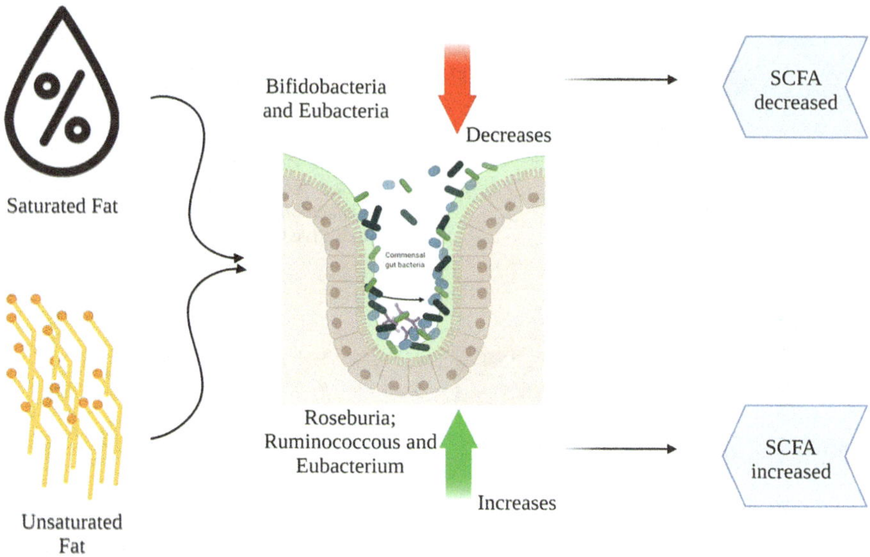


**Fig. 11.5** Effects of dietary lipids on gut microbiota: Saturated fat decreases the Bifidobacteria and Eubacteria levels; thus, SCFA concentration is reduced, giving rise to an inflammatory reaction, whereas the unsaturated fats increase the level of Roseburia, Ruminococcus, and Eubacterium, which increases the concentration of SCFA

#### 11.3.6.1 The Pulmonary Microbiota

There is a clear cross-relation between the gut and lungs, and it is necessary for training the host immune cell [110]. Several independent techniques based on amplification and analysis of 16s-rRNA show that the lungs are not as sterile as initially thought [111]. The finding that every lung has a unique microbial profile and if studied in a very subtle way it may open the scope of research in categorizing the microbial profile of healthy as well as diseased lungs and also the mechanism by which the microbial profile regulates the local as well as systemic immune response [110]. The biomass of the pulmonary microbiota is substantially lower than that of the gut microbiome [112]. The microbiotic profile depends on the colonization from oropharynges and upper respiratory tract through salivary microinhalation and host ability to eliminate (tussion and sneezing) [111]. The microbial profile of lungs is not always constant; they keep changing with the different physiological and pathological conditions, just like the changes in gut microbial profile due to various factors. The URT is mainly colonized by aerobic bacteria [113], and some anaerobes are isolated from 50% of the sputum sample [114]. The most common phyla available are Bacteroidetes, Firmicutes, Proteobacteria, Prevotella, Veillonella, Pseudomonas, Fusobacteria, and Streptococcus found by analysing the bronchoalveolar lavage of healthy adults by culture-independent technique [115].

To find the exact source of lung microbiota, culture-independent RNA/DNA sequencing methods or microarrays were used. The main challenge of this method of microbiota analysis is the low biomass of lung microbiota than the gut microbiota

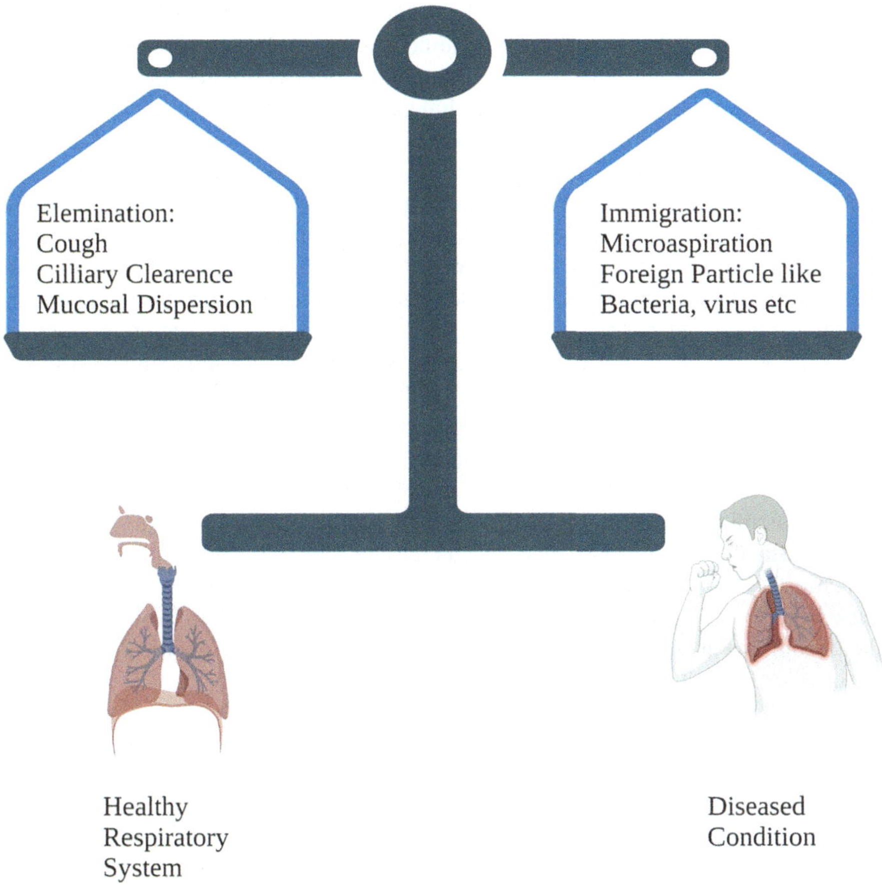


**Fig. 11.6** The First Dickson's Model: The Lung Homeostasis: Lung homeostasis is dependent on dynamics of immigration and elimination of the materials from the lung. The elimination of the foreign materials through cough, ciliary clearance, mucosal dispersion contributes to a healthy lung

and there is always a chance of contamination from the mouth [116]. By analysing the bronchoalveolar lavage fluid, it is found that the respiratory tract consists of a profile of microbiota that decreases in biomass from the URT to LRT. They also closely resemble the nose and mouth microbial profile [117]. It is also seen that the upper respiratory tract's microbial profile resembles that of the stomach [113].

Further analysis also reveals that the predominant phyla in microbial lung profile are similar to the gut microbial profile, mainly the Firmicutes as well as Bacteroidetes are proceeded by Proteobacteria and Actinobacteria [117]. Microbial lung profile also includes the fungal species, which are mainly environmental [100, 101]. Now, according to all the proofs available, it can be concluded that the lung microbial profile is dependent on both the oral and gut microbial profile.

There are three models proposed by Dickson et al. (Fig. 11.6) [118] to explain the contribution of microbes in maintaining homeostasis and also the loss in homeostatic profile in case of the disease condition. The first model postulates that the community of species that form the microbial lung profile is dependent on mainly three factors, viz. migration from one anatomical compartment to a different one, and it is at least in part attributable to microaspiration [118], elimination by natural reflexes like cough, innate and adaptive immune response and rate of reproduction of the microbes which in turn depends on the factors like, pH, blood perfusion, oxygen tension, temperature, alveolar ventilation, and the concentration and activation of host inflammatory cells. The second model suggests that the nutritional availability and other nutritional factors are crucial to reproducing the bacterial community in lung microbiota. The airway tract mainly consists of air; thus, the nutritional availability is limited; this supports the finding that the available biomass in the lung microbial profile is less than the gut or mouth [118]. As the nutritional factors are limited thus the local reproduction of the bacterial species is also limited, which again supports the fact that the lung microbial profile resembles that of the gut and the oral microbial profile. This is the more or less scenario of a lung of a healthy subject. But the resembling situation changes in any lung infection like chronic bronchitis, cystic fibrosis, asthma, and bronchiectasis as these diseases supply the dense high protein-rich growth medium of secreted mucous [119]. In some diseases like pneumonia and acute respiratory distress syndrome, the alveoli are flooded with a protein-rich medium by oedema, which causes the overgrowth of the microbes present in the lungs.

The final or the third model suggests the mechanism of signaling stress response in which tissues and cells reciprocally communicate the stress signal of the internal environment. Various signaling molecules play like the hormones (glucocorticoids, oestrogens, and androgens), neurotransmitters (e.g., catecholamines and endogenous opioids), and cytokines (e.g., TNFs, IL-1, IL-6, and IL-8) [120, 121]. Further study shows that some lung commensal microbes may adopt the signaling molecule that human cells used to communicate [118].

## 11.4 The Orchestra of Gut Microbiota, Pulmonary Microbiota, and Host Immunity: Toll-like Receptor Signaling

Gut-dysbiosis is correlated with pulmonary disorders and infection; different soluble microbial components and other microbial metabolites transported via the circulatory system play roles in the protection against lung inflammation or infections [122]. The molecules which are of main importance are peptidoglycans and lipopolysaccharides (LPS) [122]. It is found that the depletion in genus Bifidobacteria and increase of genus Clostridia in gut microbial profile cause asthma-like problems in early life [123]. In an experimental model of mice depleted of gut microbiota, if LPS is administered intrarectally, the animal shows a marked increase in the TH2 mediated response, suggesting that the gut microbial-derived LPS greatly influences the lung tissue's reaction towards the allergens (Fig. 11.7)

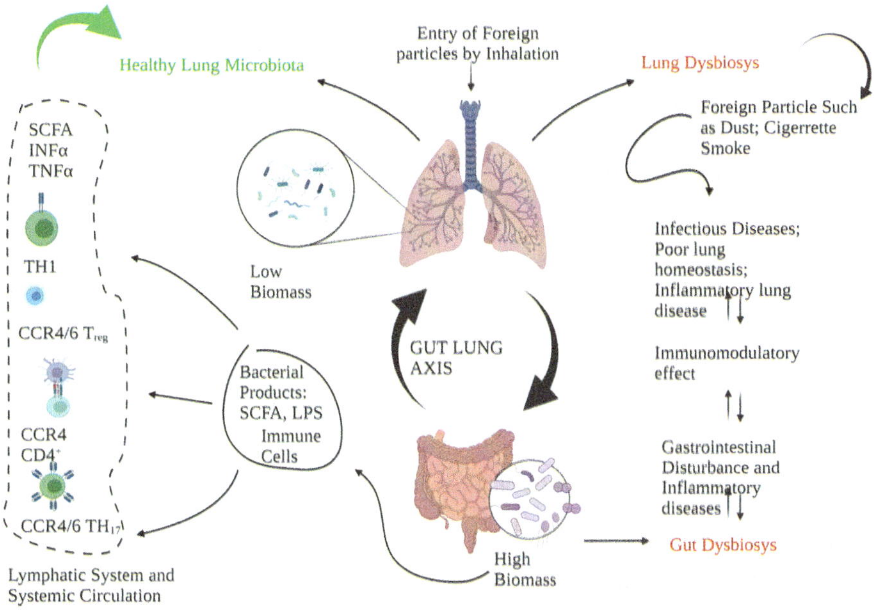


**Fig. 11.7** The Gut–Lung Axis: The Gut–Lung Axis is the prime mechanism of homeostasis of the healthy lung. The gut microbiota (generally high biomass) produces SCFA, LPS, which activates the immune cells. These immune cells keep check on the lung microbial profile. On the other hand, reduction in gut microbial richness results in gut dysbiosis, which cause the rise in inflammatory reactions, less immunomodulatory effect, thus increases in the chance to develop infections

[124]. This interrelation is bidirectional, viz. infection of lung tissue caused by *Pseudomonas aeruginosa* leads to blocking the M phase cycle and blocking the proliferation of the gut epithelial tissue [124].

The gut microbiota helps assimilate the materials that the human digestive system cannot digest, and the metabolites help in immunomodulation [125]. The SCFA (acetate, propionate, and butyrate) recruit and maturate the immune cells, thus protecting against the inflammatory response [126]. The gastric dendritic cell processes the antigens, promoting the proliferation and expansion of T cells, which then travels through the lymphatic system to the site where the antigen exposure occurs. Dysbiosis in the gut microbiota thus causes the impaired proliferation and maturation of the T cell component, thus impairing the overall systemic immune response (Fig. 11.7) [127]. The important link between the gut and the lung microbiota is the migration of immune cells from the gut to the lungs by the mucosal immune system [128]. The immune cells migrate and colonize at the mucosal sites forming the mucosa-associated lymphoid tissue, gut-associated lymphoid tissue, and nasopharyngeal tissue. The circulation occurs through the lymphatic system [110].

### 11.4.1 Gut Microbiota and Local Immunity

An extensive study has been done on potential of gut microbiota in modulating the local immune response [129]. It interacts with the mucosal immune system with pro-inflammatory and regulatory signals [130] and also interacts with neutrophils and modulates them [131]. The toll-like receptors presenting immune cells communicate with the gut microbiota and the GPR41/43 and communicate with the gut microbiomes. The link of interaction between the TLR, GPR41/43 with the gut microbiome is either MAMP or the SCFAs [132]. However, findings also discover that the bacteriobiota and the fungal symbiotic species can produce the SCFAs [133, 134]. This proves the facts that the fungi in the gut can also alter the immune system same as the bacteria and can to some extent take over each other role if somehow one is reduced due to antibiotics or antifungals or other causes [109, 135].

### 11.4.2 Lung Microbiota and Local Immunity

The study reveals that the concept of the lung microbiota and its local immunomodulatory effect is emerging through different studies and their outcomes [118]. The initial effects are the colonization of the host commensal microbiome in the lung tissue which matures the local immune cells and provides good competition to the incoming foreign bodies [136]. In a study on germ-free mice, it was observed that they exhibit an increased level of local TH2 associated CK and IgE production, which are the mediators of allergic airway inflammation [137]. Thus, the depletion of the natural lung microbiome can increase respiratory allergic reactions or acute asthma. On the other hand, it was found that the microbial profile restoration or the exposure to the commensal microbiome reduces the TH2 mediated allergic reaction and better tolerance to the foreign particles, thus decreasing acute asthma-like responses in patients [87, 136]. The microbiomes and immune cells contribute to local resident memory B-cells [138].

However, the interaction and immunomodulatory potency are a two-way communication, with a major infection like *M. Tuberculosis,* there is depletion of the normal microbial profile, which removes the protective covering the commensals providing and also there is an inverse relationship, i.e., the depletion of the normal microbial flora may give the foreign microbes to take root. The exact mechanism by which this two-way signalling pathways are working is still unclear, and extensive study is needed [139].

### 11.4.3 Long Reach and Systemic Immune Modulation by Gut–lung Axis

The knowledge of local immune response due to site-specific microbiota extends to systemic immunomodulation and long-reaching immunomodulation with further research [140]. The most important pathway for redistribution and relocation and interexchange of the immunomodulatory molecules is the Mesenteric Lymphatic System [141]. Through this Mesenteric Lymphatic system, there is interexchange of the intact bacteria, bacterial fragments and their metabolites (SCFAs) [142] through the intestinal barrier, reach the systemic circulation and modulates the immune response in the lung as well as systematically [69]. The gut microbiota metabolizes the high fibre diet into SCFA, which functions as a signaling molecule on the lung's resident antigen-presenting cell to enervate inflammatory reactions and allergic responses [143]. A study where SCFA receptor-deficient mice are used shows the increased response to inflammation and allergens and shows the acute asthmatic-like reaction [141]. The fungal population also contributes to the production of SCFA [144]. *Aspergillus fumigatus* not only produce SCFA by its own [133] but also build up biofilms that increase the bacterial formation of SCFA [134] although the bacterial SCFA cast out fungal growth. The on influence on the health and homeostasis of the host due to the commensal fungal population is still very much unclear.

Another big-league player of this immunomodulatory game is gut SFBs. These SFBs are commensal bacteria, subjugate the ileum of many animals, including humans, which account for the maturation, development, and modulation of the immune system [145]. SFBs synchronize the CD4-cell polarization into the Th17 pathway [146], which responds to the lungs' fungal infection and exemplifies autoimmune lung condition [69]. Recently it was found that concerning the inflammatory signal and in the incident of tissue damage from the lung through IL-25, the gut call up the innate lymphoid cells presupposes for the repairment of the lung tissue [147]. The activation of the toll-like receptor in the intestine activates the NF-κB dependent inflammatory response pathway accompanied with increased influenza-like response in mice [148].

Mechanisms that interconnect the lung, systemic immunity with the gut microbial profile are manifested by the surge in number of mononuclear leukocytes and phagocytic and lytic activity (results shown when treated with Bifidobacterium lactis HN019 probiotics) [149]. Diet rich in plant fibre gives the gut microbes and the probiotics fuel to metabolize and produce the SCFAs [70, 150], which alter the pulmonary immune response and thus has enormous control on the disease progression of the respiratory system [151]. The gut–lung axis and its connection to systemic and organ system-specific immunomodulation are two-way signaling. For example, the salmonella-specific gut immunization is achieved by the salmonella's nasal inoculation, which relies on the lung dendritic cells [152].

Altogether, the gut–lung axis is a synchronized system of a complex interaction between different species of bacteria and fungi that domain the various organ systems and redistribute and exchange between each other through the lymphatic and circulatory system. This complex interaction contributes to systemic and organ system-specific immunomodulation. Thus, it can be concluded that gut microbes can strongly influence lung diseases.

## 11.5 Tuberculosis Infection and Gut Microbiome

In animal studies of the gut microbiota with active TB, the study of the faecal microbial profile of the active TB patients and for cases with recurrent TB [153, 154] consistently shows low prevalence of bacterial diversity and also shows the lower count of bacteria [155, 156].

The animal studies were done on female BALB/c mice infected with aerosolized CDC1551 strain of *M. Tuberculosis* [156]. However, detailed analysis shows that there is overexpression of the Betaproteobacteria [157], under expression of Bacteroidia [157], Lachnospiraceae [156], and Ruminococcaceae [156] in the mice models studied. Another experiment was performed with a distinct strain of *M. Tuberculosis* (H37Rv; $N = 5$) and compared with a 1:1 age-matched control [156]. Faecal samples from the post-infected samples show differential clustering of bacterial prevalence and richness compared to the uninfected models (among 73 OTUs; $q < 0.01$) [156]. A study in patients with latent TB and Heliobacter pylori develops more interferon-gamma and TH1 like cytokines in contrast to the subject without pylori infection. The vaccination against TB like Bacilli Calmette–Guérin Vaccine in infants stool sample gives an increased level of Actinobacteria, which are also analogous to increase in T cell responses to vaccination [158]. Furthermore, in a study on cynomolgus macaques with active *H. pylori* infection, they are less plausible to develop an active TB infection [159]. Another study also revealed that specific enteric bacteria like *H. pylori* cause the immunomodulation systematically, thus altering *M. tuberculosis* susceptibility [160]. This may be illustrated by the fact that the Pylori causes the remodelling of the innate immune system, thus decreasing the chance for the other pathogen to get a chance to infect [160]. The phylogenic study data that can be summarized from the studies of the faecal microbial profile and the gut microbial profile reveals the fact that there are changes in the relative abundance of the bacteria of family Ruminococcaceae, Lachnospiraceae, Clostridiaceae, Streptococcaceae, Enterococcaceae, Acidaminococcaceae (only Gram-negative in firmicutes phyla) [154]. These bacterial family of phyla Firmicutes represent the most abundant Gram-positive bacteria in the human colon [161], which are directly related with maintaining the count of CD4+ cell in the peripheral blood in the newly recognized TB patients and inversely associated with the recurrent TB patients [154]. Members of the group like Lachnospira, Roseburia, Dorea, Coprococcus, Pseudobutyrivibrio, Faecalibacterium Genuses are widely known for their utilization of the plant polysaccharide and by anaerobic method providing the acetate or butyrate short-chain fatty acid (SCFA) and other anti-inflammatory molecules [162, 163]. In the light of studies done on TB infected patients, animal studies, and induced TB infected macaques, there is an increase in the genus Escherichia and Haemophilus of Phylum Proteobacteria, which are gram-negative in nature and class Gammaproteobacteria. Increase in the genus Collinsella [154] and decrease in the genus Bifidobacterium [164] of family Coriobacteriaceae and Bifidobacteriaceae of phylum Actinobacteria, these phyla are specific known for their high guanine and cytokine content in their DNA (High GC gram+); changes also seen the phylum Bacteroides (Obligate anaerobic Gram-negative) with the decrease in both the family of Bacteroidaceae [153] and Prevotellaceae [154, 164] in faecal sample of human subject [154]. The anaerobic growth and chemoautotrophic nature of the gut microbiota advocatse the formation of mixed acids, which include succinate, lactate, formate and also SCFA such as propionate and acetate. The SCFA produced by the gut microbiome is also one of the factors that are resisting the initiation as well as the progression of the active TB infection [82] as its discussed earlier that the SCFA is utilized in the body as an energy acquisition molecule and also as a signal transduction molecule for immunomodulation through downregulating proinflammatory cytokines and $T_{reg}$ cells [165]. Additionally, the structural components of the gram negatives, specifically the lipopolysaccharides obtained from the outer capsule, also act as an immunomodulatory molecule. They can provoke the immune cells to initiate the proinflammatory responses locally and distantly (occurs in case the epithelial barrier is damaged). In a study where stimulation of human peripheral blood mononuclear cell takes place by *M. tuberculosis*, it is found that the presence of SCFA butyrate significantly decreases the proinflammatory cytokine production [166]. As it is well known that the production of cell mediated response including $T_{reg}$ cells is of immense importance thus the modulation of these immune cells by SCFA is of prime importance in the link between the TB and gut microbiota [167].

Overall, dysbiosis of the gut is associated with TB infection and vice versa; the research still needs extensive attention to evaluate the more detailed profile of gut microbiota in the case of active TB patients as it is only known that the TB infection decreases the heterogeneity and richness of microbial profile of the gut and the cause–effect relationship still needs to evaluate. The mechanism of action of the SCFAs in regulating the TB infection progress still needs to be further analysed. The faecal sample did not show any traces of the mycobacterial DNA, suggesting that the cause of observed dysbiosis is not due to the appearance of the *M tuberculosis* in the gut [156]. It can also be concluded that the dysbiosis of the gut microbial profile may link the altered immune response associated with the infection of TB in the lungs, which causes the alteration in the immune-microbial synchronization in the gut, thus causing the changes observed alternatively whether these variations in gut microbiota cause the alteration in host immune response and in which mechanism is still concealed in nature.

### 11.5.1 Lung Microbiota Alterations in TB

Study shows that the richness, abundance, OTU clustering, and Shannon index all differ markedly in active TB patients than controls. Although till now the number of studies in lung microbiota is fewer than the number of studies on the gut microbial profile, still there is a rise in the number of studies in present days on the alteration of lung microbiota as a function of gut microbiota and vice versa in case of infectious lung disease (Here it is discussed about only the aspects related to infection of M. tuberculosis). As it is discussed that the microbial lung profile depends on gut as well as mouth microbial constituents [168], thus disruption in the gut microbial richness and diversity also disrupt the normal microbial flora of the lungs [169–171]. However, although there is a difference in richness based on taxonomic categories (Phyla, Genus, Species), the results show no consistent difference in particular phyla or species, indicating a far more extensive study is needed.

In the five case-control studies on active TB patients, the important consideration was the sputum study. Still, there is always a risk of contamination by the upper respiratory tract microbiome and oropharyngeal microbiomes, viz. the bacterial Genera of *Prevotella, Bulleidia, and Atopobium,* by this method. Therefore, a different approach of sample collection is required, such as throat swabs [172], bronchioalveolar lavages [171], and sputum from individuals with TB like coughing are used [169]. There is a limitation in this method, like deep cough and throat swabs containing microbiome from the upper tract rather than the lower respiratory tract and lungs. There is a relatively diverse microbial community of Streptococcus and Prevotella which are found in patients without TB infection, whereas in the case of TB infected patients, there is an abundance of mycobacteria of Phylum Actinobacteria [171, 173, 174]. The longitudinal 16s rRNA-based analysis of the sample collected from macaques infected experimentally, i.e., from the oral washes, bronchoalveolar lavage and bronchoscopy samples show that there is an increase in diversity of microbial profile early after the infection. The specific analysis of Phyla gives that Aggregatibacter, Streptococcus, and Staphylococcus genera richness is elevated and the richness of the Lachnospiraceae family is reduced at about 4 months after initial infection [175]. The genetic makeup of the individual subject and their previous record of antibiotic exposure also determine the intensity of these alterations [176]. The study on human patients for the alteration in microbial profile post-TB infections shows the increase in richness for certain species such as genus Haemophilus [177] and Acinetobacter [174] of class Gammaproteobacteria, genus Cupriavidus [171] of class Betaproteobacteria of Phylum Proteobacteria. In the case of phylum Bacteroidetes, there is overexpression of Porphyromonas [171] and shows a reduced expression of Prevotella and Alloprevotella Genuses. The results show the heterogenic expression of the Streptococcus species; some studies show over expression [177, 178] and some other shows the under expression [171, 173]. There is an increase in expression of the Lactobacillus, Staphylococcus [174], and Anoxybacillus [179] and reduce in expression of genus Gemella and Veillonella [179] of Phylum Firmicutes. Other species which also show alterations such as Fusobacterium and Cryptococcus show a reduction in richness whereas genus Thermi of Deinococcus Phylum shows overexpression [178].

## 11.6 Anti-TB Drug and Microbiome

Anti-TB drugs are classified on the basis of their clinical efficacy and tolerance [180]. Almost all of the anti-TB drugs work in the actively dividing phase of bacteria. Antibiotics kill bacteria in the concentrated phase of TB Rx, which leads to the removal of clinical symptoms and rapid sputum conversion. The continuous phase of the therapy is necessary to kill the strains of *Mycobacterium tuberculosis*, which are persistent and slow growing. For the effective treatment of TB, FLD are used, which include INH, RIF, PZA, EMB, and SM. However, due to several reasons, this first-line therapy often fails in TB treatment due to the appearance of drug-resistant bacteria. Various factors like relapse and spread of disease are involved. In MDR-TB, second-line drugs are used and they are more toxic and more expensive than first-line drugs. In MDR-TB, resistance is produced at least for isoniazid and rifampin. To prevent the occurrence of MDR-TB, an important strategy is to detect and treat single drug-resistant or drug-susceptible TB. Extensively drug-resistant TB (XDR-TB) produced because of mycobacterium TB strains have also been reported in which resistance is produced to isoniazid, rifampin, fluoroquinolones, and any other substances injectable anti-TB drug [181].

Alteration in human microbiota occurs due to antibiotic therapy, and specifically, after using broad-spectrum antibiotics, there is a fast reduction in microbiota diversity. Different antibiotics have different effects on microbiome composition, depending on the antibiotic route of administration, dosage, pharmacological properties, length of treatment, spectrum of action of an agent. Sometimes, the altered microbiome produced due to antibiotic therapy can be reversible, and recovery time may differ based on the antibiotic regimen. Even exposure to antibiotics for a short period of time can result in the production of new bacterial colonies that can live for years in a human's gut. Often the process of reversion of microbial communities to their initial state is incomplete. Alteration in the microbial environment can lead to the depletion in the population of beneficial microorganisms that have resistance to opportunistic pathogens, leading to the formation of drug-resistant colonies species. Furthermore, alteration and depletion of commensal flora may indicate the severity of infection [168].

Gut microbiota also influences the pharmacokinetics of drugs. Although the liver is the main organ for the production of primary bile acids and drug metabolism. Gut microbiota affects the drug pharmacokinetics by producing secondary bile acids. The gut microbiome also modifies expressions of enzymes and transporters involved in drug metabolism. Gut microbiota can produce drug activating or inactivating enzymes and can impact the efficacy, bioavailability, and toxicity of a drug. For example, the gut microbial enzyme transforms the sulfasalazine into its active form of 5-amino 5-salicylic acid and impacts their bioavailability by directly binding to the drug [182].

The anti-TB drug therapy impact on the diversity of the microbiome has been explored recently. Standard first-line anti-TB drugs use in multiple combinations and extended use lead to the numerous unique factors responsible for profound alteration in microbiome composition. According to the WHO, standard first-line TB treatment includes the union of four drugs, i.e., pyrazinamide, rifampin, isoniazid, and ethambutol, which are used for 2 months, isoniazid and rifampin were given for 4 months after that. The duration of antibiotic therapy exposure is the major complication of this disease, like small number of other infectious diseases need such a long duration of prophylaxis. The study available for comparison of the composition of sputum microbiota in newly formed TB, treatment failure TB, recurrent TB patients versus healthy control, does not detect the anti-TB therapy effect on lung microbiome.

Moreover, the study focuses on detecting microbiome of sputum composition in TB patients with various other diseases rather than whether sputum samples were collected before or after the antibiotic treatment. To analyse the combined TB treatment effect in lung microbiome, a well-controlled study is very much needed [168]. TB treatment is linked with the long-lasting dysbiosis of the intestine. Intestinal microbiome composition alteration caused due to antibiotics can lead to the production of novel microbiome ecological states, with initially identified but ill specified health results [183].

In paediatric TB, antibiotic exposure causes the depletion of the host-microbiome in infants, which leads to the long duration health effects. In paediatric drug-susceptible TB cases, higher dosage per kg is received by children having a body weight up to 25 kg as compared to adults of each of the anti-TB drugs: rifampin (15 versus 10 mg/kg), isoniazid (10 versus 5 mg/kg daily), ethambutol (20 versus 15 mg/kg), pyrazinamide (35 versus 25 mg/kg). Treatment duration in children is the same as in adults, i.e., intensive phase for 2 months with HRZE subsequently continuation phase for 4 months with isoniazid and rifampin. The long treatment period for paediatric TB, as compared to other bacterial diseases of childhood, likely produces any effect on a child's microbiome in comparison to chemotherapy. For childhood TB introduction of a newer fixed-dose combination (FDC) regimen has been done. These FDC regimens consist of intensive phase with isoniazid (50 mg), rifampin (75 mg), and pyrazinamide (150 mg) tablets, followed by a continuous phase with isoniazid (50 mg) and rifampin (75 mg), in per drug tablets multiple of 1 for 4–7 kg, 2 for 8–12 kg, 3 for 12–15 kg, and 4 for 16–24 kg body weight range are used. According to WHO, the duration and principles of childhood TB treatment are not changed by the introduction of FDC. In case if paediatric TB expert is not present/available for an infant having age less than 3 month and body weight below 4 kg, then treatment of infant may be done with standard childhood TB drug regimen [184].

### 11.6.1 Significance of Probiotic and Postbiotic in TB Pathogenesis

The function of macrophages could be impaired by alteration in gut microbiota and disrupt awakening of immune cells in clearance of *M. tuberculosis*. Therefore, supplementations capable of modulating the gut microbiota composition and maintaining equipoise can be given to enhance host immunity for *M. tuberculosis* and help improve the treatment results [185]. At the early twentieth century, Metchnikoff hypothesized that the Bulgarian peasants' long life span resulted from fermented milk intake in large quantities, which contains useful bacteria and leads to the generation of term probiotics [186]. Hill addressed the safety of probiotics and shown modulation in microbiota composition through inhibiting the activity and growth of damaging pathogens and stimulating the growth of beneficial bacteria. Probiotics may modulate the host system by stimulating indigenous bactericidial mechanisms and host immunoglobulins and brace up the reconciling and resistance of the host [187]. For *M. tuberculosis*, probiotic bacteria such as *Lactobacillus brevis*, *L. fermentum*, and *L. plantarum* exhibit antimicrobial activity in an in vitro study. Strong antimicrobial activity was shown by *L. casei*, *L. salivarius*, and *L. plantarum* in another in vitro study against *M. bovis* Bacillus Calmette–Guerin, and metabolite of Lactobacillus species may be the reason for this antimycobacterial activity and genes which encodes for class 2 bacterio-lysins and bacteriocins are harboured by them. Furthermore, BCG intake by phagocytes is decreased significantly by *L. plantarum*, whereas BCG intake is increased by *L. casei* and BCG intake is unaffected in case of *L. salivarius* [188]. Negi et al. in an in vivo mice model [189] found that if an antibiotic cause rise in Proteobacteria and a reduction in Firmicutes and Bacteroidetes, then it can result in the decreased macrophage-inducible calcium-dependent lectin receptor (act as pattern recognition receptor) expression and subsequently can lead to decrease in an innate immune response by recognizing and binding to the pathogen on their carbohydrate structure [190]. Alteration in gut microbiota composition causes an increase in *M. tuberculosis* stress, increases regulatory T cell number in the lung, and decreases in memory and effector T cell population. MHC-2 and macrophage-inducible calcium-dependent lectin receptor expressions could be restored on dendritic lung cells by supplementing probiotics with *Lactobacillus plantarum* MTCC 2621 and they can also decrease the burden of *M. tuberculosis* and regulatory T cells and increase in activated and effector memory CD4T cells, which exhibit CD44hiphenotype (show characteristic of activated cells in G1 stage of cell cycle) and CD62LIoCD44hi phenotype (expanded effector), respectively [189]. Further, T cells and lung dendritic cells functions in dysbiotic mice against *M. tuberculosis* could be increased by administering probiotic *L. plantarum*. Although immunoregulatory and antagonistic effects have been shown by few studies against *M. tuberculosis*, which recommend the potential role of probiotics in TB treatment as a unique strategy.

Probiotics modulate the beneficial effect of bacteria by secreted metabolites. Probiotics may be outlined as inactivated microorganism cells/components that turn out a helpful impact on host health. A benefit to the host cell could be exerted directly or indirectly by microbial cell wall fraction, exopolysaccharides, surface-associated or extracellular proteinaceous molecules, microbial metabolic such as vitamins, peptides, SCFA, amino acids, etc., and they all belong to postbiotics. Khusro et al. found that dose-dependent inhibition effect is produced by characterization and purification of the anti-tubercular protein of strain Staphylococcus hominis MANF2 having a molecular mass of 7712.3 Da [191]. Carroll et al., in their studies, found that in an in vitro condition, growth of *M. tuberculosis* H37Ra is inhibited by lacticin 3147, an antimicrobial peptide of Lactococcus lactis subsp. Cremoris MG1363, and this lacticin have MIC90value of 7.5 mg/L and great potential is exhibited by lacticin as a therapeutic agent [192]. A gut microbiota metabolite such as indole propionic acid is also known as an anti-TB agent. Negatu et al., in the Maybridge Ro3 library screened 1000 fragments primarily and in an in vivo identified 29 compounds with great anti-tubercular activity [193]. Subsequently, to know the bactericidal activity of these 29 isolated compounds against M. tuberculosis, they were co-cultured with *M. tuberculosis* and half of them can reduce the viability of *M. tuberculosis* by 100-fold. Among these compounds, the most substantial inhibition effect was shown by indole propionic acid against *M. tuberculosis*. Its direct anti-tubercular activity was indicated by testing it on a mouse model, infected through aerosol route with a small dose of *M. tuberculosis* and cause reduction of bacterial load on spleen by seven-folds. It was also shown that the indole propionic acid shows antimycobacterial activity by mimicking TrpE physiological allosteric inhibitor and blocking the biosynthesis of tryptophan in *M. tuberculosis* and it was found out after metabolic, genetic, chemical rescue, and biochemical analysis of indole propionic acid. Postbiotics anti-TB activity potential can be illustrated by these findings although further research is required to be done to explain host factors and microbiota involved in anti-TB activity [185].

## 11.7 Conclusions

Mycobacterium tuberculosis infection weigh up to approximately 1.6 billion people worldwide but most of the infection restrain by the immune system so that at any one time we get about 14.4 people have clinical symptoms. Thus, the proper functioning of the immune response specifically cellular immune response is necessary to prevent the TB infection. The dysbiosis in the gut microbiota causes alteration of the systemic immune expression negatively, also the dysbiosis in gut causes the alteration of microbial profile in other organs which in turn affects the local immune response. This whole scenario of immune alteration weakens the defence against the foreign invasion. The active M. tuberculosis infection causes alteration in gut microbiota. Anti-TB medications lead to the alteration in microbial composition, which can be of short duration and long-lasting in nature [194]. It was indicated that the microbic composition of a patient with TB was complicated than healthy volunteers and pneumonic TB patients' humor contain several foreign bacteria. The presence of those foreign bacterium could facilitate within the development of wasting disease [195]. Some specific genera of bacteria such as Rothia, Lactobacillus, Veillonella, and Leuconostoc were isolated from TB patients [196]. Probiotics and postbiotics exhibited anti-TB activity in an in vivo and in vitro conditions, which suggest that they can be used in TB treatment to overcome problems caused by using multiple antibiotics [185]. Growing evidence indicates the interaction of the microbiome with each phase of the tuberculosis spectrum. Microbiome studies should be done in clinical research of tuberculosis and trial testing of new vaccines wherever possible. At least, the microbiome will probably be associated with differential phenotypical, biological, and clinical outcomes. Hopefully, these outcomes will be changed by including the microbiome in the study of tuberculosis treatment [197].

## References

1. Lozano R, Naghavi M, Foreman K, Lim S, Shibuya K, Aboyans V, Abraham J, Adair T, Aggarwal R, Ahn SY, AlMazroa MA, Alvarado M, Anderson HR, Anderson LM, Andrews KG, Atkinson C, Baddour LM, Barker-Collo S, Bartels DH, Bell ML, Benjamin EJ, Bennett D, Bhalla K, Bikbov B, Abdulhak AB, Birbeck G, Blyth F, Bolliger I, Boufous S, Bucello C, Burch M, Burney P, Carapetis J, Chen H, Chou D, Chugh SS, Coffeng LE, Colan SD, Colquhoun S, Colson KE, Condon J, Connor MD, Cooper LT, Corriere M, Cortinovis M, de Vaccaro KC, Couser W, Cowie BC, Criqui MH, Cross M, Dabhadkar KC, Dahodwala N, De Leo D, Degenhardt L, Delossantos A, Denenberg J, Des Jarlais DC, Dharmaratne SD, Dorsey ER, Driscoll T, Duber H, Ebel B, Erwin PJ, Espindola P, Ezzati M, Feigin V, Flaxman AD, Forouzanfar MH, Fowkes FGR, Franklin R, Fransen M, Freeman MK, Gabriel SE, Gakidou E, Gaspari F, Gillum RF, Gonzalez-Medina D, Halasa YA, Haring D, Harrison JE, Havmoeller R, Hay RJ, Hoen B, Hotez PJ, Hoy D, Jacobsen KH, James SL, Jasrasaria R, Jayaraman S, Johns N, Karthikeyan G, Kassebaum N, Keren A, Khoo J-P, Knowlton LM, Kobusingye O, Koranteng A, Krishnamurthi R, Lipnick M, Lipshultz SE, Ohno SL, Mabweijano J, MacIntyre MF, Mallinger L, March L, Marks GB, Marks R, Matsumori A, Matzopoulos R, Mayosi BM, McAnulty JH, McDermott MM, McGrath J, Memish ZA, Mensah GA, Merriman TR, Michaud C, Miller M, Miller TR, Mock C, Mocumbi AO, Mokdad AA, Moran A, Mulholland K, Nair MN, Naldi L, Narayan KMV, Nasseri K, Norman P, O'Donnell M, Omer SB, Ortblad K, Osborne R, Ozgediz D, Pahari B, Pandian JD, Rivero AP, Padilla RP, Perez-Ruiz F, Perico N, Phillips D, Pierce K, Pope CA III, Porrini E, Pourmalek F, Raju M, Ranganathan D, Rehm JT, Rein DB, Remuzzi G, Rivara FP, Roberts T, De León FR, Rosenfeld LC, Rushton L, Sacco RL, Salomon JA, Sampson U, Sanman E, Schwebel DC, Segui-Gomez M, Shepard DS, Singh D, Singleton J, Sliwa K, Smith E, Steer A, Taylor JA, Thomas B, Tleyjeh IM, Towbin JA, Truelsen T, Undurraga EA, Venketasubramanian N, Vijayakumar L, Vos T, Wagner GR, Wang M, Wang W, Watt K, Weinstock MA, Weintraub R, Wilkinson JD, Woolf AD, Wulf S, Yeh P-H, Yip P, Zabetian A, Zheng Z-J, Lopez AD, Murray CJL (2012) Global and regional mortality from 235 causes of death for 20 age groups in 1990 and 2010: a systematic analysis for the Global Burden of Disease Study 2010. Lancet 380(9859):2095–2128. https://doi.org/10.1016/S0140-6736(12)61728-0
2. Organisation WH (2017) World Health Organization, Global tuberculosis report; 2017
3. Shea KM, Kammerer JS, Winston CA, Navin TR, Horsburgh CR Jr (2014) Estimated rate of reactivation of latent tuberculosis infection in the United States, overall and by population subgroup. Am J Epidemiol 179(2):216–225. https://doi.org/10.1093/aje/kwt246
4. Houben RMGJ, Dodd PJ (2016) The global burden of latent tuberculosis infection: a re-estimation using mathematical modelling. PLoS Med 13(10):e1002152. https://doi.org/10.1371/journal.pmed.1002152
5. Masjedi MR, Farnia P, Sorooch S, Pooramiri MV, Mansoori SD, Zarifi AZ, AkbarVelayati A, Hoffner S (2006) Extensively drug-resistant tuberculosis: 2 years of surveillance in Iran. Clin Infect Dis 43(7):841–847. https://doi.org/10.1086/507542
6. Zignol M, Dean AS, Falzon D, van Gemert W, Wright A, van Deun A, Portaels F, Laszlo A, Espinal MA, Pablos-Méndez A, Bloom A, Aziz MA, Weyer K, Jaramillo E, Nunn P, Floyd K, Raviglione MC (2016) Twenty years of global surveillance of antituberculosis-drug resistance. N Engl J Med 375(11):1081–1089. https://doi.org/10.1056/NEJMsr1512438
7. Frieden TR, Sterling TR, Munsiff SS, Watt CJ, Dye C (2003) Tuberculosis. Lancet (London, England) 362(9387):887–899. https://doi.org/10.1016/s0140-6736(03)14333-4
8. Hill PC, Brookes RH, Fox A, Fielding K, Jeffries DJ, Jackson-Sillah D, Lugos MD, Owiafe PK, Donkor SA, Hammond AS (2004) Large-scale evaluation of enzyme-linked immunospot assay and skin test for diagnosis of Mycobacterium tuberculosis infection against a gradient of exposure in The Gambia. Clin Infect Dis 38(7):966–973

9. Mathema B, Kurepina N, Fallows D, Kreiswirth BN (2008) Lessons from molecular epidemiology and comparative genomics. Semin Respir Crit Care Med 29(5):467–480. https://doi.org/10.1055/s-0028-1085699
10. Reed MB, Pichler VK, McIntosh F, Mattia A, Fallow A, Masala S, Domenech P, Zwerling A, Thibert L, Menzies D (2009) Major Mycobacterium tuberculosis lineages associate with patient country of origin. J Clin Microbiol 47(4):1119–1128
11. Bos KI, Harkins KM, Herbig A, Coscolla M, Weber N, Comas I, Forrest SA, Bryant JM, Harris SR, Schuenemann VJ (2014) Pre-Columbian mycobacterial genomes reveal seals as a source of New World human tuberculosis. Nature 514(7523):494–497
12. Comas I, Coscolla M, Luo T, Borrell S, Holt KE, Kato-Maeda M, Parkhill J, Malla B, Berg S, Thwaites G (2013) Out-of-Africa migration and neolithic coexpansion of Mycobacterium tuberculosis with modern humans. Nat Genet 45(10):1176–1182
13. Warner DF, Koch A, Mizrahi V (2015) Diversity and disease pathogenesis in Mycobacterium tuberculosis. Trends Microbiol 23(1):14–21
14. Reed MB, Domenech P, Manca C, Su H, Barczak AK, Kreiswirth BN, Kaplan G, Barry CE (2004) A glycolipid of hypervirulent tuberculosis strains that inhibits the innate immune response. Nature 431(7004):84–87
15. Gagneux S, DeRiemer K, Van T, Kato-Maeda M, De Jong BC, Narayanan S, Nicol M, Niemann S, Kremer K, Gutierrez MC (2006) Variable host–pathogen compatibility in Mycobacterium tuberculosis. Proc Natl Acad Sci 103(8):2869–2873
16. Albanna AS, Reed MB, Kotar KV, Fallow A, McIntosh FA, Behr MA, Menzies D (2011) Reduced transmissibility of East African Indian strains of Mycobacterium tuberculosis. PLoS One 6(9):e25075
17. Fenner L, Gagneux S, Helbling P, Battegay M, Rieder HL, Pfyffer GE, Zwahlen M, Furrer H, Siegrist HH, Fehr J (2012) Mycobacterium tuberculosis transmission in a country with low tuberculosis incidence: role of immigration and HIV infection. J Clin Microbiol 50(2):388–395
18. Lee RS, Radomski N, Proulx J-F, Levade I, Shapiro BJ, McIntosh F, Soualhine H, Menzies D, Behr MA (2015) Population genomics of Mycobacterium tuberculosis in the Inuit. Proc Natl Acad Sci 112(44):13609–13614
19. Behr M, Wilson M, Gill W, Salamon H, Schoolnik G, Rane S, Small P (1999) Comparative genomics of BCG vaccines by whole-genome DNA microarray. Science 284(5419):1520–1523
20. Forrellad MA, Klepp LI, Gioffré A, Sabio y García J, Morbidoni HR, de la Paz Santangelo M, Cataldi AA, Bigi F (2013) Virulence factors of the Mycobacterium tuberculosis complex. Virulence 4(1):3–66. https://doi.org/10.4161/viru.22329
21. Barry CE, Boshoff HI, Dartois V, Dick T, Ehrt S, Flynn J, Schnappinger D, Wilkinson RJ, Young D (2009) The spectrum of latent tuberculosis: rethinking the biology and intervention strategies. Nat Rev Microbiol 7(12):845–855
22. Esmail H, Barry C 3rd, Young D, Wilkinson R (2014) The ongoing challenge of latent tuberculosis. Philos Trans R Soc B Biol Sci 369(1645):20130437
23. Morrison J, Pai M, Hopewell PC (2008) Tuberculosis and latent tuberculosis infection in close contacts of people with pulmonary tuberculosis in low-income and middle-income countries: a systematic review and meta-analysis. Lancet Infect Dis 8(6):359–368
24. Cobat A, Gallant CJ, Simkin L, Black GF, Stanley K, Hughes J, Doherty TM, Hanekom WA, Eley B, Jaïs J-P (2009) Two loci control tuberculin skin test reactivity in an area hyperendemic for tuberculosis. J Exp Med 206(12):2583–2591
25. Orme IM, Robinson RT, Cooper AM (2015) The balance between protective and pathogenic immune responses in the TB-infected lung. Nat Immunol 16(1):57–63
26. Watford WT, Wright JR, Hester CG, Jiang H, Frank MM (2001) Surfactant protein A regulates complement activation. J Immunol 167(11):6593–6600
27. Ferguson JS, Voelker DR, McCormack FX, Schlesinger LS (1999) Surfactant protein D binds to Mycobacterium tuberculosis Bacilli and Lipoarabinomannan via carbohydrate-lectin

interactions resulting in reduced phagocytosis of the bacteria by macrophages1. J Immunol 163(1):312–321
28. Russell DG, Cardona P-J, Kim M-J, Allain S, Altare F (2009) Foamy macrophages and the progression of the human tuberculosis granuloma. Nat Immunol 10(9):943–948
29. Peters W, Ernst JD (2003) Mechanisms of cell recruitment in the immune response to Mycobacterium tuberculosis. Microbes Infect 5(2):151–158
30. Egen JG, Rothfuchs AG, Feng CG, Horwitz MA, Sher A, Germain RN (2011) Intravital imaging reveals limited antigen presentation and T cell effector function in mycobacterial granulomas. Immunity 34(5):807–819
31. Ulrichs T, Kosmiadi GA, Trusov V, Jörg S, Pradl L, Titukhina M, Mishenko V, Gushina N, Kaufmann SH (2004) Human tuberculous granulomas induce peripheral lymphoid follicle-like structures to orchestrate local host defence in the lung. J Pathol 204(2):217–228
32. Hernandez-Pando R, Jeyanathan M, Mengistu G, Aguilar D, Orozco H, Harboe M, Rook G, Bjune G (2000) Persistence of DNA from Mycobacterium tuberculosis in superficially normal lung tissue during latent infection. Lancet 356(9248):2133–2138
33. Behar S, Martin C, Booty M, Nishimura T, Zhao X, Gan H, Divangahi M, Remold H (2011) Apoptosis is an innate defense function of macrophages against Mycobacterium tuberculosis. Mucosal Immunol 4(3):279–287
34. Keane J, Remold HG, Kornfeld H (2000) Virulent Mycobacterium tuberculosis strains evade apoptosis of infected alveolar macrophages. J Immunol 164(4):2016–2020
35. Kelly DM, ten Bokum AM, O'Leary SM, O'Sullivan MP, Keane J (2008) Bystander macrophage apoptosis after Mycobacterium tuberculosis H37Ra infection. Infect Immun 76(1): 351–360
36. Hinchey J, Lee S, Jeon BY, Basaraba RJ, Venkataswamy MM, Chen B, Chan J, Braunstein M, Orme IM, Derrick SC (2007) Enhanced priming of adaptive immunity by a proapoptotic mutant of Mycobacterium tuberculosis. J Clin Invest 117(8):2279–2288
37. Velmurugan K, Chen B, Miller JL, Azogue S, Gurses S, Hsu T, Glickman M, Jacobs WR Jr, Porcelli SA, Briken V (2007) Mycobacterium tuberculosis nuoG is a virulence gene that inhibits apoptosis of infected host cells. PLoS Pathog 3(7):e110
38. Flynn JL, Goldstein MM, Chan J, Triebold KJ, Pfeffer K, Lowenstein CJ, Schreiber R, Mak TW, Bloom BR (1995) Tumor necrosis factor-alpha is required in the protective immune response against Mycobacterium tuberculosis in mice. Immunity 2(6):561–572. https://doi.org/10.1016/1074-7613(95)90001-2
39. Ladel CH, Blum C, Dreher A, Reifenberg K, Kopf M, Kaufmann SH (1997) Lethal tuberculosis in interleukin-6-deficient mutant mice. Infect Immun 65(11):4843–4849. https://doi.org/10.1128/iai.65.11.4843-4849.1997
40. MacMicking JD, North RJ, LaCourse R, Mudgett JS, Shah SK, Nathan CF (1997) Identification of nitric oxide synthase as a protective locus against tuberculosis. Proc Natl Acad Sci 94(10):5243–5248
41. Marchesi J, Ravel J (2015) The vocabulary of microbiome research. A proposal. Microbiome. 3:31
42. Turnbaugh PJ, Ley RE, Hamady M, Fraser-Liggett CM, Knight R, Gordon JI (2007) The human microbiome project. Nature 449(7164):804–810. https://doi.org/10.1038/nature06244
43. Qin J, Li R, Raes J, Arumugam M, Burgdorf KS, Manichanh C, Nielsen T, Pons N, Levenez F, Yamada T, Mende DR, Li J, Xu J, Li S, Li D, Cao J, Wang B, Liang H, Zheng H, Xie Y, Tap J, Lepage P, Bertalan M, Batto JM, Hansen T, Le Paslier D, Linneberg A, Nielsen HB, Pelletier E, Renault P, Sicheritz-Ponten T, Turner K, Zhu H, Yu C, Li S, Jian M, Zhou Y, Li Y, Zhang X, Li S, Qin N, Yang H, Wang J, Brunak S, Doré J, Guarner F, Kristiansen K, Pedersen O, Parkhill J, Weissenbach J, Bork P, Ehrlich SD, Wang J (2010) A human gut microbial gene catalogue established by metagenomic sequencing. Nature 464(7285):59–65. https://doi.org/10.1038/nature08821

44. Ley RE, Peterson DA, Gordon JI (2006) Ecological and evolutionary forces shaping microbial diversity in the human intestine. Cell 124(4):837–848. https://doi.org/10.1016/j.cell.2006.02.017
45. Bäckhed F, Ley RE, Sonnenburg JL, Peterson DA, Gordon JI (2005) Host-bacterial mutualism in the human intestine. Science 307(5717):1915–1920
46. Hasan N, Yang H (2019) Factors affecting the composition of the gut microbiota, and its modulation. PeerJ 7:e7502
47. Ranjan R, Rani A, Metwally A, McGee HS, Perkins DL (2016) Analysis of the microbiome: advantages of whole genome shotgun versus 16S amplicon sequencing. Biochem Biophys Res Commun 469(4):967–977
48. Qin J, Li R, Raes J, Arumugam M, Burgdorf KS, Manichanh C, Nielsen T, Pons N, Levenez F, Yamada T (2010) A human gut microbial gene catalogue established by metagenomic sequencing. Nature 464(7285):59–65
49. Zoetendal EG, Akkermans AD, De Vos WM (1998) Temperature gradient gel electrophoresis analysis of 16S rRNA from human fecal samples reveals stable and host-specific communities of active bacteria. Appl Environ Microbiol 64(10):3854–3859
50. Ley RE, Turnbaugh PJ, Klein S, Gordon JI (2006) Human gut microbes associated with obesity. Nature. 444(7122):1022–1023
51. Shukla SD, Budden KF, Neal R, Hansbro PMJC (2017) Microbiome effects on immunity, health and disease in the lung. Clin Transl Immunology. 6(3):e133
52. Ingenito E, Solway J, McFadden E Jr, Pichurko B, Bowman H, Michaels D, Drazen J (1987) Indirect assessment of mucosal surface temperatures in the airways: theory and tests. J Appl Physiol 63(5):2075–2083
53. Hilty M, Burke C, Pedro H, Cardenas P, Bush A, Bossley C, Davies J, Ervine A, Poulter L, Pachter L, Moffatt MF, Cookson WO (2010) Disordered microbial communities in asthmatic airways. PLoS One 5(1):e8578. https://doi.org/10.1371/journal.pone.0008578
54. Dickson RP, Martinez FJ, Huffnagle GB (2014) The role of the microbiome in exacerbations of chronic lung diseases. Lancet 384(9944):691–702
55. Lighthart B (2000) Mini-review of the concentration variations found inthe alfresco atmospheric bacterial populations. Aerobiologia 16(1):7–16
56. Munyard P, Bush A (1996) How much coughing is normal? Arch Dis Child 74(6):531–534
57. Morris A, Beck JM, Schloss PD, Campbell TB, Crothers K, Curtis JL, Flores SC, Fontenot AP, Ghedin E, Huang L (2013) Comparison of the respiratory microbiome in healthy nonsmokers and smokers. Am J Respir Crit Care Med 187(10):1067–1075
58. Segal LN, Alekseyenko AV, Clemente JC, Kulkarni R, Wu B, Chen H, Berger KI, Goldring RM, Rom WN, Blaser MJ (2013) Enrichment of lung microbiome with supraglottic taxa is associated with increased pulmonary inflammation. Microbiome 1(1):1–12
59. Venkataraman A, Bassis CM, Beck JM, Young VB, Curtis JL, Huffnagle GB, Schmidt TM (2015) Application of a neutral community model to assess structuring of the human lung microbiome. MBio 6(1):e02284–e02214
60. Bassis CM, Erb-Downward JR, Dickson RP, Freeman CM, Schmidt TM, Young VB, Beck JM, Curtis JL, Huffnagle GB (2015) Analysis of the upper respiratory tract microbiotas as the source of the lung and gastric microbiotas in healthy individuals. MBio 6(2):e00037–e00015
61. Dickson RP, Erb-Downward JR, Freeman CM, Walker N, Scales BS, Beck JM, Martinez FJ, Curtis JL, Lama VN, Huffnagle GB (2014) Changes in the lung microbiome following lung transplantation include the emergence of two distinct Pseudomonas species with distinct clinical associations. PLoS One 9(5):e97214
62. Dickson RP, Erb-Downward JR, Huffnagle GB (2013) The role of the bacterial microbiome in lung disease. Expert Rev Respir Med 7(3):245–257
63. Chioma OS, Hesse LE, Chapman A, Drake WP (2021) Role of the microbiome in interstitial lung diseases. Front Med 8:595522. https://doi.org/10.3389/fmed.2021.595522

64. Rogers GB, Zain NMM, Bruce KD, Burr LD, Chen AC, Rivett DW, McGuckin MA, Serisier DJ (2014) A novel microbiota stratification system predicts future exacerbations in bronchiectasis. Ann Am Thorac Soc 11(4):496–503
65. Molyneaux PL, Cox MJ, Willis-Owen SA, Mallia P, Russell KE, Russell A-M, Murphy E, Johnston SL, Schwartz DA, Wells AU (2014) The role of bacteria in the pathogenesis and progression of idiopathic pulmonary fibrosis. Am J Respir Crit Care Med 190(8):906–913
66. Huang YJ, Nelson CE, Brodie EL, DeSantis TZ, Baek MS, Liu J, Woyke T, Allgaier M, Bristow J, Wiener-Kronish JP (2011) Airway microbiota and bronchial hyperresponsiveness in patients with suboptimally controlled asthma. J Allergy Clin Immunol 127(2): 372–381. e373
67. Madan JC, Koestler DC, Stanton BA, Davidson L, Moulton LA, Housman ML, Moore JH, Guill MF, Morrison HG, Sogin ML, Hampton TH, Karagas MR, Palumbo PE, Foster JA, Hibberd PL, O'Toole GA (2012) Serial analysis of the gut and respiratory microbiome in cystic fibrosis in infancy: interaction between intestinal and respiratory tracts and impact of nutritional exposures. mBio 3(4). https://doi.org/10.1128/mBio.00251-12
68. Ferreira CM, Vieira AT, Vinolo MAR, Oliveira FA, Curi R, Martins FS (2014) The central role of the gut microbiota in chronic inflammatory diseases. J Immunol Res 2014:689492. https://doi.org/10.1155/2014/689492
69. McAleer JP, Kolls JK (2018) Contributions of the intestinal microbiome in lung immunity. Eur J Immunol 48(1):39–49. https://doi.org/10.1002/eji.201646721
70. Anand S, Mande SS (2018) Diet, microbiota and gut-lung connection. Front Microbiol 9:2147
71. Enaud R, Prevel R, Ciarlo E, Beaufils F, Wieërs G, Guery B, Delhaes L (2020) The gut-lung axis in health and respiratory diseases: a place for inter-organ and inter-kingdom crosstalks. Front Cell Infect Microbiol 10:9–9. https://doi.org/10.3389/fcimb.2020.00009
72. Morgan XC, Huttenhower C (2012) Chapter 12: Human microbiome analysis. PLoS Comput Biol 8(12):e1002808. https://doi.org/10.1371/journal.pcbi.1002808
73. Mori G, Morrison M, Blumenthal A (2021) Microbiome-immune interactions in tuberculosis. PLOS Pathog 17(4):e1009377. https://doi.org/10.1371/journal.ppat.1009377
74. Santiago-Rodriguez TM, Hollister EB (2019) Human virome and disease: high-throughput sequencing for virus discovery, identification of phage-bacteria dysbiosis and development of therapeutic approaches with emphasis on the human gut. Viruses 11(7). https://doi.org/10.3390/v11070656
75. Foster KR, Schluter J, Coyte KZ, Rakoff-Nahoum S (2017) The evolution of the host microbiome as an ecosystem on a leash. Nature 548(7665):43–51. https://doi.org/10.1038/nature23292
76. Spor A, Koren O, Ley R (2011) Unravelling the effects of the environment and host genotype on the gut microbiome. Nat Rev Microbiol 9(4):279–290. https://doi.org/10.1038/nrmicro2540
77. Bernard G, Pathmanathan JS, Lannes R, Lopez P, Bapteste E (2018) Microbial dark matter investigations: how microbial studies transform biological knowledge and empirically sketch a logic of scientific discovery. Genome Biol Evol 10(3):707–715. https://doi.org/10.1093/gbe/evy031
78. Faner R, Sibila O, Agustí A, Bernasconi E, Chalmers JD, Huffnagle GB, Manichanh C, Molyneaux PL, Paredes R, Pérez Brocal V, Ponomarenko J, Sethi S, Dorca J, Monsó E (2017) The microbiome in respiratory medicine: current challenges and future perspectives. Eur Respir J 49(4):1602086. https://doi.org/10.1183/13993003.02086-2016
79. Pickard JM, Zeng MY, Caruso R, Núñez G (2017) Gut microbiota: role in pathogen colonization, immune responses, and inflammatory disease. Immunol Rev 279(1):70–89. https://doi.org/10.1111/imr.12567
80. Blaser MJ (2017) The theory of disappearing microbiota and the epidemics of chronic diseases. Nat Rev Immunol 17(8):461–463. https://doi.org/10.1038/nri.2017.77
81. Wu GD, Compher C, Chen EZ, Smith SA, Shah RD, Bittinger K, Chehoud C, Albenberg LG, Nessel L, Gilroy E, Star J, Weljie AM, Flint HJ, Metz DC, Bennett MJ, Li H, Bushman FD,

Lewis JD (2016) Comparative metabolomics in vegans and omnivores reveal constraints on diet-dependent gut microbiota metabolite production. Gut 65(1):63. https://doi.org/10.1136/gutjnl-2014-308209
82. Huang R, Ju Z, Zhou P-K (2020) A gut dysbiotic microbiota-based hypothesis of human-to-human transmission of non-communicable diseases. Sci Total Environ 745:141030. https://doi.org/10.1016/j.scitotenv.2020.141030
83. Wilkins LJ, Monga M, Miller AW (2019) Defining dysbiosis for a cluster of chronic diseases. Sci Rep 9(1):12918. https://doi.org/10.1038/s41598-019-49452-y
84. Budden KF, Gellatly SL, Wood DLA, Cooper MA, Morrison M, Hugenholtz P, Hansbro PM (2017) Emerging pathogenic links between microbiota and the gut–lung axis. Nat Rev Microbiol 15(1):55–63. https://doi.org/10.1038/nrmicro.2016.142
85. Simpson HL, Campbell BJ (2015) Review article: dietary fibre–microbiota interactions. Aliment Pharmacol Ther 42(2):158–179. https://doi.org/10.1111/apt.13248
86. Gupta SS, Mohammed MH, Ghosh TS, Kanungo S, Nair GB, Mande SS (2011) Metagenome of the gut of a malnourished child. Gut Pathog 3(1):7. https://doi.org/10.1186/1757-4749-3-7
87. Russell WR, Hoyles L, Flint HJ, Dumas M-E (2013) Colonic bacterial metabolites and human health. Curr Opin Microbiol 16(3):246–254. https://doi.org/10.1016/j.mib.2013.07.002
88. Ghosh TS, Sen Gupta S, Bhattacharya T, Yadav D, Barik A, Chowdhury A, Das B, Mande SS, Nair GB (2014) Gut microbiomes of Indian children of varying nutritional status. PLoS One 9(4):e95547. https://doi.org/10.1371/journal.pone.0095547
89. Smith MI, Yatsunenko T, Manary MJ, Trehan I, Mkakosya R, Cheng J, Kau AL, Rich SS, Concannon P, Mychaleckyj JC, Liu J, Houpt E, Li JV, Holmes E, Nicholson J, Knights D, Ursell LK, Knight R, Gordon JI (2013) Gut microbiomes of Malawian twin pairs discordant for kwashiorkor. Science 339(6119):548. https://doi.org/10.1126/science.1229000
90. Fujimura Kei E, Lynch Susan V (2015) Microbiota in allergy and asthma and the emerging relationship with the gut microbiome. Cell Host Microbe 17(5):592–602. https://doi.org/10.1016/j.chom.2015.04.007
91. Bäckhed F, Roswall J, Peng Y, Feng Q, Jia H, Kovatcheva-Datchary P, Li Y, Xia Y, Xie H, Zhong H, Khan Muhammad T, Zhang J, Li J, Xiao L, Al-Aama J, Zhang D, Lee Ying S, Kotowska D, Colding C, Tremaroli V, Yin Y, Bergman S, Xu X, Madsen L, Kristiansen K, Dahlgren J, Wang J (2015) Dynamics and stabilization of the human gut microbiome during the first year of life. Cell Host Microbe 17(5):690–703. https://doi.org/10.1016/j.chom.2015.04.004
92. Bergström A, Skov Thomas H, Bahl Martin I, Roager Henrik M, Christensen Line B, Ejlerskov Katrine T, Mølgaard C, Michaelsen Kim F, Licht Tine R, Griffiths MW (2014) Establishment of intestinal microbiota during early life: a longitudinal, explorative study of a large cohort of danish infants. Appl Environ Microbiol 80(9):2889–2900. https://doi.org/10.1128/AEM.00342-14
93. Timmerman HM, Rutten NBMM, Boekhorst J, Saulnier DM, Kortman GAM, Contractor N, Kullen M, Floris E, Harmsen HJM, Vlieger AM, Kleerebezem M, Rijkers GT (2017) Intestinal colonisation patterns in breastfed and formula-fed infants during the first 12 weeks of life reveal sequential microbiota signatures. Sci Rep 7(1):8327. https://doi.org/10.1038/s41598-017-08268-4
94. Turnbaugh PJ, Ridaura VK, Faith JJ, Rey FE, Knight R, Gordon JI (2009) The effect of diet on the human gut microbiome: a metagenomic analysis in humanized gnotobiotic mice. Sci Transl Med 1(6):6ra14. https://doi.org/10.1126/scitranslmed.3000322
95. De Filippo C, Cavalieri D, Di Paola M, Ramazzotti M, Poullet JB, Massart S, Collini S, Pieraccini G, Lionetti P (2010) Impact of diet in shaping gut microbiota revealed by a comparative study in children from Europe and rural Africa. Proc Natl Acad Sci 107(33):14691. https://doi.org/10.1073/pnas.1005963107
96. Zimmer J, Lange B, Frick JS, Sauer H, Zimmermann K, Schwiertz A, Rusch K, Klosterhalfen S, Enck P (2012) A vegan or vegetarian diet substantially alters the human colonic faecal microbiota. Eur J Clin Nutr 66(1):53–60. https://doi.org/10.1038/ejcn.2011.141

97. Vemuri R, Gundamaraju R, Shastri MD, Shukla SD, Kalpurath K, Ball M, Tristram S, Shankar EM, Ahuja K, Eri R (2018) Gut microbial changes, interactions, and their implications on human lifecycle: an ageing perspective. Biomed Res Int 2018:4178607. https://doi.org/10.1155/2018/4178607
98. Kumar M, Babaei P, Ji B, Nielsen J (2016) Human gut microbiota and healthy aging: recent developments and future prospective. Nutr Healthy Aging 4:3–16. https://doi.org/10.3233/NHA-150002
99. Nagpal R, Mainali R, Ahmadi S, Wang S, Singh R, Kavanagh K, Kitzman DW, Kushugulova A, Marotta F, Yadav H (2018) Gut microbiome and aging: physiological and mechanistic insights. Nutr Healthy Aging 4:267–285. https://doi.org/10.3233/NHA-170030
100. Huffnagle GB, Noverr MC (2013) The emerging world of the fungal microbiome. Trends Microbiol 21(7):334–341. https://doi.org/10.1016/j.tim.2013.04.002
101. Huseyin CE, O'Toole PW, Cotter PD, Scanlan PD (2017) Forgotten fungi—the gut mycobiome in human health and disease. FEMS Microbiol Rev 41(4):479–511. https://doi.org/10.1093/femsre/fuw047
102. Nash AK, Auchtung TA, Wong MC, Smith DP, Gesell JR, Ross MC, Stewart CJ, Metcalf GA, Muzny DM, Gibbs RA, Ajami NJ, Petrosino JF (2017) The gut mycobiome of the human microbiome project healthy cohort. Microbiome 5(1):153. https://doi.org/10.1186/s40168-017-0373-4
103. Hoffmann C, Dollive S, Grunberg S, Chen J, Li H, Wu GD, Lewis JD, Bushman FD (2013) Archaea and fungi of the human gut microbiome: correlations with diet and bacterial residents. PLoS One 8(6):e66019. https://doi.org/10.1371/journal.pone.0066019
104. Peleg AY, Hogan DA, Mylonakis E (2010) Medically important bacterial–fungal interactions. Nat Rev Microbiol 8(5):340–349. https://doi.org/10.1038/nrmicro2313
105. Markey L, Shaban L, Green ER, Lemon KP, Mecsas J, Kumamoto CA (2018) Pre-colonization with the commensal fungus Candida albicans reduces murine susceptibility to Clostridium difficile infection. Gut Microbes 9(6):497–509. https://doi.org/10.1080/19490976.2018.1465158
106. Zhou TX, Jung JH, Zhang ZF, Kim IH (2013) Effect of dietary β-glucan on growth performance, fecal microbial shedding and immunological responses after lipopolysaccharide challenge in weaned pigs. Anim Feed Sci Technol 179(1):85–92. https://doi.org/10.1016/j.anifeedsci.2012.10.008
107. Buts J-P, Dekeyser N, Stilmant C, Delem E, Smets F, Sokal E (2006) Saccharomyces boulardii produces in rat small intestine a novel protein phosphatase that inhibits Escherichia coli endotoxin by dephosphorylation. Pediatr Res 60(1):24–29. https://doi.org/10.1203/01.pdr.0000220322.31940.29
108. Kozel TR, Castagliuolo I, Riegler Martin F, Valenick L, LaMont JT, Pothoulakis C (1999) Saccharomyces boulardii protease inhibits the effects of clostridium difficile toxins A and B in human colonic mucosa. Infect Immun 67(1):302–307. https://doi.org/10.1128/IAI.67.1.302-307.1999
109. Jiang TT, Shao T-Y, Ang WXG, Kinder JM, Turner LH, Pham G, Whitt J, Alenghat T, Way SS (2017) Commensal fungi recapitulate the protective benefits of intestinal bacteria. Cell Host Microbe 22(6):809–816.e804. https://doi.org/10.1016/j.chom.2017.10.013
110. Fabbrizzi A, Amedei A, Lavorini F, Renda T, Fontana G (2019) The lung microbiome: clinical and therapeutic implications. (1970-9366 (Electronic)). Intern Emerg Med. 14(8):1241–1250
111. Wang J, Li F, Tian Z (2017) Role of microbiota on lung homeostasis and diseases. Sci China Life Sci 60(12):1407–1415. https://doi.org/10.1007/s11427-017-9151-1
112. Segal LN, Alekseyenko AV, Clemente JC, Kulkarni R, Wu B, Chen H, Berger KI, Goldring RM, Rom WN, Blaser MJ, Weiden MD (2013) Enrichment of lung microbiome with supraglottic taxa is associated with increased pulmonary inflammation. Microbiome 1(1):19. https://doi.org/10.1186/2049-2618-1-19
113. Bassis Christine M, Erb-Downward John R, Dickson Robert P, Freeman Christine M, Schmidt Thomas M, Young Vincent B, Beck James M, Curtis Jeffrey L, Huffnagle Gary B, Ravel J

(2015) Analysis of the upper respiratory tract microbiotas as the source of the lung and gastric microbiotas in healthy individuals. mBio 6(2):e00037–e00015. https://doi.org/10.1128/mBio.00037-15
114. Tai A, Ranganath S (2008) Anaerobic bacterial infection in patients with cystic fibrosis. Am J Respir Crit Care Med 178(9):994–994. https://doi.org/10.1164/ajrccm.178.9.994
115. Beck JM, Young VB, Huffnagle GB (2012) The microbiome of the lung. Transl Res 160(4): 258–266. https://doi.org/10.1016/j.trsl.2012.02.005
116. O'Dwyer DN, Dickson RP, Moore BB (2016) The Lung Microbiome, Immunity, and the Pathogenesis of Chronic Lung Disease. J Immunol 196(12):4839. https://doi.org/10.4049/jimmunol.1600279
117. Charlson ES, Bittinger K, Haas AR, Fitzgerald AS, Frank I, Yadav A, Bushman FD, Collman RG (2011) Topographical continuity of bacterial populations in the healthy human respiratory tract. Am J Respir Crit Care Med 184(8):957–963. https://doi.org/10.1164/rccm.201104-0655OC
118. Dickson RP, Erb-Downward JR, Falkowski NR, Hunter EM, Ashley SL, Huffnagle GB (2018) The lung microbiota of healthy mice are highly variable, cluster by environment, and reflect variation in baseline lung innate immunity. Am J Respir Crit Care Med 198(4): 497–508. https://doi.org/10.1164/rccm.201711-2180OC
119. Günther A, Siebert C, Schmidt R, Ziegler S, Grimminger F, Yabut M, Temmesfeld B, Walmrath D, Morr H, Seeger W (1996) Surfactant alterations in severe pneumonia, acute respiratory distress syndrome, and cardiogenic lung edema. Am J Respir Crit Care Med 153(1):176–184. https://doi.org/10.1164/ajrccm.153.1.8542113
120. Zaborina O, Lepine F, Xiao G, Valuckaite V, Chen Y, Li T, Ciancio M, Zaborin A, Petroff E, Turner JR, Rahme LG, Chang E, Alverdy JC (2007) Dynorphin activates quorum sensing quinolone signaling in Pseudomonas aeruginosa. PLoS Pathog 3(3):e35. https://doi.org/10.1371/journal.ppat.0030035
121. Zhang X, Essmann M, Burt ET, Larsen B (2000) Estrogen effects on Candida albicans: a potential virulence-regulating mechanism. J Infect Dis 181(4):1441–1446. https://doi.org/10.1086/315406
122. Wypych TP, Wickramasinghe LC, Marsland BJ (2019) The influence of the microbiome on respiratory health. Nat Immunol 20(10):1279–1290. https://doi.org/10.1038/s41590-019-0451-9
123. Mah KW, Björkstén B, Lee BW, van Bever HP, Shek LP, Tan TN, Lee YK, Chua KY (2006) Distinct pattern of commensal gut microbiota in toddlers with eczema. Int Arch Allergy Immunol 140(2):157–163. https://doi.org/10.1159/000092555
124. Qian G, Jiang W, Zou B, Feng J, Cheng X, Gu J, Chu T, Niu C, He R, Chu Y, Lu M (2018) LPS inactivation by a host lipase allows lung epithelial cell sensitization for allergic asthma. J Exp Med 215(9):2397–2412. https://doi.org/10.1084/jem.20172225
125. Burke DG, Fouhy F, Harrison MJ, Rea MC, Cotter PD, O'Sullivan O, Stanton C, Hill C, Shanahan F, Plant BJ, Ross RP (2017) The altered gut microbiota in adults with cystic fibrosis. BMC Microbiol 17(1):58. https://doi.org/10.1186/s12866-017-0968-8
126. Vinolo MAR, Rodrigues HG, Nachbar RT, Curi R (2011) Regulation of inflammation by short chain fatty acids. Nutrients 3(10). https://doi.org/10.3390/nu3100858
127. Samuelson DR, Welsh DA, Shellito JE (2015) Regulation of lung immunity and host defense by the intestinal microbiota. Front Microbiol 6:1085
128. McGhee JR, Fujihashi K (2012) Inside the mucosal immune system. PLoS Biol 10(9): e1001397. https://doi.org/10.1371/journal.pbio.1001397
129. Elson CO, Alexander KL (2015) Host-microbiota interactions in the intestine. Dig Dis 33(2): 131–136. https://doi.org/10.1159/000369534
130. Skelly AN, Sato Y, Kearney S, Honda K (2019) Mining the microbiota for microbial and metabolite-based immunotherapies. Nat Rev Immunol 19(5):305–323. https://doi.org/10.1038/s41577-019-0144-5

131. Karmarkar D, Rock KL (2013) Microbiota signalling through MyD88 is necessary for a systemic neutrophilic inflammatory response. Immunology 140(4):483–492. https://doi.org/10.1111/imm.12159
132. Le Poul E, Loison C, Struyf S, Springael J-Y, Lannoy V, Decobecq M-E, Brezillon S, Dupriez V, Vassart G, Van Damme J, Parmentier M, Detheux M (2003) Functional characterization of human receptors for short chain fatty acids and their role in polymorphonuclear cell activation *. J Biol Chem 278(28):25481–25489. https://doi.org/10.1074/jbc.M301403200
133. Baltierra-Trejo E, Sánchez-Yáñez JM, Buenrostro-Delgado O, Márquez-Benavides L (2015) Production of short-chain fatty acids from the biodegradation of wheat straw lignin by Aspergillus fumigatus. Bioresour Technol 196:418–425. https://doi.org/10.1016/j.biortech.2015.07.105
134. Xiros C, Shahab RL, Studer MH-P (2019) A cellulolytic fungal biofilm enhances the consolidated bioconversion of cellulose to short chain fatty acids by the rumen microbiome. Appl Microbiol Biotechnol 103(8):3355–3365. https://doi.org/10.1007/s00253-019-09706-1
135. Wheeler Matthew L, Limon Jose J, Bar Agnieszka S, Leal Christian A, Gargus M, Tang J, Brown J, Funari Vincent A, Wang Hanlin L, Crother Timothy R, Arditi M, Underhill David M, Iliev Iliyan D (2016) Immunological consequences of intestinal fungal dysbiosis. Cell Host Microbe 19(6):865–873. https://doi.org/10.1016/j.chom.2016.05.003
136. Gollwitzer ES, Saglani S, Trompette A, Yadava K, Sherburn R, McCoy KD, Nicod LP, Lloyd CM, Marsland BJ (2014) Lung microbiota promotes tolerance to allergens in neonates via PD-L1. Nat Med 20(6):642–647. https://doi.org/10.1038/nm.3568
137. Herbst T, Sichelstiel A, Schär C, Yadava K, Bürki K, Cahenzli J, McCoy K, Marsland BJ, Harris NL (2011) Dysregulation of allergic airway inflammation in the absence of microbial colonization. Am J Respir Crit Care Med 184(2):198–205. https://doi.org/10.1164/rccm.201010-1574OC
138. Allie SR, Bradley JE, Mudunuru U, Schultz MD, Graf BA, Lund FE, Randall TD (2019) The establishment of resident memory B cells in the lung requires local antigen encounter. Nat Immunol 20(1):97–108. https://doi.org/10.1038/s41590-018-0260-6
139. Molyneaux PL, Mallia P, Cox MJ, Footitt J, Willis-Owen SAG, Homola D, Trujillo-Torralbo M-B, Elkin S, Kon OM, Cookson WOC, Moffatt MF, Johnston SL (2013) Outgrowth of the bacterial airway microbiome after rhinovirus exacerbation of chronic obstructive pulmonary disease. Am J Respir Crit Care Med 188(10):1224–1231. https://doi.org/10.1164/rccm.201302-0341OC
140. Chiu L, Bazin T, Truchetet M-E, Schaeverbeke T, Delhaes L, Pradeu T (2017) Protective microbiota: from localized to long-reaching co-immunity. Front Immunol 8:1678
141. Trompette A, Gollwitzer ES, Yadava K, Sichelstiel AK, Sprenger N, Ngom-Bru C, Blanchard C, Junt T, Nicod LP, Harris NL, Marsland BJ (2014) Gut microbiota metabolism of dietary fiber influences allergic airway disease and hematopoiesis. Nat Med 20(2):159–166. https://doi.org/10.1038/nm.3444
142. Bingula R, Filaire M, Radosevic-Robin N, Bey M, Berthon J-Y, Bernalier-Donadille A, Vasson M-P, Filaire E (2017) Desired turbulence? Gut-lung axis, immunity, and lung cancer. J Oncol 2017:5035371. https://doi.org/10.1155/2017/5035371
143. Cait A, Hughes MR, Antignano F, Cait J, Dimitriu PA, Maas KR, Reynolds LA, Hacker L, Mohr J, Finlay BB, Zaph C, McNagny KM, Mohn WW (2018) Microbiome-driven allergic lung inflammation is ameliorated by short-chain fatty acids. Mucosal Immunol 11(3):785–795. https://doi.org/10.1038/mi.2017.75
144. Hynes MJ, Murray SL, Khew GS, Davis MA (2008) Genetic analysis of the role of peroxisomes in the utilization of acetate and fatty acids in aspergillus nidulans. Genetics 178(3):1355–1369. https://doi.org/10.1534/genetics.107.085795
145. Yin Y, Wang Y, Zhu L, Liu W, Liao N, Jiang M, Zhu B, Yu HD, Xiang C, Wang X (2013) Comparative analysis of the distribution of segmented filamentous bacteria in humans, mice and chickens. ISME J 7(3):615–621. https://doi.org/10.1038/ismej.2012.128

146. Bradley CP, Teng F, Felix KM, Sano T, Naskar D, Block KE, Huang H, Knox KS, Littman DR, Wu H-JJ (2017) Segmented filamentous bacteria provoke lung autoimmunity by inducing gut-lung axis Th17 cells expressing dual TCRs. Cell Host Microbe 22(5):697–704.e694. https://doi.org/10.1016/j.chom.2017.10.007
147. Huang Y, Mao K, Chen X, Sun M-a, Kawabe T, Li W, Usher N, Zhu J, Urban JF, Paul WE, Germain RN (2018) S1P-dependent interorgan trafficking of group 2 innate lymphoid cells supports host defense. Science 359(6371):114. https://doi.org/10.1126/science.aam5809
148. Ichinohe T, Pang IK, Kumamoto Y, Peaper DR, Ho JH, Murray TS, Iwasaki A (2011) Microbiota regulates immune defense against respiratory tract influenza A virus infection. Proc Natl Acad Sci 108(13):5354. https://doi.org/10.1073/pnas.1019378108
149. Gill HS, Rutherfurd KJ, Cross ML, Gopal PK (2001) Enhancement of immunity in the elderly by dietary supplementation with the probiotic Bifidobacterium lactis HN019. Am J Clin Nutr 74(6):833–839. https://doi.org/10.1093/ajcn/74.6.833
150. King DE, Egan BM, Woolson RF, Mainous AG III, Al-Solaiman Y, Jesri A (2007) Effect of a high-fiber diet vs a fiber-supplemented diet on c-reactive protein level. Arch Intern Med 167(5):502–506. https://doi.org/10.1001/archinte.167.5.502
151. Varraso R, Chiuve SE, Fung TT, Barr RG, Hu FB, Willett WC, Camargo CA (2015) Alternate Healthy Eating Index 2010 and risk of chronic obstructive pulmonary disease among US women and men: prospective study. BMJ 350:h286. https://doi.org/10.1136/bmj.h286
152. Ruane D, Brane L, Reis BS, Cheong C, Poles J, Do Y, Zhu H, Velinzon K, Choi J-H, Studt N, Mayer L, Lavelle EC, Steinman RM, Mucida D, Mehandru S (2013) Lung dendritic cells induce migration of protective T cells to the gastrointestinal tract. J Exp Med 210(9): 1871–1888. https://doi.org/10.1084/jem.20122762
153. Li W, Zhu Y, Liao Q, Wang Z, Wan C (2019) Characterization of gut microbiota in children with pulmonary tuberculosis. BMC Pediatr 19(1):445. https://doi.org/10.1186/s12887-019-1782-2
154. Luo M, Liu Y, Wu P, Luo D-X, Sun Q, Zheng H, Hu R, Pandol SJ, Li Q-F, Han Y-P, Zeng Y (2017) Alternation of gut microbiota in patients with pulmonary tuberculosis. Front Physiol 8: 822
155. Dubourg G, Lagier JC, Armougom F, Robert C, Hamad I, Brouqui P, Raoult D (2013) The gut microbiota of a patient with resistant tuberculosis is more comprehensively studied by culturomics than by metagenomics. Eur J Clin Microbiol Infect Dis 32(5):637–645. https://doi.org/10.1007/s10096-012-1787-3
156. Winglee K, Eloe-Fadrosh E, Gupta S, Guo H, Fraser C, Bishai W (2014) Aerosol mycobacterium tuberculosis infection causes rapid loss of diversity in gut microbiota. PLoS One 9(5): e97048. https://doi.org/10.1371/journal.pone.0097048
157. Dumas A, Corral D, Colom A, Levillain F, Peixoto A, Hudrisier D, Poquet Y, Neyrolles O (2018) The host microbiota contributes to early protection against lung colonization by mycobacterium tuberculosis. Front Immunol 9:2656
158. Huda MN, Lewis Z, Kalanetra KM, Rashid M, Ahmad SM, Raqib R, Qadri F, Underwood MA, Mills DA, Stephensen CB (2014) Stool microbiota and vaccine responses of infants. Pediatrics 134(2):e362–e372. https://doi.org/10.1542/peds.2013-3937
159. Perry S, de Jong BC, Solnick JV, de la Luz Sanchez M, Yang S, Lin PL, Hansen LM, Talat N, Hill PC, Hussain R, Adegbola RA, Flynn J, Canfield D, Parsonnet J (2010) Infection with helicobacter pylori is associated with protection against tuberculosis. PLoS One 5(1):e8804. https://doi.org/10.1371/journal.pone.0008804
160. Arnold IC, Hutchings C, Kondova I, Hey A, Powrie F, Beverley P, Tchilian E (2015) Helicobacter hepaticus infection in BALB/c mice abolishes subunit-vaccine-induced protection against M. tuberculosis. Vaccine 33(15):1808–1814. https://doi.org/10.1016/j.vaccine.2015.02.041
161. Rinninella E, Raoul P, Cintoni M, Franceschi F, Miggiano GA, Gasbarrini A, Mele MC (2019) What is the healthy gut microbiota composition? a changing ecosystem across age,

environment, diet, and diseases. Microorganisms 7(1). https://doi.org/10.3390/microorganisms7010014
162. Atarashi K, Tanoue T, Shima T, Imaoka A, Kuwahara T, Momose Y, Cheng G, Yamasaki S, Saito T, Ohba Y, Taniguchi T, Takeda K, Hori S, Ivanov II, Umesaki Y, Itoh K, Honda K (2011) Induction of colonic regulatory T cells by indigenous Clostridium species. Science 331(6015):337. https://doi.org/10.1126/science.1198469
163. Sokol H, Pigneur B, Watterlot L, Lakhdari O, Bermúdez-Humarán LG, Gratadoux J-J, Blugeon S, Bridonneau C, Furet J-P, Corthier G, Grangette C, Vasquez N, Pochart P, Trugnan G, Thomas G, Blottière HM, Doré J, Marteau P, Seksik P, Langella P (2008) Faecalibacterium prausnitzii is an anti-inflammatory commensal bacterium identified by gut microbiota analysis of Crohn disease patients. Proc Natl Acad Sci 105(43):16731. https://doi.org/10.1073/pnas.0804812105
164. Maji A, Misra R, Dhakan DB, Gupta V, Mahato NK, Saxena R, Mittal P, Thukral N, Sharma E, Singh A, Virmani R, Gaur M, Singh H, Hasija Y, Arora G, Agrawal A, Chaudhry A, Khurana JP, Sharma VK, Lal R, Singh Y (2018) Gut microbiome contributes to impairment of immunity in pulmonary tuberculosis patients by alteration of butyrate and propionate producers. Environ Microbiol 20(1):402–419. https://doi.org/10.1111/1462-2920.14015
165. Smith PM, Howitt MR, Panikov N, Michaud M, Gallini CA, Bohlooly-Y M, Glickman JN, Garrett WS (2013) The microbial metabolites, short-chain fatty acids, regulate colonic Treg cell homeostasis. Science 341(6145):569. https://doi.org/10.1126/science.1241165
166. Lachmandas E, van den Heuvel CNAM, Damen MSMA, Cleophas MCP, Netea MG, van Crevel R (2016) Diabetes mellitus and increased tuberculosis susceptibility: the role of short-chain fatty acids. J Diabetes Res 2016:6014631. https://doi.org/10.1155/2016/6014631
167. Shafiani S, Gs T-H, Kariyone A, Takatsu K, Urdahl KB (2010) Pathogen-specific regulatory T cells delay the arrival of effector T cells in the lung during early tuberculosis. J Exp Med 207(7):1409–1420. https://doi.org/10.1084/jem.20091885
168. Hong B-Y, Maulén NP, Adami AJ, Granados H, Balcells ME, Cervantes J (2016) Microbiome changes during tuberculosis and antituberculous therapy. Clin Microbiol Rev 29(4):915–926
169. Cheung MK, Lam WY, Fung WYW, Law PTW, Au CH, Nong W, Kam KM, Kwan HS, Tsui SKW (2013) Sputum microbiota in tuberculosis as revealed by 16S rRNA pyrosequencing. PLoS One 8(1):e54574. https://doi.org/10.1371/journal.pone.0054574
170. Cui Z, Zhou Y, Li H, Zhang Y, Zhang S, Tang S, Guo X (2012) Complex sputum microbial composition in patients with pulmonary tuberculosis. BMC Microbiol 12(1):276. https://doi.org/10.1186/1471-2180-12-276
171. Zhou Y, Lin F, Cui Z, Zhang X, Hu C, Shen T, Chen C, Zhang X, Guo X (2015) Correlation between either cupriavidus or porphyromonas and primary pulmonary tuberculosis found by analysing the microbiota in patients' bronchoalveolar lavage fluid. PLoS One 10(5):e0124194. https://doi.org/10.1371/journal.pone.0124194
172. Wu J, Liu W, He L, Huang F, Chen J, Cui P, Shen Y, Zhao J, Wang W, Zhang Y, Zhu M, Zhang W, Zhang Y (2013) Sputum microbiota associated with new, recurrent and treatment failure tuberculosis. PLoS One 8(12):e83445. https://doi.org/10.1371/journal.pone.0083445
173. Hu Y, Feng Y, Wu J, Liu F, Zhang Z, Hao Y, Liang S, Li B, Li J, Lv N, Xu Y, Zhu B, Sun Z (2019) The gut microbiome signatures discriminate healthy from pulmonary tuberculosis patients. Front Cell Infect Microbiol 9:90
174. Vázquez-Pérez JA, Carrillo CO, Iñiguez-García MA, Romero-Espinoza I, Márquez-García JE, Falcón LI, Torres M, Herrera MT (2020) Alveolar microbiota profile in patients with human pulmonary tuberculosis and interstitial pneumonia. Microb Pathog 139:103851. https://doi.org/10.1016/j.micpath.2019.103851
175. Walter J, Armet AM, Finlay BB, Shanahan F (2020) Establishing or exaggerating causality for the gut microbiome: lessons from human microbiota-associated rodents. Cell 180(2):221–232. https://doi.org/10.1016/j.cell.2019.12.025

176. Cadena AM, Ma Y, Ding T, Bryant M, Maiello P, Geber A, Lin PL, Flynn JL, Ghedin E (2018) Profiling the airway in the macaque model of tuberculosis reveals variable microbial dysbiosis and alteration of community structure. Microbiome 6(1):180. https://doi.org/10.1186/s40168-018-0560-y
177. Nakhaee M, Rezaee A, Basiri R, Soleimanpour S, Ghazvini K (2018) Relation between lower respiratory tract microbiota and type of immune response against tuberculosis. Microb Pathog 120:161–165. https://doi.org/10.1016/j.micpath.2018.04.054
178. Botero LE, Delgado-Serrano L, Cepeda ML, Bustos JR, Anzola JM, Del Portillo P, Robledo J, Zambrano MM (2014) Respiratory tract clinical sample selection for microbiota analysis in patients with pulmonary tuberculosis. Microbiome 2(1):29. https://doi.org/10.1186/2049-2618-2-29
179. Hu Y, Kang Y, Liu X, Cheng M, Dong J, Sun L, Zhu Y, Ren X, Yang Q, Chen X, Jin Q, Yang F (2020) Distinct lung microbial community states in patients with pulmonary tuberculosis. Sci China Life Sci 63(10):1522–1533. https://doi.org/10.1007/s11427-019-1614-0
180. Kayukova L, Berikova E (2020) Modern anti-tuberculosis drugs and their classification. Part I: first-line drugs. Pharm Chem J 54:1–9
181. Jnawali HN, Ryoo S (2013) First-and second-line drugs and drug resistance. In: Tuberculosis-current issues in diagnosis and management, vol 20. IntechOpen, London, pp 163–180
182. Eribo OA, du Plessis N, Ozturk M, Guler R, Walzl G, Chegou NN (2020) The gut microbiome in tuberculosis susceptibility and treatment response: guilty or not guilty? Cell Mol Life Sci 77(8):1497–1509
183. Wipperman MF, Fitzgerald DW, Juste MAJ, Taur Y, Namasivayam S, Sher A, Bean JM, Bucci V, Glickman MS (2017) Antibiotic treatment for Tuberculosis induces a profound dysbiosis of the microbiome that persists long after therapy is completed. Sci Rep 7(1):1–11
184. O'Toole RF, Gautam SS (2018) The host microbiome and impact of tuberculosis chemotherapy. Tuberculosis 113:26–29
185. Liu Y, Wang J, Wu C (2021) Microbiota and tuberculosis: a potential role of probiotics, and postbiotics. Front Nutr 8:626254
186. Gibson GR, Hutkins R, Sanders ME, Prescott SL, Reimer RA, Salminen SJ, Scott K, Stanton C, Swanson KS, Cani PD (2017) Expert consensus document: the International Scientific Association for Probiotics and Prebiotics (ISAPP) consensus statement on the definition and scope of prebiotics. Nat Rev Gastroenterol Hepatol 14(8):491–502
187. Hori T, Matsuda K, Oishi K (2020) Probiotics: a dietary factor to modulate the gut microbiome, host immune system, and gut–brain interaction. Microorganisms 8(9):1401
188. Bravo M, Combes T, Martinez FO, Cerrato R, Rey J, Garcia-Jimenez W, Fernandez-Llario P, Risco D, Gutierrez-Merino J (2019) Lactobacilli isolated from wild boar (Sus scrofa) antagonize mycobacterium bovis Bacille Calmette-Guerin (BCG) in a species-dependent manner. Front Microbiol 10:1663
189. Negi S, Pahari S, Bashir H, Agrewala JN (2019) Gut microbiota regulates mincle mediated activation of lung dendritic cells to protect against Mycobacterium tuberculosis. Front Immunol 10:1142
190. Schierwagen R, Uschner FE, Ortiz C, Torres S, Brol MJ, Tyc O, Gu W, Grimm C, Zeuzem S, Plamper A (2020) The role of macrophage-inducible C-Type lectin in different stages of chronic liver disease. Front Immunol 11:1352
191. Khusro A, Aarti C, Mahizhaveni B, Dusthackeer A, Agastian P, Esmail GA, Ghilan A-KM, Al-Dhabi NA, Arasu MV (2020) Purification and characterization of anti-tubercular and anticancer protein from Staphylococcus hominis strain MANF2: in silico structural and functional insight of peptide. Saudi J Biol Sci 27(4):1107–1116
192. Carroll J, Draper LA, O'Connor PM, Coffey A, Hill C, Ross RP, Cotter PD, O'Mahony J (2010) Comparison of the activities of the lantibiotics nisin and lacticin 3147 against clinically significant mycobacteria. Int J Antimicrob Agents 36(2):132–136

193. Negatu DA, Liu JJ, Zimmerman M, Kaya F, Dartois V, Aldrich CC, Gengenbacher M, Dick T (2018) Whole-cell screen of fragment library identifies gut microbiota metabolite indole propionic acid as antitubercular. Antimicrob Agents Chemother 62(3):e01571–e01517
194. Namasivayam S, Sher A, Glickman MS, Wipperman MF (2018) The microbiome and tuberculosis: early evidence for cross talk. MBio 9(5):e01420–e01418
195. Cui Z, Zhou Y, Li H, Zhang Y, Zhang S, Tang S, Guo X (2012) Complex sputum microbial composition in patients with pulmonary tuberculosis. BMC Microbiol 12(1):1–8
196. Eshetie S, Van Soolingen D (2019) The respiratory microbiota: new insights into pulmonary tuberculosis. BMC Infect Dis 19(1):1–7
197. Naidoo CC, Nyawo GR, Wu BG, Walzl G, Warren RM, Segal LN, Theron G (2019) The microbiome and tuberculosis: state of the art, potential applications, and defining the clinical research agenda. Lancet Respir Med 7(10):892–906

# Lung Microbiome: Friend or Foe of *Mycobacterium tuberculosis* 12

Summaya Perveen and Rashmi Sharma

**Abstract**

Tuberculosis, or commonly called TB, is one of the fatal contagious diseases claiming 4000 lives daily (WHO report, 2020). Primarily a curable disease, TB has become a global health issue with emerging resistant strains. According to the WHO, an estimated 10 million individuals developed active TB with 0.5 million people with drug-resistant *Mycobacterium tuberculosis* (Mtb) strains globally in 2019. The multimodal and multidrug TB chemotherapy leads to lower patient compliance, resulting in development of multidrug-resistant strains. Microbial communities inhabit the lungs in variable niches with respect to nutrient availability and immune response. Mycobacteria, the causative agent of TB have a unique capability to adapt to the complex microenvironment of the host, acclimatize, and adapt to host microbiome and immune surveillance. There exists a crosstalk between mycobacteria, host microbiome and the host immune cells leading to a "tug of war" resulting in either clearance of mycobacteria from the host or establishment of infection. However, there is a need of further studies and characterization of the multiple players in the growth and sustenance of infection which would help in deciding the course of treatment for drug sensitive and resistant strains. This chapter would elucidate the role of host lung microbiome in pathogenesis, spread, prevention and clinical interventions in tuberculosis.

S. Perveen · R. Sharma (✉)
Infectious Diseases Division, CSIR-Indian Institute of Integrative Medicine, Jammu, India

Academy of Scientific and Innovative Research (AcSIR), Ghaziabad, India
e-mail: rashmi.sharma.09@iiim.res.in



G. Gupta et al. (eds.), *Microbiome in Inflammatory Lung Diseases*,
https://doi.org/10.1007/978-981-16-8957-4_12

**Keywords**

*Mycobacterium tuberculosis* · Lung microbiome · Multidrug resistance · Host Immune response

## 12.1 Introduction

Microbes have been known to inhabit the earth ever since its creation. The human body is the reservoir of microbes whose reduction and alteration are found to be associated with several diseases. The human microbiome is integral to the development, progression and exacerbation of a disorder. Human microbiome comprises all the microbial population in our body. It is linked to many diseases like asthma, neurodegenerative diseases, cancer, diabetes and a lot more. Thus, the knowledge of complete microbiome and its relation with disease relapse and treatment are the pressing priority [1–3].

Gut microbiota and their association with diseases have been explored extensively, but the interplay of lung microbiome with several respiratory disorders is yet to be unveiled. One such respiratory illness is tuberculosis (TB). It is one of the deadliest infectious diseases of lungs caused by *Mycobacterium tuberculosis* (*Mtb*). According to world TB report, TB killed 1.2 million people in 2019, out of which 0.2 million were co-infected with HIV [4].

The TB chemotherapy consists of the first-line drugs (rifampicin, isoniazid, ethambutol and pyrazinamide) and the second-line drugs (bedaquiline, delamanid, fluoroquinolones, clofazimine, cycloserine, para-aminosalicylic acid) [5]. However, with the long duration of chemotherapy, patient falls out of compliance, leading to advent of drug-resistant strains. With the amplified cases of drug-resistant strains, multidrug-resistant tuberculosis (MDR-TB) and extensively drug-resistant tuberculosis (XDR-TB) challenge the public health care system severely, leading to treatment failure, relapse and death. World Health Organisation (WHO) has reported about 6.8% of incident TB cases of relapse in 2020 [4]. The reasons of treatment relapse could be many, including patient non-compliance, drug resistance or toxicity [6].

Tuberculosis increases the risk of long-term lung impairment as well. It results in a myriad of lung pathologies, including cavitation and fibrosis [7]. There is a heterogeneity in the lung damage from tuberculosis, varying from mild to very severely damaged lungs. This variability could be a result of a number of factors at play, with host microbiome being one. The disorder in lung microbiome, abundance or depletion has been reported to be associated with relapse and progression of TB [8]. Thus, it is necessary to understand the lung microbiome of an individual and compare the microbiota of healthy and diseased subject for better treatment outcome and cure. This chapter would shed light on lung microbiome and plausible correlation to TB pathogenesis. We would also be discussing the impact of TB chemotherapy on lung microbiome and vice versa, pressing the need for exploring the field for better attuning the treatment with the host microbiome for better treatment outcome.

## 12.2 Lung Microbiome

The life on earth begins from a single cell to the higher order multi-cellular being. As the life developed and evolved, the single-celled microorganisms started residing each site on the earth. They can survive in the flourishing environment as well as in the extreme conditions. Several communities of microorganisms inhabited the mammals, living in symbiotic association, where the host provides a niche for their development and survival and the microbe aids in nutrients biotransformation, detoxification, educating immune response and protecting against pathogenic microbes [9]. Thus, the mammals are hybrid organism composed of diverse group of host and microbial cells functioning in continuous dynamic and mutual equilibrium.

The human microbiota forms a complex community comprising of diverse group of bacteria, fungi, protozoa and viruses that are collectively found within the host. The microbes share a similar environment and are subjected to changes during the course of life as a result to multiple external and internal factors including environmental, clinical interventions and diseased states. The newly explored field of microbiomics investigates (1) the characterisation of microbiome at diverse setting in the host body, (2) microbial gene analysis, (3) influence of microbes on the host and (4) the role of microbiome in the diseased conditions. At the dawn of next generation sequencing, the field of microbiomics has advanced with rapid amplification and sequencing of selected microbial DNA segments, their identification and diversity based on sequence similarity [10]. The human microbiome project has evaluated distinct body habitats (oral cavity, skin, lower gastrointestinal (GI) tract and vaginal samples) for the microbiome diversity determination in a group of healthy volunteers. The 16S rRNA gene analysis was performed using pyrosequencing as they are the conserved bacterial gene with hypervariable regions like V1, V2, V3, V6 and V7 that vary among species. The human microbiome project established that the diversity and richness of each habitat-specific microbes differ extensively even among healthy subjects, with solid niche specialization both within (alpha diversity) and amongst individuals (beta diversity) [11].

The microbiome of the lung was historically neglected and it was thought that the lung is devoid of microorganisms forming a sterile environment. Even the human microbiome project initially excluded the microbiota of the lungs [12]. However, recent studies have defined lung microbiota and its impact in healthy and diseased individuals and provided the researchers a new standpoint to learn about the respiratory diseases [13]. Hence, it is necessary to comprehend the changes in the lung microbiome in a diseased state, marked by microbial elimination and immigration along with the growth factors like competition for nutrients, attachment site, temperature and host–microbe interactions [14].

The lung microbiome is divided into two sections: (1) the upper respiratory tract or URT that comprised of nostrils, rhino-pharynx and oropharynx and (2) the lower respiratory tract or LRT comprising of alveoli and bronchi (Fig. 12.1) [15]. The URT and LRT colonization begins with the birth of an individual and is influenced by several modalities such as method of delivery either vaginal or caesarean,

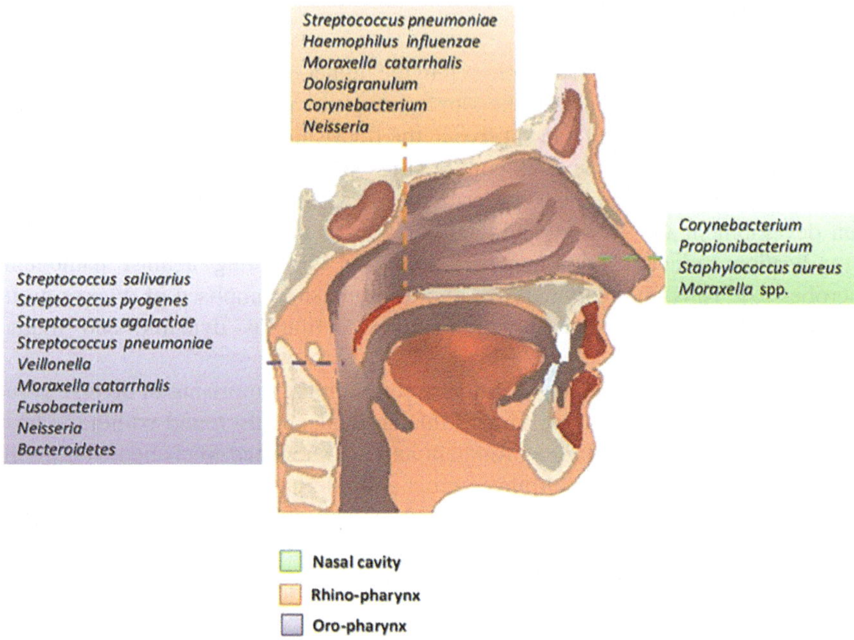


**Fig. 12.1** Lung microbiome; A representation of microbes inhabiting the upper respiratory tract in humans (Reproduced from Santacroce et al. [14])

environment, diet, age and antibiotics. The URT is majorly colonized by the microbiota belonging to the genera of *Proteobacteria Firmicutes, Actinobacteria, Bacteroidetes and Fusobacteria*. The nostril microbiome is enriched with *Propionibacterium* spp. (*Actinobacteria*) as they are able to metabolize the sebum released by glands of the nostril's epithelium. The pH decrease in the nostrils facilitates the growth of *Staphylococcus coagulase* and *Corynebacterium* belonging to *Firmicutes* and *Actinobacteria* genera, respectively. The rhino-pharynx microbiome exhibits the supremacy of *Moraxella, Dolosigranulum, Corynebacterium, Staphylococcus or Streptococcus* spp., *Haemophilus* and *Neisseria*, playing a crucial part in upholding the equilibrium of related species, preventing the development of pathogens and assisting host immunity [14, 16–18]. The oropharynx is rich in the genus *Streptococcus, Veillonella, Neisseria, Fusobacterium* and *Bacteroidetes*. The LRT is dominated by bacteria (*Pseudomonas, Streptococcus, Fusobacterium, Prevotella, Acinetobacter, Megasphaera, Sphingomonas, Staphylococcus, Veillonella*) as well as fungi (*Aspergillus, Cladosporium, Penicillium, Candida, Neosartorya, Eurotium* and *Saccharomyces*) [14]. The respiratory tract also displays the presence of several virus groups with known and unknown pathogenicity. Culture-dependent molecular assays are performed to detect viral particles in the upper and lower respiratory tract. Some of the very common virus in respiratory tract are rhinoviruses, paramyxoviruses, along with enteroviruses, orthomyxoviruses, and parvoviruses [19].

The dysbiosis of lung microbiome occurs in diseased conditions characterized by increase of certain pathogenic bacterial population and decrease of other commensal microbes [16]. The dysbiosis of *Anaerococcus, Peptoniphilus, Finegoldia, Dialister, Parvimonas, Corynebacterium, Staphylococcus* and *Propionibacterium* genera is related to chronic rhinosinusitis patients [20, 21]. The core microbiota of the patients suffering from cystic fibrosis comprised of *Streptococcus, Prevotella, Rothia, actinomyces* and *Veillonella* along with less prevalent *Pseudomonas, Burkholderia, Stenotrophomonas* and *Achromobacter* bacterial community [22]. The adult asthmatic patients displayed increased number of *Haemophilus* spp. accompanied by decreased amount of *Prevotella* spp. as compared to the control individuals [1]. Thus, the diseased conditions in human are not only associated with the genes of microbiota that resides in their body but also the expression of the microbial genes.

## 12.3 Mtb Pathogenesis

*Mycobacterium tuberculosis*, a gram positive actinobacteria causing tuberculosis to almost 10 million of individuals all around the globe and killing approximately 1.4 million of diseased individuals in 2019. Geographically, most people who developed TB belongs to South East Asia, Africa and Western Pacific regions [4, 23]. Tuberculosis is a contagious disease that spreads from one individual to another via aerosols that are produced when an individual with active pulmonary TB sneezes or coughs. When these aerosols containing bacilli are inhaled by a healthy person, the pathogenesis of *Mtb* in the new host commences (Fig. 12.2). At first, *Mtb* is phagocytosed by the alveolar macrophages of the host and apprehended in the phagosomes. The interactions between the pathogen-associated molecular patterns of mycobacteria (for instance, lipoprotein and glycolipids) and the pattern recognition receptors (like Toll-like receptors) on the macrophages cause phagocytosis of *Mtb*. Later, the phagosome fuses with lysosome forming phagolysosome to kill the bacteria in the presence of hydrolytic enzymes and acidic pH. The bactericidal activity is also accomplished by the generation of reactive oxygen and nitrogen intermediates by the macrophages [24, 25].

The bacilli that survives the harsh conditions of the macrophages replicates and divides in the phagosomes, increasing the bacterial load in the host and inducing the production of pro-inflammatory cytokines like IL-1$\alpha$, IL-6, IL-1$\beta$ and TNF-$\alpha$. The inflammatory response persuades the recruitment of immune cells including neutrophils, monocytes and dendritic cells to the infection site. Increased level of TNF-$\alpha$ results in controlled *Mtb* growth by forming granuloma, hallmark of latent tuberculosis [24, 26]. Granuloma is a compact calcified structure with a central necrotic caseum with dying macrophages surrounded by epithelioid cells, foam cells, dendritic cells, natural killer cells, neutrophils, B cells, T cells, fibroblasts and cells that secrete extracellular matrix components. Granuloma is formed to wall

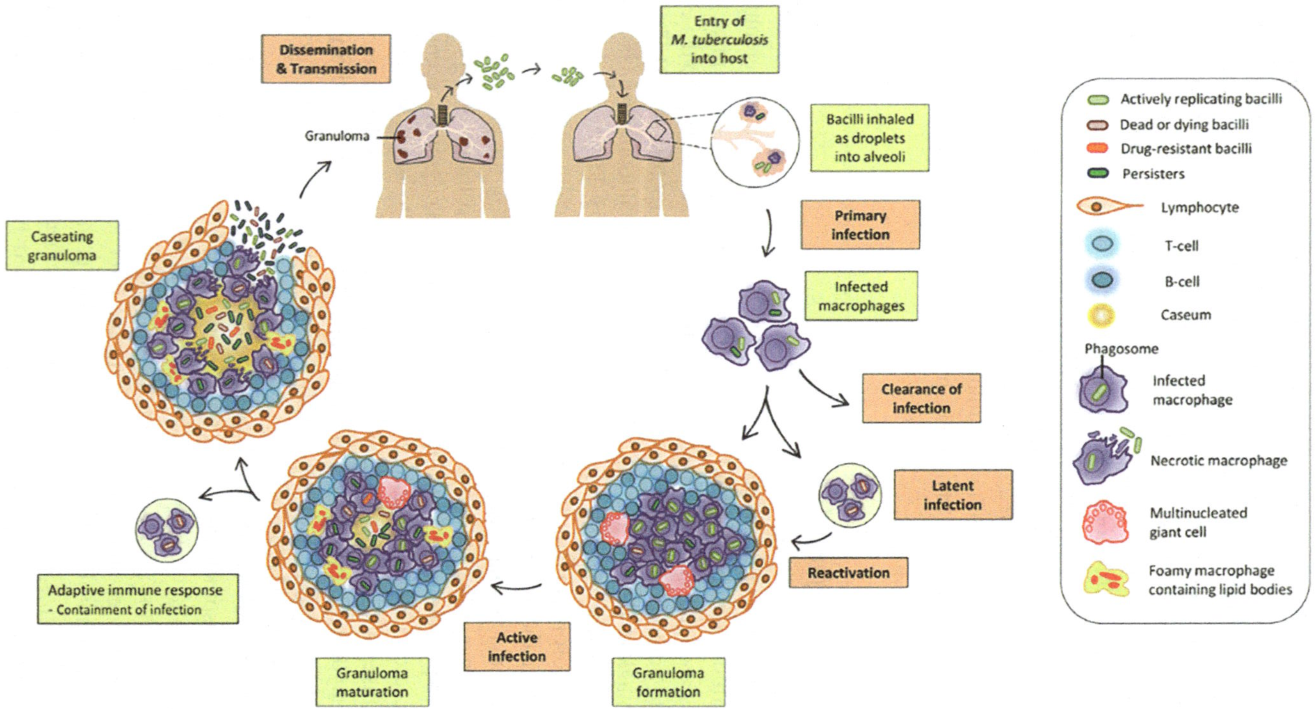


**Fig. 12.2** *Mtb* pathogenesis; Transmission and progression of tuberculosis from one infected individual to another uninfected person (Adapted from Parbhoo et al. [63] with permissions from Wiley online library)

off the pathogen from migrating to other location via lymphatic systems [27]. The production of pro-inflammatory cytokines like TNFα, IFNγ and IL-6 from macrophages along with the adaptive immune defence of the CD4+ and CD8+ T cells restricts the growth of *Mtb* in the granuloma. The bacteria in the granuloma persist in a dormant, metabolically inactive state for ages and only become active when the immune surveillance is compromised either with HIV co-infection, treatment with anti-TNF medications, or cholecalciferol deficiency [25, 26].

The WHO TB global report 2020 states that almost about 25% of world's populace has already been infected with *Mtb*. Yet, only 5–15% of the diseased individuals develops active TB in their lifetime and most of them remain dormant and develop TB only when the immune system is compromised [4]. The disequilibrium in the host immune defense causes the centre of granuloma to undergo caseation and spill active and infectious bacteria from primary location to other new site in the lungs, causing active TB. This results to formation of productive cough that further facilitates spread of bacilli to the environment, continuing the cycle of pathogenesis [23]. Naidoo et al. [28] propose a relation between how tuberculosis pathogenesis is modulated by the risk factors and its associated microbiome imbalance (Fig. 12.3). The health of host is modulated by the risk factors which makes it important to understand the interlink between the diseased microbial state and tuberculosis.

## 12.4 Lung Microbiome in TB Infection

Gut microbiota importance in human body has been studied and explored by various groups around the globe [29, 30] and their association with several disorders like neurodegenerative diseases [2, 31, 32], diabetes [33], cardiovascular diseases [34], cancer [3] and COVID-19 [35] has also been researched extensively. The lung microbiota that was previously neglected is now looked into to understand disease conditions and how the lung microbiota contributes in the treatment or the exacerbation of respiratory diseases.

Dysbiosis of lung microbiota has already been associated in several disease progression like asthma [1], cystic fibrosis [22], chronic rhinosinusitis [20] and many more. The lung microbiota effects on the pathogenesis of *Mtb* and treatment outcome are a new field to explore. In the healthy state, there is a commensal relationship between the host immunity and lung microbiota, attaining host-pathogen equilibrium. But in the immune-impaired state, i.e., infection with *Mtb*, development of either active or latent TB occurs causing heightened inflammation, hampering the resistance to pathogenic microbiota, and change in microbial ecology and diversity, leading to microbiome dysbiosis. Further the comorbidities like malnutrition, diabetes, HIV also influence the disease development and alteration in microbiota [36] (Fig. 12.4). Hence, it is necessary to comprehend the dysbiosis in lung microbiota that occurs in the diseased state. Several groups have worked upon the lung microbiome perturbation caused by tuberculosis.

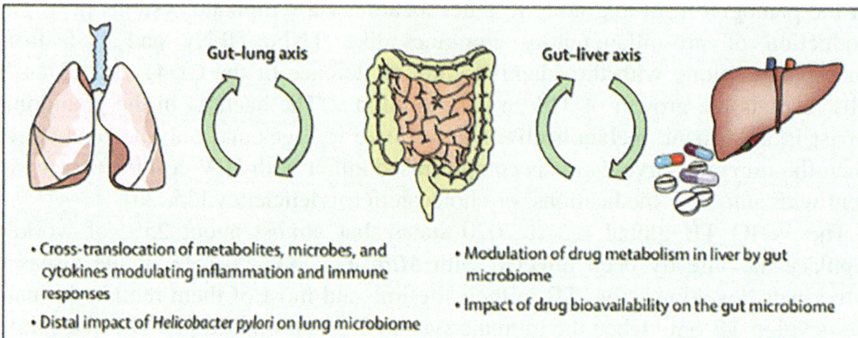


**Factors known to modulate the risk of tuberculosis disease progression, hepatic tuberculosis drug metabolism, and, separately, lung or gut microbiota**

| | Lung microbiota | Gut microbiota | Hepatic drug metabolism |
|---|---|---|---|
| Alcohol | Unknown | ↑Proteobacteria, ↓Bacteroidetes | Promotes antituberculosis drug-induced hepatotoxicity |
| HIV | ↑*Prevotella*, ↑*Veillonella*, ↑*Streptococcus* | ↓Rikenellaceae, ↓Bacteroidaceae, ↓Lachnospiraceae or ↓Ruminococcaceae | Increases risk of tuberculosis drug-induced liver injury |
| Malnutrition | Unknown | ↑*Entreococcus faecalis*, ↑*Streptococcus gallolyticus*, ↓*Faecalibacterium prauznitzii*, ↓*Bacteroides* spp, ↓*Bifidobacterium* spp | Malnutrition promotes antituberculosis-drug induced hepatotoxicity |
| Smoking | Minimal effect on microbiota | ↑Proteobacteria, ↑Bacteroidetes, ↓Actinobacteria, ↓Firmicutes | No data |
| Diabetes | Unknown | ↓Firmicutes | Unknown |
| Air pollution | ↑*Neisseria*, ↑*Streptococcus*, ↓*Tropheryma* | ↑Firmicutes, ↓Bacteroidetes | Unknown |

**Fig. 12.3** The epidemiologically important risk factors that modulates TB and their effect on lung and gut microbiota (Reproduced from Naidoo et al. [28] with approvals from the Elsevier)

Cui et al. evaluated the microbial population in the sputum sample of enrolled pulmonary TB patients [37]. Pyrosequencing was performed to analyse the total DNA extracted from 31 TB patients and 24 healthy individuals. Their study displayed that the TB patients show microbiota diversity that were classified into 24 phyla as compared to the healthy subjects that showed presence of bacteria from 17 phyla. The bacterial population unique to TB patients are *Stenotrophomonas, Thermus, Cupriavidus, Methylobacterium, Pseudomonas, Comamonas, Diaphorobacter, Sphingomonas* and *Mobilicoccus*. They concluded that these

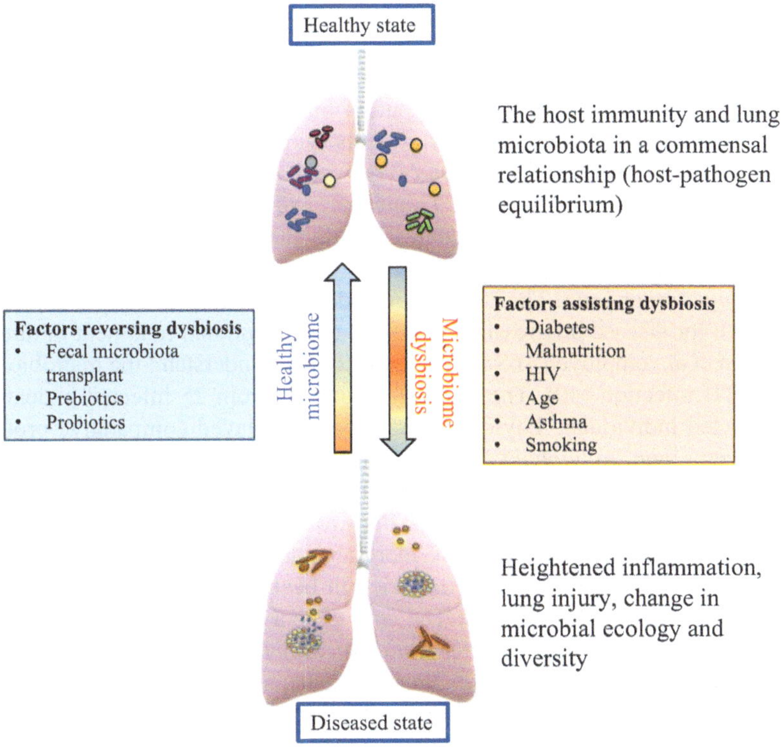


**Fig. 12.4** An overview of the balance between the healthy state and the diseased state associated with lung dysbiosis. Progression to diseased state is accelerated by the risk factor like co-infection, smoking, etc. that alter the lung microbiome, whereas the flowback to the healthy state can be achieved by the Faecal Microbiota Transplant, prebiotics and probiotics

unique foreign bacteria may have a role to play in TB development and progression [37].

Using 16S rRNA pyrosequencing technique, Cheung et al., compared sputum microbiota of 22 *Mtb*-infected individuals with 14 control sputum samples. The phyla *Proteobacteria* and *Bacteroidetes* were more signified in the TB subjects as compared to the controls. The fundamental genera found in the TB sputum microbiota is characterized by *Actinomyces, Fusobacterium, Leptotrichia, Prevotella, Streptococcus* and *Veillonella*. Along with these genera some other less represented genera found in TB samples are *Mogibacterium, Moryella* and *Oribacterium* [38]. Botero et al., collected the sputum, nasal and oropharynx samples from six TB-infected patients and compared the microbiota of lungs with the six healthy control samples. They found that the *Streptococcaceae* spp. are abundant in oropharyngeal samples of TB patients along with enriched fungal community of *Aspergillus* and *Candida* in sputum and oropharyngeal samples, whereas the control subjects showed richness in *Fusobacterium, Actinomyces, Prevotella, Leptotrichia* and *Veillonella*. Their study concluded that oropharynx samples should be analysed to study the alteration of lung microbiota in TB [39]. Recently, lung microbiome association with tuberculosis has been explored. A study conducted by Wu et al. employed 16S RNA sequencing technique to inspect association of *Mtb* infection with the sputum microbiota in TB patients with new infection, recurrent infection and TB patients with therapy collapse. They found abundance of *Granulicatella, Streptococcus* and *Pseudomonas* genera, whereas *Leptotrichia, Prevotella, Catonella, Treponema* and *Coprococcus* were reduced in TB-infected individuals than in the control healthy individuals. They also recognized the richness of *Pseudomonas* in the treatment failure patients with high *Pseudomonas/ Mycobacterium* ratio. Their study concluded that certain bacterial population is responsible for disease progression as well as reinfection and treatment failure [8].

Krishna et al., employed 16S rRNA sequencing to understand the microbiome of lungs in TB infection [40]. They collected samples from 25 infected patients and 16 uninfected individuals. Phylum-level analysis displayed comparative profusion of *Firmicutes* and *Actinobacteria* in TB-infected individuals. Their study also showed difference in the core genera in TB and in healthy person. The earlier study demonstrated the abundance of *Bacteroidetes* and *Proteobacteria* in TB patients, whereas these two phyla are more abundant in the general population in India. Further, *Firmicutes* are generally rich in healthy individual in China but in India they are high in TB patients. The core genera of TB samples found in their study are *Actinomyces, Rothia, Granulicatella, Lactobacillus, Streptococcus, Neisseria, Leptotrichia* and *Veillonella* [40]. Eshetie et al., estimated the relative percentage of lung microbiota genera in healthy as well as *Mtb*-infected individuals. *Streptococcus* (35%), *Neisseria* (27%), *Prevotella* (9%) and *Veillonella* (8%) were rich in TB-infected patients however *Prevotella* (37%), *Gammaproteobacteria* (22%), *Streptococcus* (19%) and *Haemophilus* (15%) were detected in healthy controls. The study identified the exclusive genera for both the groups. *Rothia, Veillonella, Leuconostoc* were exclusively found in TB cases, whereas *Gammaproteobacteria, Haemophilus, Lactobacillus* and *Actinobacillus* were recognized in healthy subjects only [41].

Short chain fatty acids (SCFA) are known to modulate energy metabolism, inflammatory reactions and cholesterol metabolism. It acts as an anti-inflammatory agent via blocking NF-κβ activation and IFN-γ signalling, that have an indispensable role in control of *Mtb* progression and granuloma formation and maintenance. Additionally, SCFA is also recognized to inflate regulatory T cells (Treg) proliferation and their binding with GPCR decreases allergic reaction by reducing dendritic cell-mediated Th2 responses. Treg produces anti-inflammatory cytokines like interleukin-10 to limit the activity of pro-inflammatory cytokines against *Mtb* [42–44] (Fig. 12.5). Segal et al., investigated the contribution of propionate, butyrate and other SCFA in TB progression in HIV patients medicated with antiretroviral-drug-therapy. They used 16S rRNA sequencing to scrutinize the LRT microbiome of bronchoalveolar lavage samples. They found the richness of anaerobes such as *Prevotella* in the LRT which is associated with elevated pulmonary SCFA-induced Tregs which obstructs *Mtb* control by alveolar macrophages. They concluded that antiretroviral therapy has caused pulmonary dysbiosis with the enrichment of oral

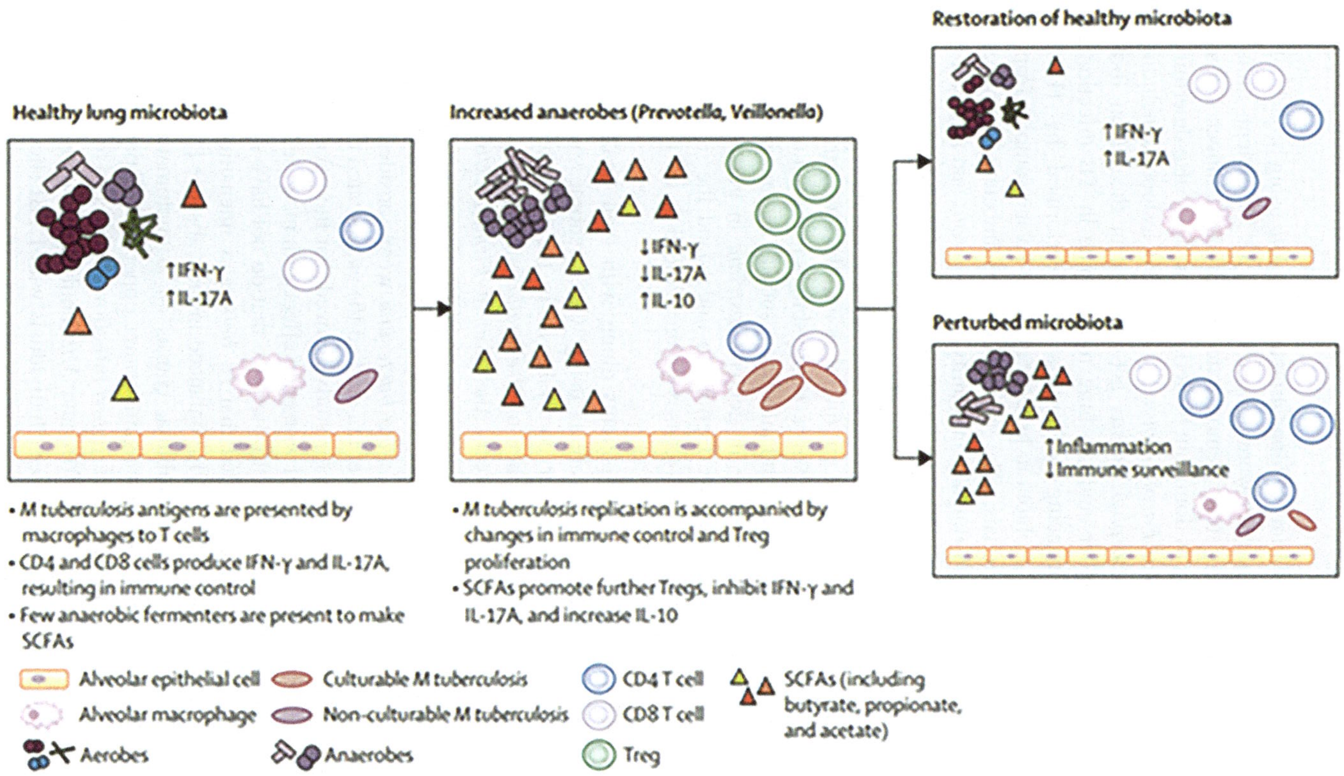


**Fig. 12.5** Overview of how the alteration in lung microbiota and production of SCFA lead to disease progression (Adapted from Naidoo et al. [28] with approval from Elsevier)

anaerobes that can increase the TB risk in HIV patients [45]. The study suggests that the metabolic profile of the microbiota may be a significant determinant for progression of active TB.

Zhou et al., studied microbiome of lungs infected with *Mtb* using bronchoalveolar lavage fluid samples. The samples were collected from 32 TB patients along with 24 healthy individuals. Their study demonstrated significant changes in the LRT microbiota with *Cupriavidus* as dominant bacterial genus. They also determined the richness of *Mycobacteria* and *Porphyromonas* within TB lesions, concluding that *Mycobacteria* along with *Porphyromonas* act together in the lesion formation [46]. The pilot study to understand the lung microbiome in TB infection using shotgun metagenomic sequencing techniques was performed by Hu et al. [47]. They analysed lung microbiota in bronchoalveolar lavage samples from *Mtb*-infected and healthy individuals. The microbiome of infected individual was enriched with *Mtb* along with fungal community, *Ascomycota* and *Basidiomycota*, whereas the uninfected individual microbiome comprised of *Streptococcus, Prevotella, Neisseria, Selenomonas* and *Bifidobacterium.*

The gut microbiota also affects the lung microbiome through the "gut-lung axis" [48]. The metabolites released by the gut microbes into the bloodstreams affect the respiratory tract microbiome and vice versa. The dysbiosis in the gut microbiome has been associated with lung disorders and infections, such as asthma where depletion in genus *Bifidobacteria* and upsurge in genus *Clostridia* in gut are linked with asthma [28, 49]. Several studies are ongoing to understand the influence of gut microbiota on lung microbiome that causes either protection or TB exacerbated infection.

Maji et al., enrolled active TB patients along with their healthy household contacts and analysed their gut microbiota using whole-genome shotgun and 16S rRNA gene sequencing. They observed dysbiosis in the gut microbiota in the infected individuals with the abundance of *Prevotella* and *Bifidobacterium* in the household contacts [42]. In TB subjects, the ratio of *Firmicutes/Bacteroidetes* is increased, directly affecting the concentration of SCFA. Propionate and butyrate-producing bacteria like *Eubacterium rectale, Faecalibacterium prausnitzii, Roseburia, Butyrivibrio* and *Phascolarcto bacterium* were considerably increased in infected individuals [42]. Majlessi et al., studied the exacerbated consequence of gut microbiota in TB progression. Their study showed that the mice gut colonized with *Helicobacter hepaticus* exhibited worsened effect with increased inflammation and lung necrosis when compared with the mice that do not have *H. hepaticus* in their gut microbiome. They concluded that *H. hepaticus* intensified the immune response against *Mtb* that was otherwise balanced in their absence [50].

*Helicobacter pylori*, a gut microbiota, causes an asymptomatic infection of stomach and may provide protection against other opportunistic pathogens. Perry et al. worked on the hypothesis that *H. pylori* infection contributes to the protection against *Mtb*. They examined *H. pylori* and *Mtb* antibody responses in enrolled subjects undergoing tuberculin skin test. Individuals with LTBI showing seropositivity to *H. pylori*- displayed increased *Mtb* specific IFN-$\gamma$ response, and an enhanced Th-1 response when compared to *H. pylori* seronegative individuals.

Their study also revealed that presence of *H. pylori* in the microbiome of Cynomolgus macaques results in less likelihood of progression of *Mtb* infection, suggesting importance of *H. pylori* in providing immune-protection against TB [51]. Negatu et al., performed whole cell screening where they screened 1000 fragments from the Maybridge Ro3 library [52]. Their screen identified indole propionic acid (IPA), a metabolite released by gut microbiota. IPA reduced bacterial load in spleen of the infected mouse by seven-fold, further also represented adequate pharmacokinetic properties. Thus, focussing on the therapeutic activity of gut microbiota against TB [52].

## 12.5 TB Treatment

Tuberculosis treatment comprised of combination of drugs that are given to patients with active TB to completely recover from diseased state. The treatment duration is of 6 months that are divided into two phases: (1) the intensive phase of 2 months and (2) the continuation phase of 4 months [53].

The intensive phase comprised of four drugs: isoniazid (H), rifampicin (R), pyrazinamide (Z) and ethambutol (E) forming HRZE regimen and the continuation phase comprised of isoniazid and rifampicin. The diseased individual completing the WHO approved 6-month guideline for treatment is thought to recover from the disease but compliance to the therapy is challenging for the patient because of the related toxicity and side-affects associated with the drugs along with the lengthy duration of the treatment. The non-compliance of patients leads to drug-resistant TB such MDR-TB or XDR-TB [6].

The WHO consolidated guidelines on drug-resistant tuberculosis categorize the medications for MDR-TB patients into three groups, group A, group B, and group C. To complete the regimen of 4 drugs to treat MDR-TB, combination of drugs from all the groups are administered. Group A consists of prioritized drugs like bedaquiline, linezolid and moxifloxacin/levofloxacin, taken in all regimen. Group B includes possibly added drugs like clofazimine and cycloserine/terizidone. Group C comprised of auxiliary drugs like pyrazinamide, ethambutol, meropenem, ethionamide, *para*-aminosalicylic acid, amikacin and delamanid that are to be given in case drugs from group A and B cannot be administered to complete the 4-drug regimen [54, 55]. The dosage, therapy duration, mode of action and bacterial spectrum of first and second-line drugs are discussed in Table 12.1.

## 12.6 The Effect of TB Treatment on Microbiome

As discussed above, the treatment for TB comprised of combination of antibiotics for a lengthy period [53]. Among these drugs some of them are specific to *Mtb* such as isoniazid, pyrazinamide and ethambutol, as these are prodrugs and only get activated by enzymes of *Mtb* [56]. But rifampicin, a broad-spectrum antibiotic, inhibits the transcription of bacterial genes. Further, treatment of MDR-TB/XDR-

**Table 12.1** Dosage, therapy duration, mechanism of action and bacterial spectrum of first and second-line drugs

| Tuberculosis drugs | | Dosage | Treatment period | Mode of action | Bacterial spectrum |
|---|---|---|---|---|---|
| First-line drugs (drug-susceptible TB) | Isoniazid (H) | 5 mg/kg | 6 months<br>• 2-month intensive phase<br>• 4-month continuation phase. | Targets InhA, of FAS II system, required for synthesis of mycolic acid layer | Narrow spectrum, bactericidal |
| | Rifampicin (R) | 10 mg/kg | | Targets rpoB gene (DNA dependent RNA polymerase), needed for transcription | Broad spectrum |
| | Pyrazinamide (Z) | 25 mg/kg | 2 months intensive phase | Targets pncA, energetics | Narrow spectrum |
| | Ethambutol (E) | 15 mg/kg | | Targets EmbB, required in synthesis of arabinogalactan layer biosynthesis | Narrow spectrum |
| Second-line drugs (group A) | Bedaquiline | 400 mg | Short regimen of 9–11 months<br>• 4–6 month of intensive phase and 5-month continuation phase<br>Long regimen of >20 months<br>• 8-month intensive phase and >12-month continuation phase) | Targets atpE of ATP synthase, obstructing energy metabolism | Narrow spectrum |
| | Linezolid | 600 mg | | Targets rplC, required for protein synthesis | Broad spectrum |
| | Fluoroquinolones | 400–1000 mg | | Targets gyrA/B subunits of topoisomerase enzyme | Broad spectrum |
| Second-line drugs (group B) | Clofazimine | 50 mg | | Primary target is outer membrane and secondary targets are respiratory chain and ion transporters | Broad spectrum |
| | Cycloserine/ terizidone | 10–15 mg/kg | | Targets cell wall biosynthesis | Broad spectrum |

| | | | | | |
|---|---|---|---|---|---|
| Second-line drugs (group C) | Delamanid | 100 mg | | Targets ddn, hindering mycolic acid biosynthesis | Narrow spectrum |
| | Pyrazinamide | 25 mg/kg | | Targets pncA, membrane energy metabolism | Narrow spectrum |
| | Ethambutol | 15 mg/kg | | Targets EmbB, required in synthesis of arabinogalactan layer biosynthesis | Narrow spectrum |
| | Meropenem | 500 mg | | Targets cell wall biosynthesis | Broad spectrum |
| | Ethionamide | 15–20 mg/kg | | Targets mycolic acid biosynthesis | Narrow spectrum |
| | *Para*-aminosalicylic acid | 8–12 g | | Targets dihydrofolate reductase, inhibits DNA precursor synthesis | Narrow spectrum |
| | Amikacin | 15 mg/kg | | Targets protein synthesis in bacteria | Broad spectrum |

TB comprised of medications that are broad-spectrum antibiotic like fluoroquinolones. Administration of these drugs for such long duration of 12 months is associated with alteration in the microbiota, thereby compromising immune response and protection against infection. Many studies are ongoing to understand the consequence of anti-TB drugs on the microbiome and this section deals with the microbiome perturbation due to anti-TB drugs [28, 57].

For identifying the effect of HRZE regimen on the intestinal microbiota, Namasivayam et al., studied the perturbation in C57BL/6 mice. They segregated mice into three groups, two groups were infected with *Mtb* (one on HRZE regimen and other left untreated) and one taken as control. They analysed the microbiota in the stool samples of the mice using 16S rRNA sequencing. In the HRZ treated animal, a significant depletion in microbes of phylum *Firmicutes*, especially *Robinsoniella, Stomatobaculum, Acetivibrio, Alkaliphilus, Acetanaerobacterium, Tyzzerella, Butyricicoccus, Peptococcus* and *Ruminococcus,* genera, followed by an increase in *Erysipelatoclostridium* and *Eggerthia* genera as compared to the control animal was observed [58]. To examine the consequence of long 3 months of anti-TB therapy on the microbiome they studied stool samples for 3 months post-cessation of treatment taking age-matched naïve animals as control. Relative decrease in *Lactobacillus*, while increase in microbes belonging to phylum *Bacteroidetes* and *Proteobacteria* was observed during post-therapy period [58]. Wipperman et al. compared gut microbiome composition of individuals undergoing the TB therapy and subjects who have recovered by the 6-month anti-TB drug treatment with the uninfected individuals as controls in a study. The microbiome was evaluated by 16S rDNA and metagenomic DNA sequencing. They concluded that the overall diversity in the infected and uninfected individual is similar but there are some specific microbiomics changes associated with the anti-TB therapy. Individuals on the HRZE regimen showed enrichment of *Fusobacterium, Prevotella* and *Erysipelatoclostridium* and depletion in *Coprococcus, Blautia, Lactobacillus, Bifidobacterium* and *Ruminococcus*. Recovered subjects showed decline in *Bacteroides* and upsurge in *Eubacterium, Ruminococcus,* and *Faecalibacterium* genera. Perturbation in the microbiome is directly affecting the immune response. *Ruminococcus* and *Coprococcus* are known to regulate peripheral cytokine production including interferons and interleukins, similarly, *Bifidobacterium* was found to conduct a Th17 immune response in mice [44].

Wang et al., described in their study about the perturbation of gut microbiota which leads to unfavourable changes in the lipid profile of individuals undergoing MDR-TB treatment. They enrolled 76 individuals in four different groups, i.e., one active MDR-TB group, one cured MDR-TB group, and two TB groups of first infection taken as control. The faecal and blood samples of the subjects were collected to analyse the gut microbiota. They observed alteration and depletion in richness (up to 26% reduction) of gut microbiome during MDR-TB therapy and even after 3–8 years after therapy success. The altered gut microbiome comprised of phylum Verrucomicrobia and Firmicutes and genera *Erysipelatoclostridium, Adlercreutzia, Butyricicoccus, Akkermansia, Coprococcus, Eubacterium,*

*Psychrobacter, Clostridioides, Fusicatenibacter, Streptococcus* and *Klebsiella*. The alteration was associated with raised low-density lipoprotein cholesterol and total cholesterol. This study focussed on the adverse side effects of the drugs caused on the gut microbiota during the course of TB therapy [59]. The study conducted by Hu et al., displayed that the anti-TB medications alter the gut microbiota within a week of administration. A significant depletion of *Ruminococcus* and *Faecalibacterium*, belonging to phylum *Firmicutes* trailed by abundance of *Bacteroides fragilis*, *Bacteroides plebeius* and *Parabacteroides distasonis*, belonging to phylum Bacteroidetes and OTU8, OTU2972 of family Erysipelotrichaceae was observed in TB patients undergoing anti-TB therapy [60].

The consequence of first- and second-line anti-TB drugs on the gut microbiome has been focussed widely but the effects of drugs on lung microbiome have not been explored a lot. It is pertinent to comprehend the perturbation of the lung microbiome caused by anti-TB drugs and their contribution in the disease progression and recurrence. This would also open a new arena to explore the contribution of microbiome in treatment of tuberculosis. Currently, researchers are conducting studies to recognize the role of altering the microbiome in favour of the host to eliminate *Mtb*. A research conducted by Cardona et al., showed the efficacy of use of heat-killed *M. manresensis* as an adjunct to treat tuberculosis. They studied the effect of $10^5$ heat-killed non-tuberculous-mycobacteria species (*M. manresensis*) on C3HeB/FeJ mice when orally administered for 14 days. Oral treatment with *M. manresensis* for 14 days induced a protein purified derivative-specific Tregs population. The therapy lessened the bacillary load in lungs, caused granulomatous infiltration and release of pro-inflammatory cytokines. Administration of *M. manresensis* orally, along with 6-month standard therapy reduced the relapse of TB. Thus, this study supports the usefulness of *M. manresensis* in TB treatment as well as control of excessive inflammatory response [61].

Another study performed by Suprapti et al., elaborated the outcome of probiotics and vitamin B (B1, B6 and B12) supplementation on pro-inflammatory cytokines release during the intensive phase of TB therapy [62]. For the study, 22 TB-infected individuals were selected and divided into two groups. One group administered HRZE regimen + B6 supplementation (control), whereas the other group received HRZE regimen + probiotics + vitamin B1, B6 and B12 supplementation (intervention). The cytokine release was measured after 2 months of intensive phase treatment using the ELISA method. The intervention group showed higher percentage of IFN-γ release as compared to the control group. This study highlighted that probiotics and vitamins B1, B6, B12 could modulate immune response through cytokine release during intensive phase therapy [62].

## 12.7 Conclusion and Way Forward

Exploring lung microbiome in relation to TB pathogenesis has garnered attention in the recent years. Growing studies and evidences indicate the interplay amongst the host microbiome and mycobacteria, influencing the immune responses in the body.

However, a better understanding of the subject can open a new field of TB diagnostics and therapeutics by targeting the microbiome in the favour of the host. The field is mostly unexplored, with a lot of unanswered questions. Future in vitro, ex vivo, in vivo and clinical studies need to be conducted to understand the interplay between host immune responses, microbiome and mycobacteria. This would further inform the healthcare on targeting microbiome as a potent therapeutic target and the use of host directed agents in clearing the mycobacteria from lungs.

## References

1. Hilty M, Burke C, Pedro H et al (2010) Disordered microbial communities in asthmatic airways. PLoS One 5(1):e8578
2. Angelucci F, Cechova K, Amlerova J et al (2019) Antibiotics, gut microbiota, and Alzheimer's disease. J Neuroinflammation 16(1):1–10
3. Lin C, Cai X, Zhang J et al (2019) Role of gut microbiota in the development and treatment of colorectal cancer. Digestion 100(1):72–78
4. Global Tuberculosis Report (2020) World Health Organisation
5. Conradie F, Diacon AH, Ngubane N et al (2020) Treatment of highly drug-resistant pulmonary tuberculosis. N Engl J Med 382(10):893–902
6. Forget EJ, Menzies D (2006) Adverse reactions to first-line antituberculosis drugs. Expert Opin Drug Saf 5(2):231–249
7. Ravimohan S, Kornfeld H, Weissman D et al (2018) Tuberculosis and lung damage: from epidemiology to pathophysiology. Eur Respir Rev 27(147):170077
8. Wu J, Liu W, He L et al (2013) Sputum microbiota associated with new, recurrent and treatment failure tuberculosis. PLoS One 8(12):e83445
9. Scotti E, Boué S, Sasso GL et al (2017) Exploring the microbiome in health and disease: implications for toxicology. Toxicol Res Appl 1:2397847317741884
10. Barko P, McMichael M, Swanson KS et al (2018) The gastrointestinal microbiome: a review. J Vet Intern Med 32(1):9–25
11. Consortium HMP (2012) Structure, function and diversity of the healthy human microbiome. Nature 486(7402):207
12. Peterson J, Garges S, Giovanni M et al (2009) The NIH human microbiome project. Genome Res 19(12):2317–2323
13. Cui L, Morris A, Huang L et al (2014) The microbiome and the lung. Ann Am Thorac Soc 11 (Supplement 4):S227–S232
14. Santacroce L, Charitos IA, Ballini A et al (2020) The human respiratory system and its microbiome at a glimpse. Biology 9(10):318
15. Adami AJ, Cervantes JL (2015) The microbiome at the pulmonary alveolar niche and its role in Mycobacterium tuberculosis infection. Tuberculosis 95(6):651–658
16. Kumpitsch C, Koskinen K, Schöpf V et al (2019) The microbiome of the upper respiratory tract in health and disease. BMC Biol 17(1):1–20
17. O'Dwyer DN, Dickson RP, Moore BB (2016) The lung microbiome, immunity, and the pathogenesis of chronic lung disease. J Immunol 196(12):4839–4847
18. Dominguez-Bello MG, Costello EK, Contreras M et al (2010) Delivery mode shapes the acquisition and structure of the initial microbiota across multiple body habitats in newborns. Proc Natl Acad Sci 107(26):11971–11975
19. Wylie KM (2017) The virome of the human respiratory tract. Clin Chest Med 38(1):11
20. Hoggard M, Biswas K, Zoing M et al (2017) Evidence of microbiota dysbiosis in chronic rhinosinusitis. Int Forum Allergy Rhinol. 7(3):230–239

21. Copeland E, Leonard K, Carney R et al (2018) Chronic rhinosinusitis: potential role of microbial dysbiosis and recommendations for sampling sites. Front Cell Infect Microbiol 8:57
22. Coburn B, Wang PW, Caballero JD et al (2015) Lung microbiota across age and disease stage in cystic fibrosis. Sci Rep 5(1):1–12
23. Russell DG (2001) Mycobacterium tuberculosis: here today, and here tomorrow. Nat Rev Mol Cell Biol 2(8):569–578
24. Zuniga J, Torres-García D, Santos-Mendoza T et al (2012) Cellular and humoral mechanisms involved in the control of tuberculosis. Clin Dev Immunol 2012:193923
25. Chai Q, Zhang Y, Liu CH (2018) Mycobacterium tuberculosis: an adaptable pathogen associated with multiple human diseases. Front Cell Infect Microbiol 8:158
26. Delogu G, Sali M, Fadda G (2013) The biology of mycobacterium tuberculosis infection. Mediterr J Hematol Infect Dis 5(1):e2013070
27. Ramakrishnan L (2012) Revisiting the role of the granuloma in tuberculosis. Nat Rev Immunol 12(5):352–366
28. Naidoo CC, Nyawo GR, Wu BG et al (2019) The microbiome and tuberculosis: state of the art, potential applications, and defining the clinical research agenda. Lancet Respir Med 7(10): 892–906
29. Adak A, Khan MR (2019) An insight into gut microbiota and its functionalities. Cell Mol Life Sci 76(3):473–493
30. Zhao J, Zhang X, Liu H et al (2019) Dietary protein and gut microbiota composition and function. Curr Protein Pept Sci 20(2):145–154
31. Grochowska M, Laskus T, Radkowski M (2019) Gut microbiota in neurological disorders. Arch Immunol Ther Exp 67(6):375–383
32. Sarkar SR, Banerjee S (2019) Gut microbiota in neurodegenerative disorders. J Neuroimmunol 328:98–104
33. Vallianou NG, Stratigou T, Tsagarakis S (2019) Metformin and gut microbiota: their interactions and their impact on diabetes. Hormones 18(2):141–144
34. Xu H, Wang X, Feng W et al (2020) The gut microbiota and its interactions with cardiovascular disease. Microb Biotechnol 13(3):637–656
35. Donati Zeppa S, Agostini D, Piccoli G et al (2020) Gut microbiota status in COVID-19: an unrecognized player? Front Cell Infect Microbiol 10:742
36. Hong B-Y, Maulén NP, Adami AJ et al (2016) Microbiome changes during tuberculosis and antituberculous therapy. Clin Microbiol Rev 29(4):915–926
37. Cui Z, Zhou Y, Li H et al (2012) Complex sputum microbial composition in patients with pulmonary tuberculosis. BMC Microbiol 12(1):1–8
38. Cheung MK, Lam WY, Fung WYW et al (2013) Sputum microbiota in tuberculosis as revealed by 16S rRNA pyrosequencing. PLoS One 8(1):e54574
39. Botero LE, Delgado-Serrano L, Cepeda ML et al (2014) Respiratory tract clinical sample selection for microbiota analysis in patients with pulmonary tuberculosis. Microbiome 2(1):1–7
40. Krishna P, Jain A, Bisen P (2016) Microbiome diversity in the sputum of patients with pulmonary tuberculosis. Eur J Clin Microbiol Infect Dis 35(7):1205–1210
41. Eshetie S, Van Soolingen D (2019) The respiratory microbiota: new insights into pulmonary tuberculosis. BMC Infect Dis 19(1):1–7
42. Maji A, Misra R, Dhakan DB et al (2018) Gut microbiome contributes to impairment of immunity in pulmonary tuberculosis patients by alteration of butyrate and propionate producers. Environ Microbiol 20(1):402–419
43. Lachmandas E, van den Heuvel CN, Damen MS et al (2016) Diabetes mellitus and increased tuberculosis susceptibility: the role of short-chain fatty acids. J Diabetes Res 2016:6014631
44. Wipperman MF, Fitzgerald DW, Juste MAJ et al (2017) Antibiotic treatment for tuberculosis induces a profound dysbiosis of the microbiome that persists long after therapy is completed. Sci Rep 7(1):1–11

45. Segal LN, Clemente JC, Li Y et al (2017) Anaerobic bacterial fermentation products increase tuberculosis risk in antiretroviral-drug-treated HIV patients. Cell Host Microbe 21(4): 530–537.e4
46. Zhou Y, Lin F, Cui Z et al (2015) Correlation between either Cupriavidus or Porphyromonas and primary pulmonary tuberculosis found by analysing the microbiota in patients' bronchoalveolar lavage fluid. PLoS One 10(5):e0124194
47. Hu Y, Cheng M, Liu B et al (2020) Metagenomic analysis of the lung microbiome in pulmonary tuberculosis-a pilot study. Emerg Microbes Infect 9(1):1444–1452
48. Dang AT, Marsland BJ (2019) Microbes, metabolites, and the gut–lung axis. Mucosal Immunol 12(4):843–850
49. Fabbrizzi A, Amedei A, Lavorini F et al (2019) The lung microbiome: clinical and therapeutic implications. Intern Emerg Med 14(8):1241–1250
50. Majlessi L, Sayes F, Bureau J-F et al (2017) Colonization with Helicobacter is concomitant with modified gut microbiota and drastic failure of the immune control of Mycobacterium tuberculosis. Mucosal Immunol 10(5):1178–1189
51. Perry S, De Jong BC, Solnick JV et al (2010) Infection with Helicobacter pylori is associated with protection against tuberculosis. PLoS One 5(1):e8804
52. Negatu DA, Liu JJ, Zimmerman M et al (2018) Whole-cell screen of fragment library identifies gut microbiota metabolite indole propionic acid as antitubercular. Antimicrob Agents Chemother 62(3):e01571–e01517
53. Sotgiu G, Centis R, D'ambrosio L et al (2015) Tuberculosis treatment and drug regimens. Cold Spring Harb Perspect Med 5(5):a017822
54. World Health Organization (2019) WHO consolidated guidelines on drug-resistant tuberculosis treatment. World Health Organization
55. Jang JG, Chung JH (2020) Diagnosis and treatment of multidrug-resistant tuberculosis. Yeungnam Univ J Med 37(4):277
56. Hameed H, Islam MM, Chhotaray C et al (2018) Molecular targets related drug resistance mechanisms in MDR-, XDR-, and TDR-Mycobacterium tuberculosis strains. Front Cell Infect Microbiol 8:114
57. O'Toole RF, Gautam SS (2018) The host microbiome and impact of tuberculosis chemotherapy. Tuberculosis 113:26–29
58. Namasivayam S, Maiga M, Yuan W et al (2017) Longitudinal profiling reveals a persistent intestinal dysbiosis triggered by conventional anti-tuberculosis therapy. Microbiome. 5(1):1–17
59. Wang J, Xiong K, Zhao S et al (2020) Long-term effects of multi-drug-resistant tuberculosis treatment on gut microbiota and its health consequences. Front Microbiol 11:53
60. Hu Y, Yang Q, Liu B et al (2019) Gut microbiota associated with pulmonary tuberculosis and dysbiosis caused by anti-tuberculosis drugs. J Infect 78(4):317–322
61. Cardona P, Marzo-Escartín E, Tapia G et al (2016) Oral administration of heat-killed Mycobacterium manresensis delays progression toward active tuberculosis in C3HeB/FeJ mice. Front Microbiol 6:1482
62. Suprapti B, Suharjono S, Raising R et al (2018) Effects of probiotics and vitamin B supplementation on IFN-γ and IL-12 levels during intensive phase treatment of tuberculosis. Indones J Pharm 29(2):80
63. Parbhoo T, Sampson SL, Mouton JM (2020) Recent developments in the application of flow cytometry to advance our understanding of Mycobacterium tuberculosis physiology and pathogenesis. https://doi.org/10.1002/cyto.a.24030

# Microbiome in Idiopathic Pulmonary Fibrosis

13

Sachchidanand Pathak, Anurag Mishra, Gaurav Gupta, Abhay Raizaday, Santosh Kumar Singh, Pramod Kumar, Sachin Kumar Singh, Neeraj Kumar Jha, Dinesh Kumar Chellappan, and Kamal Dua

**Abstract**

It is believed that Idiopathic Pulmonary Fibrosis (IPF) is an age-related chronic, progressive, and histopathologically associated fibrosing interstitial lung disorder which primarily affects the elderly. Despite tremendous progress in our knowledge of pathophysiology of diseases, we still do not know the possible causes of IPF. According to current research evidences, it is proposed that IPF may develop

S. Pathak
Kashi Institute of Pharmacy, Varanasi, India

A. Mishra (✉)
NIMS Institute of Pharmacy, Jaipur, India

G. Gupta · A. Raizaday · S. K. Singh
Suresh Gyan Vihar University, Jaipur, India

P. Kumar
Limetta Laboratories, Haridwar, India

S. K. Singh
School of Pharmaceutical Sciences, Lovely Professional University, Phagwara, India

N. K. Jha
Department of Biotechnology, School of Engineering and Technology (SET), Sharda University, Greater Noida, India

D. K. Chellappan
Department of Life Sciences, School of Pharmacy, International Medical University, Kuala Lumpur, Malaysia

K. Dua
Discipline of Pharmacy, Graduate School of Health, University of Technology Sydney, Sydney, NSW, Australia

Faculty of Health, Australian Research Centre in Complementary and Integrative Medicine, University of Technology Sydney, Ultimo, NSW, Australia



G. Gupta et al. (eds.), *Microbiome in Inflammatory Lung Diseases*,
https://doi.org/10.1007/978-981-16-8957-4_13

genotype as a result of repeated alveolar damage causing an abnormal wound-healing response. Genomic variations in epithelial integrity and host defence genes put people at risk for IPF, whereas immunosuppression and overt respiratory infection are supposed to have a high death rate. The role of infection in disease etiopathogenesis has long been suspected and its progression, or as a cause of acute aggravation, although preliminary investigations using classic culture procedures have formed inconsistent findings. Current approach of culture-independent microbiological analysis procedures to IPF patients has previously revealed various unacknowledged variations in lung microbiome and also a high microbial burden in bronchoalveolar lavage (BAL) in patients with IPF. However, connection does not always imply causation. Furthermore, lung microbiome is still incompletely defined, and more studies need to be done to explore species other than viruses and bacteria, such as fungus. The knowledge of microbiome's role in aetiology and IPF progression might leads to its modification, allowing targeted therapeutic treatment.

**Keywords**

Idiopathic Pulmonary Fibrosis (IPF) · Interstitial Lung Disease (ILD) · Microbiome · Pathogenesis · Acute Exacerbation · Infection in Lung

## 13.1 Introduction

Microbiome refers as pathogenic and symbiotic organisms and a commensal's ecological community which share human bodily space and also the intricate dealings of these microbes with the host. Various studies are conducted on the microbiome of gastrointestinal tract, with about 100 trillion microorganisms; yet, the lower respiratory tract's epithelial surface is considered to be one of least inhabited areas of human body which has been supposed as sterile in past. Identifying and isolating microorganisms were challenging due to the difficulty of physically sampling lower airways and limitations of bacterial culture, leading to the incorrect idea. To better understand the respiratory tract's microbiome, researchers switched from using culture-dependent methods to using methods independent of culture. High-throughput DNA sequencing methods use sequence similarity in extremely conserved genes like 16S ribosomal RNA gene to rapidly identify multi-faceted bacterial communities (with species which cannot be grown) [1, 2]. Because of this, scientists are now studying lung microbiome in healthy volunteers and also patients with chronic respiratory diseases as cystic fibrosis, bronchiectasis, COPD and ILD. As a result of this investigation, researchers have discovered diverse communities of fungi, bacteria and viruses [3, 4].

IPF is a debilitating, severe, fibrosing and deadly fibrotic ILD that predominantly affects elderly people and eventually causing respiratory failure with the cause of chronic dyspnoea, an inevitable reduction in the functions of lung. It is a degenerative interstitial lung disease associated with ageing, via median diagnostic age of

66 years, and a serious condition via 2.5–.5 years of survival. The IPF factors are still unknown and yet, the particular or main causing factor has not been acknowledged, the disease is supposed to be induced by abnormality in wound-healing mechanisms in genetically susceptible individuals in response to unidentified environmental triggers (such as gastric micro-aspiration, viral infections, cigarette smoke, particulate dust, etc.) [5, 6].The resulting extracellular matrix deposition & development of fibroblastic foci reasons irreparable damages in lung architecture, resulting in alveolar structure loss, impeded gas exchange and eventually causing respiratory failure. Infectious agents, such as bacteria and viruses, can cause damage in alveolar epithelial cells and apoptosis, as well as alter the host's response toward injury. Furthermore, researches involving genetic vulnerability to IPF have identified an amplified risk with genetic polymorphisms involved in characteristic host response control. A single nucleotide polymorphism in promoter region of the mucin 5B gene (MUC5B) (rs35705950), which codes for critical component of airway mucus, and single nucleotide polymorphism in the toll-interacting protein (TOLLIP) gene (rs5743890), which codes for adaptor protein that controls signaling through toll-like receptors (TLRs), are two specific examples [7, 8].

While comparing 130 IPF patients' peripheral blood transcriptomes with controls, 4 genes involved in immune defence, including alpha-defensins, were found to be upregulated. These findings propose that innate immunological vulnerability can contribute a significant role in IPF aetiology, and supports hypothesis that infection, in combination with host immune system, contributes toward an abnormal fibrosis sequence of events. This review will be looking at what we currently know about the function of the respiratory microbiome in IPF, as well as extents of debate &further research objectives and priorities [9, 10].

## 13.2 Microbiome Development and Composition in Healthy Lungs

Initially thought to be sterile, epithelial surfaces of respiratory tract have been revealed to support dynamic microbial populations utilizing various culture-independent methods. Bacterial DNA was identified in 95.7% specimens of bronchoalveolar lavage (BAL) using high-throughput sequencing of bacterial 16s-rRNA, compared with 39.1% of BAL samples using conventional standard culture methods. Healthy lungs have bacterial communities that are quite similar to those observed in the mouth, but with a bacterial load that is two to four times lower. Previous studies have reported that there are approximately 10–100 bacterial cells per 1000 human cells in lung tissues. Interestingly despite changes in temperature, pH, & oxygen concentration, level of microbiome in healthy volunteers is quite consistent [11, 12]. Firmicutes (including genera *Veillonella* sp. and *Streptococcus* sp.), Bacteroidetes (including the species *Prevotella* sp.) to a slighter extent, Actinobacteria and Proteobacteria are the most commonly found phyla in normal airways.

The microbiota composition of the lungs is largely determined by three factors: microbial immigration, which is brought on by oro-nasal cavity mucosal dispersion, micro-aspiration of gastric contents and air inhalation; microbial elimination, which is caused by cough, mucociliary clearance and immunity; and the microbiological growth environment including oxygen tension, temperature, pH and nutrition availability [13–15].

The microbiota present in lungs reflects a stable condition among microbial inflow, outflow and reproduction level, and as a result of these three variables, with the latter being primarily impacted in the event of pathological processes of chronic diseases. The microbiome of lung is changed across every lung disease examined and compared to healthy volunteers. Many studies have found pollutants samples of upper respiratory tract during sampling due to the sensitivity of molecular technologies used, resulting in an inaccurate representation of the true microbiome. The risk of oropharyngeal contamination should be considered, as the majority of published research works have utilized BAL samples to describe the lung microbiome of healthy volunteers. Furthermore, heterogeneity of the microbial composition of lung at spatially distinct lung locations within subjects has been demonstrated in healthy participants, but this variation is smaller than inter-subject community variance [16–18]. Contamination has recently been shown to have a negligible impact on microbial plethora in bronchoscopy-acquired samples, supporting utility of bronchoscopy to study microbiome of lungs. Contamination can occur at any point during a microbiome study, not only during bronchoscopy [19].

While comparing data of microbiome from very identical subject specimen utilizing distinct sequencing channel and techniques, significant variance was observed. Other significant sources of contamination include agents and extraction kits and in low biomass samples they become important factor as those obtained from the respiratory system. Recall that BAL DNA sequencing provides "instantaneous" "snapshot" in time of bacterial diversity of lower airways, but does not assess chronically changing microbial communities over time. Several research works have focused on viruses and fungi in addition to the study of lung microbiota. A new study has found that commensal fungi have an effect on both the host immune system and bacteria in the gut. This has implications for the restoration of a healthy microbiome following antibiotic therapy. Because of the wide variety of viruses that can be found in the lungs, they are thought to be a catalyst for many different types of lung disease [20–22].

## 13.3 Microbiome in Idiopathic Pulmonary Fibrosis

The lung microbiome: Previously considered to be sterile, the respiratory tract's epithelial surfaces have been demonstrated to support dynamic microbial populations utilizing culture-independent methods. With biochemical sequence analysis of the factor 16s-rRNA genomic regions, bacterial species can now be recognized; in other microbiome scientific studies, groups of bacteria with similar

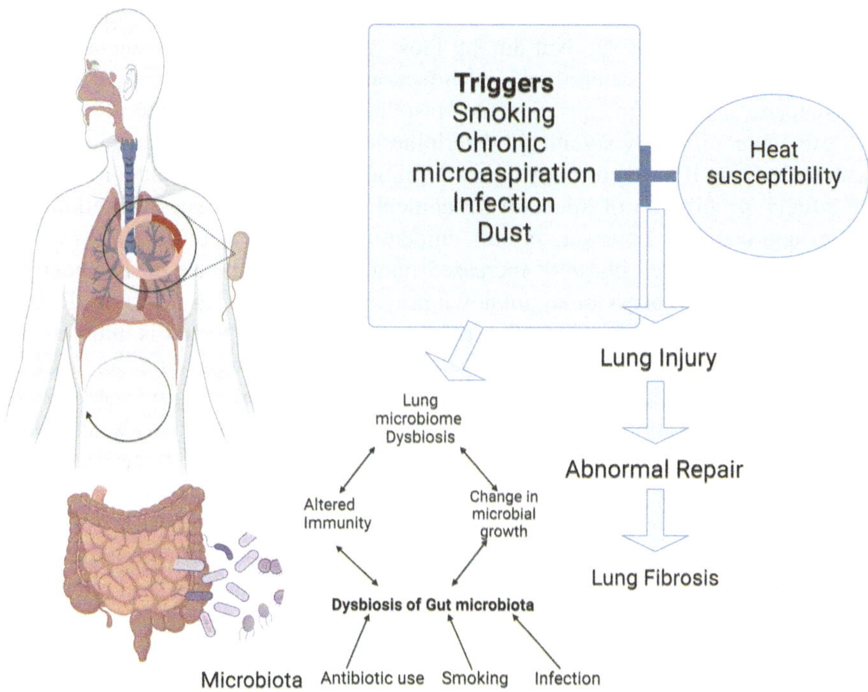


**Fig. 13.1** Microbiota interaction in idiopathic pulmonary fibrosis

genetic codes are categorized into operational taxonomic units (OTUs) and evaluated by comparing to the 16s rRNA data base. Higher microbial 16s-rRNA sequencing recognizes bacterial DNA in 95.7% of BAL specimens, compared to comparison to standard culture systems, that can locate bacteria in 39.1% of BAL samples. Using these genetic methods in characterizing microbial flora in respiratory tract of both sick population and healthy controls have shown correlations that imply the microbiome–host interaction that may be important in the aetiology and development of lung disease. Moreover, when severe asthma patients were compared with non-severe asthmatic patients and related controls, changes in the microbiome were observed, showing that the disease phenotype may be influenced by the microbial populations in the airway [23–25] (Fig. 13.1).

## 13.4 Microbiome Effect on IPF Prognosis and Exacerbation

Exacerbations are common in the IPF progression, as they are in a variety of chronic diseases of lungs. Acute episodes and exacerbations are linked to an especially bleak prognosis. Non-survivors exhibited shorter dyspnoea durations, higher C reactive protein (CRP) values, inspiratory oxygen fraction ($FiO_2$) ratios/lower arterial oxygen tension ($PaO_2$), lower proportions of lymphocytes and greater proportions of neutrophils in BALF than survivors. CRP was found to be only independently associated predictor of survival among those variables, ultimately suggesting that inflammation and/or bacterial or viral infection might be one of many pathogenic mechanisms involved in causing acute episodes and aggravations [26, 27].

"An acute, clinically significant deterioration of unidentifiable cause in a patient with underlying IPF" is currently a new definition of exacerbation and it necessitates the official prohibition of infection for clinical diagnosis. The specific aetiology of acute aggravations, however, is still unidentified, and it is uncertain whether it reflects an augmented phase of increased lung damage response or an underlying fibroproliferative process to an unknown previous or coexisting infection. Respiratory tract infections carry a mortality risk in persons with IPF, and is indistinguishable with acute aggravations is one of the factors suggesting an infection involvement in aggravation. Recent investigations involving lung microbiome during aggravations of IPF and its impact on progression of disease have also cast doubt on the definition. According to these research works, an enhanced bacterial load at time of diagnosis appears to be a biomarker for a disease that progresses more quickly and has a higher mortality risk [28, 29].

Another research including 20 patients with IPF diagnosed acute aggravations and 15 matched control subjects with constant IPF condition who undergone bronchoscopy & extraction of DNA process found that patients with IPF had a four-fold greater bacterial load during aggravations. In comparison to patients with stable IPF, their BALF included a greater number of neutrophils. This suggests the idea that bacteria have a significant role in exacerbations even if active infection is present. They are supposed to use 16S rRNA gene qPCR and pyrosequencing to investigate changes in BAL microbiota in both stable and acute exacerbation groups. There was noticeable alteration in microbiota in cases of acute exacerbation, along with a substantial increase in *Stenotrophomonas* sp. & *Campylobacter* sp., and a substantial decrease in *Campylobacter* sp. & *Veillonella* sp., despite being known best as gastrointestinal pathogen, was initially demonstrated in respiratory microbiota [30–32]. Its occurrence in respiratory microbiota is most probable due to stomach's gastric contents silent micro-aspiration. To conclude these findings, this pilot study shows that IPF acute exacerbation may be due to be a significant role of various bacteria. Micro-aspiration may play a role in the apparent transfer of bacteria that are normally restricted to the gastrointestinal system. Although a prospective longitudinal research work is needed to validate the findings, they give a justification for clinical trials including prophylactic antibiotics as a method to avoid acute aggravations in IPF patients [33, 34].

## 13.5 A Gut–Lung Axis and Regulation of Host Defence in Chronic Lung Disease Aggravations: Evidence and Implications

The exact mechanism by which bacteria influence the initial immunity which present during the birth in healthy and sick is still being researched, and a few is revealed like microbiota modulates and regulates immunity of lung or the formation of lymphoid tissue associated with bronchial related. The importance of the gut commensal microbiota as a modulator of the innate immune system is being more recognized. The intestinal microbiota in healthy individuals is dominated by three phyla: *Ruminococcus*, *Prevotella* and *Bacteroides*. Evidences reveal that the formation of the intestinal microbiome is vital for control of an adequate immune response in lungs during a critical early period of life. Alteration in the composition of the microbiota of intestine impacts the progression and vulnerability of chronic lung diseases including asthma and cystic fibrosis. Moreover, the host is more vulnerable to lung infections, such as *Klebsiella pneumoniae*, *Listeria monocytogenes* and viruses, in the absence of normal intestinal biota. This offers the intriguing hypothesis that chronic lung disease exacerbations are caused by decreased adaptive and innate immune system as a result of changes in the intestinal microbiota of host [35–37]. As previously stated, individuals with progressive IPF have an enormous burden of *Staphylococcus* and *Streptococcus* species in their lungs, and earlier researches have shown that neutrophils from microbiota depleted mice had a decreased capacity to kill *S. aureus* and *S. pneumoniae*. Recently, it has been found that microbial stimulation of gut nod-like receptor sites causes an increase in the producing of free radicals in phagocytic cells—a lung's sentinel innate immune cell. This suggests that circumstances related to loss of intestinal bacterial homeostasis (such as antibiotic use) may lead to weakened lung immunity. In COPD, viral infections can exacerbate symptoms, and the pathophysiology that follows could be linked to dysbiosis, which alters the microbiota of the airways and causes excessive inflammation [38, 39]. Despite the fact that damage of gastrointestinal commensal signaling might be responsible for impairing innate immunity of lung in this condition, cigarette smoke also makes a significant contribution to impeded lung innate immunity either directly or indirectly by altering innate immune cell phagocytosis, mucus, ciliary function and directly enhancing intestinal microbiota (e.g., enhanced formation of biofilm). These alterations may have an influence on respiratory infections' propensity to aggravate COPD. Providing viruses' proclivity for causing exacerbations in lung disease, it is worth considering the influence of respiratory viral infection on gut microbiota. According to Wang and associates, influenza infection can cause abnormalities in the gut microbiota, including an increase in Enterobacteriaceae and decrease in *Lactococcus* & *Lactobacillus*. As stated earlier, this might result in a reduction of beneficial bacteria, which could contribute to smoking-related illness. According to the scientists, these changes in gut microbiome were not caused by lytic influenza intestinal infection [40, 41]. Th17 cells were involved in the damage, and neutralization of IL-17 reduced the severity of the injury. In addition, reduction of the intestinal flora caused by antibiotics

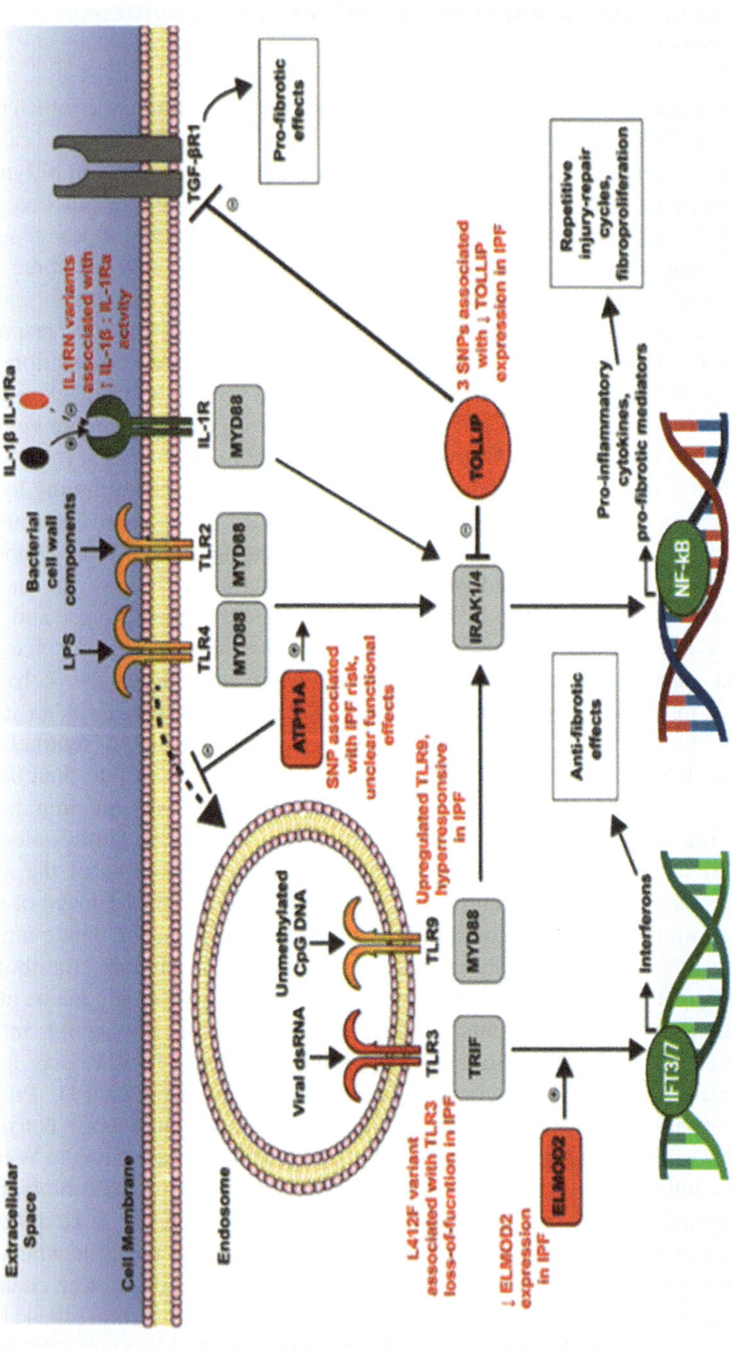


**Fig. 13.2** Risk factor for idiopathic pulmonary fibrosis

resulted in less intestinal damage. The relevance of effector T cell developed in the lung after infection and subsequently moved to small intestine, produced IFN-g and modified the gut microbiota, was also highlighted in the research work by scientists. Finally, Th17 response was aided via triggered epithelial-derived IL-15 due to changes in the gut microbiota. It is conceivable that the responses of IL-17 in intestine contribute to the progression of lung diseases. Certain microorganisms are eradicated by IL-17, which has been linked to the pathophysiology of sarcoidosis, asthma, cystic fibrosis, necrotizing bronchial asthma and bone marrow transplant-related pneumonitis [42, 43] (Fig. 13.2).

IL-17 might potentially have an impact in the dynamic changes that occur in pulmonary microbiome in COPD patients. In emphysema animal model, Yadava & their colleagues studied the effects of experimental changes on the lung microbiome. LPS/elastase was given to pathogen-free & axenic mice for 4 weeks. Through an excess of *Lactobacillus*, *Pseudomonas* and a reduction of *Prevotella*, microbiota diversity & abundance were reduced in LPS/elastase model. The loss of bacterial load was linked to a reduction in IL-17 production. Axenic mice were given microbiota-enriched fluid intranasally, which increased IL-17 production. In mice with microbiota, inhibition of IL-17 resulted in decreased inflammation and disease load. IL-17 has been linked to hepatic fibrosis in several investigations, and several experimental models of pulmonary fibrosis are IL-17A–dependent [44, 45]. In addition, research works looking into the onset of intestinal fibrosis have found a link between changes in the microbiome and Th17 responses. Through the adhesion of segmented filamentous bacteria on intestinal epithelial cells, gut is a recognized source of Th17 cells. In lungs, the situation may be identical. In animal models, Gauguet and his colleagues revealed that intestine segmented filamentous bacteria can enhance pulmonary innate immunity by inducing IL-17, resulting in resistance to *S. aureus* pneumonia. This adds to the growing body of data that suggests the gut–lung microbiome axis is important in regulating the lung's innate immune response [46, 47].

## 13.6 Limitation

Han and colleagues were limited to naming the progression-related bacteria Staphylococcus OTU 1348 & *Streptococcus* OTU 1345, as 16S rRNA sequencing can never be utilized for genetic markers. More research is required to fully characterize these bacteria, either in form of microbe-specific sequencing or sequencing specific to a particular culture. Despite the fact that the cohort contained multiple *Streptococcus* and *Staphylococcus* species, only two OTUs were linked to progression of diseases. In any disease related to lung, there are certain general limits to microbiome research. Particularly in the context of the molecular technologies used, infection of samples from upper respiratory tract during taking specimen is an apparent problem in many research, yielding a misleading depiction of the real microbiome [48, 49]. Kits including reagents and extraction agents can potentially be a source of contamination, which is especially critical when working with small biomass samples like those from respiratory system. Contamination can occur at any point during a microbiome study, not particularly during the bronchoscopy stage. While corelating the data of microbiome from same patient samples using various sequencing techniques and platforms, significant variance was observed. There are a number of biases in primer design that can favour or penalize certain bacteria, resulting in the exclusion of entire genera [50, 51].

While studies on the IPF microbiome have used high-throughput molecular technologies to identify bacterial species and loads, they have not yet demonstrated a causal, mechanistic link to the disease process and advancement. In the investigational studies of IPF, it is not clear if changes to the lungs' microbiome reflect the disease's aetiology or are due to a lack of underpinning immune defences in this patient population. The information gleaned from this research is probable to be less significant and more irrelevant because it does not reveal how the various bacterial colonies interact with one another [52–54].

This study provides a "snapshot" of the lower respiratory microbiome by DNA sequencing from a BAL sample, but it does not look at longitudinal modifications. Because it is unrealistic to perform bronchoscopies on a regular basis, other approaches of tracking the lower airway microbiome over time should be explored. BAL taken from one lobe of lung may not be representative of microbiota in the other lobes, especially because histological hallmark of IPF and UIP shows spatial variability through fibrosis alongside typical parenchyma. This is the case for IPF and UIP. Ex-planted lung tissue sections via cystic fibrosis patient were sequenced using 16s rDNA to uncover variances in microbial communities within lungs [55, 56]. As per our consideration of IPF microbiome improves, sample & sequencing techniques improve and composition of patient's microbiome might serve as a biomarker to help with prognosis & therapy stratification. A key question for future IPF research is whether or not prophylactic antibiotics should be used to target specific microbiome "signatures" in patients in order to improve survival, based on the results of a trial testing co-trimoxazole in patients with IPF [57, 58].

## 13.7 Conclusion

As per various studies and researches, alteration in microbiome load, composition and diversity have been linked to aetiology of disease, acute exacerbation, progression and death in idiopathic pulmonary fibrosis. Lung microbiome dysbiosis will be linked to IPF development and progression, according to the study's findings. When it comes to IPF, microbiome manipulation could soon be a treatment modality to restore a "healthy" microbiome culture. However, a comprehensive method to account for various factors driving development of disease, advancement, & episodic exacerbating is more probable. It is unclear if antibiotics, probiotics (extrinsic microorganisms given for health purposes) or prebiotics (molecules that encourage growth and development of specific bacteria) should be used to control the lung microbiome. However, modification of microbiome should focus on pathogenic microbes while leaving the rest of the microbial population intact but that would be considerably more difficult to achieve. All of these studies suggest that anti-biotherapy may have a significant role in IPF patients, and they establish a justification for long-term anti-biotherapy related clinical trials, that acts as a modulator of immunity and anti-bioprophyl axis to avoid acute exacerbations. Future research on lung microbiome dynamics could aid in the selection of suitable, targeted & more customized anti-biotherapy over the course of disease, particularly in cases of IPF aggravations. These studies require more advanced metagenomic techniques to determine functional relevance of particular microbial species & populations in the development of IPF.

## References

1. Anand S, Mande SS (2018) Diet, microbiota and gut-lung connection. Front Microbiol 9:2147
2. Bacci G, Taccetti G, Dolce D, Armanini F, Segata N, Di Cesare F et al (2020) Untargeted metagenomic investigation of the airway microbiome of cystic fibrosis patients with moderate-severe lung disease. Microorganisms 8(7):1003
3. Beck JM, Young VB, Huffnagle GB (2012) The microbiome of the lung. Transl Res 160(4): 258–266
4. Bhagirath AY, Li Y, Somayajula D, Dadashi M, Badr S, Duan K (2016) Cystic fibrosis lung environment and Pseudomonas aeruginosa infection. BMC Pulm Med 16(1):174
5. Blanchard AC, Waters VJ (2019) Microbiology of cystic fibrosis airway disease. Semin Respir Crit Care Med 40(6):727–736
6. Budden KF, Shukla SD, Rehman SF, Bowerman KL, Keely S, Hugenholtz P et al (2019) Functional effects of the microbiota in chronic respiratory disease. Lancet Respir Med 7(10): 907–920
7. Carney SM, Clemente JC, Cox MJ, Dickson RP, Huang YJ, Kitsios GD et al (2020) Methods in lung microbiome research. Am J Respir Cell Mol Biol 62(3):283–299
8. Caverly LJ, Zhao J, LiPuma JJ (2015) Cystic fibrosis lung microbiome: opportunities to reconsider management of airway infection. Pediatr Pulmonol 50(Suppl 40):S31–S38
9. Chmiel JF, Aksamit TR, Chotirmall SH, Dasenbrook EC, Elborn JS, LiPuma JJ et al (2014) Antibiotic management of lung infections in cystic fibrosis. I. the microbiome, methicillin-resistant Staphylococcus aureus, gram-negative bacteria, and multiple infections. Ann Am Thorac Soc 11(7):1120–1129
10. Chunxi L, Haiyue L, Yanxia L, Jianbing P, Jin S (2020) The gut microbiota and respiratory diseases: new evidence. J Immunol Res 2020:2340670
11. Cribbs SK, Beck JM (2017) Microbiome in the pathogenesis of cystic fibrosis and lung transplant-related disease. Transl Res 179:84–96
12. Cuthbertson L, Walker AW, Oliver AE, Rogers GB, Rivett DW, Hampton TH et al (2020) Lung function and microbiota diversity in cystic fibrosis. Microbiome. 8(1):45
13. de Almeida OGG, Capizzani C, Tonani L, Grizante Barião PH, da Cunha AF, De Martinis ECP et al (2020) The lung microbiome of three Young Brazilian patients with cystic fibrosis colonized by fungi. Front Cell Infect Microbiol 10:598938
14. Dickson RP (2016) The microbiome and critical illness. Lancet Respir Med 4(1):59–72
15. Dmitrijeva M, Kahlert CR, Feigelman R, Kleiner RL, Nolte O, Albrich WC et al (2021) Strain-resolved dynamics of the lung microbiome in patients with cystic fibrosis. MBio 12(2):e02863–e02820
16. Drakopanagiotakis F, Wujak L, Wygrecka M, Markart P (2018) Biomarkers in idiopathic pulmonary fibrosis. Matrix Biol 68-69:404–421

17. Fastrès A, Felice F, Roels E, Moermans C, Corhay JL, Bureau F et al (2017) The lung microbiome in idiopathic pulmonary fibrosis: a promising approach for targeted therapies. Int J Mol Sci 18(12):2735
18. Flume PA, Chalmers JD, Olivier KN (2018) Advances in bronchiectasis: endotyping, genetics, microbiome, and disease heterogeneity. Lancet 392(10150):880–890
19. Françoise A, Héry-Arnaud G (2020) The microbiome in cystic fibrosis pulmonary disease. Genes 11(5):536
20. Han MK, Huang YJ, Lipuma JJ, Boushey HA, Boucher RC, Cookson WO et al (2012) Significance of the microbiome in obstructive lung disease. Thorax 67(5):456–463
21. Han MK, Zhou Y, Murray S, Tayob N, Noth I, Lama VN et al (2014) Lung microbiome and disease progression in idiopathic pulmonary fibrosis: an analysis of the COMET study. Lancet Respir Med 2(7):548–556
22. Heirali AA, Acosta N, Storey DG, Workentine ML, Somayaji R, Laforest-Lapointe I et al (2019) The effects of cycled inhaled aztreonam on the cystic fibrosis (CF) lung microbiome. J Cyst Fibros 18(6):829–837
23. Héry-Arnaud G, Boutin S, Cuthbertson L, Elborn SJ, Tunney MM (2019) The lung and gut microbiome: what has to be taken into consideration for cystic fibrosis? J Cyst Fibros 18(1): 13–21
24. Huang YJ, LiPuma JJ (2016) The microbiome in cystic fibrosis. Clin Chest Med 37(1):59–67
25. Invernizzi R, Wu BG, Barnett J, Ghai P, Kingston S, Hewitt RJ et al (2021) The respiratory microbiome in chronic hypersensitivity pneumonitis is distinct from that of idiopathic pulmonary fibrosis. Am J Respir Crit Care Med 203(3):339–347
26. Kanda T, Goto T, Hirotsu Y, Masuzaki R, Moriyama M, Omata M (2020) Molecular mechanisms: connections between nonalcoholic fatty liver disease, steatohepatitis and hepatocellular carcinoma. Int J Mol Sci 21(4):1525
27. Kitsios GD, Rojas M, Kass DJ, Fitch A, Sembrat JC, Qin S et al (2018) Microbiome in lung explants of idiopathic pulmonary fibrosis: a case-control study in patients with end-stage fibrosis. Thorax 73(5):481–484
28. Lipinski JH, Moore BB, O'Dwyer DN (2020) The evolving role of the lung microbiome in pulmonary fibrosis. Am J Physiol Lung Cell Mol Physiol 319(4):L675–L182
29. Lira-Lucio JA, Falfán-Valencia R, Ramírez-Venegas A, Buendía-Roldán I, Rojas-Serrano J, Mejía M et al (2020) Lung microbiome participation in local immune response regulation in respiratory diseases. Microorganisms. 8(7):1059
30. Lynch SV (2016) The lung microbiome and airway disease. Ann Am Thorac Soc 13 Suppl 2 (Suppl 5):S462–S4s5
31. Mammen MJ, Scannapieco FA, Sethi S (2020) Oral-lung microbiome interactions in lung diseases. Periodontology 2000 83(1):234–241
32. Mendez R, Banerjee S, Bhattacharya SK, Banerjee S (2019) Lung inflammation and disease: a perspective on microbial homeostasis and metabolism. IUBMB Life 71(2):152–165
33. Metwally AA, Ascoli C, Turturice B, Rani A, Ranjan R, Chen Y et al (2020) Pediatric lung transplantation: dynamics of the microbiome and bronchiolitis obliterans in cystic fibrosis. J Heart Lung Transplant 39(8):824–834
34. Moffatt MF, Cookson WO (2017) The lung microbiome in health and disease. Clin Med (Lond) 17(6):525–529
35. Monsó E (2020) Look at the wood and not at the tree: the microbiome in chronic obstructive lung disease and cystic fibrosis. Arch Bronconeumol 56(1):5–6
36. Morris A, Gibson K, Collman RG (2014) The lung microbiome in idiopathic pulmonary fibrosis What does it mean and what should we do about it? Am J Respir Crit Care Med 190(8):850–852
37. Morton JT, Aksenov AA, Nothias LF, Foulds JR, Quinn RA, Badri MH et al (2019) Learning representations of microbe-metabolite interactions. Nat Methods 16(12):1306–1314
38. Muhlebach MS, Zorn BT, Esther CR, Hatch JE, Murray CP, Turkovic L et al (2018) Initial acquisition and succession of the cystic fibrosis lung microbiome is associated with disease progression in infants and preschool children. PLoS Pathog 14(1):e1006798

39. Mur LA, Huws SA, Cameron SJ, Lewis PD, Lewis KE (2018) Lung cancer: a new frontier for microbiome research and clinical translation. Ecancermedicalscience 12:866
40. O'Dwyer DN, Ashley SL, Gurczynski SJ, Xia M, Wilke C, Falkowski NR et al (2019) Lung microbiota contribute to pulmonary inflammation and disease progression in pulmonary fibrosis. Am J Respir Crit Care Med 199(9):1127–1138
41. O'Dwyer DN, Garantziotis S (2021) The lung microbiome in health, hypersensitivity pneumonitis, and idiopathic pulmonary fibrosis: a heavy bacterial burden to bear. Am J Respir Crit Care Med 203(3):281–283
42. Paudel KR, Dharwal V, Patel VK, Galvao I, Wadhwa R, Malyla V et al (2020) Role of lung microbiome in innate immune response associated with chronic lung diseases. Front Med 7:554
43. Permall DL, Pasha AB, Chen XQ, Lu HY (2019) The lung microbiome in neonates. Turk J Pediatr 61(6):821–830
44. Quinn RA, Adem S, Mills RH, Comstock W, DeRight GL, Humphrey G et al (2019) Neutrophilic proteolysis in the cystic fibrosis lung correlates with a pathogenic microbiome. Microbiome 7(1):23
45. Ramsey KA, Schultz A, Stick SM (2015) Biomarkers in paediatric cystic fibrosis lung disease. Paediatr Respir Rev 16(4):213–218
46. Rogers GB, Bruce KD, Hoffman LR (2017) How can the cystic fibrosis respiratory microbiome influence our clinical decision-making? Curr Opin Pulm Med 23(6):536–543
47. Saint-Criq V, Lugo-Villarino G, Thomas M (2021) Dysbiosis, malnutrition and enhanced gut-lung axis contribute to age-related respiratory diseases. Ageing Res Rev 66:101235
48. Salisbury ML, Han MK, Dickson RP, Molyneaux PL (2017) Microbiome in interstitial lung disease: from pathogenesis to treatment target. Curr Opin Pulm Med 23(5):404–410
49. Scialo F, Amato F, Cernera G, Gelzo M, Zarrilli F, Comegna M et al (2021) Lung microbiome in cystic fibrosis. Life. 11(2):94
50. Shukla SD, Budden KF, Neal R, Hansbro PM (2017) Microbiome effects on immunity, health and disease in the lung. Clin Transl Immunol 6(3):e133
51. Spagnolo P, Molyneaux PL, Bernardinello N, Cocconcelli E, Biondini D, Fracasso F et al (2019) The role of the Lung's microbiome in the pathogenesis and progression of idiopathic pulmonary fibrosis. Int J Mol Sci 20(22):5618
52. Takahashi Y, Saito A, Chiba H, Kuronuma K, Ikeda K, Kobayashi T et al (2018) Impaired diversity of the lung microbiome predicts progression of idiopathic pulmonary fibrosis. Respir Res 19(1):34
53. Tan JY, Tang YC, Huang J (2020) Gut microbiota and lung injury. Adv Exp Med Biol 1238: 55–72
54. Tojo R, Suárez A, Clemente MG, de los Reyes-Gavilán CG, Margolles A, Gueimonde M et al (2014) Intestinal microbiota in health and disease: role of bifidobacteria in gut homeostasis. World J Gastroenterol 20(41):15163–15176
55. Tong X, Su F, Xu X, Xu H, Yang T, Xu Q et al (2019) Alterations to the lung microbiome in idiopathic pulmonary fibrosis patients. Front Cell Infect Microbiol 9:149
56. Torrisi SE, Kahn N, Vancheri C, Kreuter M (2020) Evolution and treatment of idiopathic pulmonary fibrosis. Presse Med. 49(2):104025
57. Tracy M, Cogen J, Hoffman LR (2015) The pediatric microbiome and the lung. Curr Opin Pediatr 27(3):348–355
58. Trivedi R, Barve K (2020) Gut microbiome a promising target for management of respiratory diseases. Biochem J 477(14):2679–2696

# SARS-CoV-2 and Microbiota

# 14

Edda Russo, Lavinia Curini, Alessio Fabbrizzi, and Amedeo Amedei


**Abstract**

Today, the globe is dealing with the COVID-19 epidemic, which poses a pandemic danger.

This infectious disease is triggered by severe acute respiratory syndrome coronavirus-2 (SARS-CoV-2) and at lung level, like as intestine, there are different bacterial populations, named microbiota, that have a significant influence on the host's immunological homeostasis.

But, to date, our understanding of bacterial flora and its symbiotic connection with immunological processes in the context of SARS-CoV-2 is, however, incomplete. A disturbed bacterial flora combined with too many opportunistic infections can trigger a cascade of inflammatory reactions dysregulating the immune system and resulting in multi-organ damage.

The involvement of the lung and intestinal microbiota in immune regulation of SARS-CoV-2 infection via multiple pathways will be discussed in this chapter. In addition, diet and lifestyle have a huge impact on the microbiota-inflammation



E. Russo · L. Curini
Department of Clinical and Experimental Medicine, University of Florence, Florence, Italy
e-mail: edda.russo@unifi.it; lavinia.curini@unifi.it

A. Fabbrizzi
SOD of Interdisciplinary Internal Medicine, Azienda Ospedaliera Universitaria Careggi (AOUC), Florence, Italy
e-mail: alessio.fabbrizzi@unifi.it

A. Amedei (✉)
Department of Clinical and Experimental Medicine, University of Florence, Florence, Italy

SOD of Interdisciplinary Internal Medicine, Azienda Ospedaliera Universitaria Careggi (AOUC), Florence, Italy
e-mail: amedeo.amedei@unifi.it







G. Gupta et al. (eds.), *Microbiome in Inflammatory Lung Diseases*,
https://doi.org/10.1007/978-981-16-8957-4_14

crosstalk affecting both frequency and timing of this viral infection. So, in the last section we focus on the microbiota function in COVID-19, with an emphasis on immunological activation by the lung and gut bacteria and potential future therapies based on probiotic administration, able to regulate the microbiota–immunity axis.

**Keywords**

SARS-CoV-2 · Covid-19 · Microbiota · Lung microbiota · Gut microbiota · Microbiota–immune axis · Lung–gut axis · Probiotics

## 14.1 SARS Covid-19

Coronaviruses belong to RNA viruses. These kind of virus are divided into four genera, notably alpha, beta, gamma, and delta-coronavirus. Both animals and humans can be infected and major symptoms are respiratory, neurological, hepatic, and gastrointestinal [1].

Considering the coronaviruses' high prevalence and global distribution, their genetic diversity, and common genome recombination, and the increasing human–animal interface activities, novel coronaviruses with the potential to infect people are commonly detected [1]. As we know, several local health officials in Wuhan-China reported clusters of subjects with pneumonia of unclear cause, in late December 2019. Local hospitals using a surveillance method for "pneumonia of uncertain cause" identified the virus of infection as new coronavirus [2]. As is well known, the advent of the new coronavirus has been considered a risk to world public health, so much so as to cause a global lockdown, which began in China and rapidly extended throughout the world. COVID-19 is a pandemic triggered by the severe acute respiratory syndrome coronavirus 2 (SARS-CoV-2), as previously reported, it belongs to RNA coronavirus family. Approximately 40% of infected people are asymptomatic and hence are not included in clinical testing results. Furthermore, despite improvements in the mass kit manufacture, diagnostic testing has remained in high demand, creating fears of a shortage. The rapid and unexpected emergence of positive COVID-19 cases in numerous countries has been attributed to convert community broadcasts, cases imported by incoming visitors or unintentional distribution by asymptomatic. For the first time in human history, this has resulted in multiple cycles of lockdown-open-lockdown of cities, and in extreme cases, entire nations, while a clear perception of the situation is lacking.

### 14.1.1 Epidemiology

As previously reported, multiple cases of unexplained pneumonia have been recorded in some hospitals in Wuhan city (Hubei province—China), since December 2019, with a history of exposure to a huge seafood market. On January 30, 2020, the

World Health Organization named the novel coronavirus 2019-nCoV a Public Health Emergency of International Concern. The cause of the acute respiratory infection has been identified as a new coronavirus [3]. So far, the sickness has spread fast from Wuhan to other China countries and 66 nations. Then, as the epidemic progressed, clustering instances and confirmed patients without a history of travel to Wuhan appeared. In addition, confirmed cases with no obvious link to the Wuhan seafood market have been reported in a few international countries. In accordance to the National Health Commission of the People's Republic of China, a total of 80,302 CoVID-19 cases has been confirmed in China as of 24:00 on March 2, 2020, in 31 provinces, and Xinjiang PRC. Moreover, SARS-CoV-2 infection has caused 219 million COVID-19 cases worldwide. Because only a minority of acute infections are detected and reported, the real totality of cases is underestimated. Seropositivity indicated that the incidence of past SARS-CoV-2 exposure surpasses the incidence of reported cases by a factor of two after accounting for possible false positives or negatives [4].

### 14.1.2 Genomic and Viral Elements

The entire genome of Wuhan-Hu-1 coronavirus (WHCV), a strain of SARSCoV-2, was identified in a worker in a Wuhan seafood market and measures 29.9 kb [5]. The positive-sense RNA genomes of SARS-CoV and MERS-CoV are 27.9 kb and 30.1 kb, respectively. The genomes of CoVs have been discovered to have a variety of open reading frames (ORFs). The first ORF (ORF1a/b) encodes 16 non-structural proteins (NSP), the polyproteins "pp1a" and "pp1ab" and contains viral RNA. The remaining ORFs codify accessory and structural proteins. Spike (S) glycoprotein, small envelope (E) protein, matrix (M) protein, and nucleocapsid (N) protein are the four structural proteins that influence the host's innate immune response (IR), as well as several accessory proteins [6].

### 14.1.3 Current Covid-19 Variants of Concern

SARS-CoV-2, like other viruses, changes over time. Most SARS-CoV-2 genetic changes do not exert any kind of influence on the ability of the virus to function. Several variants have attracted a lot of attention because of their fast development within populations, transmission, and clinical implications [7]. The World Health Organization (WHO) has also established names for significant variants using the Greek alphabet. Each variant has numerous names based on the nomenclature used by different evolutionary categorization schemes [8].

Lineage B.1.1.7 (Alpha)—This variant, also known as 20I/501Y.V1, was first detected in the United Kingdom in late 2020 and was connected to a growth in local infections. This strain contains more than a dozen differences from other circulating strains, including those in the spike protein. It has since been detected in a number of other states. The B.1.1.7 variant outperforms wild-type strains in terms of transmission; early estimations suggested that the variant was 50–75% more transmissible. According to certain research, the B.1.1.7 variation is linked to a higher risk of illness severity. There has been no evidence that the B.1.1.7 variant is connected to clinically meaningful immunological escape thus far. Finally, serum from COVID-19 vaccine recipients has been found to have neutralizing activity against the B.1.1.7 mutant, indicating that some immunizations are still effective against it [9].

Lineage B.1.617.2 (Delta)—The 20A/S:478K lineage was discovered in India in December 2020 and has become one of the most common forms. B.1.617.2 is more contagious than B.1.1.7, according to UK data: the proportion of SARS-CoV-2 infections caused by B.1.617.2 increased while the proportion of SARS-CoV-2 infections caused by B.1.1.7 decreased, and the secondary household infection rate associated with B.1.617.2 infection was 13.6% compared to 9.0% for B.1.1.7. B.1.617.2 is linked to a higher incidence of hospitalization than B.1.1.7 according to the same study [10].

Lineage B.1.351 (Beta)—In late 2020, in South Africa, this variety, also known as 20H/501Y.V2, was found. It is not phylogenetically related to B.1.1.7, however, it does share several mutations, particularly the spike protein mutation N501Y. A report from South Africa suggests that this strain swiftly became the most prevalent [11].

P.1 lineage (Gamma)—This variant, also known as 20J/501Y.V3, was originally discovered in four Brazilian visitors in Japan in December 2020 and was later discovered to account for 42% of 31 sequenced specimens in Brazil's Amazonas state. Other countries have since detected it [9].

Lineages B.1.427 and B.1.429 (Epsilon) are two closely similar variants, also known as 20C/S452R or CAL.20C. Only four global instances were detected in October 2020, all in Southern California; by January 2021, the variation had been identified in other nations and accounted for 35% of viral samples sequenced in California. Several spike protein mutations are present in the variation, including L452R, which has been linked to enhance cell entrance and lower sensitivity to neutralization in vitro by convalescent and vaccine recipient plasma.

Lineage B.1.1.529 (Omicron) was first reported to the World Health Organization (WHO) from South Africa on 24 November 2021. Omicron multiplies around 70 times faster than the Delta variant in the bronchi (lung airways) but it is less severe than previous strains, especially compared to the Delta variant. Omicron might be less able to penetrate deep lung tissue.

### 14.1.4 Coronavirus Replication and Pathogenesis

The angiotensin-converting enzyme 2 (ACE2) is the human cell receptor for SARS-CoV and it is almost ubiquitarian in human organs but particularly present in the lower respiratory tract and governs both cross-species and human-to-human transmission. After extracting SARS-CoV-2 from the bronchoalveolar lavage fluid (BALF) of a COVID-19 patient with interstitial pneumonia, a recent study validated that the virus uses the same cellular entry receptor, ACE2, as SARS-CoV [12]. The surface coronavirus S-glycoprotein can bind to the ACE2 receptor on the surface of human cells. S1 and S2 are two subunits of the S glycoprotein. S1 uses the key function domain RBD to determine the virus's host range and cellular tropism, whereas S2 uses two tandem domains, heptad repeats 1 (HR1) and HR2, to mediate virus-cell membrane fusion. After the membrane fusion, SARS COV2 RNA enters the cytoplasm and translates two polyproteins, pp1a and pp1ab (which encode non-structural proteins), to assemble the replication-transcription complex (RTC) in the double-membrane vesicle. On a continual basis, RTC replicates and generates a tiered series of sub-genomic RNAs that encode auxiliary and structural proteins. In the endoplasmic reticulum (ER) and golgi, newly synthesized genomic RNA, nucleocapsid proteins, and envelope glycoproteins mix to create viral particle buds. Finally, the virus is released when the virion-containing vesicles merge with the plasma membrane.

### 14.1.5 Host and Reservoir

Natural reservoir hosts, such as wild animals (including bats), play an important role in the transmission of many viruses, such as Ebola, Nipah, Coronavirus, and others. SARS-CoV-2 is the seventh member of the coronavirus family, and it is a beta-CoV with a genomic sequence that is nearly identical to SRAS-nCoV. SARS-CoV-2, like SARS-CoV, MERS-CoV, and many other coronaviruses, is thought to have originated in bats, but more research is needed to determine if SARS-CoV-2-infected pneumonia is directly transmitted from bats or perhaps through an intermediary host.

Notably, the virus is 96% identical to a bat coronavirus at the whole-genome level, indicating that this species are the most likely hosts of the SARS-CoV-2. In addition, Ji and colleagues have shown that snack can serve as a virus reservoir for human infection. According to Zhu et al., bats and minks could be the two possible hosts of coronavirus, with minks perhaps serving as intermediary hosts. Pangolins have since been identified as probable intermediate hosts in research, however intermediate hosts can have many hosts in general [13].

### 14.1.6 Transmission Route

The predominant source of infection was SARS-CoV-2-infected pneumonia patients. The major modes of transmission is respiratory droplet transmission and touch; and asymptomatic instances are crucial in the transmission process. The capability of the virus to multiply in the digestive tract has now been confirmed, implying the possibility of fecal-oral transmission. Furthermore, the novel corona-virus has the potential to cause neonatal infection through mother-to-child transmission [14].

### 14.1.7 Clinical Manifestations

COVID-19 has an estimated incubation duration of up to 14 days after contact, with an incubation period generally of 4–5 days. Patients experience a wide spectrum of clinical symptoms, including asymptomatic, mild, and severe disease that is rapidly progressing and fulminant [15].

- *Asymptomatic or Presymptomatic patients*: this group include all patients tested positive for SARS-CoV-2 by virological swab (such as a nucleic acid amplification test [NAAT] or an antigen test) yet do not show the typical COVID-19 symptoms.
- *Mild Illness:* any of the COVID-19 symptoms like fever, sore throat, dry cough, asthenia, headache, myalgias, gastroenteric symptoms, loss of taste and smell, but no shortness of breath with augmented respiratory frequency, dyspnea, or abnormal chest imaging.
- *Moderate Illness*: desaturation with generally SpO2 less than 94% with indications of involved lower respiratory illness during clinical examination or imaging.
- *Severe Illness:* Patients with PaO2 /FiO2 ratio of 300 mmHg, respiratory frequency >30 breaths/min, or lung infiltrates of more than 50% at CT scan.
- *Critical Illness*: In these patients' septic shock, acute respiratory distress syndrome (ARDS), cardiac dysfunction, exaggerated inflammatory response, and/or exacerbation of underlying comorbidities are almost all possible. Especially patients recovered in intensive care units could also have hepatic, cardiac, renal, central nervous system (CNS), or thrombotic disease in addition to pulmonary disfunction.

The majority of SARS-CoV-2 patients were healthy and mild, and their mortality was lower than that of MERS-CoV and SARS-CoV [16]. The most prevalent symptoms are fever (98%), cough (76%), myalgia or weariness (44%), whereas atypical symptoms included are sputum (28%), headache (8%), diarrhea (5%), and hemoptysis (5%). Dyspnea was present in about half of the individuals (the median from onset to dyspnea was 8 days). Lymphocytic leukemia was found in 63% of the individuals. All of the patients were infected with pneumonia. ARDS (29%) was the most common complication, followed by acute cardiac injury (12%) and secondary infections (10%); 32% of patients needed to be admitted to the intensive care unit. In China, 81% of COVID-19 cases were mild (defined as no pneumonia or mild pneumonia in this study), while 14% were severe (defined as dyspnea, respiratory frequency of 30 breaths/min, SpO2: 93%, and a ratio of arterial partial pressure of oxygen to fraction of inspired oxygen [PaO2 /FiO2] 50% within 24–48 h) and 5% were critical (defined as respiratory failure, septic shock, and/or multiorgan dysfunction or failure) [17]. Diarrhea, disorientation, rhinorrhea, anosmia, dysgeusia, sore throat, abdominal pain, anorexia, and vomiting are among the other symptoms mentioned. The prevalence of these gastrointestinal symptoms varied widely between studies, ranging from 5–61% [18].

SARS-CoV-2 has been found in COVID-19 patients' feces, even in those who show no signs or symptoms of gastrointestinal illness [19]. Furthermore, fecal viral shedding has been documented to last for weeks after first diagnosis and even after PCR negativity. Detection of COVID-19 patients using anal swabs has been employed in various Chinese areas, and it is now used in many countries. The discovery of viral RNA in COVID-19 patients' anal swabs or stool samples could be used frequently to make decisions about hospitalization, recovery, and discharge. In other words, continuation of transmission-based measures for patients may be considered until viral RNA transformation in stool is negative [20].

According to research, subjects over the 60 age are at a higher risk than youngsters, and those may be less likely to become infected or have milder symptoms or even asymptomatic infection [20].

Several clinical research have highlighted the relationship between diabetes mellitus (DM) and infectious illnesses [21]. As a result of their reduced immunity, DM sufferers are more susceptible to infections than individuals without the disease. Notably, pneumonia caused by SARS-CoV, Middle East respiratory syndrome coronavirus (MERS-CoV), and the 2009 influenza A (H1N1), have been shown to increase the vulnerability of DM patients. The same could be in COVID 19 patients [22].

#### 14.1.7.1 SARS-CoV-2 Reinfection

Reinfection is thought to be much more frequent in people who had a weaker IR during the early contact with the virus, as commonly occurs in the case of patients who had a mild disease. As early IRs fade over time, reinfection may occur. Nonetheless, in three cases, SARS-CoV-2 reinfection occurred despite previous severe sickness and already 3 weeks after diagnosis of the first contagion, according to one recent review of the literature. Although there are fears that reinfection may occur more frequently when new variants circulate, the real prevalence of reinfection is unknown [23].

### 14.1.8 Diagnosis

People with COVID-19-like symptoms, as well as subjects who have had a known high-risk SARS-CoV-2 exposures, have to do theSARS-CoV-2 molecular swab or the antigen test [24].

Although nasopharyngeal specimens are still the best bet for diagnosing SARS-CoV-2, nasal or oropharyngeal swabs are also possible options. Bronchoalveolar lavage and other lower respiratory tract samples are more accurate than upper respiratory tract samples despite the fact that they are infrequently acquired due to fears about viral aerosolization during sample collection techniques. In fact, reverse transcriptase polymerase chain reaction (RT-PCR)-based diagnostic assays are the gold standard for identifying SARS-CoV-2 infection. NAATs have lately expanded to encompass a number of new platforms (for example, reverse transcriptase loop mediated isothermal amplification [RT-LAMP]) [25]. In clinical practice, there may be a 5-day window after exposure before viral nucleic acids can be detected. When a mutation happens in the region of the virus' genome that is examined by that test, some NAATs can generate false negative results. BAL and sputum induction are aerosol-producing process that should only be carried out after a comprehensive evaluation of the risk of infecting health workers. Endotracheal aspiration appears to be less likely to generate aerosols than BAL, and many authors believe that the sensitivity and specificity of those procedures are equivalent. Antigen-based diagnostic tests are less sensitive than RT-PCR-based diagnostics yet with comparable specificity. When the viral load is assumed to be at its greatest, antigen testing function is best in the early stage of symptomatic SARS-CoV-2 infection. Antigen-based diagnostics have the advantages of being inexpensive and having a quick turnaround time. Due to the availability of speedy results, they are an appealing solution in high-risk congregate settings where avoiding transmission is critical. Repeat testing is also possible with antigen-based tests to quickly identify people SARS-CoV-2 positive [26]. Nothing like antigen tests and NAATs for SARS-CoV-2, which only identify the virus's existence, serologic and antibody tests for SARS-CoV-2 can detect recent or past infection. Because seroconversion (the formation of measurable immunoglobulin M and-or IgG antibodies to SARS-CoV-2) may take 21 days or longer following symptom onset, 21–26 guidelines does not suggest serologic testing [27].

#### 14.1.8.1 Physical Examination

Positive indicators may not be seen in patients with moderate symptoms. Dyspnea, wet rales at the objective exam of the lungs, attenuated breath sounds, and dullness in percussion are all symptoms of a severe disease.

#### 14.1.8.2 Chest-X Ray and CT Imaging Examination

When an imaging examination is performed, the age of patient, the immunological condition, the disease stage at the time of scanning, and pharmacological interventions may all be significant aspects to consider. Chest imaging may show several tiny patchy shadows and interstitial alterations in the first stages of SARS COV2 interstitial pneumonia, notably in the lung periphery [28]. In severe cases, ground-glass opacity and segmental consolidation in bilateral lungs, especially in the lung periphery, are shown more clearly on CT than on X-ray scanning. Generally, the imaging findings for CoVID-19 are comparable to those for SARS and MERS, which is not surprising given that the responsible viruses are both coronaviruses [29].

#### 14.1.8.3 Laboratory Diagnosis

The imaging findings for CoVID-19 are comparable to those for MERS and SARS, and are not surprising that the viruses responsible are both coronaviruses. As a result, a laboratory diagnosis is required. CoVID-19 is identified primarily through virus isolation and viral nucleic acid testing. The isolation of SARS COV2 is the "gold standard" for laboratory diagnosis, according to classic Koch's postulates. The most essential feature about viral nucleic acids is that they can be used for early detection. As a result, we should look for SARS-CoV-2 nucleic acid: detailed SARS-CoV-2 RNA detection has diagnostic value [30].

### 14.1.9 Current Anti-COVID-19 Treatments

Infection prevention and control methods, as well as supportive care, such as supplemental oxygen and mechanical ventilatory, are being used in the treatment of COVID-19 [31]. A high SARS-CoV-2 viral load was independently correlated indicator of disease gravity and mortality in most investigations and may be beneficial in susceptible persons like the elderly, patients with co-existing medical diseases like diabetes or heart disease, and immunocompromised people. IL-2, IL-4, IL-6, IL-10, TNF, IFN, and C-reactive protein levels are all linked to a high viremia, leading in a hyper-inflammatory state and, as a result, a severe infection. However, because of the wide heterogeneity in fluid samples and diverse disease stages, these data should be cautiously interpreted and solely as trends [32].

Supportive care, actions to decrease the potential of SARS-CoV-2 transmission such as the isolation of patients and educating individuals on when to call a health care practitioner and request a person evaluation should all be included in the management of non-hospitalized patients with acute infection. Before getting in-person care, individuals with COVID-19 symptoms should be triaged using telehealth consultations whenever possible. Patients with dyspnea should be referred to a health care practitioner for an in-person evaluation, and they should be thoroughly monitored in the days after dyspnea beginning to check for worsening respiratory status. The patient's physical exam results, risk factors for severe illness development, and the accessibility of health-care resources should be considered at the same time while developing a management plan. After being diagnosed with COVID-19, a patient's usual treatment and supplement regimen should be maintained. ACE enzyme inhibitors, nonsteroidal anti-inflammatory drugs, statin medication, oral inhalation, and intranasal corticosteroids should all be maintained as prescribed, especially for concomitant conditions [32].

#### 14.1.9.1 Therapeutic Management of non-hospitalized COVID-19+ Adults

Fever, headache, myalgias, and cough can all be treated with over-the-counter antipyretics, analgesics, or antitussives. Dyspnea patients may find it more comfortable to rest in the prone rather than the supine position. Because severe breathlessness might create anxiety, health care practitioners should consider educating patients about breathing exercises. In addition, patients should be encouraged to take water on a regular basis to avoid becoming dehydrated. Additional therapy measures are suggested when needed during the acute COVID-19 phase, and ambulation and other kinds of activity should be improved according to the patient's tolerance [33].

##### Dexamethasone

Generally, patients with mild-moderate COVID-19 who do not require to be hospitalized or supplementary oxygen should not be treated with dexamethasone or other systemic glucocorticoids, according to the guidelines. Dexamethasone was proven in recovery trial to lower mortality in hospitalized COVID-19 patients who needed supplementary oxygen. However, no benefit of Dexamethasone was found in hospitalized patients who did not receive oxygen support [34].

**Remdesivir**

To date, the only medicine licensed by the FDA for the management of COVID-19 is remdesivir. It is suggested in hospitalized patients who require supplementary oxygen. The remdesivir safety and efficacy were investigated in clinical trials that ended after patients were discharged from the hospital [35]. Remdesivir should not be continued in hospitalized stable COVID-19 patients who do not need supplemental oxygen.

## 14.1.10 Anti-SARS-CoV-2 and Monoclonal Antibodies

In subjects with moderate symptoms of COVID-19 and some risk factors for developing the disease, a single monoclonal antibody (sotrovimab) and two anti-SARS-CoV-2 monoclonal antibodies (casirivimab plus imdevimab and bamlanivimab plus etesevimab) minimized hospitalization and death risk in the outpatient setting. As a result, the FDA has granted these medications Emergency Use Authorizations (EUAs) for the COVID-19 treatment in these patients, as well as in people with other risk factors for progression revealed in population-based research [36]. Guidelines recommend treating outpatients with mild to moderate COVID-19 who are at high risk of clinical progression, as defined by the EUAs criteria and the guidelines' statement, with one of the anti-SARS-CoV-2 monoclonal antibodies listed below casirivimab in combination with imdevimab, or sotrovimab. Because of an increase in the number of gamma (P.1) and beta (B.1.351) variants, that have decreased susceptibility to both etesevimab and bamlanivimab, the guidelines currently advise against using bamlanivimab plus etesevimab (AIII). Within 10 days of the onset of symptoms, treatment should begin immediately after the patient has tested positive for a SARS-CoV-2 antigen test or an NAAT [36].

### 14.1.10.1 Baricitinib

Patients with COVID-19 were enrolled in the pivotal safety and effectiveness trials for baricitinib, and treatment was ceased at the time of hospital discharge. Guidelines recommend against continuing baricitinib in COVID-19 patients who do not require additional oxygen administration (AIIa).

### 14.1.10.2 Anticoagulants and Antiplatelet Medication

Those medication should not be started in the outpatient setting to prevent arterial thrombosis and venous thromboembolism unless the patient has other risk factors or is enrolled in a clinical study.

### 14.1.10.3 Tocilizumab

This is an anti-IL-6 monoclonal drug and compared to dexametazone, it has a considerably greater effect on survival, clinical improvement, and hospital discharge rate in patients with a severe COVID-19 course, particularly those who develop cytokine storm. When compared to the administration of tocilizumab alone, the combination of tocilizumab and dexametazone does not improve therapy success in subjects with a severe infection [37].

### 14.1.11 Long-Term Symptoms (Long Covid)

Covid-19 has had an unprecedented impact thus far, and long-term symptoms could have much more serious consequences [38]. Recent data suggests that in many people who have had Covid-19, a variety of symptoms can persist after the acute infection has cleared, a condition known as persistent or long covid. The National Institute for Health and Care Excellence (NICE) describes long Covid as symptoms that persist or increase from 4 to 12 weeks after an acute covid-19 infection and are not due to another illness. On the other hand, the National Institutes of Health (NIH) uses the CDC's (Centers for Disease Control and Prevention) definition of long Covid, which classifies the illness as lasting more than 4 weeks following infection [39]. The structure and function of numerous organs are involved and impaired in people with extended Covid. Long-term symptoms after covid-19 have been reported across the disease severity spectrum. Long Covid can occur in any patient with Covid-19, regardless of the severity of their infection or the intensity of their treatment. The frequency of long-term symptoms linked with Covid-19 is similar in patients treated on wards and in intensive care units. The percentage of people who have long Covid symptoms is almost the same whether they are treated with oxygen, continuous positive airway pressure (CPAP), or invasive ventilation [39]. On the other, hand many patients with mild acute clinical manifestations develop long Covid symptoms, [39] and studies show that the prevalence of long Covid symptomatology differs little between non-hospitalized and hospitalized Covid-19 positive patients. Asthenia, myalgia, dyspnea, heart anomalies, cognitive decline, sleep disturbances, post-traumatic stress disorder symptoms, concentration issues, and headache are the most common symptoms [40].

### 14.1.12 Prevention and Prophylaxis

SARS-CoV-2 is assumed to be spread mostly through inhalation of air carrying aerosol particles and very fine droplets passed from an infected individual to those within about 6 ft of the infected person. after prolonged exposure (>15 min) in an enclosed location and with poor ventilation. Covering coughs and sneezes and keeping at least 6 ft away from others can help reduce the risk of SARS-CoV-2 transmission. Face coverings may help to prevent the spread of infectious droplets from people infected with SARS-CoV-2 when regular distance is not possible. Handwashing frequently also minimizes the chance of infection. Vaccination is another essential approach to avoid SARS-CoV-2 infection.

### 14.1.13 Vaccines

The FDA granted Emergency Use Authorizations (EUAs) for two mRNA vaccines, BNT162b2 (Pfizer-BioNTech) and mRNA-1273, in December 2020 (Moderna) [41]. The FDA granted an EUA for Ad26.COV2.S (Johnson & Johnson/Janssen), a human adenovirus type 26 (Ad26) vectored vaccination, in February 2021. After a two-dose series, the mRNA-1273 and BNT162b2 vaccines were found to be >90% effective in avoiding symptomatic, laboratory-confirmed COVID-19 and >95% effective in preventing severe COVID-19 in large, placebo-controlled trials. The single-dose of Ad26.COV2.S vaccine was found to be 66% effective in preventing moderate to severe laboratory-confirmed COVID-19 infection. SARS-CoV-2 vaccines approved for use seem to protect against asymptomatic illness, transmission, and infection by presently circulating or emergent SARS-CoV-2 variants. These vaccines have a lot of side effects, both local and systemic, in particular after the second dosage of a SARS-CoV-2 mRNA vaccine. The majority of vaccine-related side effects were from mild to moderate severity. There are a few reports of severe allergic reactions, including anaphylaxis, after receiving the SARS-CoV-2 mRNA vaccine. Thrombosis with thrombocytopenia has been found to occur in roughly three vaccinated people per million in the USA. Almost all cases of this dangerous disease have been reported in vaccinated women between the ages of 18 and 49. This adversity is considerably more uncommon among women over 50 and males of all ages. Thrombosis can occur in unusual places, such as the cerebral and abdominal veins; lower extremities thrombosis and pulmonary emboli are also possible. Thrombocytopenia and venous thrombosis have been reported in patients who received vaccines with an adenoviral vector. The original vaccine studies did not include pregnant or breastfeeding women. According to a study looking at data from three US vaccination safety reporting systems, the prevalence of adverse events among 35,691 pregnant vaccine participants was similar to that seen in non-pregnant patients [42].

## 14.2 The Microbiome

SARS-CoV-2 were identified in infected patients' mid-nasal, nasopharyngeal, rectal, and stool swabs [43–45]. Indeed, multiple organs, including the respiratory and gastrointestinal systems, express the ACE2 receptors [46]. Impaired ACE2 expression is linked to viral infection and today immunological imbalance and lung and intestinal microbiota disequilibrium are well documented in SARS-CoV-2 [47].

In general, microflora, microbiota, and normal flora are terms used to describe colonies of microorganisms living in close proximity to their hosts. In the other hand, the term "microbiome" defines the combined genetic material of the microorganisms in a particular environment. The human microbiota might include up to 1014 microorganisms (more than the number of human cells). The human body was once thought to be a self-sustaining organism capable of controlling all of its

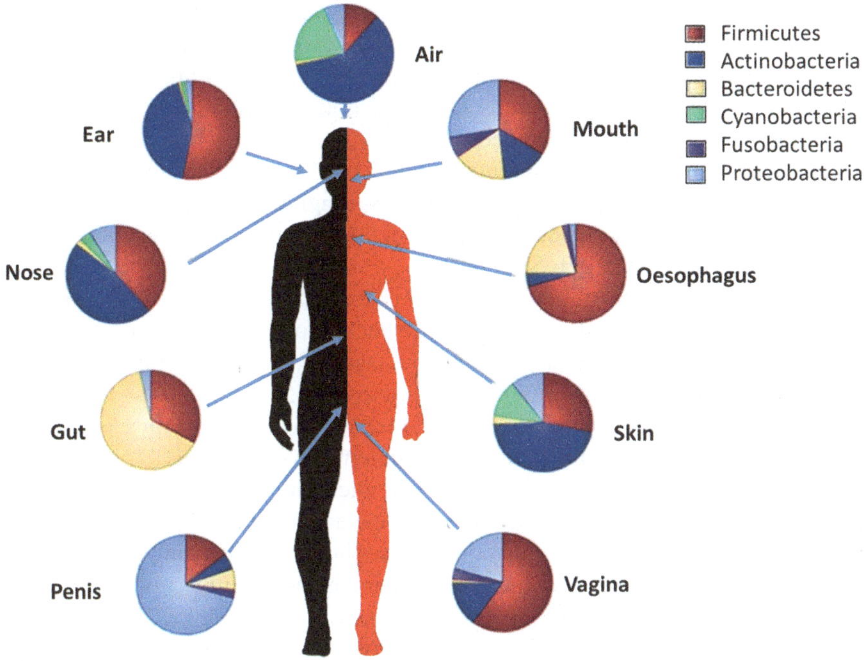


**Fig. 14.1** Surfaces of the human body inhabited from microbiome. The microbiota is resident in every body surface exposed to the external environment such as skin and mucosa (from the gastrointestinal, to respiratory and urogenital tract)

metabolic processes, however nowadays scientists have demonstrated that it can be considered as an ecosystem with billions of microbes.

The microbiota is normally found on every part of the body that is accessible to the external environment, such as the skin and mucosa (from the gastrointestinal, to respiratory and urogenital tract) (Fig. 14.1). The gastrointestinal tract (GIT) has the greatest number of microorganisms that produce compounds that may be utilized as nutrients, making it a prime location for colonization; in fact, the colon contains approximately 70% of all bacteria in the body. This human GIT environment is the product of an evolutionary process of microflora and body coexistence. The microbiota has a big impact on physiological processes including digestion and immune system activation [48]. In addition, microorganisms from the *Archaea*, *Bacteria*, and *Eukarya* domains (fungi, and protozoa, as well as their viruses), make up the "human microbiota" mainly composed of stringent anaerobes bacteria, however facultative anaerobes and aerobes are in the minority. The commensal bacteria are symbiotic, but they may trigger a pathogenic state following translocation through the mucosa or in certain compromised conditions, such as immunodeficiency. In general, the human microbiota composition is very individual, although there is greater variety in the organization of the bacterial community among body locations than there is between people. This shows that the human microflora is a complex ecosystem with a wide range of microbiological components [49, 50]. However, within different body locations, there is a conserved microbial community of a healthy microbiota, called "core." Interestingly, a large percentage of the human microbiota, around 70%, is composed by bacteria that cannot be grown using conventional microbiological procedures [51]. The host microflora's protective role in a number of pathological situations has been emphasized in studies using animals treated with broad-spectrum antibiotics, ranging from metabolic disorders [52] to infectious and inflammatory diseases in the intestine and at distal body sites, such as the lungs and skin [53–55].

The host-microbiota symbiotic balance is extremely susceptible to different biological factors, such as the host's genetic background, antibiotic use, nutrition, and the availability of allergens or infectious agents, resulting in an alteration of microbiota architecture, and thus in a "dysbiosis" [56]. Dysbiosis can worsen existing conditions or make people more vulnerable to new ones, for example, the emergence of potentially pathogenic endosymbionts or pathobionts.

Today, in order to analyze the bacterial microbiota profile, genomic Next-Generation Sequencing (NGS) technique and metagenomics techniques (16S ribosomal RNA profiling, as well as the more precise shotgun-sequencing technique) are widely used. NGS is able to give more information about the impact of microflora in host metabolic reaction, diseases progression, and inflammation [57, 58]. The existence, distribution, and relative abundance of microbial commensals in previously thought-to-be-sterile body areas, such as the lungs, have been detected using metagenomic techniques. Since 2010, research has shown that changes in the lung microbiota (LM) are linked to a plethora diseases, including cystic fibrosis, chronic obstructive pulmonary disease, and asthma [59–61] suggesting that the microflora of the lungs has an impact on both illness and respiratory health. Furthermore, the intestinal microflora via the gut–lung axis may affect also pulmonary immunity [62, 63] (Fig. 14.2) (see paragraph 3).

### 14.2.1 The Lung Microbiome-Covid-19 Link

Because they are continually exposed to a broad variety of external environments, the lungs are at the forefront of immunity. As previously reported, until recently, the lung was thought to be a sterile, bacteria-free environment [64]. Due to a favorable moisture, temperature, and mucus environment, and to the possibility of being in touch with the external milieu, the lung is considered as a highly populated bacteria location [65]. Because of the bidirectional flow of air and mucus, the LM is dynamic and transitory [66, 67]. The lung microbiota of a healthy individual has a moderate density but as a result in increase of interacting microbes. The most frequent phyla are *Bacteroidetes*, *Firmicutes*, and *Proteobacteria*, whereas the most common bacteria at the genus level are *Veillonella*, *Prevotella*, and *Streptococcus* [64, 68]. The LM profile is influenced by microbial movement, removal, and relative growth rates of its components. Furthermore, these variables may alter in lung

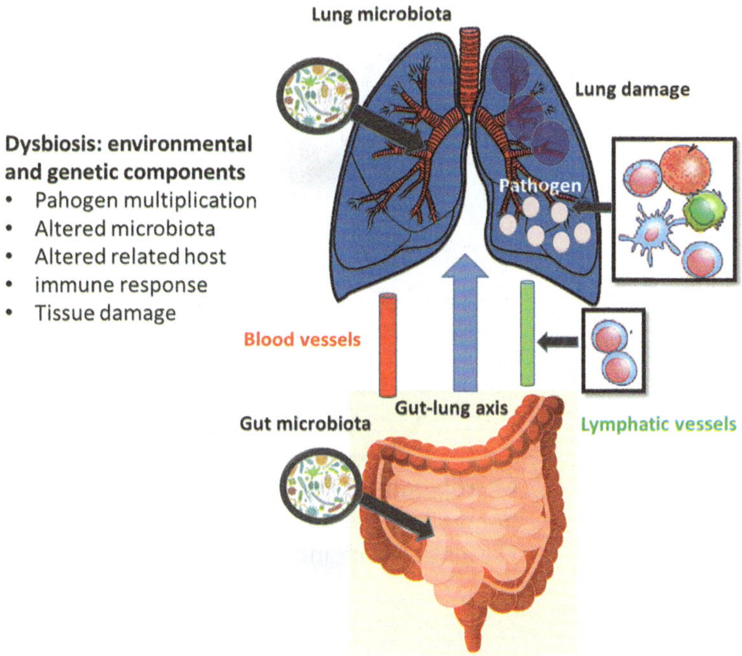


**Fig. 14.2** The gut–lung axis in a dysbiotic status. When the commensal bacterial diversity is disrupted by a respiratory pathogen, pathobionts can develop in the gut and/or lungs, resulting in dysbiosis. Dysbiosis causes a disruption in the number and activity of leucocytes, which can lead to lung injury

illnesses, resulting in overgrowth of certain species and a decrease in microbial diversity [69] (Fig. 14.3).

However, during lung pathology, the local environment dramatically changes, producing an ideal milieu for bacteria development. Bronchoscopy samples from one of the most prevalent lung diseases, such as asthma, have revealed an increased number of *Haemophilus*, *Neisseria*, *Fusobacterium*, and *Porphyromonas* [59, 70]. *Lactobacillus*, *Fusobacteria*, *Leptotrichia*, and *Fusobacterium* were found in large quantities in another prevalent lung illness, as chronic obstructive pulmonary disease (COPD) [60].

New data suggests a link between lung bacteria and pneumonia risk [71]. The most frequent bacterial causative agents of pneumonia are *Streptococcus pneumoniae* and *Haemophilus influenzae type b*, whereas the most frequent viral agent is respiratory syncytial virus (RSV) [72]. The bacterial load was higher in patients with ventilator-associated pneumonia, but the abundance of the *Bacilli* in their endotracheal aspirates was lower [73]. *Rothia, Pseudomonas,* and *Corynebacterium* were more common in intubated pneumonia patients than in those without pneumonia, whereas *Prevotella* and *Streptococcus* were less common [74]. In patients with interstitial pneumonia, the abundance of *Firmicutes*' phylum, as well

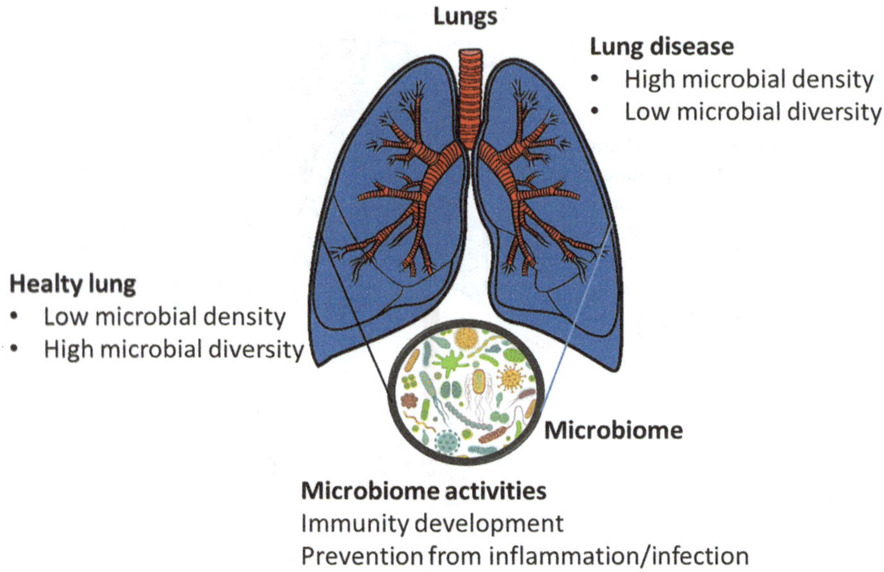


**Fig. 14.3** Lung microbiome activities

as the total phyla richness, was lower than in the healthy group. In contrast to healthy subjects, the quantity of *Prevotella* and *Veillonella* was greatly enriched, whereas *Streptococcus* was considerably decreased in interstitial pneumonia patients [75]. Furthermore, the microbial diversity and composition tended to alter depending on the pneumonia causative agent, indicating that the LM might vary depending on the pneumonia kind [76]. In addition, *Moraxella*, *Haemophilus*, *Streptococcus*, *Dolosigranulum,* and *Corynebacterium* dominated the upper respiratory tract microbiota of rhinovirus and RSV infected children [77]. In addition, rhinovirus infection increased the relative abundance of *Neisseria* and *Haemophilus*, two bacteria linked to subsequent lung infections [78].

In conclusion, the LM is very dynamic, and bacterial and viral infection can cause fast changes in the microbiome composition, with microbial communities playing a key role in those diseases. The alveolar surface is constantly exposed to invading microorganisms because it is in close contact with the external environment. As a result, there is a mutual interplay between the microbiome and the IR, with several defence lines against potential infections. For inhaled pathogens, a continuous layer of pulmonary epithelial cells acts as barrier. On the surface and in the fluids surrounding alveoli, the pulmonary epithelium's mucus layer, defence proteins (e.g., lactoferrin, immunoglobulins, defensins), lysozymes, and proteolytic enzymes defend against infection [65]. In addition, there is an intimate interaction between pulmonary epithelial cells and immune system cells producing a wide spectrum of chemokines and cytokines. They have pattern-recognition receptors (PRRs) on their cell surfaces, such as toll-like receptors (TLRs) that allow them to identify pathogen-associated molecular patterns (PAMPs) from microorganisms [79].

Immune cells such as lymphocytes, dendritic cells, innate lymphoid cells, macrophages, and neutrophils promote adaptive and innate immune processes. All of the immune cells are engaged in phagocytosis, or the activation of antigen-removing effector molecules, which remove antigens from the respiratory tract. Regulatory T cells (Tregs) in the lungs are critical for maintaining immunological tolerance to airborne particle [79]. In addition, the lung resident memory T cells represent another sort of T cells' subsets. Apart from their involvement in cytokines' production to prevent viral and bacterial infection, T cells provide a prompt IR at barrier surfaces, recalling antigens that were previously exposed via the lung mucosa [80]. However, SARS-CoV-2 causes lung infection evading the immune system.

The respiratory tract microbiota of COVID-19 patients has been recently studied. When compared to subjects with other pneumonia types, a meta-transcriptomic study of sputum samples and nasopharyngeal swabs from pneumonia patients displayed a lower alpha diversity in patients infected with SARS-CoV-2. Other symptoms in these individuals might explain this result, such as an augmented susceptibility to pulmonary infections and activation of various immunological pathways associated with the production of cytokines [81, 82]. COVID-19 patients showed a higher rate of parallel infections with different bacteria, viruses, and fungi than subjects with non-COVID-19 pneumonia, whose cytokine distribution indicated Gram-negative infections. In the same way, a meta-transcriptomic study of BAL fluid from both COVID-19 and non-COVID-19 patients revealed a substantial growth of bacterial and other pathogens, highlighting the existence of LM dysbiosis in COVID-19 patients [83]. In addition, *Acinetobacter*, a bacterium typically linked with lung infections that can lead to pneumonia was found to dominate the lung microbiome of 20 deceased COVID-19 patients [84]. Analyzing the lung mycobiome (fungi microbiome) of dead patients, it was found that opportunistic infections such as *Cladosporium*, *Cryptococcus*, *Alternaria*, *Issatchenkia*, *Candida,* and *Aspergillus* were also prevalent. Some of these infections are lethal, notably *Cryptococcus* infection, which is associated with a high risk of morbidity and fatality [84]. The nasopharyngeal microbiota of patients with acute respiratory disease with COVID-19 suspicion, on the other hand, showed no differences in the composition comparing patients who were COVID-19+ to those who were negative [85]. Analogously, the nasopharyngeal microbiome of COVID-19 patients was assessed comparing sample procedures and swab types. The result confirmed that microbiota architecture was not changed by swab type, but rather by sampling methodology [86]. In the same study, *Bacteroidetes*, *Firmicutes*, *Proteobacteria,* and *Actinobacteria* were the most predominant taxa in the nasopharynx of COVID-19 patients. Furthermore, after whole-genome sequencing of BALF samples, the microbiota in COVID-19 subjects was comparable to that in CAP (community-acquired pneumonia) patients. However, the LM composition of COVID-19 and CAP groups was substantially dissimilar from healthy controls, who showed enrichment with recognized oral and upper respiratory commensal bacteria; this result suggests dysbiosis in the COVID-19 patients' lung microbiota [83].

Recently, COVID-19 co-infection with another virus, such as influenza A/B, rhino- or enteroviruses, or respiratory syncytial virus, was discovered in 11 of

20 patients, utilizing next-generation sequencing of nasopharyngeal swabs [87]. *Haemophilus parainfluenzae*, *Streptococcus mitis*, *Leptotrichia buccalis*, *Streptococcus bovis*, *Neisseria cinerea*, and *Rothia mucilaginosa* were among the bacteria found in COVID-19 throat samples, according to another study. Despite possessing COVID-19, the pharyngeal microbiota diversity of all investigated bacterial phyla were reduced in older individuals compared to younger ones. This might explain the severity disparities [88]. Although there is no conclusive evidence that the lung microbiome impacts the risk of SARS-CoV-2 infection or the severity of disease, more evidence suggest that bacteria in the lungs play an essential pathologic role. As a result, it indicates that the lung microbiome can influence the risk of SARS-CoV infection. Recent findings linking COVID-19 illness severity to the gut microbiota appear to be very promising ([89, 90] #4029).

### 14.2.2 The Role of Gut Microbiome in COVID-19 Infection

Despite the fact that over 50 bacterial phyla have been identified, just two dominate the typical human flora of the human gut: *Bacteroidetes* and *Firmicutes*, with *Actinobacteria*, *Proteobacteria*, *Fusobacteria*, *Verrucomicrobia*, and *Cyanobacteria* appearing in small proportions [91]. The number of microbial species identified in the intestine varies greatly depending on the study type, but it is generally estimated to be in the range of 500–1000 species. According to a new research involving a large number of subjects, the human gut microbiota (GM) comprises over 35,000 different bacterial species [92].

The mouth cavity, stomach, small intestine, and colon are the four separate areas that make up the human digestive system. The mucosa is the body's biggest surface that is constantly exposed to bacterial and food antigens. The bacterial phyla present on Earth are more than 50, but the most common human gut-associated microbiota is composed, as previously reported, of four phyla: *Firmicutes*, 30.6–83% (*Ruminococcus*, *Clostridium*, *Peptococcus*, *Eubacterium*, *Dorea*, *Lactobacillus* - L, *Peptostreptococcus*); *Bacteroidetes*, 8–48% (*Bacteroides*); 16.7% (*Bifidobacterium*), *Actinobacteria*, 0.7-and *Proteobacteria*; 0.1–26.6% (*Enterobacteriacee*; [91, 93]). However, the GM structure is not homogeneous. The amount of bacteria in the human GIT rises from the mouth (fewer than 200 species) to the colon (bacteria exceeding 1010–1012/g of luminal composition, with prevalence of anaerobe bacteria) [94]. In addition, the bacterial structure varies among the different GIT compartments. Various microbial strains are concentrated at different regions, when comparing biopsy samples of the small intestine and colon from healthy controls. Bacilli class of the *Firmicutes* and *Actinobacteria* is increased in the specimens of the small intestine. On the contrary, *Bacteroidetes* and the *Lachnospiraceae* families of the *Firmicutes* were more dominant in colonic samples, [92]. A thick mucus layer separates the intestinal mucosa from the lumen, resulting in substantial latitudinal microbial variety. The microbiota in the intestinal lumen differs substantially from the microbiota entrenched in the mucus layer, as well as the bacterial population living in the near epithelium area. The mucus layer and epithelial crypts were not accessible to many bacterial strains found in the intestinal lumen. *Streptococcus*, *Bacteroides*, *Bifidobacterium*, members of *Clostridium*, *Enterobacteriaceae*, *Enterococcus*, *Lactobacillus*, and *Ruminococcus* were all observed in fecal samples, whereas only *Lactobacillus Clostridium*, and *Enterococcus* were observed in the epithelial crypts of the small intestine and in the gut mucus layer [95]. Different variables, such as bacterial factors (enzymes, metabolic activity, adhesion capacity), host elements (bile acids, mucus pH, digestive enzymes, transit time), and non-host features, might all contribute to the diversifications over the length of the GI tract (medication, nutrients, environmental factors) [96]. Due to the abundance of nutrients, the human oral cavity represents the ideal habitat for microorganisms. At least six billion microorganisms take place in mouth belonging to the *Bacteroidetes*, *Firmicutes* (Gram positive), *Proteobacteria* (Gram negative), *Fusobacteria* (Gram negative), and *Actinobacteria* (Gram positive; [97]). The gastric microbiota is composed mostly of *Actinobacteria* but, due to the acidic environment, *Helicobacter* (e.g., *H. pylori*) is also present [98]. The microbiota composition of the small intestine is comparable to that of the colon, but the latter includes a greater number of bacteria. The small intestine hosts few bacteria in its proximal part, the microbiota is composed of *Enterococcus faecalis* and Gram+ *Lactobacillus*. More bacteria are found in the distal part, e.g., *Bacteroides* and coliforms. In the colon, *Firmicutes* and *Bacteroidetes* were dominant and, at the genus level, anaerobic lactic acid bacteria, e.g., *Bifidobacterium bifidum* and anaerobic *Bacteroides*, prevailed [97]. The GIT microbiota is critical to human physiology because it produces metabolites interacting with the host and performs key metabolic activities. The bacteria in the gut microbiota, in particular, act as a first defence line against pathogen colonization, breaking down indigestible dietary components [99], promoting angiogenesis, supporting fat metabolism, synthesizing vitamins, maintaining homeostasis, and especially assisting immune system development [100]. A stratum of epithelial cells separates the bacteria population from the interior gut milieu, providing a physical and chemical barrier that balances crosstalk between the immunological host system and the external environment. Furthermore, epithelial surfaces have acquired antimicrobial capabilities to combat microorganism invasion. The mucosa and the internal environment of the human body are protected by adaptive and innate IRs. The mucosal-associated immune system induces almost the 80% of active immunological cells, mostly located in the GI tract, where the quantity of immunogenic dietary components and bacterial flora is highest compared to other body sites. The immune system usually tolerates commensal bacteria and maintains homeostasis, so the bacterial flora does not evoke a pro-inflammatory response; however, when new pathogenic bacteria enter into this well-balanced system or the eubiosis is disrupted (e.g., antibiotic usage, immunodeficiency, and poor diets), the immune system responds to the microbiota, causing inflammation and cancer development in the gut [101]. Several studies have linked an GM imbalance and metabolic processes to the onset and progression of GIT diseases such as colorectal cancer, functional dyspepsia, severe diarrhea, inflammatory bowel disease (IBD), celiac disease, and irritable bowel syndrome (IBS) [102, 103]. Extrinsic (e.g., stress, genetics, and age) and intrinsic (e.g., stress, genetics, and aging) variables can both trigger the GM dysbiosis (e.g., appendectomy, diet, and antibiotic use).

Current results, such as the enteric microbiota dysbiosis ([89, 90] #3997, [104] #3998) and persistent detection of viral RNA in fecal samples ([44] #3995, [45] #3996), suggest a significant role for the GIT tract also in SARS-CoV-2 infected individuals. Moreover, in a subset of patients, has been observed that the SARS-CoV-2 virus infects and actively replicates in enterocytes ([105] #3990, [106] #3991, [12] #3992], resulting in symptomatic GIT disease ([107] #3993). These results indicate that SARS-CoV-2 may interact with the commensal bacteria in the GIT. The mechanisms that lead to the manifestation of gastrointestinal symptoms are still partially vague but it is reasonable to consider the hypothesis that ACE2 receptors can be involved in the biochemical mechanism, since these kinds of receptors are highly expressed on the enterocyte membrane [108, 109]. Different recent investigations on SARS-CoV-2 + patients have found fecal microbiome dysbiosis with an increase of opportunistic pathogens. Commonly, opportunistic pathogens are members of the commensal microflora that can turn pathogenic in the presence of compromised host immune system or a host disturbance, such as dysbiosis [82, 83, 89, 90, 110]. Two different studies found that COVID-19 patients with chronic dysbiosis also had a rise in opportunistic microorganisms in their enteric system [90, 111]. Moreover, during SARS-CoV-2 infection, the spread of opportunistic and commensal pathogens was observed in a small group of 15 patients, showing that the severe form of COVID-19 disease was correlated with more *Coprobacillus, Clostridium ramosum,* and *Clostridium hathewayi,* while the presence of *Faecalibacterium prausnitzii* was related to the milder form of the SARS-CoV-2 [90, 112]. Even though *Faecalibacterium* has been linked to a mechanism that reduces intestinal inflammation [113], one study found that it was adversely connected with severe COVID-19 ([89, 90] #4006).

However, although recent data showed a link between opportunistic infections and the gut microbiome, the fine nature of opportunistic pathogen enrichment and pathogenicity remained unknown. When the host immune system is compromised, these pathogens may play a role in secondary bacterial infection [90]. Analogously, COVID-19 + patients showed a GM dysbiosis with an excess of opportunistic infections [89, 90, 110, 111, 114]. Moreover, even though the mechanism of GM alteration underlying severe diseases is unclear, it has been found to be a predisposing factor for pro-inflammatory settings such as sepsis ([113] #4008). Furthermore, COVID-19 severity was shown to be linked to the baseline fecal microbiota in a recent observational research [90]. Indeed, a recent cohort study investigated how the GM of COVID-19 patients links to disease severity and associated inflammatory markers [115]. When the microbiota structure of hospitalized COVID-19 patients was compared to that of non-COVID-19 subjects, *Bacteroidetes* were found to be more numerous in positive patients, whereas *Actinobacteria* were found to be more abundant in non–COVID-19 subjects. In comparison to non-COVID-19 patients, the GM of COVID-19 patients was predominantly enriched with taxa including *Ruminococcus torques, Ruminococcus gnavus,* and *Bacteroides dorei*, and deficient with *Bacteroides adolescentis, Faecalibacterium prausnitzii*, and *Eubacterium rectale*. Regardless of whether they had taken antibiotics, the intestinal microflora of recovered patients was augmented in *Lactobacillus ruminis* and *Bifidobacterium dentium* and reduced in *Bifidobacterium longum, Eubacterium rectale, Faecalibacterium prausnitzii, Ruminococcus bromii* [115]. In addition, in the same study, the enteric microbiome dysbiosis remained after SARS-CoV-2 clearance, which might be a cause in the development of chronic symptoms or multisystem inflammatory syndromes reported in certain patients after the virus was cleared (Long Covid-19) [115].

In contrast to previous finding, a recent report found no changes in GM composition in relation to COVID-19 severity or gut inflammatory markers, suggesting that only antibiotic-treated individuals had significant microbiome alterations with limited GM diversity [116].

However, it is unknown how much the human gut microbiota has a role in COVID-1. More research into the bacterial microflora is needed, particularly the convalescent phase of COVID-19 and the antibody generation phase against SARS-CoV-2. Identifying a putative link between intestinal microflora and COVID-19 might lead to the discovery of microbial species implicated in disease pathogenesis and/or microbial biomarkers for disease severity, which could be used as a predictor of disease progression. Furthermore, early GM manipulation (e.g., by symbiotics, probiotics, fermented foods, and fecal transplant) might be helpful in terms of prevention and treatments.

## 14.3 The Microbiota–Inflammation Axis in Covid-19 Disease

As previously reported, during the COVID-19 disease, pulmonary and extrapulmonary symptoms, particularly related to the immunological response, were observed. SARS -Cov-2 mediates a significant damage of lung tissue interfering with lung microbiota, indeed, if the host IR does not stop the virus duplication, the effects can result in a great lung damage (Fig. 14.4) [117, 118].

All the available evidence, pointing to the interaction between COVID-19 and the host microbiota, involves the activation of inflammatory cascade and of the innate and adaptive IR [108, 109, 119]. Indeed, the IR activated is characterized by the secretion a vast number of chemokines and especially cytokines, such as TNF-α, IFN-α, IFN-γ, IL-6, IL-1β, and IL-12, creating a critical hyperinflammation status (named cytokine storm) and a life-threatening outcomes [120].

It is therefore evident the existence of a continuous and reciprocal communication between the immune system and microbiota (inflammatory cascade), reported as "microbiota-inflammation or microbiota-immunity axis" [121]. In detail, the microbiota is crucial not only in the activation of systemic inflammation and IR through inflammatory cytokine modulation, as previously stated, but also through the transfer of lipopolysaccharides (LPS) from gram-negative bacteria to the systemic circulation, thus inducing a strong inflammatory tone (Fig. 14.5).

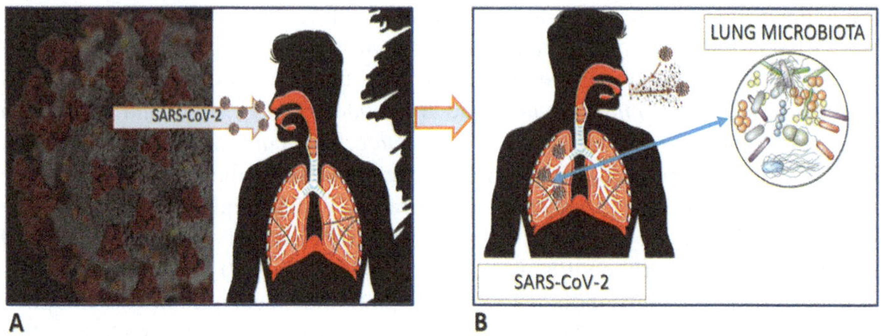


**Fig. 14.4** SARS-CoV-2 infection. COVID-19 patients can be infected and consequently replicate and spread the virus by infecting other people (**a**). The interaction between SARS-CoV-2 and lung microbiota can be very damaging, as it induces a great activation of the inflammatory cascade and an uncontrolled IR that helps the virus spread. If the body's IR cannot stop SARS-CoV-2 from multiplying, the consequence is severe lung damage (**b**)

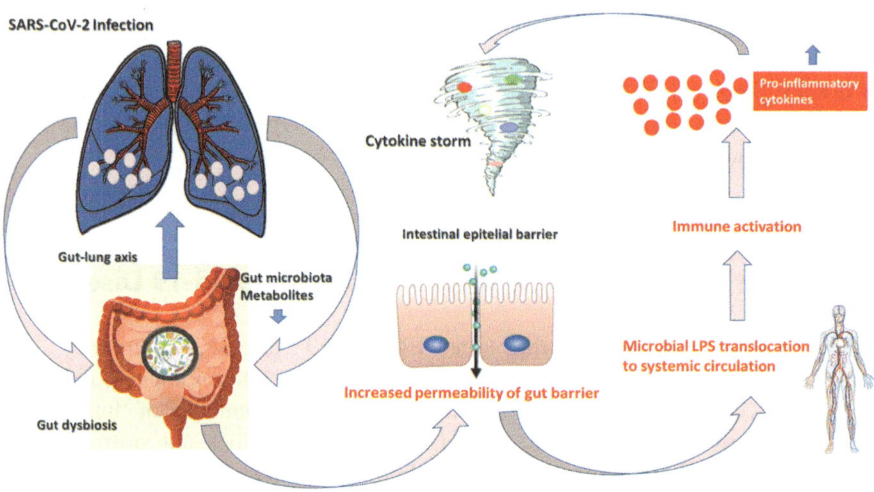


**Fig. 14.5** The perturbation of the gut microbiota–inflammation axis leading to severe COVID-19 by cytokine storm

Furthermore, a very recent work demonstrated how the gut microbiota affects lung health, through the "gut-lung axis," involving immunoregulatory mechanisms [44]. In addition, immune cells or gut bacteria play a major role in this mutual interaction, suggesting the existence of a tight crosstalk between enteric and lung microbiota [44]. In other words, the IR in the lung environment can be influenced by the gut microbiota and vice versa, involving also intestinal microbial metabolites, such as endotoxins damaging the lung tissue. As a result, a changed microbiota promotes the production of pro-inflammatory cytokines and the presence of opportunistic microbial species, both of which are known to aggravate the severity of SARS-CoV-2 infection [122, 123].

The impact of the gut microbiome–inflammation axis on the SARS CoV-2 was documented in several studies, as briefly summarized in Table 14.1, some of them using in vivo models [124, 125].

In one of these studies, the antimalarial drug Hydroxychloroquine (HCQ) was proposed to treat COVID, and the disease was tested by studying the direct effects on gut microbiome–inflammation axis of 15 female C57BL/6J mice. The findings revealed that a high dose of HCQ alters gut microbiota composition but has no effect on gut integrity or IRs [125].

In addition, the gut virome and bacteriome were examined in a group of 13 COVID-19 patients and compared to healthy controls in a recent study. The researchers discovered: (1) a decrease in intestinal microbial diversity; (2) a higher prevalence of pathogenic bacterial strains; and (3) a significant decrease in SCFAs (short chain fatty acids), particularly butyrate (crucial in modulating the immune and inflammatory responses). In conclusion these data suggest an intestinal gut bacterial dysbiosis [124]. In order to unravel the gut microbiota–inflammation axis, the results obtained on humans were later replicated using hACE2 transgenic mice infected with SARS-CoV-2, observing also that during infection, related genes were expressed differently in gut epithelial cells during infection, which could explain the virome and bacteriome dynamics. The outcomes confirmed what previously reported in human patients, highlighting how microbiome, virome with their associated immunological profile can influence COVID-19 disease development, treatment, and healing processes of patients [124].

In the study of Yeoh et al., in some hospitalized COVID-19 subjects, inflammatory indicators such as TNF-α, CXCL10, lactate dehydrogenase, aspartate transaminase, IL-10, γ-glutamyl transferase, C-reactive protein, erythrocyte sedimentation rate, and N-terminal pro-B-type natriuretic peptide were related with changes in the gut microbiota profile, six species in particular, resulted deficient in the COVID-19 cohort were negatively correlated with CXCL10, five species negatively correlated with IL-10, and two species anti-correlated with CCL2 and TNF- α.

Moreover, they investigated the differences in GM composition between hospitalized COVID-19 and non-COVID-19 patients. The results showed that *Bacteroidetes* were found to be more prevalent in COVID-19 patients (mean 23.9% versus 12.8%), whereas *Actinobacteria* were found to be more abundant in non-COVID-19 subjects (26.1% vs 19.0%; P.05).

Another relevant study examined the hypothesis if the low COVID-19 distribution in Africa and Asia is linked to protective immunity directed against galactose-1,3-galactose (α-Gal), triggering an immunological response by the activation of various systems implicated in SARS-CoV-2 neutralization. Anti-SARS-CoV-2 immunity and the risk of developing life-threatening lung inflammation could be another factor to consider in lowering SARS-CoV-2 transmission and also the risk of developing life-threatening lung inflammation. Consequently, an experimental research using GGTA1-knockout mice infected with SARS-CoV-2 generated in

**Table 14.1** A summarizing table of the significant results and aims of studies regarding the microbiota–inflammation axis role in SARS CoV-2 infection

| References | Organism (human or mice) | Aim | Result |
|---|---|---|---|
| Cao et al. [124] | Mice | Replicate the results observed in a cohort of 13 COVID-19 in gut virome and bacteriome in mouse models. | Mice COVID-19 model confirmed that SARS-CoV-2 infections change microbiota components such as the bacteriome and virome; their compositional signatures may reflect or perhaps contribute to the severity of the disease and the healing processes. |
| Pan et al. [125] | Mice | Investigate the effects of hydroxychloroquine (HCQ) an antimalarial drug considered to treat COVID-19 in the gut microbiota of mice | A short-term high dose of HCQ alters intestinal microbiota but not gut integrity or IRs. Therapeutic trials should be made to control the effects of HCQ on gut microecology. |
| Yeoh et al. [115] | Human | Compare the microbiota composition and inflammation markers of hospitalized COVID-19 patients with non-COVID-19 subjects | Inflammatory parameters were shown to be substantially linked to alterations in the gut microbiome profile in COVID-19 hospitalized patients. The authors also found that the diversity of the gut microbiota is linked to the intensity of the immunological response to COVID-19, and thus to the severity of clinical illness. |
| Hodžić et al. [126] | Human | Evaluate the possibility that the lower distribution of COVID-19 in Africa and Asia may depend on the protective immunity direct against the galactose-α-1,3-galactose (α-Gal), inducing an IR through the activation of different mechanisms involved in SARS-CoV-2 neutralization | α-Gal immunity produced by the gut microbiota may be useful in preventing COVID-19, limiting SARS-CoV-2 infection through virus neutralization, downregulating the ACE2 receptor, and reducing illness severity by downregulating the inflammatory response. It is acceptable to think about -Gal immunity as another element that could help to reduce SARS-CoV-2 transmission |
| Donati Zeppa et al. [127] | Human | Discuss the gut microbiota's biology, physiopathological, and clinical implications in COVID-19, as well as measures to improve/maintain its healthy status as a simple and supplementary strategy to | Gut microbiota can influence IR, thereby affecting the disease progression and IR possibly associated with the gut microbiota status leading to serious clinical complications in COVID-19. Prescription of |

(continued)

**Table 14.1** (continued)

| References | Organism (human or mice) | Aim | Result |
|---|---|---|---|
| | | minimize COVID-19 virulence and socio-sanitary burden. | prebiotics and probiotics should be considered as either an adjunctive treatment to limit COVID-19 progression |
| Merenstein et al. [128] | Human | Describe the COVID-19 respiratory tract microbiome and the link between illness severity, systemic immunologic characteristics, and outcomes | COVID-19 patients had a dysbiotic respiratory microbiome. Integrated characteristics of the microbiome at early sampling sites have strong capacity to discriminate the ultimate level of COVID-19 severity, according to machine learning analysis |
| Zhang et al. [82] | Human | Analyzed meta transcriptomics in 187 patients (62 cases with COVID-19 and 125 controls with lung disease) and evaluate transcriptional aspects, pathogens, microbiome and host responses, building a host gene classifier and examined its potential for diagnosing COVID-19 | The airway microbiome of COVID-19 patients had lower alpha diversity, as well as taxa with varying abundances and harmful microorganisms. A transcriptional signature of 36 differentially expressed genes substantially related with immunological pathways, such as cytokine signaling, was discovered using host gene analysis. COVID-19 could be diagnosed with the help of a host gene classifier based on this model |

human cells that express the antigen B enzyme is needed to establish the potential protective impact of -Gal immunity against COVID-19 [126].

Furthermore, bacterial LPS can aggressively activate inflammatory system cells. Moreover, the LPS levels in plasma correspond with the level of intestinal absorption in various situations. Several research have found a link between LPS and T cell activation, as well as increased pro-inflammatory response [129].

In fact, LPS levels have been found to be higher in cases of severe lung injury, suggesting that LPS may play a role in the cytokine cascade and COVID-19-related microvascular consequences. In some cases of COVID-19, GM dysbiosis may enable LPS translocation into the systemic circulation, which would further excite Kupffer cells in the liver, culminating in NF-B pathway activation and release of TNF- and IFN- [90].

In addition, another interesting study demonstrated a significant increase in the absorption of the intestinal epithelium in individuals with severe COVID-19 disease, indicating a leaky gut condition. The concentration of zonulin, a protein that works as a permeability modulator in the digestive tract, increased considerably [130].

It is also worth noting that the crosstalk among the intestine and the lung can indicate the influence of SARS-CoV-2 infection on the GM structure. COVID-19 patients in fact had a different composition of fecal bacteria than healthy controls, and the intestinal microbiota composition pattern was positively linked with an increased expression of IL-18, a renowned pro-inflammatory cytokine [114, 131].

Remarkably, Giron et al. recently examined the amounts of 50 gut-associated plasma metabolites, discovering that when compared to controls—most of these metabolites were found to be altered in severe forms of COVID-19 disease., as well as individuals who have a minor infection-correlated score disease. Moreover, citrulline, an amino acid that has long been used as a marker of intestinal functionality, was found to be significantly lower in the study and in the same time, the levels of succinic acid (used as indicator of gut dysbiosis) were shown to rise [130].

Even if the incidence of SARS-CoV-2 RNA in the stool was not assumed to be linked to digestive symptoms, a recent study compared COVID-19 subjects with mild disease and digestive symptoms with or without respiratory difficulties, showing that these kind of patients are more prone to have a severe COVID infection. and a detectable presence of the virus in stool compared to patients who show only respiratory signs [132, 133].

The individual susceptibility to COVID-19 is likely to be influenced by the microbiome–inflammation axis 's pre-existing health condition and changes after SARS-CoV-2 infection [127]. In fact, COVID-19 is responsible for the great majority of the severe clinical disorders and the virus replication triggers the release of chemokines and cytokines. This process brings to the development of a severe acute IR with an increasing gut permeability that bring to the activation of pro-inflammatory bacteria, thus triggering the inflammatory response [134, 135]. The increased inflammation can result in a leaky gut, allowing microbial pathogens and metabolites to enter the bloodstream. COVID-19 patients' septic states may get much worse as a result of this. Regarding lung microbiota–inflammation axis, in a recent study, oropharyngeal, nasopharyngeal, and endotracheal samples from COVID-19 patients, non-COVID patients, and healthy controls were evaluated. COVID-19 patients had an upper respiratory microbiome alteration that changed more over time than critically sick individuals who did not have COVID-19 [128].

In addition, the diversity of the oropharyngeal microbiota was inversely linked to systemic immune parameters and immune profiling of peripheral blood mononuclear cells during hospitalization, and the microbiota profile was linked to systemic immune parameters and immune profiling of peripheral blood mononuclear cells. Intubated patients had patient-specific lung microbiome communities that were often highly dynamic, with *Staphylococcus* being prominent. In severe illness, *Anelloviridae* and *Redondoviridae* demonstrated more frequent colonization and higher titers. Finally, machine learning research revealed that integrated microbiota traits at early sampling points had strong power to distinguish COVID-19 severity levels at the final level [128].

Furthermore, in a current report metatranscriptomics analyses were performed in cases with COVID-19 and with non-COVID-19 pneumonia [82]. Pathogens, the microbiota, and host responses together with their respectively transcriptional features have been investigated. A host gene classifier was developed based on the host transcriptional signature, and its potential for diagnosing COVID-19 and identifying disease severity was investigated. With 18 taxa of differential abundance, the airway microbiome of COVID-19 patients demonstrated a lower alpha diversity. In 47% of the COVID-19 cases, potentially pathogenic microorganisms were found, with respiratory viruses accounting for 58%. A transcriptional signature of 36 differentially expressed substantially related with immunological pathways, such as cytokine signaling, was discovered using host gene analysis. COVID-19 might be diagnosed and illness severity could be predicted using a host gene classifier based on such a signature. COVID-19 immune-associated host transcriptional markers have the potential to improve COVID-19 diagnosis while also showing disease severity [82].

As previously reported, the SARS-CoV-2 infects many organs presenting the ACE2 receptor, mainly expressed in enterocytes and in epithelial cells in the lungs. Normally, the renin–angiotensin–aldosterone system (RAAS), which modulates blood pressure and liquid balance, requires the ACE2 enzyme, which, additionally modulates the intestinal activities and protects organs from inflammatory damage. SARS-CoV-2 infection causes ACE2 dysfunction, this leads to organs epithelium inflammation and breakdown. On the other hand, inflammation has been shown to increase ACE2 levels establishing a vicious circle [136].

The study conducted by Zou et al. concerns COVID patients who did not show any respiratory symptoms but suffering from different disturbs, such as kidney failure [137].

Single-cell RNA sequencing datasets derived from cardiovascular, respiratory, digestive, and urinary systems were evaluated in order to create a risk map of distinct affected human organs, considered exposed to SARS-CoV-19 infection. Real-risk organs have been identified, including lungs, heart, esophagus, kidney, bladder, and ileus, as well as certain cell types (e.g., alveolar cells, myocardial cells, kidney cells) were assessed through these data analyses. The outcome confirmed the ability of COVID to invade other organs, in addition to the lungs [137].

In a very recent study, Xiao et al. [138] described in COVID patient's intestinal tract a significant infiltration of plasma cells and lymphocytes, associated with an important ACE2 downregulation of, leading to reduced tryptophan (Trp) uptake, which physiologically plays a very important role in intestinal microbiota maintenance, and a consequent decrease in the release of antimicrobial peptides that normally help to convert SARS-CoV-2- lesions in the gut and improve systemic conditions. All these events of evasion from the normal control mechanisms are the basis of the survival of the virus [138].

Moreover, ACE2 gene expression has been shown to increase with age and is potentially responsible for the augmented susceptibility and the more severe disease course in older people elderly [139].

For these reasons, the idea of using ACE inhibitors as a therapy treatment to decrease lung inflammation and benefit COVID-19 patients has been considered [140].

Additionally, nowadays the most persistent global health challenges are the human immunodeficiency, such as HIV or hepatitis C virus (HCV) together with COVID-19 and are estimated that populations infected with these virus could be exposed to a high risk for a high responses if simultaneously suffering from COVID 19 disease [141]. As a result, people with lung lesions who also have COVID-19 could be at a serious risk of developing HIV-HCV-COVID-19 co-infection.

### 14.3.1 The Effect of Diet Lifestyle and Probiotic Treatment on Microbiota–Inflammation Axis in Covid-19 Disease

Importance of lacking of nutrients in diet and consequently in the metabolic wellbeing resulting in wrong lifestyle behavior have been connected with the possibility and seriousness of COVID-19 disease [142]. In fact, due to micronutrient deficiencies and incorrect assumptions, a low nutritional state can intensify the infection's immunological response. Malnutrition causes a reduction in immune cells, particularly T- and B-cells, resulting in leukopenia and an ineffective IR [143, 144].

On the other hand, a diet rich in prebiotics, anti-inflammatory, antioxidant phytonutrients found in colorful vegetables, fermented foods, and beverages supports the healthy GM composition and function and as a consequence, of the whole organism [145].

In the past, it has been proven that addressing social health determinants, such as poor nutrition, reduced the gravity of some infectious illnesses [146, 147].

Although evidence of the association between food intake quality and COVID-19 risk or severity needs deeper investigation, prioritizing social health determinants in the public health response to COVID-19 could pave the way to a new medical approach [148]. Previous studies, for examples, linked the COVID-19 to lifestyle patterns, shown that a person lifestyle is an important component in preventing diseases [149]. Lifestyle is a complex notion that combines a person's life consciousness, conduct, and attitude, and can be classified based on people life patterns. Several research works have emphasized the relevance of leading a healthy lifestyle in sustaining and improving personal quality of life [150]. Lifestyle outcomes in people during the COVID-19 pandemic resulted in a considerable drop in several important activities, such as the body training and the daily life occupations [151].

The most recurrent GI symptoms are tight connected to the enterocytes infected by SARS-CoV-2 and damaging the intestinal epithelium, as previously mentioned [152]. As a result, a variety of GM manipulation treatments have been proposed, including a specific diet high in probiotics, prebiotics or resort to interventions like the fecal microbiota transplantation (FMT), or the use of bacterial components to cure or mitigate COVID-19-correlated disorders [153]. Nowadays, there are no experimental data to confirm the role of probiotics, as a valid method to treat

COVID-19, although the remarkable role in helping regulate respiratory tract immunity through the gut–lung axis is well documented [154].

However, normalizing intestinal dysbiosis using probiotics could be one of the methods to treating COVID-19, since this strategy has been effective in the treatment of other viral respiratory tract infections [155]. Currently, the most common strains commercially available belong to the *Lactobacillus* spp. and *Bifidobacterium spp* [156].

A meta-analysis conducted by Kang EJ, et al. showed how these two bacterial strains can reduce the common cold, one of the symptoms of SARS-CoV-2 infection [157]. Probiotic treatment reduced the disease severity or shortened its duration in this trial, whereas 2–47% of COVID-19 subjects required the mechanical ventilation.

In a meta-analysis of randomized controlled studies, probiotics were found to reduce ventilator-associated pneumonia (VAP) and the duration of antibiotic treatment for VAP [158]. Consequently, *Lactobacillus spp.* and *Bifidobacterium spp.* are known to play a role in many physiological processes, including immunological activation, prevention of pathogenic bacterial colonization, synthesis of SCFA, catabolism of cancerous cells and vitamin synthesis, conferring a benefit for host's health [159].

In particular, *Lactobacillus spp.*, generally found in healthy intestine, can impact on IR both in the respiratory and in the intestinal tract, protecting against viral respiratory infections by stimulating Th (T helper)1 mediated IR, Natural Killer (NK) cells and increasing IgA-mediated mucosal immunity [160].

Furthermore, oral administration of L. plantarum causes the activation of cytotoxic CD8+ T lymphocytes as well as an increase in granulocyte phagocytic activity. In addition, *L. plantarum* administration induces a decrease in pro-inflammatory cytokines IFN-$\gamma$ and TNF-$\alpha$ that may contribute to creation of an immune over reaction during COVID-19 disease, and an increase in anti-inflammatory cytokines IL-4 and IL-10 [161, 162]. As a result, it is hypothesized that boosting probiotics in high-risk and critically infected patients, as well as frontline healthcare workers, may help to reduce infection and flatten the COVID-19 curve. Additionally, probiotics have been shown to increase vaccination responses against respiratory viral infections, and recent studies have suggested that preserving the balance of the intestinal microbiota may be advantageous to COVID-19 + patients and aid in recovery due to enhanced immunological state [163].

Currently, several experimental trials are being performed throughout the world to determine the effect of probiotics in COVID-19 prevention and treatment [164].

However, there are no current randomized controlled trials (RCTs) available to provide conclusive data. So, in addition to the best medical treatments available, it is also necessary to consider an additional pathophysiology-based options to cure and prevent COVID-19 disease, addressing intestinal dysbiosis. In this context, probiotic supplements can be a good option to consider [164, 165]. However, probiotic strains of different species displayed varied physiology and metabolism, therefore their effects on the human body vary, causing different health consequences. Finally, it is well documented that probiotics taken at a greater dose may not be as effective as one consumed at the recommended level. Likewise, depending on the host, different doses of the same probiotic strain can have different impacts. Consequently, probiotic effects must be demonstrated at the strain level to ensure efficiency, and probiotic strains must be carefully chosen to achieve maximum benefit [164].

## 14.4 Limitation of the Current Research on Covid-19 and Microbiome

The fine details of the relationship between human microbiota and COVID-19 are currently unknown. Our chapter brought up several important points that call for more microbiome research in COVID-19. To begin with, longitudinal data, including post-admission data, are scarce. Aside from a lack of understanding of COVID-19 pathophysiology, there are little information of the clinical consequences that may persist in patients after the absence of viral contamination proved with negative RT-PCR test [166].

So, it is critical to build a long-term cohort of people who have healed from COVID-19 and look at the link between their microbiota and clinical characteristics of acute respiratory infections.

In addition, Chinese reports are the most prevalent in this sector. Data from other countries is required, as ethnicity is known to play a substantial role in microbiome diversity.

Moreover, studying the microbiome of non-intestinal organs including the skin, oral cavity, and urine tract can help researchers to better understand the microbiota in COVID-19 patients.

More microbiome data on their relationship to COVID-19 severity is required. If, in the future the microbiome might serve as a predictive indicator for disease development. Furthermore, early GM management (e.g., by symbiotics, probiotics, fermented foods, and fecal transplant) might be helpful in terms of prevention and treatments in such situation. The consequences of changing the gut microbiota are unknown at this time, and the findings of ongoing experiments are needed.

## 14.5 Conclusion

In our chapter, we have summarized and discussed the current studies that may provide light on the microbioma function in SARS-CoV-2 infection. In detail, we highlight lung and gut microbiota and their consequences in relation to COVID-19, with an emphasis on immunomodulation. There are hypotheses that imply a link, such as the so-called gut–lung axis, in which the intestinal microbiota influences the lungs, or immunomodulatory signals generated by the microbiota. By promoting the differentiation of a big amount of immune cells, the healthy gut microbiota can control the SARS-CoV-2- infection, comparing to a gut and lung dysbiosis condition characterized by a smaller number of immune cells.

Finally, dietary probiotics are important regulators of the gut microbial ecology, and they may be useful in maintaining microbiome homeostasis, influencing COVID-19 infection.

## References


1. Artika IM, Dewantari AK, Wiyatno A (2020) Molecular biology of coronaviruses: current knowledge. Heliyon 6(8):e04743
2. Jin YH, Cai L, Cheng ZS, Cheng H, Deng T, Fan YP, Fang C, Huang D, Huang LQ, Huang Q, Han Y, Hu B, Hu F, Li BH, Li YR, Liang K, Lin LK, Luo LS, Ma J, Ma LL, Peng ZY, Pan YB, Pan ZY, Ren XQ, Sun HM, Wang Y, Wang YY, Weng H, Wei CJ, Wu DF, Xia J, Xiong Y, Xu HB, Yao XM, Yuan YF, Ye TS, Zhang XC, Zhang YW, Zhang YG, Zhang HM, Zhao Y, Zhao MJ, Zi H, Zeng XT, Wang YY, Wang XH, for the Zhongnan Hospital of Wuhan University Novel Coronavirus Management and Research Team, Evidence-Based Medicine Chapter of China International Exchange and Promotive Association for Medical and Health Care (CPAM) (2020) A rapid advice guideline for the diagnosis and treatment of 2019 novel coronavirus (2019-nCoV) infected pneumonia (standard version). Mil Med Res 7(1):4
3. Lau SK, Chan JF (2015) Coronaviruses: emerging and re-emerging pathogens in humans and animals. Virol J 12:209
4. Cascella M, Rajnik M, Aleem A, Dulebohn SC, Di Napoli R (2021) Features, evaluation, and treatment of coronavirus (COVID-19). StatPearls, Treasure Island, F)
5. Mittal A, Manjunath K, Ranjan RK, Kaushik S, Kumar S, Verma V (2020) COVID-19 pandemic: insights into structure, function, and hACE2 receptor recognition by SARS-CoV-2. PLoS Pathog 16(8):e1008762
6. Mousavizadeh L, Ghasemi S (2021) Genotype and phenotype of COVID-19: their roles in pathogenesis. J Microbiol Immunol Infect 54(2):159–163
7. Almubaid Z, Al-Mubaid H (2021) Analysis and comparison of genetic variants and mutations of the novel coronavirus SARS-CoV-2. Gene Rep 23:101064
8. Aleem A, Akbar Samad AB, Slenker AK (2021) Emerging variants of SARS-CoV-2 and novel therapeutics against coronavirus (COVID-19). StatPearls, Treasure Island (FL)
9. Duong D (2021) Alpha, beta, delta, gamma: what's important to know about SARS-CoV-2 variants of concern? CMAJ 193(27):E1059–E1060
10. Lustig Y, Zuckerman N, Nemet I, Atari N, Kliker L, Regev-Yochay G, Sapir E, Mor O, Alroy-Preis S, Mendelson E, Mandelboim M (2021) Neutralising capacity against Delta (B.1.617.2) and other variants of concern following Comirnaty (BNT162b2, BioNTech/Pfizer) vaccination in health care workers, Israel. Euro Surveill 26(26):2100557
11. Yaniv K, Ozer E, Shagan M, Lakkakula S, Plotkin N, Bhandarkar NS, Kushmaro A (2021) Direct RT-qPCR assay for SARS-CoV-2 variants of concern (alpha, B.1.1.7 and Beta, B.1.351) detection and quantification in wastewater. Environ Res 201:111653
12. Zhou F, Yu T, Du R, Fan G, Liu Y, Liu Z, Xiang J, Wang Y, Song B, Gu X, Guan L, Wei Y, Li H, Wu X, Xu J, Tu S, Zhang Y, Chen H, Cao B (2020) Clinical course and risk factors for mortality of adult inpatients with COVID-19 in Wuhan, China: a retrospective cohort study. Lancet 395(10229):1054–1062
13. Akhmetzhanov AR, Jung SM, Cheng HY, Thompson RN (2021) A hospital-related outbreak of SARS-CoV-2 associated with variant Epsilon (B.1.429) in Taiwan: transmission potential and outbreak containment under intensified contact tracing, January-February 2021. Int J Infect Dis 110:15–20
14. Harrison AG, Lin T, Wang P (2020) Mechanisms of SARS-CoV-2 transmission and pathogenesis. Trends Immunol 41(12):1100–1115
15. Fu L, Wang B, Yuan T, Chen X, Ao Y, Fitzpatrick T, Li P, Zhou Y, Lin YF, Duan Q, Luo G, Fan S, Lu Y, Feng A, Zhan Y, Liang B, Cai W, Zhang L, Du X, Li L, Shu Y, Zou H (2020)

Clinical characteristics of coronavirus disease 2019 (COVID-19) in China: a systematic review and meta-analysis. J Infect 80(6):656–665
16. Oran DP, Topol EJ (2020) Prevalence of asymptomatic SARS-CoV-2 infection : a narrative review. Ann Intern Med 173(5):362–367
17. Costa VO, Nicolini EM, da Costa BMA, Teixeira FM, Ferreira JP, Moura MA, Montessi J, Campos RL, Guaraldo AN, Costa PM (2021) Evaluation of the risk of clinical deterioration among inpatients with COVID-19. Adv Virol 2021:6689669
18. Cabrera Martimbianco AL, Pacheco RL, Bagattini AM, Riera R (2021) Frequency, signs and symptoms, and criteria adopted for long COVID-19: a systematic review. Int J Clin Pract 75: e14357
19. Jones DL, Baluja MQ, Graham DW, Corbishley A, McDonald JE, Malham SK, Hillary LS, Connor TR, Gaze WH, Moura IB, Wilcox MH, Farkas K (2020) Shedding of SARS-CoV-2 in feces and urine and its potential role in person-to-person transmission and the environment-based spread of COVID-19. Sci Total Environ 749:141364
20. Wang C, Wang Z, Wang G, Lau JY, Zhang K, Li W (2021) COVID-19 in early 2021: current status and looking forward. Signal Transduct Target Ther 6(1):114
21. Magdy Beshbishy A, Oti VB, Hussein DE, Rehan IF, Adeyemi OS, Rivero-Perez N, Zaragoza-Bastida A, Shah MA, Abouelezz K, Hetta HF, Cruz-Martins N, Batiha GE (2021) Factors behind the higher COVID-19 risk in Diabetes: a critical review. Front Public Health 9: 591982
22. Grant MC, Geoghegan L, Arbyn M, Mohammed Z, McGuinness L, Clarke EL, Wade RG (2020) The prevalence of symptoms in 24,410 adults infected by the novel coronavirus (SARS-CoV-2; COVID-19): a systematic review and meta-analysis of 148 studies from 9 countries. PLoS One 15(6):e0234765
23. SeyedAlinaghi S, Oliaei S, Kianzad S, Afsahi AM, MohsseniPour M, Barzegary A, Mirzapour P, Behnezhad F, Noori T, Mehraeen E, Dadras O, Voltarelli F, Sabatier JM (2020) Reinfection risk of novel coronavirus (COVID-19): a systematic review of current evidence. World J Virol 9(5):79–90
24. Shenoy S (2021) SARS-CoV-2 (COVID-19), viral load and clinical outcomes; lessons learned one year into the pandemic: a systematic review. World J Crit Care Med 10(4):132–150
25. Li M, Wei R, Yang Y, He T, Shen Y, Qi T, Han T, Song Z, Zhu Z, Ma X, Lin Y, Yuan Y, Zhao K, Lu H, Zhou X (2021) Comparing SARS-CoV-2 testing in anterior nasal vestibular swabs vs. Oropharyngeal Swabs. Front Cell Infect Microbiol 11:653794
26. Dankova Z, Novakova E, Skerenova M, Holubekova V, Lucansky V, Dvorska D, Brany D, Kolkova Z, Strnadel J, Mersakova S, Janikova K, Samec M, Pokusa M, Petras M, Sarlinova M, Kasubova I, Loderer D, Sadlonova V, Kompanikova J, Kotlebova N, Kompanikova A, Hrnciarova M, Stanclova A, Antosova M, Dzian A, Nosal V, Kocan I, Murgas D, Krkoska D, Calkovska A, Halasova E (2021) Comparison of SARS-CoV-2 detection by rapid antigen and by three commercial RT-qPCR tests: a Study from Martin University Hospital in Slovakia. Int J Environ Res Public Health 18(13):7037
27. Leli C, Matteo LD, Gotta F, Cornaglia E, Vay D, Megna I, Pensato RE, Boverio R, Rocchetti A (2021) Performance of a SARS CoV-2 antigen rapid immunoassay in patients admitted to the emergency department. Int J Infect Dis 110:135–140
28. Ashtari S, Vahedian-Azimi A, Shojaee S, Pourhoseingholi MA, Jafari R, Bashar FR, Zali MR (2021) Computed tomographic features of coronavirus disease-2019 (COVID-19) pneumonia in three groups of Iranian patients: a single center study. Radiologia (Engl Ed) 63(4):314–323
29. Greffier J, Hoballah A, Sadate A, de Oliveira F, Claret PG, de Forges H, Loubet P, Mauboussin JM, Hamard A, Beregi JP, Frandon J (2021) Ultra-low-dose chest CT performance for the detection of viral pneumonia patterns during the COVID-19 outbreak period: a monocentric experience. Quant Imaging Med Surg 11(7):3190–3199
30. Stumpf J, Siepmann T, Lindner T, Karger C, Schwobel J, Anders L, Faulhaber-Walter R, Schewe J, Martin H, Schirutschke H, Barnett K, Huther J, Muller P, Langer T, Pluntke T, Anding-Rost K, Meistring F, Stehr T, Pietzonka A, Escher K, Cerny S, Rothe H, Pistrosch F,

Seidel H, Paliege A, Beige J, Bast I, Steglich A, Gembardt F, Kessel F, Kroger H, Arndt P, Sradnick J, Frank K, Klimova A, Mauer R, Grahlert X, Anft M, Blazquez-Navarro A, Westhoff TH, Stervbo U, Tonn T, Babel N, Hugo C (2021) Humoral and cellular immunity to SARS-CoV-2 vaccination in renal transplant versus dialysis patients: a prospective, multicenter observational study using mRNA-1273 or BNT162b2 mRNA vaccine. Lancet Reg Health Eur 9:100178
31. Bohman JK, Nei SD, Mellon LN, Ashmun RS, Guru PK (2021) Physical therapy and sedation while on extracorporeal membrane oxygenation for COVID-19-associated acute respiratory distress syndrome. J Cardiothorac Vasc Anesth. https://doi.org/10.1053/j.jvca.2021.06.030
32. Tobaiqy M, Qashqary M, Al-Dahery S, Mujallad A, Hershan AA, Kamal MA, Helmi N (2020) Therapeutic management of patients with COVID-19: a systematic review. Infect Prev Pract 2(3):100061
33. Hausmann JS, Kennedy K, Simard JF, Liew JW, Sparks JA, Moni TT, Harrison C, Larche MJ, Levine M, Sattui SE, Semalulu T, Foster G, Surangiwala S, Thabane L, Beesley RP, Durrant KL, Mateus EF, Mingolla S, Nudel M, Palmerlee CA, Richards DP, Liew DFL, Hill CL, Bhana S, Costello W, Grainger R, Machado PM, Robinson PC, Sufka P, Wallace ZS, Yazdany J, Sirotich E, Alliance C-GR (2021) Immediate effect of the COVID-19 pandemic on patient health, health-care use, and behaviours: results from an international survey of people with rheumatic diseases. Lancet Rheumatol. https://doi.org/10.1016/S2665-9913(21)00175-2
34. Chen F et al (2021) Potential adverse effects of dexamethasone therapy on COVID-19 patients: review and recommendations. Infect Dis Ther 10(4):1907–1931
35. Du W, Yu J, Liu X, Chen H, Lin L, Li Q (2020) Persistence of SARS-CoV-2 virus RNA in feces: a case series of children. J Infect Public Health 13(7):926–931
36. Pogue JM, Lauring AS, Gandhi TN, Marshall VD, Eschenauer GA, Nagel JL, Baang JH, Zhou S, Valesano AL, Petty LA (2021) "monoclonal antibodies for early treatment of COVID-19 in a world of evolving SARS-CoV-2 mutations and variants." open forum. Infect Dis 8(7):ofab268
37. Zarebska-Michaluk D, Jaroszewicz J, Rogalska M, Martonik D, Pabjan P, Berkan-Kawinska-A, Bolewska B, Oczko-Grzesik B, Kozielewicz D, Tudrujek-Zdunek M, Kowalska J, Moniuszko-Malinowska A, Klos K, Rorat M, Leszczynski P, Piekarska A, Polanska J, Flisiak R (2021) Effectiveness of tocilizumab with and without dexamethasone in patients with severe COVID-19: a retrospective study. J Inflamm Res 14:3359–3366
38. Taribagil P, Creer D, Tahir H (2021) 'Long COVID' syndrome. BMJ Case Rep 14(4):e241485
39. Baig AM (2021) Chronic COVID syndrome: need for an appropriate medical terminology for long-COVID and COVID long-haulers. J Med Virol 93(5):2555–2556
40. Humphreys H, Kilby L, Kudiersky N, Copeland R (2021) Long COVID and the role of physical activity: a qualitative study. BMJ Open 11(3):e047632
41. Kumar A (2021) COVID-19 gripped the globe with some unnoticed facts and too many questions. Virus 32:1–4
42. Pressman AR, Lockhart SH, Shen Z, Azar KMJ (2021) Measuring and promoting SARS-CoV-2 vaccine equity: development of a COVID-19 vaccine equity index. Health Equity 5(1):476–483
43. Huang C, Wang Y, Li X, Ren L, Zhao J, Hu Y, Zhang L, Fan G, Xu J, Gu X, Cheng Z, Yu T, Xia J, Wei Y, Wu W, Xie X, Yin W, Li H, Liu M, Xiao Y, Gao H, Guo L, Xie J, Wang G, Jiang R, Gao Z, Jin Q, Wang J, Cao B (2020) Clinical features of patients infected with 2019 novel coronavirus in Wuhan, China. Lancet 395(10223):497–506
44. Wolfel R, Corman VM, Guggemos W, Seilmaier M, Zange S, Muller MA, Niemeyer D, Jones TC, Vollmar P, Rothe C, Hoelscher M, Bleicker T, Brunink S, Schneider J, Ehmann R, Zwirglmaier K, Drosten C, Wendtner C (2020) Virological assessment of hospitalized patients with COVID-2019. Nature 581(7809):465–469

45. Xu Y, Li X, Zhu B, Liang H, Fang C, Gong Y, Guo Q, Sun X, Zhao D, Shen J, Zhang H, Liu H, Xia H, Tang J, Zhang K, Gong S (2020) Characteristics of pediatric SARS-CoV-2 infection and potential evidence for persistent fecal viral shedding. Nat Med 26(4):502–505
46. Roncon L, Zuin M, Rigatelli G, Zuliani G (2020) Diabetic patients with COVID-19 infection are at higher risk of ICU admission and poor short-term outcome. J Clin Virol 127:104354
47. Sajdel-Sulkowska EM (2021) A dual-route perspective of SARS-CoV-2 infection: lung- vs. gut-specific effects of ACE-2 deficiency. Front Pharmacol 12:684610
48. Ackerman J (2012) The ultimate social network. Sci Am 306(6):36–43
49. Proctor LM (2011) The human microbiome project in 2011 and beyond. Cell Host Microbe 10(4):287–291
50. Ursell LK, Metcalf JL, Parfrey LW, Knight R (2012) Defining the human microbiome. Nutr Rev 70 Suppl 1:S38–S44
51. Fraher MH, O'Toole PW, Quigley EM (2012) Techniques used to characterize the gut microbiota: a guide for the clinician. Nat Rev Gastroenterol Hepatol 9(6):312–322
52. Turnbaugh PJ, Ley RE, Mahowald MA, Magrini V, Mardis ER, Gordon JI (2006) An obesity-associated gut microbiome with increased capacity for energy harvest. Nature 444(7122): 1027–1031
53. From the American Association of Neurological Surgeons (AANS), American Society of Neuroradiology (ASNR), Cardiovascular and Interventional Radiology Society of Europe (CIRSE), Canadian Interventional Radiology Association (CIRA), Congress of Neurological Surgeons (CNS), European Society of Minimally Invasive Neurological Therapy (ESMINT), European Society of Neuroradiology (ESNR), European Stroke Organization (ESO), Society for Cardiovascular Angiography and Interventions (SCAI), Society of Interventional Radiology (SIR), Society of NeuroInterventional Surgery (SNIS), and World Stroke Organization (WSO), Sacks D, Baxter B, BCV C, Carpenter JS, Cognard C, Dippel D, Eesa M, Fischer U, Hausegger K, Hirsch JA, Shazam Hussain M, Jansen O, Jayaraman MV, Khalessi AA, Kluck BW, Lavine S, Meyers PM, Ramee S, Rüfenacht DA, Schirmer CM, Vorwerk D (2018) Multisociety consensus quality improvement revised consensus statement for endovascular therapy of acute ischemic stroke. Int J Stroke 13(6):612–632
54. Samuelson DR, Shellito JE, Maffei VJ, Tague ED, Campagna SR, Blanchard EE, Luo M, Taylor CM, Ronis MJJ, Molina PE, Welsh DA (2017) Alcohol-associated intestinal dysbiosis impairs pulmonary host defense against Klebsiella pneumoniae. PLoS Pathog 13(6):e1006426
55. Zhang M, Jiang Z, Li D, Jiang D, Wu Y, Ren H, Peng H, Lai Y (2015) Oral antibiotic treatment induces skin microbiota dysbiosis and influences wound healing. Microb Ecol 69(2): 415–421
56. Levy M, Kolodziejczyk AA, Thaiss CA, Elinav E (2017) Dysbiosis and the immune system. Nat Rev Immunol 17(4):219–232
57. Human Microbiome Project C (2012) Structure, function and diversity of the healthy human microbiome. Nature 486(7402):207–214
58. Kau AL, Ahern PP, Griffin NW, Goodman AL, Gordon JI (2011) Human nutrition, the gut microbiome and the immune system. Nature 474(7351):327–336
59. Hilty M, Burke C, Pedro H, Cardenas P, Bush A, Bossley C, Davies J, Ervine A, Poulter L, Pachter L, Moffatt MF, Cookson WO (2010) Disordered microbial communities in asthmatic airways. PLoS One 5(1):e8578
60. Pragman AA, Kim HB, Reilly CS, Wendt C, Isaacson RE (2012) The lung microbiome in moderate and severe chronic obstructive pulmonary disease. PLoS One 7(10):e47305
61. Willner D, Haynes MR, Furlan M, Schmieder R, Lim YW, Rainey PB, Rohwer F, Conrad D (2012) Spatial distribution of microbial communities in the cystic fibrosis lung. ISME J6(2): 471–474
62. Budden KF, Gellatly SL, Wood DL, Cooper MA, Morrison M, Hugenholtz P, Hansbro PM (2017) Emerging pathogenic links between microbiota and the gut-lung axis. Nat Rev Microbiol 15(1):55–63

63. Trompette A, Gollwitzer ES, Yadava K, Sichelstiel AK, Sprenger N, Ngom-Bru C, Blanchard C, Junt T, Nicod LP, Harris NL, Marsland BJ (2014) Gut microbiota metabolism of dietary fiber influences allergic airway disease and hematopoiesis. Nat Med 20(2):159–166
64. Dickson RP, Erb-Downward JR, Martinez FJ, Huffnagle GB (2016) The microbiome and the respiratory tract. Annu Rev Physiol 78:481–504
65. Invernizzi R, Lloyd CM, Molyneaux PL (2020) Respiratory microbiome and epithelial interactions shape immunity in the lungs. Immunology 160(2):171–182
66. Dickson RP, Huffnagle GB (2015) The lung microbiome: new principles for respiratory bacteriology in health and disease. PLoS Pathog 11(7):e1004923
67. Huffnagle GB, Dickson RP, Lukacs NW (2017) The respiratory tract microbiome and lung inflammation: a two-way street. Mucosal Immunol 10(2):299–306
68. Dickson RP, Erb-Downward JR, Freeman CM, McCloskey L, Beck JM, Huffnagle GB, Curtis JL (2015) Spatial variation in the healthy human lung microbiome and the adapted island model of lung biogeography. Ann Am Thorac Soc 12(6):821–830
69. Dickson RP, Martinez FJ, Huffnagle GB (2014) The role of the microbiome in exacerbations of chronic lung diseases. Lancet 384(9944):691–702
70. Durack J, Lynch SV, Nariya S, Bhakta NR, Beigelman A, Castro M, Dyer AM, Israel E, Kraft M, Martin RJ, Mauger DT, Rosenberg SR, Sharp-King T, White SR, Woodruff PG, Avila PC, Denlinger LC, Holguin F, Lazarus SC, Lugogo N, Moore WC, Peters SP, Que L, Smith LJ, Sorkness CA, Wechsler ME, Wenzel SE, Boushey HA, Huang YJ, National Heart L, A. Blood Institute's (2017) Features of the bronchial bacterial microbiome associated with atopy, asthma, and responsiveness to inhaled corticosteroid treatment. J Allergy Clin Immunol 140(1):63–75
71. Wu BG, Segal LN (2018) The lung microbiome and its role in pneumonia. Clin Chest Med 39(4):677–689
72. Gadsby NJ, Russell CD, McHugh MP, Mark H, Conway Morris A, Laurenson IF, Hill AT, Templeton KE (2016) Comprehensive molecular testing for respiratory pathogens in community-acquired pneumonia. Clin Infect Dis 62(7):817–823
73. Emonet S, Lazarevic V, Leemann Refondini C, Gaia N, Leo S, Girard M, Nocquet Boyer V, Wozniak H, Despres L, Renzi G, Mostaguir K, Dupuis Lozeron E, Schrenzel J, Pugin J (2019) Identification of respiratory microbiota markers in ventilator-associated pneumonia. Intensive Care Med 45(8):1082–1092
74. Woo S, Park SY, Kim Y, Jeon JP, Lee JJ, Hong JY (2020) The dynamics of respiratory microbiota during mechanical ventilation in patients with pneumonia. J Clin Med 9(3):638
75. Mori G, Morrison M, Blumenthal A (2021) Microbiome-immune interactions in tuberculosis. PLoS Pathog 17(4):e1009377
76. Wang H, Dai W, Qiu C, Li S, Wang W, Xu J, Li Z, Wang H, Li Y, Yang Z, Feng X, Zhou Q, Han L, Li Y, Zheng Y (2016) Mycoplasma pneumoniae and Streptococcus pneumoniae caused different microbial structure and correlation network in lung microbiota. J Thorac Dis 8(6):1316–1322
77. Rosas-Salazar C, Shilts MH, Tovchigrechko A, Schobel S, Chappell JD, Larkin EK, Shankar J, Yooseph S, Nelson KE, Halpin RA, Moore ML, Anderson LJ, Peebles RS Jr, Das SR, Hartert TV (2016) Differences in the nasopharyngeal microbiome during acute respiratory tract infection with human rhinovirus and respiratory syncytial virus in infancy. J Infect Dis 214(12):1924–1928
78. Hofstra JJ, Matamoros S, van de Pol MA, de Wever B, Tanck MW, Wendt-Knol H, Deijs M, van der Hoek L, Wolthers KC, Molenkamp R, Visser CE, Sterk PJ, Lutter R, de Jong MD (2015) Changes in microbiota during experimental human rhinovirus infection. BMC Infect Dis 15:336
79. Lloyd CM, Marsland BJ (2017) Lung homeostasis: influence of age, microbes, and the immune system. Immunity 46(4):549–561
80. Cheng M, Hu S (2017) Lung-resident gammadelta T cells and their roles in lung diseases. Immunology 151(4):375–384

81. Zhang F, Gan R, Zhen Z, Hu X, Li X, Zhou F, Liu Y, Chen C, Xie S, Zhang B, Wu X, Huang Z (2020) Adaptive immune responses to SARS-CoV-2 infection in severe versus mild individuals. Signal Transduct Target Ther 5(1):156
82. Zhang H, Ai JW, Yang W, Zhou X, He F, Xie S, Zeng W, Li Y, Yu Y, Gou X, Li Y, Wang X, Su H, Zhu Z, Xu T, Zhang W (2021) Metatranscriptomic characterization of coronavirus disease 2019 identified a host transcriptional classifier associated with immune signaling. Clin Infect Dis 73(3):376–385
83. Shen Z, Xiao Y, Kang L, Ma W, Shi L, Zhang L, Zhou Z, Yang J, Zhong J, Yang D, Guo L, Zhang G, Li H, Xu Y, Chen M, Gao Z, Wang J, Ren L, Li M (2020) Genomic diversity of severe acute respiratory syndrome-coronavirus 2 in patients with coronavirus disease 2019. Clin Infect Dis 71(15):713–720
84. Fan J, Li X, Gao Y, Zhou J, Wang S, Huang B, Wu J, Cao Q, Chen Y, Wang Z, Luo D, Zhou T, Li R, Shang Y, Nie X (2020) The lung tissue microbiota features of 20 deceased patients with COVID-19. J Infect 81(3):e64–e67
85. De Maio F, Posteraro B, Ponziani FR, Cattani P, Gasbarrini A, Sanguinetti M (2020) Nasopharyngeal microbiota profiling of SARS-CoV-2 infected patients. Biol Proced Online 22:18
86. Minich JJ, Ali F, Marotz C, Belda-Ferre P, Chiang L, Shaffer JP, Carpenter CS, McDonald D, Gilbert JA, Allard SM, Allen EE, Knight R, Sweeney DA, Swafford AD (2020) Feasibility of using alternative swabs and storage solutions for paired SARS-CoV-2 detection and microbiome analysis in the hospital environment. medRxiv
87. Ai JW, Zhang H, Xu T, Wu J, Zhu M, Yu YQ et al (2020) Optimizing diagnostic strategy for novel coronavirus pneumonia, a multi-center study in Eastern China. medRxiv
88. Budding A, Sieswerda E, Wintermans B, Bos M (2020) An age dependent pharyngeal microbiota signature associated with SARS-CoV-2 infection (4/21/2020). Available at SSRN: https://ssrn.com/abstract=3582780 or http://dx.doi.org/10.2139/ssrn.3582780
89. Zuo T, Zhan H, Zhang F, Liu Q, Tso EYK, Lui GCY, Chen N, Li A, Lu W, Chan FKL, Chan PKS, Ng SC (2020) Alterations in fecal fungal microbiome of patients with COVID-19 during time of hospitalization until discharge. Gastroenterology 159(4):1302–1310 e1305
90. Zuo T, Zhang F, Lui GCY, Yeoh YK, Li AYL, Zhan H, Wan Y, Chung ACK, Cheung CP, Chen N, Lai CKC, Chen Z, Tso EYK, Fung KSC, Chan V, Ling L, Joynt G, Hui DSC, Chan FKL, Chan PKS, Ng SC (2020) Alterations in gut microbiota of patients with COVID-19 during time of hospitalization. Gastroenterology 159(3):944–955.e948
91. Eckburg PB, Bik EM, Bernstein CN, Purdom E, Dethlefsen L, Sargent M, Gill SR, Nelson KE, Relman DA (2005) Diversity of the human intestinal microbial flora. Science 308(5728): 1635–1638
92. Frank DN, St Amand AL, Feldman RA, Boedeker EC, Harpaz N, Pace NR (2007) Molecular-phylogenetic characterization of microbial community imbalances in human inflammatory bowel diseases. Proc Natl Acad Sci U S A 104(34):13780–13785
93. Mahowald MA, Rey FE, Seedorf H, Turnbaugh PJ, Fulton RS, Wollam A, Shah N, Wang C, Magrini V, Wilson RK, Cantarel BL, Coutinho PM, Henrissat B, Crock LW, Russell A, Verberkmoes NC, Hettich RL, Gordon JI (2009) Characterizing a model human gut microbiota composed of members of its two dominant bacterial phyla. Proc Natl Acad Sci U S A 106(14):5859–5864
94. Ottman N, Smidt H, de Vos WM, Belzer C (2012) The function of our microbiota: who is out there and what do they do? Front Cell Infect Microbiol 2:104
95. Swidsinski A, Loening-Baucke V, Lochs H, Hale LP (2005) Spatial organization of bacterial flora in normal and inflamed intestine: a fluorescence in situ hybridization study in mice. World J Gastroenterol 11(8):1131–1140
96. McConnell EL, Fadda HM, Basit AW (2008) Gut instincts: explorations in intestinal physiology and drug delivery. Int J Pharm 364(2):213–226
97. Dave M, Higgins PD, Middha S, Rioux KP (2012) The human gut microbiome: current knowledge, challenges, and future directions. Transl Res 160(4):246–257

98. Schirmer M, Smeekens SP, Vlamakis H, Jaeger M, Oosting M, Franzosa EA, Ter Horst R, Jansen T, Jacobs L, Bonder MJ, Kurilshikov A, Fu J, Joosten LAB, Zhernakova A, Huttenhower C, Wijmenga C, Netea MG, Xavier RJ (2016) Linking the human gut microbiome to inflammatory cytokine production capacity. Cell 167(4):1125–1136.e1128
99. Sonnenburg JL, Angenent LT, Gordon JI (2004) Getting a grip on things: how do communities of bacterial symbionts become established in our intestine? Nat Immunol 5(6):569–573
100. Holmes E, Li JV, Athanasiou T, Ashrafian H, Nicholson JK (2011) Understanding the role of gut microbiome-host metabolic signal disruption in health and disease. Trends Microbiol 19(7):349–359
101. Schwabe RF, Jobin C (2013) The microbiome and cancer. Nat Rev Cancer 13(11):800–812
102. Mukherjee PK, Sendid B, Hoarau G, Colombel JF, Poulain D, Ghannoum MA (2015) Mycobiota in gastrointestinal diseases. Nat Rev Gastroenterol Hepatol 12(2):77–87
103. Rautava S, Luoto R, Salminen S, Isolauri E (2012) Microbial contact during pregnancy, intestinal colonization and human disease. Nat Rev Gastroenterol Hepatol 9(10):565–576
104. Gu S et al (2020) Alterations of the gut microbiota in patients with coronavirus disease 2019 or H1N1 Influenza. Clin Infect Dis 71(10):2669–2678
105. Lamers MM et al (2020) SARS-CoV-2 productively infects human gut enterocytes. Science 369(6499):50–54
106. Zhang Y, Ma ZF (2020) Impact of the COVID-19 pandemic on mental health and quality of life among local residents in Liaoning Province, China: a cross-sectional study. Int J Environ Res Public Health 17(7)
107. Sultan S et al (2020) Low vitamin D and its association with cognitive impairment and dementia. J Aging Res 2020:6097820
108. Zhang D, Li S, Wang N, Tan HY, Zhang Z, Feng Y (2020) The cross-talk between gut microbiota and lungs in common lung diseases. Front Microbiol 11:301
109. Zhang H, Li HB, Lyu JR, Lei XM, Li W, Wu G, Lyu J, Dai ZM (2020) Specific ACE2 expression in small intestinal enterocytes may cause gastrointestinal symptoms and injury after 2019-nCoV infection. Int J Infect Dis 96:19–24
110. Zuo T, Liu Q, Zhang F, Lui GC, Tso EY, Yeoh YK, Chen Z, Boon SS, Chan FK, Chan PK, Ng SC (2021) Depicting SARS-CoV-2 faecal viral activity in association with gut microbiota composition in patients with COVID-19. Gut 70(2):276–284
111. Gu S, Chen Y, Wu Z, Chen Y, Gao H, Lv L, Guo F, Zhang X, Luo R, Huang C, Lu H, Zheng B, Zhang J, Yan R, Zhang H, Jiang H, Xu Q, Guo J, Gong Y, Tang L, Li L (2020) Alterations of the gut microbiota in patients with coronavirus disease 2019 or H1N1 influenza. Clin Infect Dis 71(10):2669–2678
112. Penninger JM, Grant MB, Sung JJY (2021) The role of angiotensin converting enzyme 2 in modulating gut microbiota, intestinal inflammation, and coronavirus infection. Gastroenterology 160(1):39–46
113. Adelman MW, Woodworth MH, Langelier C, Busch LM, Kempker JA, Kraft CS, Martin GS (2020) The gut microbiome's role in the development, maintenance, and outcomes of sepsis. Crit Care 24(1):278
114. Tao W, Zhang G, Wang X, Guo M, Zeng W, Xu Z, Cao D, Pan A, Wang Y, Zhang K, Ma X, Chen Z, Jin T, Liu L, Weng J, Zhu S (2020) Analysis of the intestinal microbiota in COVID-19 patients and its correlation with the inflammatory factor IL-18. Med Microecol 5:100023
115. Yeoh YK, Zuo T, Lui GC, Zhang F, Liu Q, Li AY, Chung AC, Cheung CP, Tso EY, Fung KS, Chan V, Ling L, Joynt G, Hui DS, Chow KM, Ng SSS, Li TC, Ng RW, Yip TC, Wong GL, Chan FK, Wong CK, Chan PK, Ng SC (2021) Gut microbiota composition reflects disease severity and dysfunctional immune responses in patients with COVID-19. Gut 70(4):698–706
116. Britton GJ, Alice C-L, Cossarini F, Livanos AE, Spindler MP, Plitt T (2020) SARS-CoV-2-specific IgA and limited inflammatory cytokines are present in the stool of select patients with acute COVID-19. medRxiv
117. Ahlawat S, Asha, Sharma KK (2020) Immunological co-ordination between gut and lungs in SARS-CoV-2 infection. Virus Res 286:198103

118. Li X, Geng M, Peng Y, Meng L, Lu S (2020) Molecular immune pathogenesis and diagnosis of COVID-19. J Pharm Anal 10(2):102–108
119. Derrien M, van Hylckama Vlieg JE (2015) Fate, activity, and impact of ingested bacteria within the human gut microbiota. Trends Microbiol 23(6):354–366
120. Coperchini F, Chiovato L, Croce L, Magri F, Rotondi M (2020) The cytokine storm in COVID-19: an overview of the involvement of the chemokine/chemokine-receptor system. Cytokine Growth Factor Rev 53:25–32
121. Jiao Y, Wu L, Huntington ND, Zhang X (2020) Crosstalk between gut microbiota and innate immunity and its implication in autoimmune diseases. Front Immunol 11:282
122. Barcik W, Boutin RCT, Sokolowska M, Finlay BB (2020) The role of lung and gut microbiota in the pathology of asthma. Immunity 52(2):241–255
123. Keely S, Talley NJ, Hansbro PM (2012) Pulmonary-intestinal cross-talk in mucosal inflammatory disease. Mucosal Immunol 5(1):7–18
124. Cao J, Wang C, Zhang Y, Lei G, Xu K, Zhao N, Lu J, Meng F, Yu L, Yan J, Bai C, Zhang S, Zhang N, Gong Y, Bi Y, Shi Y, Chen Z, Dai L, Wang J, Yang P (2021) Integrated gut virome and bacteriome dynamics in COVID-19 patients. Gut Microbes 13(1):1–21
125. Pan ZY, Chang YX, Han N, Hou FY, Lee BJY, Zhi FC, Yang RF, Bi YJ (2021) Short-term high-dose gavage of hydroxychloroquine changes gut microbiota but not the intestinal integrity and immunological responses in mice. Life Sci 264:118450
126. Hodzic A, de la Fuente J, Cabezas-Cruz A (2020) COVID-19 in the developing world: is the immune response to alpha-gal an overlooked factor mitigating the severity of infection? ACS Infect Dis 6(12):3104–3108
127. Donati Zeppa S, Agostini D, Piccoli G, Stocchi V, Sestili P (2020) Gut microbiota status in COVID-19: an unrecognized player? Front Cell Infect Microbiol 10:576551
128. Merenstein C, Liang G, Whiteside SA, Cobian-Guemes AG, Merlino MS, Taylor LJ, Glascock A, Bittinger K, Tanes C, Graham-Wooten J, Khatib LA, Fitzgerald AS, Reddy S, Baxter AE, Giles JR, Oldridge DA, Meyer NJ, Wherry EJ, McGinniss JE, Bushman FD, Collman RG (2021) Signatures of COVID-19 severity and immune response in the respiratory tract microbiome. mBio 12:e0177721
129. Santos-Oliveira JR, Regis EG, Leal CR, Cunha RV, Bozza PT, Da-Cruz AM (2011) Evidence that lipopolisaccharide may contribute to the cytokine storm and cellular activation in patients with visceral leishmaniasis. PLoS Negl Trop Dis 5(7):e1198
130. Giron LB, Dweep H, Yin X, Wang H, Damra M, Goldman AR, Gorman N, Palmer CS, Tang HY, Shaikh MW, Forsyth CB, Balk RA, Zilberstein NF, Liu Q, Kossenkov A, Keshavarzian A, Landay A, Abdel-Mohsen M (2021) Plasma markers of disrupted gut permeability in severe COVID-19 patients. Front Immunol 12:686240
131. He LH, Ren LF, Li JF, Wu YN, Li X, Zhang L (2020) Intestinal Flora as a potential strategy to fight SARS-CoV-2 infection. Front Microbiol 11:1388
132. El Ouali S, Achkar JP, Lashner B, Regueiro M (2021) Gastrointestinal manifestations of COVID-19. Cleve Clin J Med. https://doi.org/10.3949/ccjm.87a.ccc049
133. Han C, Duan C, Zhang S, Spiegel B, Shi H, Wang W, Zhang L, Lin R, Liu J, Ding Z, Hou X (2020) Digestive symptoms in COVID-19 patients with mild disease severity: clinical presentation, stool viral RNA testing, and outcomes. Am J Gastroenterol 115(6):916–923
134. Aktas B, Aslim B (2020) Gut-lung axis and dysbiosis in COVID-19. Turk J Biol 44(3): 265–272
135. Zhao Y, Cao Y, Wang S, Cai K, Xu K (2020) COVID-19 and gastrointestinal symptoms. Br J Surg 107(10):e382–e383
136. Beyerstedt S, Casaro EB, Rangel EB (2021) COVID-19: angiotensin-converting enzyme 2 (ACE2) expression and tissue susceptibility to SARS-CoV-2 infection. Eur J Clin Microbiol Infect Dis 40(5):905–919
137. Zou X, Chen K, Zou J, Han P, Hao J, Han Z (2020) Single-cell RNA-seq data analysis on the receptor ACE2 expression reveals the potential risk of different human organs vulnerable to 2019-nCoV infection. Front Med 14(2):185–192

138. Xiao L, Sakagami H, Miwa N (2020) ACE2: the key molecule for understanding the pathophysiology of severe and critical conditions of COVID-19: demon or angel? Viruses 12(5):491
139. Baker SA, Kwok S, Berry GJ, Montine TJ (2021) Angiotensin-converting enzyme 2 (ACE2) expression increases with age in patients requiring mechanical ventilation. PLoS One 16(2): e0247060
140. Meng J, Xiao G, Zhang J, He X, Ou M, Bi J, Yang R, Di W, Wang Z, Li Z, Gao H, Liu L, Zhang G (2020) Renin-angiotensin system inhibitors improve the clinical outcomes of COVID-19 patients with hypertension. Emerg Microbes Infect 9(1):757–760
141. Tamuzi JL, Ayele BT, Shumba CS, Adetokunboh OO, Uwimana-Nicol J, Haile ZT, Inugu J, Nyasulu PS (2020) Implications of COVID-19 in high burden countries for HIV/TB: a systematic review of evidence. BMC Infect Dis 20(1):744
142. The Lancet Diabetes E (2021) Metabolic health: a priority for the post-pandemic era. Lancet Diabetes Endocrinol 9(4):189
143. Silverio R, Goncalves DC, Andrade MF, Seelaender M (2021) Coronavirus disease 2019 (COVID-19) and nutritional status: the missing link? Adv Nutr 12(3):682–692
144. Suardi C, Cazzaniga E, Graci S, Dongo D, Palestini P (2021) Link between viral infections, immune system, inflammation and diet. Int J Environ Res Public Health 18(5):2455
145. Gasmi A, Tippairote T, Mujawdiya PK, Peana M, Menzel A, Dadar M, Benahmed AG, Bjorklund G (2021) The microbiota-mediated dietary and nutritional interventions for COVID-19. Clin Immunol 226:108725
146. Storm I, den Hertog F, van Oers H, Schuit AJ (2016) How to improve collaboration between the public health sector and other policy sectors to reduce health inequalities? - a study in sixteen municipalities in the Netherlands. Int J Equity Health 15:97
147. Willett WC, Stampfer MJ (2013) Current evidence on healthy eating. Annu Rev Public Health 34:77–95
148. Drew DA, Nguyen LH, Steves CJ, Menni C, Freydin M, Varsavsky T, Sudre CH, Cardoso MJ, Ourselin S, Wolf J, Spector TD, Chan AT, C. Consortium (2020) Rapid implementation of mobile technology for real-time epidemiology of COVID-19. Science 368(6497):1362–1367
149. Hamer M, O'Donovan G, Stamatakis E (2019) Lifestyle risk factors, obesity and infectious disease mortality in the general population: linkage study of 97,844 adults from England and Scotland. Prev Med 123:65–70
150. Park KH, Park JH (2020) Development of an elderly lifestyle profile: a Delphi survey of multidisciplinary health-care experts. PLoS One 15(6):e0233565
151. Park KH, Kim AR, Yang MA, Lim SJ, Park JH (2021) Impact of the COVID-19 pandemic on the lifestyle, mental health, and quality of life of adults in South Korea. PLoS One 16(2): e0247970
152. Megyeri K, Dernovics A, Al-Luhaibi ZII, Rosztoczy A (2021) COVID-19-associated diarrhea. World J Gastroenterol 27(23):3208–3222
153. Ngo VL, Gewirtz AT (2021) Microbiota as a potentially-modifiable factor influencing COVID-19. Curr Opin Virol 49:21–26
154. de Oliveira GLV, Oliveira CNS, Pinzan CF, de Salis LVV, Cardoso CRB (2021) Microbiota modulation of the gut-lung Axis in COVID-19. Front Immunol 12:635471
155. Park MK, Ngo V, Kwon YM, Lee YT, Yoo S, Cho YH, Hong SM, Hwang HS, Ko EJ, Jung YJ, Moon DW, Jeong EJ, Kim MC, Lee YN, Jang JH, Oh JS, Kim CH, Kang SM (2013) Lactobacillus plantarum DK119 as a probiotic confers protection against influenza virus by modulating innate immunity. PLoS One 8(10):e75368
156. Fijan S (2014) Microorganisms with claimed probiotic properties: an overview of recent literature. Int J Environ Res Public Health 11(5):4745–4767
157. Kang EJ, Kim SY, Hwang IH, Ji YJ (2013) The effect of probiotics on prevention of common cold: a meta-analysis of randomized controlled trial studies. Korean J Fam Med 34(1):2–10

158. Su M, Jia Y, Li Y, Zhou D, Jia J (2020) Probiotics for the prevention of ventilator-associated pneumonia: a meta-analysis of randomized controlled trials. Respir Care 65(5):673–685
159. Ceccarelli G, Statzu M, Santinelli L, Pinacchio C, Bitossi C, Cavallari EN, Vullo V, Scagnolari C, d'Ettorre G (2019) Challenges in the management of HIV infection: update on the role of probiotic supplementation as a possible complementary therapeutic strategy for cART treated people living with HIV/AIDS. Expert Opin Biol Ther 19(9):949–965
160. Kikuchi Y, Kunitoh-Asari A, Hayakawa K, Imai S, Kasuya K, Abe K, Adachi Y, Fukudome S, Takahashi Y, Hachimura S (2014) Oral administration of Lactobacillus plantarum strain AYA enhances IgA secretion and provides survival protection against influenza virus infection in mice. PLoS One 9(1):e86416
161. Auld SC, Caridi-Scheible M, Blum JM, Robichaux C, Kraft C, Jacob JT, Jabaley CS, Carpenter D, Kaplow R, Hernandez-Romieu AC, Adelman MW, Martin GS, Coopersmith CM, Murphy DJ, C.-Q. and the Emory and C. Clinical Research (2020) ICU and ventilator mortality among critically ill adults with coronavirus disease 2019. Crit Care Med 48(9):e799–e804
162. Azad MAK, Sarker M, Wan D (2018) Immunomodulatory effects of probiotics on cytokine profiles. Biomed Res Int 2018:8063647
163. Wang L, Zhu L, Qin S (2019) Gut microbiota modulation on intestinal mucosal adaptive immunity. J Immunol Res 2019:4735040
164. Kurian SJ, Unnikrishnan MK, Miraj SS, Bagchi D, Banerjee M, Reddy BS, Rodrigues GS, Manu MK, Saravu K, Mukhopadhyay C, Rao M (2021) Probiotics in prevention and treatment of COVID-19: current perspective and future prospects. Arch Med Res 52:582–594
165. Spagnolello O, Pinacchio C, Santinelli L, Vassalini P, Innocenti GP, De Girolamo G, Fabris S, Giovanetti M, Angeletti S, Russo A, Mastroianni CM, Ciccozzi M, Ceccarelli G, d'Ettorre G (2021) Targeting microbiome: an alternative strategy for fighting SARS-CoV-2 infection. Chemotherapy 66(1–2):24–32
166. Samrah SM, Al-Mistarehi AH, Kewan T, Al-Khatib SM, Ibnian AM, Samrah RS, Khassawneh BY (2021) Viral clearance course of COVID-19 outbreaks. J Multidiscip Healthc 14:555–565

# Microbiome in SARS-CoV-2 (Covid-19) 15

Subha Manoharan, Lakshmi Thangavelu,
Mallineni Sreekanth Kumar, Gaurav Gupta, Kamal Dua,
and Dinesh Kumar Chellappan

**Abstract**

Coronavirus was first identified in the year 1931 as avian infectious bronchitis virus (IBV). In 1966 and 1967, two new strains of virus causing human infection were identified (HCoV-229E and HCoV-OC43), respectively. Coronavirus is positive sense, single-stranded RNA virus with an envelope. It belongs to the family Coronaviridae, suborder Cornidovirineae and order Nidovirales. The initial havoc was due to the lack of understanding with the viral pathogenesis. As the pathogenesis became clear, the management protocol was designed with many tailor-made combinations suitable for the individual patient requirement.

S. Manoharan
Department of Oral Medicine & Radiology, Madha Dental College & Hospital, Chennai, India

L. Thangavelu (✉)
Department of Pharmacology, Saveetha Dental College & Hospital, Saveetha Institute of Medical and Technical Sciences, Saveetha University, Chennai, India
e-mail: lakshmi@saveetha.com

M. S. Kumar
Department of Pediatric & Preventive Dental Sciences, Majmaah University, Al Zulfi, Saudi Arabia

G. Gupta
School of Pharmacy, Suresh Gyan Vihar University, Jaipur, India

K. Dua
Discipline of Pharmacy, Graduate School of Health, University of Technology, Sydney, NSW, Australia

D. K. Chellappan
Department of Life Sciences, School of Pharmacy, International Medical University, Kuala Lumpur, Malaysia



G. Gupta et al. (eds.), *Microbiome in Inflammatory Lung Diseases*,
https://doi.org/10.1007/978-981-16-8957-4_15

**Keywords**

Microbiome · COVID-19 · Virus · Interleukin · MERS

## 15.1 Introduction

Corona, means halo or crown in Latin, is the name of the coronavirus due to its crown-like projection. Living true to the name it ruled the world for over 2 years now. The World Health Organization reported 185,291,530 cases and 4,010,834 mortalities till June 2021. Though the virus was identified in 1931, this century faces the third pandemic by coronavirus. The transformation in the virus itself is the vital cause and it is not sure if the pandemic ends here. The huge burden of this disease is mainly due to the lack of understanding of the viral structure, pathogenesis, and its genomic transformation. Understanding of these phenomena came to light by extensive research. This challenge to humankind and the technology had been overcome with proper understanding of the viral pathogenesis. The management protocol and vaccines were formulated at a rapid pace to save mankind. This chapter gives an overview on the origin, evolution, structure, mode of transmission, pathogenesis, clinical features, investigation, and management. Understanding them is essential to handle the infection better.

## 15.2 Origin and Evolution

Coronavirus was first identified in the year 1931 as avian infectious bronchitis virus (IBV) [1]. In 1966 and 1967, two new strains of virus causing human infection were identified (HCoV-229E and HCoV-OC43), respectively [1]. Guangdong province, China was reported to face the coronavirus pandemic in 2002. The identified virus was named as SARS-CoV, correlating with the clinical disease, severe acute respiratory syndrome [2]. Almost 8000 humans were infected with around 800 deaths [3]. In 2012, there is yet another outbreak emerged in the Middle East, the MERS-CoV. First reported in Saudi Arabia, infecting 2500 humans with around 800 fatalities [4–6]. The episode repeated itself in South Korea in the year 2015 as a larger outbreak. In 2019, novel coronavirus emerged from Wuhan, initially named by China as SARS-CoV-2 and later the WHO renamed it as Covid-19. World pandemic was confirmed by WHO in early 2020 and a second wave of the disease caused massive destruction of human life in 2021 too. The total confirmed cases as of June 2021 reported by WHO were 181,930,736 with a fatality of 3,945,832 [7].

## 15.3 Structure

Coronavirus is positive sense, single-stranded RNA virus with an envelope. It comes from the family Coronaviridae, suborder Cornidovirineae belonging to order Nidovirales [8, 9]. It is further divided into three groups. It is further classified into the subfamily Orthocoronavirinae. Orthocoronavirinae has the following four genera: alphacoronavirus, betacoronavirus, gammacoronavirus, and deltacoronavirus. Alphacoronavirus and betacoronavirus exclusively involve mammalian hosts, whereas gammacoronavirus and deltacoronavirus affect avian hosts [10]. SARS-CoV belongs to betacoronavirus with 7 different strains: HCoV-229E, HCoV-NK63, HCoV-OC43, HCoVHKU1, SARS-CoV, MERS-CoV, and SARS-CoV-2 [11].

Coronavirus has 29.9 kb genomic size with four structural proteins and 16 non-structural proteins. The four structural proteins are S protein, E protein, N protein and M protein [12]. The nucleocapsid protein covers the genome which in turn is covered by an envelope which has three structural proteins; spike (S), membrane (M), envelope (E) [13, 14]. The spike protein determines the transmission capacity of the virus. It has three parts: ectodomain (S1 and S2), a transmembrane anchor, and an intercellular tail [15]. The S1 ectodomain has two terminals: the NTD and RBD which plays a vital role in binding to the host receptor. S2 ectodomain has FP, HR1, CH, CD, HR2, TM, and CT which aids in fusion to the host cell membrane [16, 17]. At the junction of S1 and S2 site is the cleavage site known as S1/S2 protease cleavage site. S2 protein can be in an open gate state and closed gate state. They need to be in an open gate state to adhere to the host cell. Binding to the host ACE 2 receptors is aided by the S protein [18].

The M protein has numerous amino acids and is abundant in the virus. There are two terminus guarding the three transmembrane domain. The two terminus includes a short amino terminus and a long carboxy terminus lying outside and inside, respectively [19]. It binds to the nucleocapsid, maintaining the shape and integrity of coronavirus [20, 21].

The E protein is the smallest polypeptide that acts as a viroporin. This also has 3 domains: a short amino terminal, a large transmembrane domain, which are hydrophilic hydrophobic, respectively, and an efficient C-terminal domain [22]. Organizing and releasing of the virus are the main function of E protein [23]. Any alteration in viroporin alters the virulence of the virus [24].

The N protein has three specific domains: an NTD, an RNA-binding domain, and a CTD. NTD binds to the viral genome, RNA-binding domain also known as linker domain containing serine and arginine is responsible for cell signaling [25, 26]. Viral genomic complex formation, interaction with M protein, and viral transcription are managed by the N protein [25].

## 15.4 Mode of Transmission

The virus gets transmitted physical contact either from an infected person or an asymptomatic carrier [27]. It is transmitted through droplet infection by sneezing, coughing. The virus stays in air as droplets can be contacted by a person staying within 1 m from a patient or carrier while coughing or sneezing. The virus is transmitted by oral, nasal, and conjunctival secretions. The virus can survive in dry surfaces too hence they can be transmitted by fomites also. It could survive in low temperature and low humid surfaces and be infective till 48 h. It can survive for 7 days in plastic, metal, ceramic, glass, wood, gloves, and mask and cause transmission of the disease [28]. Other means of transmission include airborne, fecal-oral, sexual, and breastfeed.

Fecal discharge of the virus is identified in patients. Studies show that even patients who had recovered and expressed negative nasopharyngeal swabs have discharged the virus until 33 days and 47 day of the positive results [29]. The presence of ACE2 protein in the rectal epithelium [30] and positive gastrointestinal symptoms during the course of the disease explains the possibility of fecal route of transmission [31–33]. Reproductive organ also has ACE2 receptors, hence vertical transmission from infected mother to baby during labor may occur. Presence of the virus in placenta and breast milk requires further studies to establish the cause [34, 35]. ACE2 receptors are present in ovaries, vagina, and male reproductive organs hence direct sexual transmission can occur [36].

## 15.5 Pathogenesis

The virus encounters a host cell with an ACE 2 receptor and undergoes sequential changes [37]. First, it attaches itself to cell surface ACE2 receptor, then enters the cell membrane by endocytosis or membrane fusion. The spike protein S1 helps in adhesion of the virus and S2 enhances the fusion of the cell membrane. The S1/S2 junctional protease cleavage activates the spike protein. Furin cleavage is prevalent while it has transmembrane protease serine 2 and cathepsin L also acting as cleavage proteins. The virus releases its contents into the host cell, the viral nucleic content, the RNA enters the nucleus of the host cell and replicates. Viral protein biosynthesis occurs with the viral mRNA. Maturation of the viral particles occurs and is released [38]. The SARS-CoV-2 receptor is predominantly seen in the lung; however, it is also seen in heart, kidney, liver, and ileum which is attributed to the clinical signs of the organs [39] (Fig. 15.1).

The innate immune system gets activated as the virus paves way into the body and starts replicating. It detects the viral RNA and secretes antiviral interferons and chemokines. Though the immune system prevents the host by secreting numerous inflammatory products, an excessive secretion of cytokines and chemokines would affect the host system itself. This excess secretion of cytokines is known as the cytokine storm. Plasma leakage, increased vascular permeability, and disseminated vascular coagulation are consequences of cytokine storm. Acute lung injuries are

**Fig. 15.1** Sequential events in viral replication

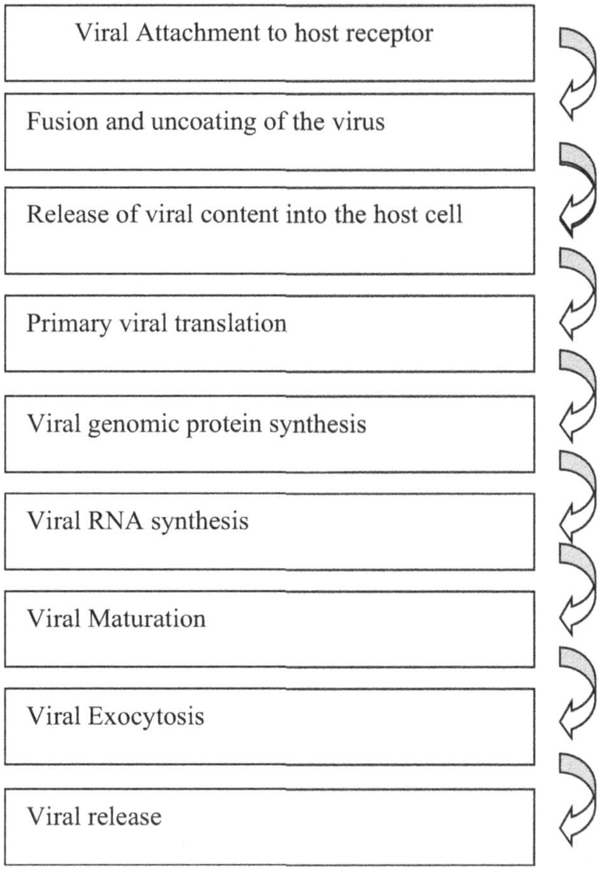


consequences of excessive proinflammatory host response. The cardiovascular system is affected due to availability of ACE2 receptors in the myocardial cells and the hypoxic state acquired due to the respiratory distress. Similarly renal and neurological damage occurs in Covid-19 infected patients. Even though the viral load reduces after 14 days of infection the damage due to host response continues to progress causes endotheliitis leading to microvascular thrombosis. This leads to intravascular coagulation and organ failure of the affected person [40, 41].

## 15.6 Clinical Signs and Symptoms

SARS-CoV-2 has an incubation period of 14 days but can vary between 6 and 14 days depending on the host immunity [5]. Coronavirus mainly infects the older adults, there seems to be no gender differences. Any underlying comorbidity like hypertension, diabetes, cardiovascular disorder, lung disorder is considered as a risk factor for coronavirus infection [38].

**Table 15.1** Classification of Covid-19 patients (clinical)

| Clinical type | Clinical signs and symptoms | Nucleic acid test | Chest imaging | Recovery period |
|---|---|---|---|---|
| Asymptomatic | Nil | Negative | Normal | Not applicable |
| Mild | Acute upper respiratory infection, gastrointestinal symptoms | Positive | Lung lesions may or may not be positive | 2 weeks |
| Moderate | Pneumonia without hypoxemia | Positive | Lung lesions positive | 2–4 weeks |
| Severe | Pneumonia with hypoxemia | Positive | Severe lung lesions | 4 weeks |
| Critical | ARDS, shock, encephalopathy, myocardial injury, heart failure, coagulation dysfunction and acute kidney injury | Positive | Massive lung involvement<br>Onset of acute respiratory distress syndrome<br>Mild:<br>200 mmHg $< PaO_2/FiO_2 \leq$ 300 mmHg<br>Moderate:<br>100 mmHg $< PaO_2/FiO_2 \leq$ 200 mmHg<br>Severe: $PaO_2/FiO_2 \leq$ 100 mmHg | Longer duration |

The disease is characterized by high grade fever and respiratory infection. Other features include pyrexia, cough, sore throat, fatigue, headache, body pain, dysgeusia, lack of taste and smell, nausea, vomiting, diarrhea, dyspnea, chest pain. Based on the severity, the disease is categorized as mild, moderate, and severe (Table 15.1) [42].

Extrapulmonary disorders also occur with covid 19 infection. It affects the major organs in gastrointestinal system (GI), hepatobiliary system, CVS, renal system, and CNS. The possible causes attributed to this multiorgan involvement include, viral toxicity, ischemic injury due to thrombosis or vasculitis, immune dysregulation, and renin-angiotensin-aldosterone dysregulation. (Table 15.2).

## 15.7 Diagnosis

Molecular testing of SARS-CoV-2 includes real time PCR assay of nasopharyngeal swab for SARS-CoV-2. Oropharyngeal swab, anterior/mid-turbinate nasal swabs, nasopharyngeal aspirates, BAL for patients in ventilators and saliva can be considered to obtain samples for investigation. SARS-CoV-2 antigen testing is also available but less sensitive than real time PCR. Antibody testing by serological evaluation is also a choice. Usually used in epidemiological studies for checking the effectiveness of the vaccine and assessing the immunity acquired post-infection. CBC, LFT, RFT, coagulation test are mandatory for hospitalized patients.

**Table 15.2** Extrapulmonary disorders

| System | Disease | Pathology | Laboratory findings |
|---|---|---|---|
| Cardiovascular system [43] | Myocardial ischemia/ infarction (MI) and myocarditis arrhythmias, cardiomyopathy, and cardiogenic shock | ACE2 receptors also exhibited by myocardial cells allows the virus to have direct effect | Elevated troponin, interleukins, cytokines |
| Renal system [44] | Acute kidney injury | Hypervolemia (fluid overload), injury due to drugs and vascular changes, and direct cytotoxicity of the virus | Proteinuria, hematuria, electrolyte abnormalities, altered acid-base balance like metabolic acidosis |
| Gastrointestinal system [45] | Dysgeusia, nausea, vomiting, diarrhea, and abdominal pain | Viral cytotoxicity on the ACE2 receptors in the intestinal mucosa, mucosal inflammation induced by cytokines, disrupted intestinal flora, and vascular changes. Acute mesenteric ischemia and portal vein thrombosis | Electrolyte disturbances |
| Hepatobiliary system [46] | Liver failure | Viral replication in the ACE2 receptors in the liver, virus mediated cytotoxicity, damage due to vasculitis | Elevated levels of AST and ALT |
| Central nervous system [47] | Anosmia and ageusia, headache, stroke, impairment of consciousness, seizure disorder, and toxic metabolic encephalopathy | Transsynaptic transfer across infected neurons via the olfactory nerve, vasculitis, or migration of leukocytes through the blood–brain barrier | Elevated inflammatory products |
| Cutaneous manifestation [48] | Erythematous maculopapular rash, vesicular rashes, urticarial rashes, vascular rashes, erythema multiforme-like eruptions [49] | | |
| Endocrine manifestations [50] | Diabetes mellitus | | Elevated blood glucose levels, metabolic acidosis and ketonuria with normal blood glucose levels, and diabetic ketoacidosis |

(continued)

**Table 15.2** (continued)

| System | Disease | Pathology | Laboratory findings |
|---|---|---|---|
| Hematological manifestations [51] | Thromboembolic events ischemic strokes, and arterial thrombosis | Hypercoagulability, ACE 2 mediated lymphocyte destruction, lymphocyte apoptosis | Thrombocytopenia, leukopenia, increased ESR levels, CRP, LDH, and leukocytosis, elevated D-dimer, fibrinogen levels, prolonged PT, and aPTT |
| Oral manifestations [52–54] | Dysgeusia/hypogeusia, erosion and ulcers, xerostomia, geographic tongue, erythema multiforme, candidiasis, mucormycosis | Presence of ACE 2 receptors in salivary glands and tongue, immunocompromised state, drug induced | Elevated cytokines and inflammatory products |

**Table 15.3** Chest radiograph severity grading

| | |
|---|---|
| 1. | Normal appearance |
| 2. | Patchy atelectasis and/or hyperinflation and/or bronchial wall thickening of the lungs |
| 3. | Focal consolidation in the lungs |
| 4. | Multifocal consolidation in the lungs |
| 5. | Diffuse alveolar change in the lungs |

Inflammatory status can be evaluated by ESR, CRP, ferritin, lactate dehydrogenase, D-dimer, and procalcitonin [55].

Imaging modalities include chest radiograph, lung ultrasound, and chest computed tomography. Chest radiograph does not detect the early stages of the disease, in advanced stage multifocal alveolar opacities can be seen which can involve the entire lung. Pleural effusion can be identified. Taylor et al. (2015) in British Medical Imaging presented a scoring system for chest radiographs (Table 15.3). Ultrasonographic examination can reveal the lung involvement of focal interstitial pattern to white lung. The features include irregular thick pleural line with small consolidations or nodules which progresses to B line, coalescent white lung which on further progression shows thickening and consolidation followed by pleural effusion.

Chest Computed Tomography is recommended for routine initial imaging or screening by The American College of Radiology. High resolution CT reveals ground glass patterns, consolidations commonly in the posterior region of the lower lobe, intra and interlobular septal thickening. Characteristic reverse halo signs can be seen, which are patchy areas of ground glass pattern with a peripheral halo. Cavitation, calcification, lymphadenopathy, and pleural effusion can also be detected. Li et al. (2020) prescribed a severity grading in CT (Table 15.4).

**Table 15.4** CT severity grading

| Score | Percentage of lung involvement |
|---|---|
| 0 | Nill lung involvement or 0% |
| 1 | Less than 5% of lung involvement |
| 2 | 5–25% of lung involvement |
| 3 | 26–49% of lung involvement |
| 4 | 50–75% of lung involvement |
| 5 | Greater than 75% of lung involvement |

## 15.8 Management

Researchers have identified drugs to restrict the virus from entering into the host cell and to inhibit replication. Chloroquine and hydroxychloroquine, the antimalarial drugs, prevent viral entry. Chloroquine modifies the glycosylation of ACE-2 and decreases the adhering capacity of the spike protein. It increases the endosomal pH and inhibits fusion. It also regulates proinflammatory signaling pathways and gives symptomatic relief to the patient. Recommended dose is 500 mg twice daily for 10 days. However, doses beyond 5 g may cause ventricular arrhythmia and hypokalemia [56]. Recombinant human angiotensin converting enzyme 2, which is still under trial, blocks the viral entry by blocking the spike proteins. No serious adverse effect has been reported so far [57]. Meplazumab, blocks the spike protein adhesion by blocking the CD147 which is essential for the spike protein to adhere to the host cell [58]. Human monoclonal antibody (MAB) which is yet another drug under trial blocks the spike protein ectodomain thereby preventing the adhesion of the virus [58, 59].

Remdesivir, a broad spectrum antiviral drug is the first line of drug used in management of SARS-CoV-2. In target cells it becomes active by transforming into triphosphate form. It terminates the viral replication by getting incorporated into nascent virus [60, 61]. Favipiravir, a guanine analogue is a promising antiviral drug under trial [62]. Umefenovir is another broad spectrum antiviral drug, under trial against SARS-CoV-2. Lopinavir and ritonavir act against viral proteases essential for maturation of the viral protein thereby inhibiting their progress [56], however on a trial they had shown adverse effects like headache, vomiting, and diarrhea and does not show any difference in mortality [63]. N3, a mechanism based, computer aided drug designed blocks the main protein [64]. Camostat mesylate blocks the serine protease and inhibits viral fusion [65]. Carmofur, an antineoplastic drug blocks the protease essential for fusion [64, 66].

Azithromycin, a macrolide antibiotic, is used as an adjuvant antibacterial for sore throat in SARS-CoV-2. It inhibits bacterial protein synthesis and modulates the host immune system. It regulates the cytokine release by decreasing the respiratory syncytial virus (RSV) release [56]. ACE inhibitors and stimulators block the availability of the host receptor thereby blocking the viral entry [17]. Convalescent plasma also known as pooled plasma containing immunoglobulins extracted from patients recovered from SARS-CoV-2 infection [67]. Though all the mentioned

**Table 15.5** Therapeutic management of Covid 19 infection recommended by NIH

| Severity | Recommendation |
|---|---|
| Hospitalized, does not require oxygen supplement | Remdesivir for high risk patients<br>Corticosteroids are not prescribed |
| Hospitalized and require oxygen supplement | For patients with minimal oxygen requirement: Remdesivir<br>For patients with increase oxygen requirement: Dexamethasone and Remdesivir<br>For patients to whom the above combination is contraindicated or non-availability of remdesivir: Dexamethasone |
| Hospitalized and require oxygen delivery by high flow device or non-invasive ventilation | Dexamethasone or Dexamethasone plus remdesivir<br>In patients with systemic inflammation and rapidly increasing oxygen need: Baricitinib or tocilizumab is added |
| Hospitalized and requires IMV or ECMO | Dexamethasone or Dexamethasone plus tocilizumab |

drugs give a logical explanation to act against SARS-CoV-2 virus most of it is under different phases of trial. Remdesivir is the only antiviral introduced in the management protocol. Azithromycin is used along with it and rarely plasma transfusion is provided for patients. The standard protocol for SARS-CoV-2 management is given in Table 15.5.

## 15.9 Conclusion

A pandemic in any century is a challenge to mankind. Covid 19 began at the end of 2019 and almost a year and a half has strangulated mankind physically, emotionally, and economically too. The initial havoc was due to the lack of understanding with the viral pathogenesis. As the pathogenesis became clear, the management protocol was designed with many tailor-made combinations suitable for the individual patient requirement. This virus is going to exist among us forever like its predecessors SARS-CoV and MERS. Hence, mass awareness about the virus, its mode of transmission, pathogenesis, initial signs and symptoms, and personal care among the population is a must to avoid further disaster to mankind. On the research front, scientists should be able to formulate a definitive treatment protocol and vaccine. This demands further research on the genotypic and phenotypic expression of the virus. Natural disasters are a heavy burden on a nation's shoulder. That too pandemic like SARS-CoV-2 curbs the normal functioning of the whole world. This adds to the economic downfall too which will have an impact on human life for a few years even after the pandemic ends. Hence, preparedness is essential. All national and international health care organizations and systems should be prepared for any such occurrence in future.

## References


1. V'kovski P, Kratzel A, Steiner S, Stalder H, Thiel V (2021) Coronavirus biology and replication: implications for SARS-CoV-2. Nat Rev Microbiol 19:155–170
2. Zhong NS, Zheng BJ, Li YM, Poon LLM, Xie ZH, Chan KH et al (2003) Epidemiology and cause of severe acute respiratory syndrome (SARS) in Guangdong, People's Republic of China, in February, 2003. Lancet 362:1353–1358. https://doi.org/10.1016/s0140-6736(03)14630-2
3. Ksiazek TG, Erdman D, Goldsmith CS, Zaki SR, Peret T, Emery S et al (2003) A novel coronavirus associated with severe acute respiratory syndrome. N Engl J Med 348:1953–1966. https://doi.org/10.1056/nejmoa030781
4. Zaki AM, van Boheemen S, Bestebroer TM, Osterhaus ADM, Fouchier RAM (2012) Isolation of a novel coronavirus from a man with pneumonia in Saudi Arabia. N Engl J Med 367:1814–1820. https://doi.org/10.1056/nejmoa1211721
5. Lu R, Zhao X, Li J, Niu P, Yang B, Wu H et al (2020) Genomic characterisation and epidemiology of 2019 novel coronavirus: implications for virus origins and receptor binding. Lancet 395:565–574. https://doi.org/10.1016/s0140-6736(20)30251-8
6. Kelly-Cirino C, Mazzola LT, Chua A, Oxenford CJ, Van Kerkhove MD (2019) An updated roadmap for MERS-CoV research and product development: focus on diagnostics. BMJ Glob Health 4:e001105. https://doi.org/10.1136/bmjgh-2018-001105
7. WHO Coronavirus (COVID-19) Dashboard n.d.. https://covid19.who.int. Accessed 2 July 2021
8. Payne S (2017) Family coronaviridae. Viruses:149–158. https://doi.org/10.1016/b978-0-12-803109-4.00017-9
9. Channappanavar R, Zhao J, Perlman S (2014) T cell-mediated immune response to respiratory coronaviruses. Immunol Res 59:118–128. https://doi.org/10.1007/s12026-014-8534-z
10. Corman VM, Muth D, Niemeyer D, Drosten C (2018) Hosts and sources of endemic human coronaviruses. Adv Virus Res:163–188. https://doi.org/10.1016/bs.aivir.2018.01.001
11. Liu DX, Liang JQ, Fung TS (2021) Human coronavirus-229E, -OC43, -NL63, and -HKU1 (Coronaviridae). In: Encyclopedia of virology. Elsevier, Amsterdam, pp 428–440
12. Xiong C, Jiang L, Chen Y, Jiang Q (n.d.) Evolution and variation of 2019-novel coronavirus. https://doi.org/10.1101/2020.01.30.926477
13. Bosch BJ, van der Zee R, de Haan CAM, Rottier PJM (2003) The coronavirus spike protein is a class I virus fusion protein: structural and functional characterization of the fusion core complex. J Virol 77:8801–8811
14. Brian DA, Baric RS (2005) Coronavirus genome structure and replication. Curr Top Microbiol Immunol:1–30. https://doi.org/10.1007/3-540-26765-4_1
15. Xu X, Chen P, Wang J, Feng J, Zhou H, Li X et al (2020) Evolution of the novel coronavirus from the ongoing Wuhan outbreak and modeling of its spike protein for risk of human transmission. Sci China Life Sci 63:457–460. https://doi.org/10.1007/s11427-020-1637-5
16. Li F (2012) Evidence for a common evolutionary origin of coronavirus spike protein receptor-binding subunits. J Virol 86:2856–2858. https://doi.org/10.1128/jvi.06882-11
17. Wan Y, Shang J, Graham R, Baric RS, Li F (2020) Receptor recognition by the novel coronavirus from Wuhan: an analysis based on decade-long structural studies of SARS coronavirus. J Virol 94. https://doi.org/10.1128/JVI.00127-20
18. Walls AC, Park Y-J, Alejandra Tortorici M, Wall A, McGuire AT, Veesler D (2020) Structure, function, and antigenicity of the SARS-CoV-2 spike glycoprotein. Cell 181:281–92.e6. https://doi.org/10.1016/j.cell.2020.02.058
19. Arndt AL, Larson BJ, Hogue BG (2010) A conserved domain in the coronavirus membrane protein tail is important for virus assembly. J Virol 84:11418–11428. https://doi.org/10.1128/jvi.01131-10
20. Nal B, Chan C, Kien F, Siu L, Tse J, Chu K et al (2005) Differential maturation and subcellular localization of severe acute respiratory syndrome coronavirus surface proteins S, M and E. J Gen Virol 86:1423–1434. https://doi.org/10.1099/vir.0.80671-0

21. Neuman BW, Kiss G, Kunding AH, Bhella D, Fazil Baksh M, Connelly S et al (2011) A structural analysis of M protein in coronavirus assembly and morphology. J Struct Biol 174:11–22. https://doi.org/10.1016/j.jsb.2010.11.021
22. Schoeman D, Fielding BC (2019) Coronavirus envelope protein: current knowledge. Virol J 16: 69
23. Nieto-Torres JL, DeDiego ML, Verdiá-Báguena C, Jimenez-Guardeño JM, Regla-Nava JA, Fernandez-Delgado R et al (2014) Severe acute respiratory syndrome coronavirus envelope protein ion channel activity promotes virus fitness and pathogenesis. PLoS Pathog 10:e1004077
24. DeDiego ML, Álvarez E, AlmazÁn F, Rejas MT, Lamirande E, Roberts A et al (2007) A severe acute respiratory syndrome coronavirus that lacks the E gene is attenuated in vitro and in vivo. J Virol 81:1701–1713. https://doi.org/10.1128/jvi.01467-06
25. McBride R, van Zyl M, Fielding B (2014) The coronavirus nucleocapsid is a multifunctional protein. Viruses 6:2991–3018. https://doi.org/10.3390/v6082991
26. Fan H, Ooi A, Tan YW, Wang S, Fang S, Liu DX et al (2005) The Nucleocapsid protein of coronavirus infectious bronchitis virus: crystal structure of its N-terminal domain and Multimerization properties. Structure 13:1859–1868. https://doi.org/10.1016/j.str.2005.08.021
27. Gao M, Yang L, Chen X, Deng Y, Yang S, Xu H et al (2020) A study on infectivity of asymptomatic SARS-CoV-2 carriers. Respir Med 169:106026
28. van Doremalen N, Bushmaker T, Morris DH, Holbrook MG, Gamble A, Williamson BN et al (n.d.) Aerosol and surface stability of HCoV-19 (SARS-CoV-2) compared to SARS-CoV-1. https://doi.org/10.1101/2020.03.09.20033217
29. Wu Y, Guo C, Tang L, Hong Z, Zhou J, Dong X et al (2020) Prolonged presence of SARS-CoV-2 viral RNA in faecal samples. Lancet Gastroenterol Hepatol 5:434–435
30. Kumar A, Faiq MA, Pareek V, Raza K, Narayan RK, Prasoon P et al (2020) Relevance of SARS-CoV-2 related factors ACE2 and TMPRSS2 expressions in gastrointestinal tissue with pathogenesis of digestive symptoms, diabetes-associated mortality, and disease recurrence in COVID-19 patients. bioRxiv. https://doi.org/10.1101/2020.04.14.040204
31. Amirian ES, Susan AE (2020) Potential fecal transmission of SARS-CoV-2: current evidence and implications for public health. Int J Infect Dis 95:363–370. https://doi.org/10.1016/j.ijid.2020.04.057
32. Patel KP, Vunnam SR, Patel PA, Krill KL, Korbitz PM, Gallagher JP et al (2020) Transmission of SARS-CoV-2: an update of current literature. Eur J Clin Microbiol Infect Dis 39:2005–2011
33. Xiao F, Tang M, Zheng X, Liu Y, Li X, Shan H (2020) Evidence for gastrointestinal infection of SARS-CoV-2. Gastroenterology 158:1831–3.e3
34. Wang S, Guo L, Chen L, Liu W, Cao Y, Zhang J et al (2020) A case report of neonatal 2019 coronavirus disease in China. Clin Infect Dis 71:853–857
35. Jing Y, Run-Qian L, Hao-Ran W, Hao-Ran C, Ya-Bin L, Yang G et al (2020) Potential influence of COVID-19/ACE2 on the female reproductive system. Mol Hum Reprod 26:367–373
36. Candotto V, Lauritano D, Nardone M, Baggi L, Arcuri C, Gatto R et al (2017) HPV infection in the oral cavity: epidemiology, clinical manifestations and relationship with oral cancer. Oral Implantol 10:209–220
37. Li W, Moore MJ, Vasilieva N, Sui J, Wong SK, Berne MA et al (2003) Angiotensin-converting enzyme 2 is a functional receptor for the SARS coronavirus. Nature 426:450–454
38. Yuki K, Fujiogi M, Koutsogiannaki S (2020) COVID-19 pathophysiology: a review. Clin Immunol 215:108427
39. Zou X, Chen K, Zou J, Han P, Hao J, Han Z (2020) Single-cell RNA-seq data analysis on the receptor ACE2 expression reveals the potential risk of different human organs vulnerable to 2019-nCoV infection. Front Med 14:185–192
40. Alipoor SD, Mortaz E, Jamaati H, Tabarsi P, Bayram H, Varahram M et al (2021) COVID-19: molecular and cellular response. Front Cell Infect Microbiol 11. https://doi.org/10.3389/fcimb.2021.563085

41. Streicher F, Jouvenet N (2019) Stimulation of innate immunity by host and viral RNAs. Trends Immunol 40:1134–1148
42. Oh H-LJ, Gan SK-E, Bertoletti A, Tan Y-J (2012) Understanding the T cell immune response in SARS coronavirus infection. Emerg Microbes Infect 1:1–6. https://doi.org/10.1038/emi.2012.26
43. Huang C, Wang Y, Li X, Ren L, Zhao J, Hu Y et al (2020) Clinical features of patients infected with 2019 novel coronavirus in Wuhan, China. Lancet 395:497–506
44. Martinez-Rojas MA, Vega-Vega O, Bobadilla NA (2020) Is the kidney a target of SARS-CoV-2? Am J Physiol Renal Physiol 318:F1454–F1462. https://doi.org/10.1152/ajprenal.00160.2020
45. Tariq R, Saha S, Furqan F, Hassett L, Pardi D, Khanna S (2020) Prevalence and mortality of COVID-19 patients with gastrointestinal symptoms: a systematic review and meta-analysis. Mayo Clin Proc 95:1632–1648
46. Cascella M, Rajnik M, Aleem A, Dulebohn SC, Di Napoli R (2021) Features, evaluation, and treatment of coronavirus (COVID-19). StatPearls Publishing, Treasure Island, FL
47. Zubair AS, McAlpine LS, Gardin T, Farhadian S, Kuruvilla DE, Spudich S (2020) Neuropathogenesis and neurologic manifestations of the coronaviruses in the age of coronavirus disease 2019. JAMA Neurol 77:1018. https://doi.org/10.1001/jamaneurol.2020.2065
48. Daneshgaran G, Dubin DP, Gould DJ (2020) Cutaneous manifestations of COVID-19: an evidence-based review. Am J Clin Dermatol 21:627–639
49. Sachdeva M, Gianotti R, Shah M, Bradanini L, Tosi D, Veraldi S et al (2020) Cutaneous manifestations of COVID-19: report of three cases and a review of literature. J Dermatol Sci 98:75–81. https://doi.org/10.1016/j.jdermsci.2020.04.011
50. Gupta A, Madhavan MV, Sehgal K, Nair N, Mahajan S, Sehrawat TS et al (2020) Extrapulmonary manifestations of COVID-19. Nat Med 26:1017–1032
51. Coopersmith CM, Antonelli M, Bauer SR, Deutschman CS, Evans LE, Ferrer R et al (2021) The surviving sepsis campaign: research priorities for coronavirus disease 2019 in critical illness. Crit Care Med 49:598–622. https://doi.org/10.1097/ccm.0000000000004895
52. Wang H, Zhou M, Brand J, Huang L (2009) Inflammation and taste disorders: mechanisms in taste buds. Ann N Y Acad Sci 1170:596–603
53. Takeda N, Takaoka T, Ueda C, Toda N, Kalubi B, Yamamoto S (2004) Zinc deficiency in patients with idiopathic taste impairment with regard to angiotensin converting enzyme activity. Auris Nasus Larynx 31:425–428. https://doi.org/10.1016/s0385-8146(04)00142-7
54. Chaux-Bodard A-G, Deneuve S, Desoutter A (2020) Oral manifestation of Covid-19 as an inaugural symptom? J Oral Med Oral Surg 26:18. https://doi.org/10.1051/mbcb/2020011
55. Gandhi RT, Lynch JB, Del Rio C (2020) Mild or moderate Covid-19. N Engl J Med 383:1757–1766
56. Kumar A, Singh R, Kaur J, Pandey S, Sharma V, Thakur L et al (2021) Wuhan to world: the COVID-19 pandemic. Front Cell Infect Microbiol 11:596201
57. Groß S, Jahn C, Cushman S, Bär C, Thum T (2020) SARS-CoV-2 receptor ACE2-dependent implications on the cardiovascular system: from basic science to clinical implications. J Mol Cell Cardiol 144:47–53. https://doi.org/10.1016/j.yjmcc.2020.04.031
58. Tian X, Li C, Huang A, Xia S, Lu S, Shi Z et al (2020) Potent binding of 2019 novel coronavirus spike protein by a SARS coronavirus-specific human monoclonal antibody. Emerg Microbes Infect 9:382–385. https://doi.org/10.1080/22221751.2020.1729069
59. Wang M-Y, Zhao R, Gao L-J, Gao X-F, Wang D-P, Cao J-M (2020) SARS-CoV-2: structure, biology, and structure-based therapeutics development. Front Cell Infect Microbiol 10:587269
60. Gurwitz D (2020) Angiotensin receptor blockers as tentative SARS-CoV-2 therapeutics. Drug Dev Res 81:537–540. https://doi.org/10.1002/ddr.21656
61. Siegel D, Hui HC, Doerffler E, Clarke MO, Chun K, Zhang L et al (2017) Discovery and synthesis of a phosphoramidate prodrug of a Pyrrolo[2,1-f][triazin-4-amino] adenine C-nucleoside (GS-5734) for the treatment of Ebola and emerging viruses. J Med Chem 60:1648–1661. https://doi.org/10.1021/acs.jmedchem.6b01594

62. Furuta Y, Komeno T, Nakamura T (2017) Favipiravir (T-705), a broad spectrum inhibitor of viral RNA polymerase. Proc Jpn Acad Ser B 93:449–463. https://doi.org/10.2183/pjab.93.027
63. Cao Y-C, Deng Q-X, Dai S-X (2020) Remdesivir for severe acute respiratory syndrome coronavirus 2 causing COVID-19: an evaluation of the evidence. Travel Med Infect Dis 35: 101647
64. Jin Z, Du X, Xu Y, Deng Y, Liu M, Zhao Y et al (2020) Structure of Mpro from SARS-CoV-2 and discovery of its inhibitors. Nature 582:289–293. https://doi.org/10.1038/s41586-020-2223-y
65. Bittmann S (2020) COVID 19: Camostat and the role of serine protease entry inhibitor TMPRSS2. J Regen Biol Med. https://doi.org/10.37191/mapsci-2582-385x-2(2)-020
66. Jin Z, Zhao Y, Sun Y, Zhang B, Wang H, Wu Y et al (2020) Structural basis for the inhibition of SARS-CoV-2 main protease by antineoplastic drug carmofur. Nat Struct Mol Biol 27:529–532. https://doi.org/10.1038/s41594-020-0440-6
67. Rojas M, Rodríguez Y, Monsalve DM, Acosta-Ampudia Y, Camacho B, Gallo JE et al (2020) Convalescent plasma in Covid-19: possible mechanisms of action. Autoimmun Rev 19:102554. https://doi.org/10.1016/j.autrev.2020.102554

# Microbiome in Influenza-A Virus Infection 16

Suhas Suresh Awati, Santosh Kumar Singh, Abhay Raizaday, Pramod Kumar, Yogendra Singh, Mohammad Arshad Javed Shaikh, and Gaurav Gupta

**Abstract**

Influenza is an infective virus infection. The indications range from mild-to-severe and sometimes consist of pyrexia, coughing, runny-nose, muscle pain, headache, swelling in throat, and weakness. Influenza virus infection continues to be a major global health threat. Influenza-A virus may create pandemic flu, i.e., worldwide epidemic illness. Avoidance and treatment of influenza viral infection remain restricted, and alternative host protection policies are desperately required. The microbiomes play a vital position in immunomodulatory and in tissue homeostasis. The objective of current work is to emphasize the modern approaches into the regulatory function of microbiome in influenza-A viral infection and present a fresh fact of the connections and fundamental means of the bonding among the microbiome and management of influenza-A viral infection.

**Keywords**

Influenza-A · Microbiome · Viral infection · Flu · Microbiota

S. S. Awati
School of Pharmacy, Suresh Gyanvihar University, Jaipur, India

Dr. Shivajirao Kadam College of Pharmacy, Sangli, India

S. K. Singh (✉) · A. Raizaday · M. A. J. Shaikh · G. Gupta
School of Pharmacy, Suresh Gyanvihar University, Jaipur, India
e-mail: santosh.krsingh@mygyanvihar.com

P. Kumar
Limetta Laboratories, Haridwar, India

Y. Singh
Maharishi Arvind College of Pharmacy, Jaipur, India





G. Gupta et al. (eds.), *Microbiome in Inflammatory Lung Diseases*,
https://doi.org/10.1007/978-981-16-8957-4_16

## 16.1 Introduction

Influenza is an infective virus infection. The indications range from mild-to-severe and sometimes consist of pyrexia, coughing, runny-nose, muscle pain, headache, swelling in throat, and weakness. Influenza virus infection continues to be a major global health threat. Influenza-A virus may create pandemic flu, i.e., worldwide epidemic illness. Avoidance and treatment of influenza viral infection remain restricted, and alternative host protection policies are desperately required. The microbiomes play a vital position in immunomodulatory and in tissue homeostasis. The objective of current work is to emphasize the modern approaches into the regulatory function of microbiome in influenza-A viral infection and present a fresh fact of the connections and fundamental means of the bonding among the microbiome and management of influenza-A virus infection. Respiratory viral infectivity results in harsh morbid and mortal effects in humans as well as in animals globally. Influenza viruses are original sources of rigorous respiration system diseases that cause 3–5 millions of infective cases. There is steady risk of such dangers to society, like latest pandemics of swine-flu and Covid-19 clearly demonstrated [1]. These symptoms ordinarily start 1–4 days subsequent to viral contact and remain around 3–8 days. Diarrheal and vomit conditions can ensue, especially in kids. Influenza to pneumonia progression is a viral to bacterial complications journey. Various else problems of disease like meningitis worsen the original pathological conditions like asthma. Influenza virus infection continues to be a major global health threat [2].

Influenza-A/B/C and D are major four categories. Water bird is the main host of influenza A that may infect human and pig which cause seasonal epidemic. Influenza B and C basically contaminate people with mild infections, while Influenza D infects steer and pig that are less infective to humans. In human, these viruses are mainly transferred through respiration droplet of cough and sneeze [3]. Successive hand wash and mouth/nose masking during cough and sneeze decrease transmissions. Yearly vaccines are able to secure in opposition to flu. Influenza viruses, especially IAV, evolve rapidly, so influenza immunizations are updated routinely to go with circulation strains. H1N1 and H3N2 subtypes are vaccine protected currently. Many lab tests are available to recognize such viruses. Antiviral drugs and support services can cure such infections. Such infections are risky in comorbidity [4]. Every year, almost 5–15% population indentures such infections. Many severe cases and deaths are due to influenza worldwide. Children and elder patients are more prone to death. Winter is more susceptible season for influenza infections compared to summer. Pandemic threats are occurring after every 20–50 years due to influenza viruses like Spanish flu or recent COVID-19 [5].

Influenza-A virus is single-stranded RNA virus. It is divided into two types depending on hemagglutinin/neuraminidase ratio. There are 198 influenza-A sub-type combinations which are available in nature. Time to time these viruses undergo genetic and antigenic properties [6] (Fig. 16.1).

Microbiome includes all microorganisms living in digestive tract of any human, animal, or insect. Even from birth many body parts are prone to expose with various

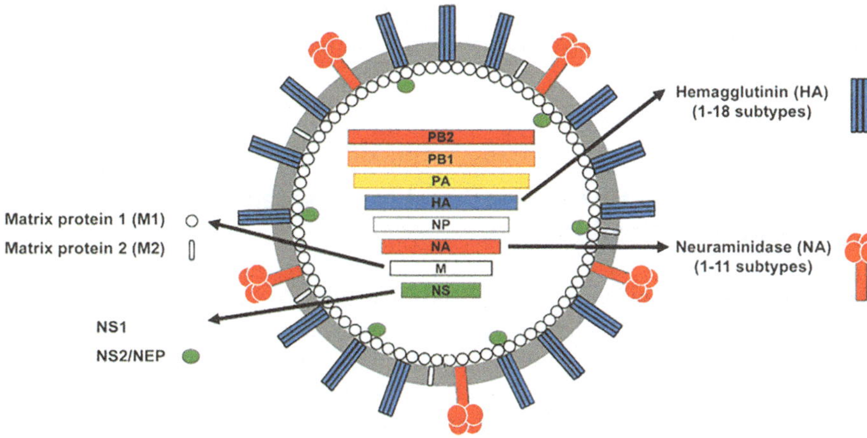


**Fig. 16.1** Influenza-A virus

microbes, collectively called as microbiome [7–9]. During routine life they affect their host at different stages like digestion [9–11]. All immunity reactions are intensely influenced by GIT-microbiome [12]. Avoidance and treatment of influenza viral illness stay restricted, and substitute host protective guidelines are desperately required.

## 16.2 Influenza Viruses

Influenza virus enters the body from first parts of respiratory tract and changes the microbiota extensively subsequent to illness. It reduces healthy microbes and increases pathogenic microbial load [13–15]. According to literature, it is found that nasopharyngeal and oropharyngeal microbiota show discrete alterations after influenza infection. Ramos et al. found that throat microbiota was flexible to such infections, with extremely stable compositions subsequent to influenza infections [16, 17]. Interruptions in microbe–host homeostasis, created by infections that affect normal physiological mechanisms. According to research, such changes are due to over-production of interferon. The troubled gut microbiota additionally enthused Interleukin-15 generation from intestinal epithelium [18–20].

## 16.3 Host Factors Effect on of Influenza-A Viral Infection

Influenza pathogenesis has two chapters. Initial chapter is for 1–3 days with high intensive viral load and inflammations. The later chapter is either controlled for virus or severe and mortal. Death is associated with deregulated immune response. Obese and old patients with comorbidity are prone for more inflammation and severe diseases. Extreme inflammations affect tissue repair and cell mediated immunity.

Various host factors like age, weight, sex, etc. decide the severity. Progress of effectual and largely defensive vaccines and therapy are major points in future to control such threats [21–23].

## 16.4 The Bi-directional Relationship Between Influenza and the Microbiome

The association of influenza virus and the microbiome is believed to be bi-directional. First, the microbiome may influence influenza virus infection. Although not yet examined in human populations, murine studies suggest that the microbiome can influence influenza virus infection through immunomodulation [24, 25]. It is still unclear which attributes of the microbiome may be driving this relationship. However, recent randomized controlled trials report substantial reductions in the incidence of RTIs among infants known synbiotic treatment compared to placebo [26, 27]. Further exploration in human populations could enhance synbiotic approaches for preventing influenza virus infection.

Second, influenza virus may influence the microbiome. Man et al. fittingly describe the upper respiratory tract (URT) as the "gatekeeper to respiratory health" [28] as colonization at this site is a necessary precursor of respiratory infection for certain bacterial pathogens [29, 30]. Influenza can perturb the microbiota, enhancing the acquisition [31, 32] and overgrowth [33, 34] of microbes. This perturbation increases danger of persistent diseases [35, 36], mainly through a deep lung [37]. But, no longitudinal studies among human populations have yet examined how influenza virus infection alters the microbiome. As the majority of influenza deaths are attributed to inferior bacterial infections due to common bacterial residents of the URT, characterizing these alterations in the microbiome would be a first step towards designing synbiotic methods for improving microbiome resilience and reducing disease severity [38].

## 16.5 Inducing an Immuno-regulatory Microenvironment

The significance chore of the varied microorganism occupying a major indispensable responsibility has been markedly suggested in modifying the improvement of immunity of host, together within and outer side the intestine [39–42]. The establishment of an immune tolerant microenvironment contributes immensely in generating immuno-regulatory cells by growth, demarcation, and instigation of T cells which bring out of effector immune response and maintain homeostasis, as well as by encouraging the making of various pro-inflammatory cytokines like interferon (IFN)-gamma throughout infection [43–45]. Consequently, it potentiates the commensal microbiota persuade Treg cells which impel cytokines and thereby relating to limit the degrees of antiviral immune responses [44].

## 16.6 Influenza Viral Infection Suppression by the Gut Microbiota

### 16.6.1 Suppression (Direct) of Viral Infection

The existence of commensal microbiomes at places where some viruses use as get access into the host is considered as a significant interactions slot for the attacking viruses and commensal microbiota that possibly have suppressive effects for infection of virus. Besides, *Enterococcus faecium* is capable of averting infection by trapping those influenza viruses by direct adsorptive as well as they build a variety of metabolites possessing anti-microbial property against virus infestation [46]. Mechanistic approach may be put forward by commensal microbiota-derived LPS that have potential to bind to and subvert morphological characteristic of influenza virions, thus diminishing the overall viral stability and later by production of extra-cellular matrix binding protein formed by Gram-positive bacteria *Staphylococcus epidermidis* present in rhinal cavity as a commensal and has capacity to tightly bind to influenza virus thereby obstructing additional viral contagion [47, 48].

### 16.6.2 Suppression (Indirect) of Viral Infection

A significant role of commensal microbiota stands influential for the host immune responses guaranteeing effectual exclusion of viral invasion and has been supported by numerous studies highlighting with the intention that intact healthy commensal microbiota benefits to preserve healthy immunity against virus whilst microbiota disturbance augment infection of virus. Steed et al. in his findings illustrated that *Clostridium orbiscindens* an exact human-related gut microbiome produces metabolite desamino-tyrosine that amplify loop of IFN (Type-I) signalling thereby protects from influenza [49]. Yitbarek et al. in their research finding signified modulation of type-I interferon as well as antibody mediated immune responses by gut microbiota against influenza virus [50]. Steady with this discovering, it has been shown that during respiratory flu infection disease, antibiotic exposure led to a faulty generation of virus specific CD4 and CD8 T cells and antibodies due to a damaged inflammasome reliant movement of antigen-presenting cells (APC) from the lung to the depleting lymph nodes [51] (Fig. 16.2).

Safe affirmation of the gut microbiota is significant for the age and commencement of immuno-regulatory cells to debilitate local/systemic stimulation of immunity. Indeed, Rosshart et al. discovered that reconstruction of the gut microbiome from wild mice confers compelling defending impacts to laboratory mice (GF) during deadly flu infections, an impact chiefly intervened through the avoidance of extreme inflammation by means of IL-10 and IL-13 creation in the mice affected by virus by the regular gut microbiota [52] (Table 16.1).

On infestation of the influenza virus, commensal microbiota, in addition to their components (like various TLR-ligands) or else metabolites (like desamino-tyrosine) trigger the inflammasome thereby significantly producing two cytokines (IL-1b and

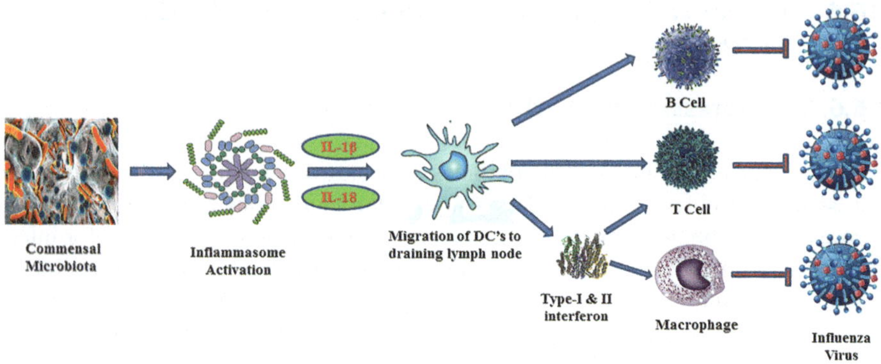


**Fig. 16.2** The mechanism of flu infection suppression by the commensal microbiome

**Table 16.1** The mechanism of flu infection suppression by the commensal microbiome

| Type of suppression | Mechanisms | Virus types | References |
|---|---|---|---|
| Direct suppression | Adsorptive trapping of viruses | Influenza-A viruses | [46] |
| | Binding to and destabilizing virion morphology | Influenza-A viruses | [47] |
| | Binding to and blocking further infections | Influenza-A viruses | [48] |
| Indirect suppression | Enhancing type-I IFN signalling | Influenza-A virus | [49, 50] |
| | Promoting APC migration and T cell activation | Influenza-A virus | [51] |
| | Encouraging TLR mediated humeral and cellular immune responses against virus | Influenza-A virus | [51] |
| | Preventing excessive inflammation and inflammation-associated pathology | Influenza-A virus | [52] |

IL-18) that influence the immigration of dendritic cells commencing lung to the lymph nodes, wherein acting as an antigen-presenting cells (APC) that prime virus specific B-cells, macrophages, CD4+T cells, and CD8+T cells. Accordingly, these effector cells produce antibodies against specific virus or inflammatory cytokines or exercise direct killing of virus effects to suppress the flu infection [53].

### 16.6.3 Dysbiosis of Gut Microbiota in Flu Infection

Yearly the more figures of influenza-A virus (IAV)-associated death and morbidness, ensuing deaths in large numbers has raised questions on accomplishment in fighting virus infections, thus imposing the advancement of novel remedies having potency to decrease the severity of related infections of IAV. Intestinal complications during respiratory diseases have exemplified the relationship involving the intestinal and respiratory tract [54].

Wang et al. demonstrated that gut microbiota dysbiosis resulting from flu infection is arbitrated by IFN-ϒ (INF-I and II) that has been created by lung derived CCR9 and CD4+T cells which are employed into the small bowel, and interceded by the CCL25–CCR9 cytokine axis.

Wang et al. in their research findings signified the induction of immune injury in the intestine by microbiota mediated Th-17 cell reliant inflammation furnish by H1N1 virus triggered the intestine immune injury, and declined following neutralizing of IL-17A [55].

The consequences of cellular immune reply to flu infection resulted in immune mediated in appetence and loss of weight was put forward by Groves et al. [56]

Influenza virus infestation results in dysbiotic microenvironment of intestine and leading to dynamic reduction of numerous essential groups of bacteria, mucosal layers distraction and making of more numbers of anti-microbial peptides in Paneth cells.

Sencio et al. reported dysbiosis of gut microbiota contributed to pulmonary pneumococcal infection in lungs by changed production of short-chain fatty acid [57].

### 16.6.4 Respiratory Tract Microbiome and Influenza-A Viral Infection

An examination Leung et al. showed significantly augmentation of pathogen Firmicutes and Proteobacteria by Influenza-A (H1N1) 2009 pandemic viral infection on examination of oropharyngeal microbiome in patients of pneumonia [58]. The studies revealed by Chaban et al. and Hanada et al. on patients harbouring pandemic H1N1 signified and characterized the prime phyla of the URT were Actinobacteria, Firmicutes, and Proteobacteria nevertheless, the authors recommended that influenza is allied with a development of Proteobacteria which is usually less abundant in healthy hosts [59, 60].

Greninger et al. revealed that Enterobacter and Moraxella spp. (which are categorized as Proteobacteria) were found to be the most significantly present bacteria in samples of nasopharyngeal found in pandemic H1N1 influenza patients with [61]. Though the studies demonstrated significant inter-subject variability, subsequently emphasize on the necessity for longitudinal studies to decipher changes post-viral infection. Subsequent viral exposure alters respiratory tract microbiome given that bacterial pneumonia every now and again emerges because of suctioned bacterial microorganisms; an expected system by which viral diseases may build the danger of auxiliary bacterial contaminations is through expanded colonization of the URT by bacterial microbes.

In human subjects, live attenuated influenza vaccine (LAIV) and human rhinovirus (hRV) have exposed to disturb the local host bacterial community, with amplified relative richness of potential pathogens (or pathobionts), such as Neisseria and Staphylococcal species.

Respiratory viruses not only modify the bacterial flora in the URT but in addition endorse bacterial colonization of the LRT through the different mechanisms that harm bacterial clearance. Making of mucus in the respiratory tract is augmented to facilitate virus clearance throughout infections. Nevertheless, extreme production of mucus can lead to airway hindrance by impeding mucociliary clearance [60, 62].

### 16.6.5 Microbiome Disturbance of the URT in Influenza-A Virus Infection in Humans and Ferrets

Bacterial co-infections aggravate the influenza infection whilst resulting in syndrome exacerbation owing to host replies and cell damage. The core URT microbiome is agitated by IAV infection through uncharacterized indirect and direct processes, in turn facilitating co-infections with bacteria type pathogens triggering raised number of hospitalizations and morbidity related with infection of IAV. Examining temporal dynamics of uncontaminated and flu virus infected in humans and ferrets for URT microbiomes, it was observed that both the uninfected humans and ferret had stable "healthy ecostate" for URT microbiomes for both within and among the individuals. On the contrary, infected patients and ferrets revealed huge alteration in bacterial community structure over time and between individuals. The "unhealthy" ecostates of contaminated persons developed in the direction of "healthy ecostate" eventually, subsequently showing clearance of virus infection. *Pseudomonas* blooms were observed and measured recurrently in the concerned microbiomes in the infected individuals. Antiviral responses of the host may be contributed to microbiome perturbation in a dynamic way which necessitates host's microbiome metatranscriptomics or metaproteomics measurements in precise experiments attentive at the beginning of viral infestation. A close relation was suggested among dynamic infection, sickness, and disturbance of the microbiome, showed high disturbance which is correlated amid high viral loads and loss in weight in the models of ferret, wherein the kinetics are comparable to the antiviral reply that are induced in infection of IAV. The rationality of potential therapeutic target to avert IAV related co-infections of bacteria can be significantly related to the dynamic and flexible form of the microbiome during flu infection for the preservation of the homeostasis of microbiome [63, 64].

## 16.7 Conclusions and Future Perspectives

Aforementioned discussion highlights recent consideration of modulation of infection of virus by the commensal microbiota of the host and the fundamental mechanisms in this regulation as well as illustrated the role of viral infectivity towards the disturbances of homeostasis of microbiota in the host. The complete understanding of the degree to which commensal microbiota may perhaps establish the competence of viral reproduction, transmission, perseverance, and in many cases may not be comprehensible clear as well as the reports suggested significant mechanistic approaches primarily influencing the microbiota of host by invading viruses are uncertain.

The information establishes to assist a close interaction of continuous commands and relation between the commensal microbiota with infecting viruses, an interface that the result of an infection. The speculation of drugs against virus proposed for the modulating virus–microbiota interactions has fascinated particularly due to its effectiveness in monitoring the action of numerous virus infections. The medicinal claim of FMT and probiotic supplement has on date established its usefulness in decreasing the severity of numerous diseases in human and non-human primate-based studies, while these efforts may end up being ineffective in specific conditions and could even bring about undesirable complications [65–68]. Subsequently, there remain a breach in our perception of the interactions amongst the viruses and commensal microbiota. A continuous amendment of these possible treatment approaches is apparently desirable to enhance the regulation of viral infections during the modulation of commensal microbiota. Besides, it is currently well acknowledged that overuse of antibiotic leads to the development of antibiotic-resistant bacteria or even super bacteria that may bring about severe life threatening infections. In view of that, we do not advocate the support for the usage of antibiotics in treatment or prevention of viral infections. On the other hand, considering how commensal microbiota augment virus infection, particularly the molecular necessities for the microbiota mediated encouragement of virus infections, may possibly direct to the progress of novel, practicable strategies of antiviral. The mechanism consideration that underlines the modulation of commensal microbiota by viral infections is not completely clear, thus the intriguing question certainly warrants further investigations.

Respiratory infestation of infective virus can instigate a surge of host immunological response altering microbial development environment in the gut, LRT, and URT. Stimulation of flu virus induced antiviral interferon path may lead to insufficient responses of innate immune cells through host defense against secondary infections of bacteria, resulting in the increase of potentially species of pathogenic bacteria. Attending alteration in microbiome of gut sought to bring about by the underlying infection of virus can correspondingly modify immune cell priming beside secondary challenge of bacteria, though still it is not explored scientifically so far.

Though the image is incomplete, the current literature of microbiome conveys further insights showing keen interest on the pathogenesis of dysregulated immune might support the progress of secondary bacterial pneumonias. Thereby making clear variances and dynamics of microbiota of respiratory tract in both healthy and cases of chronic infection of lung and in acute type of viral infections of respiratory tract revealing pathogenesis of virus-bacteria interactions wherein providing a base in the emerging novel tactics for the anticipation, and the managing acute type of viral infections of respiratory tract and worsening of chronic diseases of the lung.

## References


1. Takeshi IC, Iris KP, Yosuke K, David R, Ho JH, Thomas SM, Akiko I (2011) Microbiota regulates immune defense against respiratory tract influenza A virus infection. PNAS 108(13): 5354–5359
2. Flu Symptoms & Diagnosis. Centers for Disease Control and Prevention (CDC). 10 July 2019. Retrieved 24 January 2020
3. Flu Symptoms & Complications. Centers for Disease Control and Prevention (CDC). 26 February 2019
4. Call SA, Vollenweider MA, Hornung CA, Simel DL, McKinney WP (2005) Does thispatient have influenza? JAMA 293(8):987–997
5. Dharmapalan D (2020) Influenza. Indian J Pediatr 87(10):828–832
6. Tong S et al (2013) New world bats harbor diverse Influenza-A viruses. PLoS Pathog 9(10): e1003657
7. Ley RE, Peterson DA, Gordon JI (2006) Ecological and evolutionary forces shaping microbial diversity in the human intestine. Cell 124:837–848
8. Dethlefsen L, McFall-Ngai M, Relman DA (2007) An ecological and evolutionary perspective on human-microbe mutualism and disease. Nature 449:811–818
9. Round JL, Mazmanian SK (2009) The gut microbiota shapes intestinal immune responses during health and disease. Nat Rev Immunol 9:313–323
10. Kamada N, Seo SU, Chen GY, Nunez G (2013) Role of the gut microbiota in immunity and inflammatory disease. Nat Rev Immunol 13:321–335
11. Hooper LV, Gordon JI (2001) Commensal host-bacterial relationships in the gut. Science 292: 1115–1118
12. Macpherson AJ, Harris NL (2004) Interactions between commensal intestinal bacteria and the immune system. Nat Rev Immunol 4:478–485
13. Edouard S, Million M, Bachar D, Dubourg G, Michelle C, Ninove L et al (2018) The nasopharyngeal microbiota in patients with viral respiratory tract infections is enriched in bacterial pathogens. Eur J Clin Microbiol Infect Dis 37:1725–1733. https://doi.org/10.1007/s10096-018-3305-8
14. Li Y, Ding J, Xiao Y, Xu B, He W, Yang Y et al (2017) 16S rDNA sequencing analysis of upper respiratory tract flora in patients with influenza H1N1 virus infection. Front Lab Med 1:16–26. https://doi.org/10.1016/j.flm.2017.02.005
15. Wen Z, Xie G, Zhou Q, Qiu C, Li J, Hu Q et al (2018) Distinct nasopharyngeal and oropharyngeal microbiota of children with influenza a virus compared with healthy children. Bio Med Res Int:6362716. https://doi.org/10.1155/2018/6362716
16. Ramos-Sevillano E, Wade WG, Mann A, Gilbert A, Lambkin-Williams R, Killingley B et al (2019) The effect of influenza virus on the human oropharyngeal microbiome. Clin Infect Dis 68:1993–2002. https://doi.org/10.1093/cid/ciy821
17. Yildiz S, Mazel-Sanchez B, Kandasamy M, Manicassamy B, Schmolke M (2018) Influenza A virus infection impacts systemic microbiota dynamics and causes quantitative enteric dysbiosis. Microbiome:6–9. https://doi.org/10.1186/s40168-017-0386-z
18. Groves HT, Cuthbertson L, James P, Moffatt MF, Cox MJ, Tregoning JS (2018) Respiratory disease following viral lung infection alters the murine gut microbiota. Front Immunol 9:182. https://doi.org/10.3389/fimmu.2018.00182
19. Zhao N, Wang SP, Li HY, Liu SL, Li M, Luo J et al (2018) Influence of novel highly pathogenic avian influenza A (H5N1) virus infection on migrating whooper swans fecal microbiota. Front Cell Infect Microbiol 8:46. https://doi.org/10.3389/fcimb.2018.00046
20. Wang J, Li FQ, Wei HM, Lian ZX, Sun R, Tian ZG (2014) Respiratory influenza virus infection induces intestinal immune injury via microbiota-mediated Th17 cell-dependent inflammation. J Exp Med 211:2397–2410. https://doi.org/10.1084/jem.20140625

21. Yitbarek A, Weese JS, Alkie TN, Parkinson J, Sharif S (2018) Influenza A virus subtype H9N2 infection disrupts the composition of intestinal microbiota of chickens. FEMS Microbiol Ecol 94:165. https://doi.org/10.1093/femsec/fix165
22. Deriu E, Boxx GM, He XS, Pan C, Benavidez SD, Cen LJ et al (2016) Influenza virus affects intestinal microbiota and secondary salmonella infection in the gut through type I interferons. PLoS Pathog 12:e1005572. https://doi.org/10.1371/journal.ppat.1005572
23. Gounder AP, Boon ACM (2019) Influenza pathogenesis: the effect of host factors on severity of disease. J Immunol 202:341–350
24. Ichinohe T, Pang IK, Kumamoto Y et al (2011) Microbiota regulates immune defense against respiratory tract influenza A virus infection. Proc Natl Acad Sci U S A 108:5354
25. Abt MC, Osborne LC, Monticelli LA et al (2012) Commensal bacteria calibrate the activation threshold of innate antiviral immunity. Immunity 37:158–170
26. Luoto R, Ruuskanen O, Waris M, Kalliomäki M, Salminen S, Isolauri E (2014) Prebiotic and probiotic supplementation prevents rhinovirus infections in preterm infants: a randomized, placebo-controlled trial. J Allergy Clin Immunol 133:405–413
27. Panigrahi P, Parida S, Nanda NC et al (2017) A randomized synbiotic trial to prevent sepsis among infants in rural India. Nature 548(7668):407–412
28. Man WH, de SteenhuijsenPiters WA, Bogaert D (2017) The microbiota of the respiratory tract: gatekeeper to respiratory health. Nat Rev Microbiol 15:259–270
29. Bogaert D, De Groot R, Hermans PWM (2004) Streptococcus pneumoniae colonisation: the key to pneumococcal disease. Lancet Infect Dis 4:144–154
30. Simell B, Auranen K, Käyhty H et al (2012) The fundamental link between pneumococcal carriage and disease. Expert Rev Vaccines 11:841–855
31. Grijalva CG, Griffin MR, Edwards KM et al (2014) The role of Influenza and parainfluenza infections in nasopharyngeal pneumococcal acquisition among young children. Clin Infect Dis 58:1369–1376
32. De Lastours V, Malosh R, Ramadugu K et al (2016) Co-colonization by Streptococcus pneumoniae and Staphylococcus aureus in the throat during acute respiratory illnesses. Epidemiol Infect 22:1–13
33. McCullers JA, McAuley JL, Browall S, Iverson AR, Boyd KL, HenriquesNormark B (2010) Influenza enhances susceptibility to natural acquisition of and disease due to Streptococcus pneumoniae in ferrets. J Infect Dis 202:1287–1295
34. Fan RR, Howard LM, Griffin MR et al (2016) Nasopharyngeal pneumococcal density and evolution of acute respiratory illnesses in young children, Peru, 2009–2011. Emerg Infect Dis 22:1996–1999
35. Atarashi K, Tanoue T, Shima T et al (2011) Induction of colonic regulatory T cells by indigenous clostridium species. Science 331:337–341
36. Buffie CG, Pamer EG (2013) Microbiota-mediated colonization resistance against intestinal pathogens. Nat Rev Immunol 13:790
37. Candela M, Perna F, Carnevali P et al (2008) Interaction of probiotic Lactobacillus and Bifidobacterium strains with human intestinal epithelial cells: adhesion properties, competition against enteropathogens and modulation of IL-8 production. Int J Food Microbiol 125:286–292
38. Morens DM, Taubenberger JK, Fauci AS (2008) Predominant role of bacterial pneumonia as a cause of death in pandemic influenza: implications for pandemic influenza preparedness. J Infect Dis 198:962–970
39. Tedesco D, Thapa M, Chin CY, Ge Y, Gong M, Li J et al (2018) Alterations in intestinal microbiota lead to production of interleukin 17 by intrahepatic gamma delta T-cell receptor-positive cells and pathogenesis of cholestatic liver disease. Gastroenterology 154:2178–2193. https://doi.org/10.1053/j.gastro.2018.02.019
40. Yu H, Gagliani N, Ishigame H, Huber S, Zhu S, Esplugues E et al (2017) Intestinal type 1 regulatory T cells migrate to periphery to suppress diabetogenic T cells and prevent diabetes development. Proc Natl Acad Sci U S A. 114:10443–10448. https://doi.org/10.1073/pnas.1705599114

41. Zhao Q, Elson CO (2018) Adaptive immune education by gut microbiota antigens. Immunology 154:28–37. https://doi.org/10.1111/imm.12896
42. Sefik E, Geva-Zatorsky N, Oh S, Konnikova L, Zemmour D, McGuire AM et al (2015) Individual intestinal symbionts induce a distinct population of RORgamma (+) regulatory T cells. Science 349:993–997. https://doi.org/10.1126/science.aaa9420
43. Jakobsson HE, Abrahamsson TR, Jenmalm MC, Harris K, Quince C, Jernberg C et al (2014) Decreased gut microbiota diversity, delayed Bacteroidetes colonisation and reduced Th1 responses in infants delivered by caesarean section. Gut 63:559–566. https://doi.org/10.1136/gutjnl-2012-303249
44. Tanoue T, Atarashi K, Honda K (2016) Development and maintenance of intestinal regulatory T cells. Nat Rev Immunol 16:295–309. https://doi.org/10.1038/nri.2016.36
45. Round JL, Mazmanian SK (2010) Inducible Foxp3+ regulatory T-cell development by a commensal bacterium of the intestinal microbiota. Proc Natl Acad Sci U S A 107:12204–12209. https://doi.org/10.1073/pnas.0909122107
46. Wang ZY, Chai WD, Burwinkel M, Twardziok S, Wrede P, Palissa C et al (2013) Inhibitory influence of Enterococcus faecium on the propagation of swine influenza A virus in vitro. PLoS One 8:e53043. https://doi.org/10.1371/journal.pone.0053043
47. Bandoro C, Runstadler JA (2017) Bacterial lipopolysaccharide destabilizes influenza viruses. mSphere 2:17. https://doi.org/10.1128/mSphere.00267-17
48. Chen HW, Liu PF, Liu YT, Kuo S, Zhang XQ, Schooley RT et al (2016) Nasal commensal Staphylococcus epidermidis counteracts influenza virus. Sci Rep 6:27870. https://doi.org/10.1038/srep27870
49. Steed AL, Christophi GP, Kaiko GE, Sun LL, Goodwin VM, Jain U et al (2017) The microbial metabolite desaminotyrosine protects from influenza through type I interferon. Science 357:498–502. https://doi.org/10.1126/science.aam5336
50. Yitbarek A, Alkie T, Taha-Abdelaziz K, Astill J, Rodriguez-Lecompte JC, Parkinson J et al (2018) Gut microbiota modulates type I interferon and antibody mediated immune responses in chickens infected with influenza virus subtype H9N2. Benef Microbes 9:417–427. https://doi.org/10.3920/BM2017.0088
51. Ichinohe T, Pang IK, Kumamoto Y, Peaper DR, Ho JH, Murray TS et al (2011) Microbiota regulates immune defense against respiratory tract influenza A virus infection. Proc Natl Acad Sci U S A 108:5354–5359. https://doi.org/10.1073/pnas.1019378108
52. Rosshart SP, Vassallo BG, Angeletti D, Hutchinson DS, Morgan AP, Takeda K et al (2017) Wild mouse gut microbiota promotes host fitness and improves disease resistance. Cell 171:1015–1028. https://doi.org/10.1016/j.cell.2017.09.016
53. Na L, Wen-Tao M, Ming P, Qin-Lei F, Jin-Lian H (2019) The commensal microbiota and viral infection: a comprehensive review. Front Immunol 10:1–16
54. Minodier L, Masse S, Capai L, Blanchon T, Ceccaldi P-E, van der Werf S, Hanslik T, Charrel R, Falchi A (2017) Clinical and virological factors associated with gastrointestinal symptoms in patients with acute respiratory infection: a two year prospective study in general practice medicine. BMC Infect Dis 1:729
55. Wang J, Li F, Wei H, Lian ZX, Sun R, Tian Z (2011) Respiratory influenza virus infection induces intestinal immune injury via microbiota-mediated Th17 cell dependent inflammation. J Exp Med 12:2397–2410
56. Groves HT, Higham SL, Moffatt MF, Cox MJ, Tregoning JS (2020) Respiratory viral infection alters the gut microbiota by inducing inappetence. MBio 11(1):e03236–e03219
57. Sencio V et al (2020) Gut dysbiosis during influenza contributes to pulmonary pneumococcal superinfection through altered short-chain fatty acid production. Cell Rep 30(9):2934–2947
58. Leung RK-K, Zhou J-W, GuanW LS-K, Yang Z-F, Tsui SK-W (2013) Modulation of potential respiratory pathogens by pH1N1 viral infection. Clin Microbiol Infect 19:930–935. https://doi.org/10.1111/1469-0691.12054

59. Chaban B, Albert A, Links MG, Gardy J, Tang P, Hill JE (2013) Characterization of the upper respiratory tract microbiomes of patients with pandemic H1N1 influenza. PLoS One 8:e69559. https://doi.org/10.1371/journal.pone.0069559
60. Hanada S, Pirzadeh M, Carver KY, Deng JC (2018) Respiratory viral infection-induced microbiome alterations and secondary bacterial pneumonia. Front Immunol 9(2640):1–15
61. Greninger AL, Chen EC, Sittler T, Scheinerman A, Roubinian N, Yu G et al (2010) A metagenomic analysis of pandemic influenza A (2009 H1N1) infection in patients from North America. PLoS One 5:e13381. https://doi.org/10.1371/journal.pone.0013381
62. Vareille M, Kieninger E, Edwards MR, Regamey N (2011) The airway epithelium: soldier in the fight against respiratory viruses. Clin Microbiol Rev 24:210–229. https://doi.org/10.1128/CMR.00014-10
63. Killip MJ, Fodor E, Randall RE (2015) Influenza virus activation of the interferon system. Virus Res 209:11–22
64. Kaul D et al (2020) Microbiome disturbance and resilience dynamics of the upper respiratory tract during influenza A virus infection. Nat Commun 11:2537–2548. https://doi.org/10.1038/s41467-020-16429-9
65. O'Toole PW, Marchesi JR, Hill C (2017) Next-generation probiotics: the spectrum from probiotics to live biotherapeutics. Nat Microbiol 2:57
66. Zuo T, Wong SH, Lam K, Lui R, Cheung K, Tang W et al (2018) Bacteriophage transfer during faecal microbiota transplantation in Clostridium difficile infection is associated with treatment outcome. Gut 67:634–643. https://doi.org/10.1136/gutjnl-2017-313952
67. Suez J et al (2018) Post-antibiotic gut mucosal microbiome reconstitution is impaired by probiotics and improved by autologous FMT. Cell 174:1406–1423. https://doi.org/10.1016/j.cell.2018.08.047
68. Zmora N, Zilberman-Schapira G, Suez J, Mor U, Dori-Bachash M, Bashiardes S et al (2018) Personalized gut mucosal colonization resistance to empiric probiotics is associated with unique host and microbiome features. Cell 174:1388–1405. https://doi.org/10.1016/j.cell.2018.08.041

# Microbiome in Upper Respiratory Tract Infections

17

Piyush Mittal, Manjari Mittal, Ujjawal Rawat, and Ambika

**Abstract**

Infections of the respiratory tract are result of microbiome dysbiosis. Microbiome dysbiosis is regularly portrayed by a deficiency of useful bacteria, which secures host from the abundance of opportunistic pathogenic microbes. Microbiome occurs in various body areas such as GIT, skin, respiratory tract and occurs in symbiotic relationship with human micro-habitats. The nasal cavity, anterior nares, nasopharynx, sinuses, oral cavity, Eustachian tube, oropharynx, middle ear cavity, and larynx comprise the upper respiratory tract. The middle meatus, anterior nares, and nasopharynx are the favored destinations for sampling. The URT microbiome starts developing right after birth. Also, the type of microbiome formed depends on the type of delivery whether vaginal or cesarean. It also has role in protecting the host against the foreign substances and infectious agents. The microbiome of upper respiratory tract keeps on developing and altering with age. Multiple factors like cigarette smoking, environmental conditions, use of medications like corticosteroids, use of probiotics also have their effects on the microbiome. Probiotics and bovine colostrum have been proved to be of great importance in microbiome dysbiosis. Multiple diseases like chronic rhinosinusitis, otitis media with effusion, and acute otitis media are some diseases which have been discussed in this review related to dysbiosis of upper respiratory tract.

P. Mittal (✉) · U. Rawat · Ambika
Teerthanker Mahaveer College of Pharmacy, Teerthanker Mahaveer University, Moradabad, India

M. Mittal
MET Faculty of Pharmacy, Moradabad, India



G. Gupta et al. (eds.), *Microbiome in Inflammatory Lung Diseases*,
https://doi.org/10.1007/978-981-16-8957-4_17

**Keywords**

URTI · Microbiome · Microbiome dysbiosis · Rhinosinusitis

## 17.1 Introduction

An intricate group of microorganisms which coexists in a harmonious symbiotic relationship in human micro-habitats is referred to as the human microbiome. Microbiome encompasses all surfaces of human body. Human health is a result of complex interactions between human and its microbiome [1]. Microbial composition and functionality differ depending on the human body site such as GIT, skin, airways, owning this difference to microbiome particularity [2, 3]. The upper respiratory tract (URT) microbiome is a supreme component of others microbiome occurring in the human body. The URT performs a variety of important physiological functions, which includes filtering, warming, humidification of inhaled air, and also immediate pathogen detection by olfactory sensors [4–8]. The anatomy of the URTI along with certain factors lead to development of a particular microbiome of the URT such as sinuses, nasal cavity, oropharynx and nasopharynx microbiomes [4, 9, 10]. As a grown-up inhales around 7000 L of air on daily basis, the upper respiratory tract is continuously washed with air stream from the external environment. In the process of inhalation not only air but $10^4$–$10^6$ bacterial cells/cubic meter of air are breathed in on daily basis. Along with these natural particulates, the URT is also exposed to certain other factors like humidity, oxygen, nutrients, immunological factors, or chemicals. Like other microbiomes in humans, the microbiome of the upper RT starts developing right after birth. Later in life, as the child starts growing, the adult URT microbiome emerges from this first microbial population, which is thinner and more diverse. According to many studies, the phyla Firmicutes, Actinobacteria, Proteobacteria, and Bacteroidetes, representatives of genera *Corynebacterium, Bifidobacterium, Streptococcus, Staphylococcus, Moraxella,* and *Dolosigranulum* dominate the nasal microbiome [11–14].

## 17.2 Anatomy of Upper Respiratory Tract (URT)

The URT is comprised of the nasal cavity, nasopharynx, anterior nares, sinuses, Eustachian tube, oral cavity, middle ear cavity, oropharynx, and larynx. The nasal cavity is a link between the lower respiratory tract, the GI tract, and the external environment. it is lined by various sorts of epithelium, which gives distinctive micro niches. The three nasal turbinates further split the nasal cavity into the superior, middle, and inferior meatus [4, 15]. The non-keratinized skin-like epithelium of the anterior nares gives way to stratified squamous epithelial cells lacking microvilli, which is followed by transitional epithelium with small microvilli, which leads to the pseudostratified columnar epithelium of the center meatus [16]. Adjacent to the nasal vestibule, lies the middle meatus. It is a great area of interest for researchers as it collects drainage from the anterior ethmoids, frontal and maxillary sinuses [17]. In comparison to other URT regions, the surface in the nasal vestibule and anterior nares are somewhat dry and are most exposed to the external habitat and consist of vibrissae (hair) and sebaceous glands. Large particles (>3 μm) from inhaled air are trapped by these hairs, whereas little particulate (0.5–3 μm, including organisms) is caught by flowing mucus layer that covers the entire nasal cavity [16, 18]. The nasopharynx has many crypts and folds and the keratinized and non-keratinized stratified squamous epithelium, as well as pseudostratified ciliated epithelia dominates its wall [19]. Lying within the facial skeleton are the air filled, paired cavities—frontal, ethmoid, sphenoid, and maxillary sinuses. Their function is warming and humidification of the air inhaled by the individual. They are covered by ciliated columnar epithelium which is responsible for generating mucus that drains in the nasal cavity [20] and here a local microbiome with its specific microbiome is created [21]. Another fascinating specialty for microbiome is the olfactory region, which has a link with the nasal cavity [22]. It is situated at the ceiling of the nasal cavity. Though the research focused on bacterial species in the human nasal cavity, viruses, fungi, and archaea also populate the nasal cavity [23]. The nasal cavity contains a distinctive, highly diversified archaeal community in addition to bacterial and viral components. Archaea are microorganisms that are distinct from the bacteria because of their biology. They are significant segments of the human microbiome colonizing the oral cavity, skin, GI tract, and many other parts of body [24]. The archaeal microbiota of nasal cavity is similar to that of the skin and the gastrointestinal tract. Methanogenic Euryarchaeota (*Methanosphaera, Methanobrevibacter*) and also Thaumarchaeota (*Nitrososphaera*) are the predominating species. Out of all other body sites, the hotspot of the nasal cavity is the archaeal community with high archaeal 16S rRNA content [25].

## 17.3 Dysbiosis of Upper Respiratory Tract microbiome

The role of microbial populations in the URT on human health has long been debated. Blooming in the URT is the initial step for majority of respiratory microorganisms such as *Streptococcus pneumoniae, staphylococcus aureus,* and *klebsiella pneumoniae* to cause respiratory disease via dysbiosis of the respiratory microbiome [26, 27]. A lack of beneficial, commensal bacteria, which protects the host from opportunistic pathogenic germs, is a frequent problem of microbiome dysbiosis [28–30]. As previously discussed, being a connection between the external habitat and the lungs, the nasal cavity especially the anterior nares comes in contact with more than 7000 L of air which is inhaled [31]. Human health has been depicted as the consequence of complex interactions between humans and their microbiomes, and dysbiosis of the microbiome leads to a variety of diseases [32], for, e.g., upper respiratory tract infections, chronic rhinosinusitis as a result of URT dysbiosis [33–36]. As a result, in addition to the GI tract, the nasal cavity has been proposed as a major entry point for foreign particles such as pathogens, pollutants, pollens which causes dysbiosis of the nasal microbiome. Dysbiosis of the microbiome is thought to

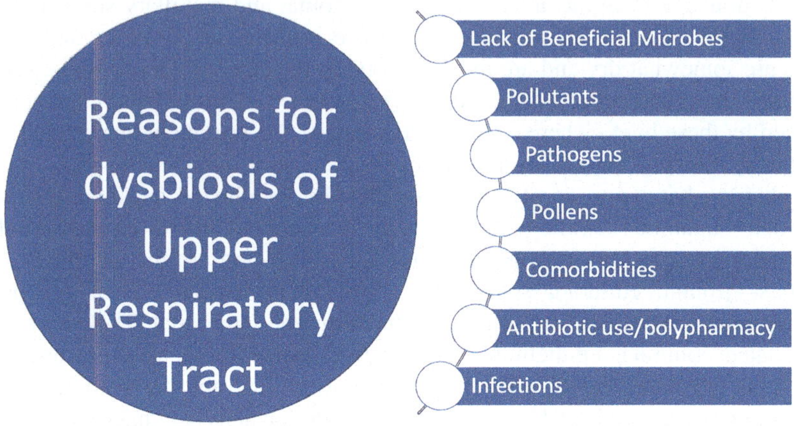


**Fig. 17.1** Reasons for dysbiosis of URT

be an important biomarker for human illness like chronic rhinosinusitis. As a result of dysbiosis, inflammatory URTIs develop for which a few distinct treatments have been proposed [37–40]. Intranasal corticosteroids and antibiotics which have anti-microbial and anti-inflammatory properties are used, microbial diversity is disoriented as a result of these treatments, which could lead to rise in gram negative bacteria in the nose [41–43]. Not only the affected microbiome but also in the nearby microbiome's alterations are evident [17]. Microorganisms that cause infections of the lower tract have been considered to enter through URT [44–46]. Studies depict that upper and lower respiratory tract pathogenesis are closely related. The lungs were perceived to be sterile traditionally but using recent molecular techniques it was concluded that even healthy lungs have bacterial microbiome, though at lower levels than the upper respiratory tract [45, 47, 48]. Microorganisms that cause infections of the lower tract have been considered to enter through URT. The reasons for dysbiosis are summarized in Fig. 17.1.

## 17.4 Analysis of Nasal Microbiome

For nasal microbiome analysis, the most commonly chosen sampling sites are the nasopharynx, the middle meatus (MM), and the anterior nares (AN) [1, 47, 48]. The nasopharynx, middle meatus, and anterior nares are the favored destinations for taking samples because other areas are difficult to reach [47–49]. From the research conducted by Hoping wang et al., it was derived that anterior nares and nasopharynx have similarity in their microbiome niche but oropharynx had different microbiome composition and characteristics. This difference of composition could be attributed to the factors such as food ingestion and esophageal reflux [50–52]. URT microbiome research deals with different methodological issues, including deciding the sample procedures (like swabs, nasal rinses, dry filter paper) and inspecting or sampling destinations [53]. This brings about error of exploration in research and study For, e.g., the center meatus is tested rather than the sinuses when chronic sinusitis is studied. Therefore, choosing the sampling site is an important part of the research.

## 17.5 Protection of Upper Respiratory Tract Microbiome

To protect the host against harmful inhaled microbes, the respiratory tract has innate and adaptive mechanisms. The mucus layer, the epithelium, nasopharyngeal associated lymphoid tissue (NALT) all have a role in URT defense [1]. The mucus layer released by goblet cells, glands, and ciliated cells has a secretory role in addition to humidification. It captures inhaled germs or microparticles. This mucus is subsequently evacuated from the nasal cavity via the esophagus [54]. This complete process is termed as mucociliary clearance [55, 56]. Also, the epithelium of the respiratory tract secretes antimicrobial components including lysozyme, lactoferrin, or defensins [57–61]. NALT associated cells downregulate the immune responses by excreting cytokines and chemokines [62–64]. The olfactory and trigeminal system plays its role in protection by sensing the foreign substances [65]. Therefore, through these mechanisms the host is protected from the inhaled harmful foreign substances including microbes.

## 17.6 Development of Upper Respiratory Tract Microbiome

The URT microbiome of an infant starts developing right after birth which with time upgrades into the adult URT microbiome. The microbiome in adult is comparatively more diverse and less thick as compared to a child. Within the elderly, the microbiome of particular microenvironment ends up being more similar and the diversity decreases [66]. Therefore, the URT microbiome keeps on changing with age and had different dominating species at different periods of life.

The type of delivery whether vaginal or cesarean is a major determinant in the colonization of URT and will decide which species will form the majority [67]. During the last few decades, the total number of cesarean sections (C-sections) deliveries has skyrocketed. It is a cause for concern because C-sections are linked to both short- and long-term respiratory illness [68].

### 17.6.1 URT Microbiome in Infants

The first nasopharyngeal assemblage occurs shortly after the birth and the nasopharyngeal microbiome of the infants look like the maternal vaginal or skin microbiome in case of cesarean delivery [69]. *Dolosigranulum*, *Streptococcus*, *Moraxella*, *Corynebacterium*, *Hemophilus*, and *Staphylococcus* are the six dominant genera of the nares and nasopharyngeal microbiome in infants, of which a couple typically overwhelms [70, 71]. Except *Moraxella catarrhalis*, which has been linked to wheezing in 1 month old infants, along with *S. pneumoniae* and *H. influenza*, children with Moraxella species dominated profile had protection from the URTI. Also, nasopharyngeal *Streptococcus* was found to be a solid indicator for asthma in around 2-month-old kids [72]. Breastfeeding during 1–5 months of age maintains the original microbiome makeup, resulting in consistent *Corynebacterium* and *Dolosigranulum* profiles. This is distinctive from infants fed with formula in whom *S. aureus* predominated. In the nares and nasopharynx of these 1.5-month-old babies, *Staphylococcus, Moraxella, Streptococcus, Corynebacterium*, and *Dolosigranulum* are the most common bacteria [73]. Overall, Dolosigranulum and Moraxella, coupled with Corynebacterium, constitute a more stable microbiome in children's first 2 years of life than the *Streptococcus* and *Haemophilus* dominating profiles, which have been linked to respiratory viruses and a high risk of bronchiolitis in infancy [74].

### 17.6.2 URT Microbiome in Adults

When it comes to children's microbiomes, it is more dense but less diverse, whereas on the other hand, adults have a less dense but more diverse microbiome [75]. When compared to infants, the URT microbiome of adults is quite similar but the niche characteristics are comparable. Actinobacteria, firmicutes, and small number of anaerobic bacteria are the dominating species in adult's anterior nares [75–77]. When compared to the anterior nares microbiome, the sphenoethmoidal recess (SR) and middle meatus (MM) have less proteobacteria, but the MM and SR have more firmicutes and actinobacteria, and the SR and MM microbiomes are likewise similar.

### 17.6.3 URT Microbiome of Elderly

The microbiota of the nasopharynx, tongue, buccal mucosa, oropharynx, and other sampling site in elderly widely varies from the anterior nares microbiota in adults (18–40 years). As a result of process of aging, changes in the microbiome of URT are observed and keep on shifting with age. Within the elderly, the microbiome of particular microenvironment ends up being more similar and the diversity decreases. In a middle-aged adult, i.e., age 40–65 years the majority is of *Staphylococcus, Corynebacterium,* and *Cutibacterium* in nasal cavity which changes to more of oropharyngeal population in the people aged above 65 years [78].

## 17.7 Effects of Smoking on Upper Respiratory Tract Microbiome

Exposure to cigarette smoke, whether direct or indirect is linked to increased risk of not only cardiovascular disease, periodontitis, malignancy but also putting patient to more risk of developing acute as well as chronic respiratory tract diseases [79]. Cigarette smoke comes into direct association with nasal surfaces and consequently has effect on the microbiome by oxygen hardship, antimicrobial action, and other mechanisms [80, 81]. Through multiple mechanisms the functionality of microbiome is disrupted. The mucociliary clearance is impaired in upper as well as lower respiratory tract infections by toxic agents produced by smoking, also bacterial colonization and attachment to airway endothelial are also enhanced via biofilm formation [53, 82, 83]. The URT of non-smoker harbors especially *Peptostreptococcus* species, alpha- hemolytic *Streptococci, Prevotella* species which appear to relate adversely with microbial presence. Also, on the one hand, the normal healthy microbiome is disrupted, whereas pathogenic bacterial (*H. Influenza, M. Catarrhalis, Campylobacter* species, *Streptococcus Pneumoniae, Streptococcus pyogenes*) growth increases in the other [84]. In comparison to non-smokers, URT microbiome of smokers was discovered to be more diverse but less robust. The probability of conveying gram-positive anaerobic ancestries (Eggerrthella, Erysipelotrichaceae, Dorea, Eubacterium) and Enterovirus species is expanded in the nasopharynx of the smokers, incorporating microbes related to URT contamination and endocarditis. Surprisingly, after a year of quitting smoking, the URT microbiome appears to rejuvenate and looks similar to microbiome of non-smoker. Smoking is not only injurious for grownups yet in addition for babies when they are presented to passive smoking. *S. pneumoniae* has been found to be elevated in babies with smoking parents, according to several research. Smoking parents' 2-year-old children are also more likely to have meningococcal meningitis, otitis media, and lower respiratory tract infections [1, 84].

## 17.8 Upper Respiratory Tract Infections (URTI)

### 17.8.1 Chronic Rhinosinusitis

In persistent rhinosinusitis, URT microbiome diversity and beneficial microbes are reduced. Chronic rhinosinusitis (CRS) is a frequent and long-lasting condition in which the human paranasal sinuses are inflamed for more than 3 months [85, 86]. Despite the fact that CRS is classified as an inflammatory rather than an infectious condition, it is suggested that it will be treated as such, the role of bacteria in the commencement and advancement of inflammation is important [87]. The conducted research depict that CRS is caused by involvement of multiple microbes [88]. Streptococcaceae, Staphylococceae, Pseudomonadaceae, and Corynebacterium were found to dominate in a study of CRS patients to explore the microbiome of the sinus. Corynebacterium and Staphylococcus Tuberculostearicum enrichment was discovered in the sinuses, as well as Pseudomonas, Staphylococcus, and

H. influenza were spotted [89, 90]. A decline in microbial richness, variety, and evenness, all of which are common hallmarks of other chronic inflammatory diseases, was observed in multiple investigations [91, 92]. These changes are attributed to the presence of large number of anaerobic bacteria flourishing in biofilms [93, 94]. Anaerobic genera which predominated in middle meatus of CRS patients included *Anaerococcus, Lactobacillus, Finegoldia,* and *Peptoniphilus*. Through their study, Hoggard et al. *Staphylococcus, Peptoniphilus, Finegoldia, Corynebacterium, Propionibacterium,* and *Anaerococcus,* which have been identified as typical health-related URT bacteria, have been found to be exhausted in CRS patients [91]. This depletion of healthy microbial community results in more serious inflammatory response as well as clinical severity [95]. In the middle meatus, Copeland et al. observed a negative connection between CRS disease state and six OTUs from the genera *Dolosigranulum, Corynebacterium,* and *Staphylococcus*.

CRS is divided into two types: CRS without nasal polyps (CRPsNP) and CRS with nasal polyps (CRPsNP) (CRPwNP). Inflammation causes nasal polyps, which are fleshy swellings. Comorbidities like asthma and aspirin intolerance are particularly common in CRSwNP patients [96]. Both subtypes' microbiomes blooming in the inferior and middle meatus were studied. *Alloiococcus, Staphylococcus,* and *Corynebacterium* species were present in CRSwNP samples while on the other hand, CRSsNP samples had anaerobes such as *Fusobacteria, Streptococcus,* and *Hemophilus* species. Because sinus cavities are generally not anaerobic, presence of anaerobic microbiota is a hallmark of disease progression and pathology [97]. Another intriguing feature is that CRS patient's responses to taste molecules are modified. The responsiveness towards bitter taste decreases, whereas increases towards sweet molecules [98]. The most usual treatment for CRS is nasal washes, corticosteroids, and sinus surgery which ultimately leads to dysbiosis of the URT microbiome.

### 17.8.2 Otitis Media

Existence of fluid in the middle ears having related sign and symptoms is called acute otitis media. Majority of children are affected in first 3 years of life and out of this, 50% cases become recurrent [99]. When 4 or more acute otitis media episodes occur in 12 months or more than 3 episodes in 6 months, it is defined as recurrent acute otitis media [100, 101]. Laufer et al. performed the earliest studies in which he composed a microbiome in children suffering from AOM (acute otitis media) to healthy children. He found out that the higher abundance of *Dolosigranulum* and *Corynebacterium* along with *Staphylococcus, Lactobacillus,* and *Propionibacterium* leads to low incidence of AOM [102, 103]. These findings of bacterial colonization during URTI are consistent with the findings of Hilty et al., who claimed that the NP bacterial density is lower in children with AOM episodes than in healthy children [104]. By taking 139 neonates Chonmaitree in 2017 collected 971 swabs performed monthly, since birth of the first to 12 months of life. He looked at the characteristics of the NP microbiota as the patient progressed from URTI to AOM, as URTI frequently precedes AOM. Data revealed that the domination of otopathogens leads to progression of URTI to AOM along with symptomatic viral infection [105]. On the Australian population, a complicated investigation with a larger sample size was done. To uncover putative defense species, the NP microbiota of 103 healthy children and 93 otitis-prone children were compared. The importance of *Dolosigranulum* and *Corynebacterium* in the NP microbiota in healthy children versus otitis-prone children is demonstrated in this study. The author next compared the MEF microbiome (middle ear fluid) taken from children having RAOM surgery to the NP microbiome of the same individual. In comparison to the NP microbiome, *alloiococcus* and *turicella* were shown to be more numerous in MEF [106]. Seventy-nine subjects aged 5–42 months were investigated further by taking MEF microbiota during an AOM event. According to reports, otopathogens dominated in MEF during AOM [107]. Spontaneous tympanic membrane perforation is a major complication of AOM [108]. However, research suggests that STMP is deficient in children who have had RAOM in the past [109].

#### 17.8.2.1 Otitis Media with Effusion

Otitis media with effusion (OME) is defined as the presence of middle ear fluid with signs or symptoms, as well as an acute infection. It could last for more than 3 months [99]. Liu et al. in his first study looked at the microbiota of the adenoid, tonsils, and middle ear of an 8-year-old child with chronic middle ear effusion. He discovered that the *Pseudomonadaceae* microbiota dominated the middle ear, while the *Streptococcaceae* microbiota dominated the tonsil. Adenoid microbiota was more complex as it includes *Streptococcaceae, Pseudomonadaceae, Fusobacteriaceae,* and *Pasteurellaceae.* He concluded that adenoid could act as wellspring of all those pathogens [110].

Fago-olsen conducted an experiment by investigating palatine tonsil microbiota and adenoids of children undergoing adenoid hyperplasia vs subject having SOM and likely to be treated with surgery. He concluded that palatine tonsil acts as wellspring but not adenoid. However, we must keep in mind that the microbiota of MEFs was not examined in this work [111].

A study of ten children following adenotonsillectomy and grommet implantation for OME revealed concerns regarding the difference between the NP and MEF microbiota. He discovered that the tonsil and adenoid microbiotas were very similar, and that the adenoid and ME microbiotas were also very similar [112].

Ani et al. later verified this finding in a wider group of children with OME. *Alloiococcus otitidis* (44%), *turicella otitidis* (6%), and *staphylococcus auricularis* are all found in the ME microbiome (30%). *Granulicatella, Staphylococcus*, and *Rothia* make up the adenoid microbiome.

In a Chinese hospital, researchers compared ME and adenoid microbiota from children receiving surgery for OME and adenoid hypertrophy to subjects of adenoid microbiota undergoing adenotonsillectomy for OSA without ear illness. Researchers discovered that *Haemophillus* (14.75%), *Staphylococcus* (9.37%), and *Halomonas* (7.85%) are the most common bacteria in ME, with low levels of *Alloiococcus otitidis* (3.75%) and *turicella*. In the OME group, 4 infections were identified to be in abundance between ME and adenoid (*Streptococcus, Neisseria, Alloprevotella, Actinobacillus*). Adenoid microbiota was dominated with *Haemophillus* (15.96%), *Streptococcus* (13.37%), and *Moraxella* (12.28%). Researchers concluded that difference in microbial composition between these two challenges the PRH in OME [113]. Hemophilus is another major otopathogen in OME, as discovered by examining the ME microbiota of 55 children with chronic middle ear effusion. Hemophilus was found to be prominent in children with hearing loss, and it was linked to MEF, which had MUC5B and MUC5A, implying a link. The existence of the NP microbiota is rare in children with OME, according to two case control studies [114, 115].

## 17.9 Methods for Microbiome Redevelopment

### 17.9.1 Bovine Colostrum

As compared to human colostrum, the bovine colostrum (BC) has many folds higher immunoglobulins. It is hundred times higher in bovine milk [116–118]. Additionally, the bovine milk has multiple components responsible for development of acquired and innate immunity [119, 120]. BC has been investigated for its potential as a passive immunotherapeutic agent to combat several infections. BC was proven to be viable in the anticipation of URTI in a study conducted by Ahmad Alsayed et al., and it was suggested that BC should be suggested as a therapeutic option for persons with recurrent RTI. The nasal swab microbiota was largely impacted by BC [121].

### 17.9.2 Probiotic Therapy

In many respiratory infections, microbial dysbiosis is a result of increased pathogens and decrease in healthy microbiome. Probiotics (live beneficial bacteria) seems to be a beneficial option and is intended to provide health advantages to the host [1, 122, 123]. Also, it was helpful in decreasing symptoms of URTI in obese as well as aged patients. Through multiple mechanisms, probiotics can act as a pioneer in microbiome dysbiosis [124]. It can also be used as a supplement in daily life.

## 17.10 Conclusion

This review provides information regarding different factors which have role in microbiome dysbiosis. An attempt was made to summarize the aspect of microbiome in maintaining the health and also the outcomes of microbiome dysbiosis on the host. Also, the upper respiratory tract infections like pharyngitis, chronic rhinosinusitis can be an etiology for microbiome dysbiosis. Upper respiratory tract microbiome can alter the nearby microbiomes. The upper respiratory tract infections can also cause otitis media. Continuous researches are going on to find the accurate relation between microbiome of an individual and the corresponding health. Also, studies are going on to find a method to rejuvenate the disrupted microbiomes like probiotic therapy and the bovine colostrum.

## References

1. Kumpitsch C, Koskinen K, Schöpf V et al (2019) The microbiome of the upper respiratory tract in health and disease. BMC Biol 17:87. https://doi.org/10.1186/s12915-019-0703-z
2. Costello EK, Lauber CL, Hamady M, Fierer N, Gordon JI, Knight R (2009) Bacterial community variation in human body habitats across space and time. Science. 326:1694–1697. https://doi.org/10.1126/science.1177486
3. Lloyd-Price J, Mahurkar A, Rahnavard G, Crabtree J, Orvis J, Hall AB et al (2017) Strains, functions and dynamics in the expanded Human Microbiome Project. Nature. 550:61. https://doi.org/10.1038/nature23889
4. de SteenhuijsenPiters WAA, Sanders EAM, Bogaert D (2015) The role of the local microbial ecosystem in respiratory health and disease. Philos Trans R Soc B Biol Sci. 370:20140294
5. Jones N (2001) The nose and paranasal sinuses physiology and anatomy. Adv Drug Deliv Rev. 51:5–19. https://doi.org/10.1016/S0169-409X(01)00172-7
6. Jimenez PN, Koch G, Thompson JA, Xavier KB, Cool RH, Quax WJ (2012) The multiple signaling systems regulating virulence in *Pseudomonas aeruginosa*. Am Soc Microbiol. 76: 46–65
7. Li Z, Nair SK (2012) Quorum sensing: how bacteria can coordinate activity and synchronize their response to external signals? Protein Sci. 21:1403–1417
8. Patel NN, Workman AD, Cohen NA (2018) Role of taste receptors as sentinels of innate immunity in the upper airway. J Pathog. 2018:9541987
9. Dickson R, Erb-Downward J, Martinez F, Huffnagle G (2017) The microbiome and the respiratory tract. Annu Rev Physiol 78:481–504. HHS Public Access
10. Lighthart B (2000) Mini-review of the concentration variations found in the alfresco atmospheric bacterial populations. https://link.springer.com/content/pdf/10.1023%2FA%3A1007694618888.pdf. Accessed 22 Oct 2018.
11. Whelan FJ, Verschoor CP, Stearns JC, Rossi L, Luinstra K, Loeb M et al (2014) The loss of topography in the microbial communities of the upper respiratory tract in the elderly. Ann Am Thorac Soc. 11:513–521
12. Bassis CM, Tang AL, Young VB, Pynnonen MA (2014) The nasal cavity microbiota of healthy adults. Microbiome. 2:27
13. Shilts MH, Rosas-Salazar C, Tovchigrechko A, Larkin EK, Torralba M, Akopov A et al (2016) Minimally invasive sampling method identifies differences in taxonomic richness of nasal microbiomes in young infants associated with mode of delivery. Microb Ecol. 71:233–242
14. Stearns JC, Davidson CJ, Mckeon S, Whelan FJ, Fontes ME, Schryvers AB et al (2015) Culture and molecular-based profiles show shifts in bacterial communities of the upper respiratory tract that occur with age. ISME J. 9:1246–1259. https://doi.org/10.1038/ismej.2014.250
15. Sahin-Yilmaz A, Naclerio RM (2011) Anatomy and physiology of the upper airway. Proc Am Thorac Soc. 8:31–39. https://doi.org/10.1513/pats.201007-050RN
16. Geurkink N (1983) Nasal anatomy, physiology, and function. J Allergy Clin Immunol. 72: 123–128
17. Yan M, Pamp SJ, Fukuyama J, Hwang PH, Cho D-Y, Holmes S et al (2013) Nasal microenvironments and interspecific interactions influence nasal microbiota complexity and *S. aureus* carriage. Cell Host Microbe. 14:631–640

18. Cohen N (2006) Sinonasal mucociliary clearance in health and disease. Ann Otol Rhinol Laryngol Suppl. 196:20–26. www.ncbi.nlm.nih.gov/pubmed/17040014
19. Ali M (1965) Histology of the human nasopharyngeal mucosa. J Anat. 99:657–672. https://europepmc.org/backend/ptpmcrender.fcgi?accid=PMC1270703&blobtype=pdf. Accessed 5 Nov 2018
20. Bell GW, Joshi BB, Macleod RI (2011) Maxillary sinus disease: diagnosis and treatment. Br Dent J. 210:113–118. https://doi.org/10.1038/sj.bdj.2011.47
21. Proctor DM, Relman DA, Section D, Alto P (2018) The landscape ecology and microbiota of the human nose, mouth and throat. Cell Hot Microbe. 21:421–432
22. Koskinen K, Reichert JL, Hoier S, Schachenreiter J, Duller S, Moissl-Eichinger C et al (2018) The nasal microbiome mirrors and potentially shapes olfactory function. Sci Rep. 8:1–11
23. Koskinen K, Pausan MR, Perras AK, Bang MBC, Mora M, Schilhabel A et al (2017) First insights into the diverse human archaeome: Specific detection of archaea in the gastrointestinal tract. MBio. 8:1–17
24. Mahnert A, Blohs M, Pausan M-R, Moissl-Eichinger C (2018) The human archaeome: methodological pitfalls and knowledge gaps. Emerg Top Life Sci. 2:469. https://doi.org/10.1042/ETLS20180037
25. Pausan MR, Csorba C, Singer G, Till H, Schoepf V, Santigli E et al (2018) Measuring the archaeome: detection and quantification of archaea signatures in the human body. bioRxiv. 334748. https://doi.org/10.1101/334748.
26. Marik PE, Kaplan D (2003) Aspiration pneumonia and dysphagia in the elderly. Chest. 124(1): 328–336. https://doi.org/10.1378/chest.124.1.328
27. Bogaert D, De Groot R, Hermans PW. *Streptococcus pneumoniae* colonisation: the key to pneumococcal disease. The Lancet infectious diseases. 2004; 4(3):144–154. Epub 2004/03/05. doi: https://doi.org/10.1016/S1473-3099(04)00938-7. PMID: 14998500.
28. Copeland E, Leonard K, Carney R, Kong J, Forer M, Naidoo Y et al (2018) Chronic rhinosinusitis: Potential role of microbial dysbiosis and recommendations for sampling sites. Front Cell Infect Microbiol. 8:57. https://doi.org/10.3389/Fcimb.2018.00057
29. Petersen C, Round JL (2014) Microreview Defining dysbiosis and its influence on host immunity and disease. Cell Microbiol. 16:1024–1033
30. Hoggard M, Waldvogel-Thurlow S, Zoing M, Chang K, Radcliff FJ, Wagner Mackenzie B et al (2018) Inflammatory endotypes and microbial associations in chronic rhinosinusitis. Front Immunol. 9:2065. https://doi.org/10.3389/Fimmu.2018.02065
31. Van der Schans CP (2007) Bronchial mucus transport. Respir Care. 52:1150–1156.; discussion 1156–8. https://doi.org/10.1016/0952-8180(93)90100-S
32. Vayssier-Taussat M, Albina E, Citti C, Cosson J-F, Jacques M-A, Lebrun M-H et al (2014) Shifting the paradigm from pathogens to pathobiome: new concepts in the light of meta-omics. Front Cell Infect Microbiol. 4:29. https://doi.org/10.3389/fcimb.2014.00029
33. Abreu NA, Nagalingam NA, Song Y, Roediger FC, Pletcher SD, Goldberg AN et al (2012) Sinus microbiome diversity depletion and *Corynebacterium tuberculostearicum* enrichment mediates rhinosinusitis. Sci Transl Med. 4:151ra124. https://doi.org/10.1126/scitranslmed.3003783
34. Coburn B, Wang PW, Diaz Caballero J, Clark ST, Brahma V, Donaldson S et al (2015) Lung microbiota across age and disease stage in cystic fibrosis. Sci Rep. 5:10241. https://doi.org/10.1038/srep10241
35. Gevers D, Kugathasan S, Denson LA, Vázquez-Baeza Y, Van Treuren W, Ren B et al (2014) The treatment-naive microbiome in new-onset Crohn's disease. Cell Host Microbe. 15:382–392. https://doi.org/10.1016/j.Chom.2014.02.005
36. Hartstra AV, Bouter KEC, Bäckhed F, Nieuwdorp M (2015) Insights into the role of the microbiome in obesity and type 2 diabetes. Diabetes Care. 38:159–165. https://doi.org/10.2337/dc14-0769
37. Bernstein JA (2013) Characterizing rhinitis subtypes. Am J Rhinol Allergy. 27:457–460. https://doi.org/10.2500/ajra.2013.27.3983

38. Ginat DT (2015) Posttreatment imaging of the paranasal sinuses following endoscopic sinus surgery. Neuroimaging Clin N Am. 25:653–665. https://doi.org/10.1016/J.NIC.2015.07.008
39. Principi N, Esposito S (2017) Nasal irrigation: an imprecisely defined medical procedure. Int J Environ Res Public Health. 14. https://doi.org/10.3390/ijerph14050516
40. Ramakrishnan VR, Holt J, Nelson LF, Ir D, Robertson CE, Frank DN (2018) Determinants of the nasal microbiome: pilot study of effects of intranasal medication use. Allergy Rhinol (Providence). 9:2152656718789519. https://doi.org/10.1177/2152656718789519
41. Feazel LM, Robertson CE, Ramakrishnan VR, Frank DN (2012) Microbiome complexity and *Staphylococcus aureus* in chronic rhinosinusitis. Laryngoscope. 122:467–472. https://doi.org/10.1002/lary.22398
42. Prevaes SMPJ, De Winter-De Groot KM, Janssens HM, De SteenhuijsenPiters WAA, Tramper-Stranders GA, Wyllie AL et al (2016) Development of the nasopharyngeal microbiota in infants with cystic fibrosis. Am J Respir Crit Care Med. 193:504–515
43. Teo SM, Mok D, Pham K, Kusel M, Serralha M, Troy N et al (2015) The infant nasopharyngeal microbiome impacts severity of lower respiratory infection and risk of asthma development. Cell Host Microbe. 17:704–715. https://doi.org/10.1016/j.chom.2015.03.008
44. Man WH, de SteenhuijsenPiters WA, Bogaert D. The microbiota of the respiratory tract: gatekeeper to respiratory health. Nat Rev Microbiol. 2017; 15(5):259–270. Epub 2017/03/21. doi: https://doi.org/10.1038/nrmicro.2017.14. PMID: 28316330.
45. Dickson RP, Huffnagle GB (2015) The lung microbiome: new principles for respiratory bacteriology in health and disease. PLoS Pathogens. 11(7):e1004923. https://doi.org/10.1371/journal.ppat.1004923
46. Cookson W, Cox MJ, Moffatt MF (2018) New opportunities for managing acute and chronic lung infections. Nat Rev Microbiol 16(2):111–120. Epub 2017/10/25. https://doi.org/10.1038/nrmicro.2017.122
47. Lal D, Keim P, Delisle J, Barker B, Rank MA, Chia N et al (2017) Mapping and comparing bacterial microbiota in the sinonasal cavity of healthy, allergic rhinitis, and chronic rhinosinusitis subjects. Int Forum Allergy Rhinol. 7:561–569. https://doi.org/10.1002/alr.21934
48. Zhou Y, Mihindukulasuriya KA, Gao H, La Rosa PS, Wylie KM, Martin JC et al (2014) Exploration of bacterial community classes in major human habitats. Genome Biol. 15:R66
49. Luna PN, Hasegawa K, Ajami NJ, Espinola JA, Henke DM, Petrosino JF et al (2018) The association between anterior nares and nasopharyngeal microbiota in infants hospitalized for bronchiolitis. Microbiome. 6:1–14
50. Wang H, Dai W, Feng X, Zhou Q, Wang H, Yang Y, Li S, Zheng Y (2018) Microbiota composition in upper respiratory tracts of healthy children in Shenzhen, China, differed with respiratory sites and ages. Biomed Res Int. 2018:6515670. https://doi.org/10.1155/2018/6515670
51. Charlson ES, Bittinger K, Haas AR et al (2011) Topographical continuity of bacterial populations in the healthy human respiratory tract. Am J Respir Crit Care Med 184(8):957–963
52. Bassis CM, Erb-Downward JR, Dickson RP et al (2015) Analysis of the upper respiratory tract microbiotas as the source of the lung and gastric microbiotas in healthy individuals. MBio 6(2):e00037–e00015
53. Garmendia J, Morey P, Bengoechea JA (2012) Impact of cigarette smoke exposure on host-bacterial pathogen interactions. Eur Respir J. 39:467–477
54. Casado B, Pannell LK, Iadarola P, Baraniuk JN (2005) Identification of human nasal mucous proteins using proteomics. Proteomics. 5:2949–2959
55. Wanner A, Salathé M, O'Riordan T (1996) Mucociliary clearance in the airways. Am J Respir Crit Care Med. 154:1868–1902
56. Schenck LP, Surette MG, Bowdish DME (2016) Composition and immunological significance of the upper respiratory tract microbiota. FEBS Lett. 590:3705–3720

57. Ganz T (2002) Antimicrobial polypeptides in host defense of the respiratory tract. J Clin Invest. 109:693–697
58. Sanchez L, Calvo M, Brock JH (1992) Biological role of lactoferrin. Arch Dis Child. 67:657–661
59. Klebanoff SJ (2005) Myeloperoxidase: friend and foe. J Leukoc Biol. 77:598–625
60. Nauseef WM (2004) Assembly of the phagocyte NADPH oxidase. Histochem Cell Biol. 122: 277–291
61. Thomas EL, Aune TM (1978) Lactoperoxidase, peroxide, thiocyanate antimicrobial system: correlation of sulfhydryl oxidation with antimicrobial action. Infect Immun. 20:456–463
62. Patel NN, Kohanski MA, Maina IW, Triantafillou V, Workman AD, Tong CCL et al (2018) Solitary chemosensory cells producing interleukin-25 and group-2 innate lymphoid cells are enriched in chronic rhinosinusitis with nasal polyps. Int Forum Allergy Rhinol. 8:900–906
63. Kato A, Schleimer RP (2007) Beyond inflammation: airway epithelial cells are at the interface of innate and adaptive immunity. Curr Opin Immunol. 19:711–720
64. Liao B, Cao P, Zeng M, Zhen Z, Wang H, Zhang Y et al (2015) Interaction of thymic stromal lymphopoietin, IL-33, and their receptors in epithelial cells in eosinophilic chronic rhinosinusitis with nasal polyps. Eur J Allergy Clin Immunol. 70:1169–1180
65. Finger TE, Böttinger B, Hansen A, Anderson KT, Alimohammadi H, Silver WL (2003) Solitary chemoreceptor cells in the nasal cavity serve as sentinels of respiration. Proc Natl Acad Sci U S A. 100:8981–8986
66. de SteenhuijsenPiters WAA, Huijskens EGW, Wyllie AL, Biesbroek G, Van Den Bergh MR, Veenhoven RH et al (2016) Dysbiosis of upper respiratory tract microbiota in elderly pneumonia patients. ISME J. 10:97–108
67. Dominguez-Bello MG, Costello EK, Contreras M, Magris M, Hidalgo G, Fierer N et al (2010) Delivery mode shapes the acquisition and structure of the initial microbiota across multiple body habitats in newborns. Proc Natl Acad Sci U S A. 107:11971–11975. https://doi.org/10.1073/pnas.1002601107
68. Dominguez-Bello MG, Costello EK, Contreras M, Magris M, Hidalgo G, Fierer N, Knight R (2010) Delivery mode shapes the acquisition and structure of the initial microbiota across multiple body habitats in newborns. Proc Natl Acad Sci U S A 107(26):11971–11975. https://doi.org/10.1073/pnas.1002601107. Epub 2010 Jun 21. PMID: 20566857; PMCID: PMC2900693
69. Biesbroek G, Tsivtsivadze E, Sanders EAM, Montijn R, Veenhoven RH, Keijser BJF et al (2014) Early respiratory microbiota composition determines bacterial succession patterns and respiratory health in children. Am J Respir Crit Care Med. 190:1283–1292
70. Frayman KB, Armstrong DS, Grimwood K, Ranganathan SC (2017) The airway microbiota in early cystic fibrosis lung disease. Pediatr Pulmonol. 52:1384–1404
71. Teo SM, Mok D, Pham K, Kusel M, Serralha M, Troy N et al (2015) The infant airway microbiome in health and disease impacts later asthma development. Cell Host Microbe. 17: 704–715. https://doi.org/10.1016/j.chom.2015.03.008
72. Von Linstow ML, Schønning K, Hoegh AM, Sevelsted A, Vissing NH, Bisgaard H (2013) Neonatal airway colonization is associated with troublesome lung symptoms in infants. Am J Respir Crit Care Med. 188:1041–1042
73. Biesbroek G, Bosch AATM, Wang X, Keijser BJF, Veenhoven RH, Sanders EAM et al (2014) The impact of breastfeeding on nasopharyngeal microbial communities in infants. Am J Respir Crit Care Med. 190:298–308
74. Vissing NH, Chawes BLK, Bisgaard H (2013) Increased risk of pneumonia and bronchiolitis after bacterial colonization of the airways as neonates. Am J Respir Crit Care Med. 188:1246–1252
75. Camarinha-Silva A, Jáuregui R, Chaves-Moreno D, Oxley APA, Schaumburg F, Becker K et al (2014) Comparing the anterior nare bacterial community of two discrete human populations using Illumina amplicon sequencing. Environ Microbiol. 16:2939–2952

76. Zhou Y, Gao H, Mihindukulasuriya KA, Rosa PSL, Wylie KM, Vishnivetskaya T et al (2013) Biogeography of the ecosystems of the healthy human body. Genome Biol. 14:R1. https://doi.org/10.1186/gb-2013-14-1-r1
77. Oh J, Byrd AL, Deming C, Conlan S, Program NCS, Kong HH et al (2014) Biogeography and individuality shape function in the human skin metagenome. Nature. 514:59–64. https://doi.org/10.1038/nature13786
78. Franceschi C, Bonafè M, Valensin S, Olivieri F, De Luca M, Ottaviani E et al (2006) Inflamm-aging: an evolutionary perspective on immunosenescence. Ann N Y Acad Sci. 908:244–254. https://doi.org/10.1111/j.1749-6632.2000.Tb06651.x
79. Stämpfli MR, Anderson GP (2009) How cigarette smoke skews immune responses to promote infection, lung disease and cancer. Nat Rev Immunol. 9:377EP
80. Macgregor I (1989) Effects of smoking on oral ecology. A review of the literature. Clin Prev Dent. 11:3–7
81. Yu G, Phillips S, Gail MH, Goedert JJ, Humphrys MS, Ravel J et al (2017) The effect of cigarette smoking on the oral and nasal microbiota. Microbiome. 5:1–6. https://doi.org/10.1186/s40168-016-0226-6
82. Bagaitkar J, Demuth DR, Daep CA, Renaud DE, Pierce DL, Scott DA (2010) Tobacco upregulates *P. gingivalis*fimbrial proteins which induce TLR2 hyposensitivity. PLoS One. 5: e9323
83. Goldstein-Daruech N, Cope EK, Zhao K-Q, Vukovic K, Kofonow JM, Doghramji L et al (2011) Tobacco smoke mediated induction of sinonasal microbial biofilms. PLoS One. 6: e15700
84. Brook I, Gober AE (2005) Recovery of potential pathogens in the nasopharynx of healthy and otitis media-prone children and their smoking and nonsmoking parents. Ann Otol Rhinol Laryngol. 117:727–730
85. Fokkens WJ, Lund VJ, Mullol J, Bachert C, Alobid I, Baroody F et al (2012) EPOS 2012: European position paper on rhinosinusitis and nasal polyps 2012. A summary for otorhinolaryngologists. Rhinol J. 50:1–12. https://doi.org/10.4193/Rhino50E2
86. Foreman A, Jervis-Bardy J, Wormald P-J (2011) Do biofilms contribute to the initiation and recalcitrance of chronic rhinosinusitis? Laryngoscope. 121:1085–1091. https://doi.org/10.1002/lary.21438
87. Mahdavinia M, Keshavarzian A, Tobin MC, Landay A, Schleimer RP (2016) A comprehensive review of the nasal microbiome in chronic rhinosinusitis (CRS). Clin Exp Allergy. 46:21–41. https://doi.org/10.1111/cea.12666
88. Cope EK, Goldberg AN, Pletcher SD, Lynch SV (2017) Compositionally and functionally distinct sinus microbiota in chronic rhinosinusitis patients have immunological and clinically divergent consequences. Microbiome. 5:53. https://doi.org/10.1186/s40168-017-0266-6
89. Aurora R, Chatterjee D, Hentzleman J, Prasad G, Sindwani R, Sanford T (2013) Contrasting the microbiomes from healthy volunteers and patients with chronic rhinosinusitis. JAMA Otolaryngol Neck Surg. 139:1328. https://doi.org/10.1001/jamaoto.2013.5465
90. Chalermwatanachai T, Vilchez-Vargas R, Holtappels G, Lacoere T, Jáuregui R, Kerckhof FM, Pieper DH, Van de Wiele T, Vaneechoutte M, Van Zele T, Bachert C (2018) Chronic rhinosinusitis with nasal polyps·is characterized by dysbacteriosis of the nasal microbiota. Sci Rep 8(1):7926. https://doi.org/10.1038/s41598-018-26327-2. PMID: 29784985; PMCID: PMC5962583
91. Wagner Mackenzie B, Waite DW, Hoggard M, Douglas RG, Taylor MW, Biswas K (2017) Bacterial community collapse: a meta-analysis of the sinonasal microbiota in chronic rhinosinusitis. Environ Microbiol. 19:381–392. https://doi.org/10.1111/1462-2920.13632
92. Psaltis AJ, Wormald P-J (2017) Therapy of sinonasal microbiome in CRS: a critical approach. Curr Allergy Asthma Rep. 17:59. https://doi.org/10.1007/s11882-017-0726-x
93. Dlugaszewska J, Leszczynska M, Lenkowski M, Tatarska A, Pastusiak T, Szyfter W (2016) The pathophysiological role of bacterial biofilms in chronic sinusitis. Eur Arch Otorhinolaryngol. 273:1989–1994. https://doi.org/10.1007/s00405015-3650-5

94. Stephenson M-F, Mfuna L, Dowd SE, Wolcott RD, Barbeau J, Poisson M et al (2010) Molecular characterization of the polymicrobial flora in chronic rhinosinusitis. J Otolaryngol Head Neck Surg. 39:182–187
95. Hirschberg A, Kiss M, Kadocsa E, Polyanka H, Szabo K, Razga Z et al (2016) Different activations of toll-like receptors and antimicrobial peptides in chronic rhinosinusitis with or without nasal polyposis. Eur Arch Otorhinolaryngol. 273:1779–1788. https://doi.org/10.1007/s00405-015-3816-1
96. Chalermwatanachai T, Vilchez-Vargas R, Holtappels G, Lacoere T, Jáuregui R, Kerckhof F-M et al (2018) Chronic rhinosinusitis with nasal polyps is characterized by dysbacteriosis of the nasal microbiota. Sci Rep. 8:7926. https://doi.org/10.1038/s41598-018-26327-2
97. Brook I (2006) The role of anaerobic bacteria in sinusitis. Anaerobe. 12:5–12. https://doi.org/10.1016/J.ANAEROBE.2005.08.002
98. Naraghi M, Deroee AF, Ebrahimkhani M, Kiani S, Dehpour A (2007) Nitric oxide: a new concept in chronic sinusitis pathogenesis B. Am J Otolaryngol. 28:334–337
99. Schilder AG, Chonmaitree T, Cripps AW, Rosenfeld RM, Casselbrant ML, Haggard MP, Venekamp RP (2016) Otitis media. Nat. Rev. Dis. Primers. 2:16063
100. Teele DW, Klein JO, Rosner B (1989) Epidemiology of otitis media during the first seven years of life in children in greater Boston: a prospective, cohort study. J. Infect. Dis. 160:83–94
101. Dirain CO, Silva RC, Collins WO, Antonelli PJ (2017) The Adenoid Microbiome in Recurrent Acute Otitis Media and Obstructive Sleep Apnea. J Int Adv Otol 13:333–339
102. Laufer AS, Metlay JP, Gent JF, Fennie K, Kong Y, Pettigrew MM (2011) Microbial communities of the upper respiratory tract and otitis media in children. mBio 2:e00245–e00310
103. Bomar L, Brugger SD, Yost BH, Davies SS, Lemon KP (2016) Corynebacterium accolens releases antipneumococcal free fatty acids from human nostril and skin surface triacylglycerols. mBio 7:01725
104. Hilty M, Qi W, Brugger SD, Frei L, Agyeman P, Frey PM, Aebi S, Mühlemann K (2012) Nasopharyngeal microbiota in infants with acute otitis media. J Infect Dis. 205:1048–1055
105. Chonmaitree T, Jennings K, Golovko G, Khanipov K, Pimenova M, Patel JA, McCormick DP, Loeffelholz MJ, Fofanov Y (2017) Nasopharyngeal microbiota in infants and changes during viral upper respiratory tract infection and acute otitis media. PLoS ONE 12:e0180630
106. Lappan R, Imbrogno K, Sikazwe C, Anderson D, Mok D, Coates H, Vijayasekaran S, Bumbak P, Blyth C, Jamieson SE et al (2018) A microbiome case-control study of recurrent acute otitis media identified potentially protective bacterial genera. BMC Microbiol. 18:13
107. Sillanpää S, Kramna L, Oikarinen S, Sipilä M, Rautiainen M, Aittoniemi J, Laranne J, Hyöty H, Cinek O (2017) Next-generation sequencing combined with specific PCR assays to determine the bacterial 16S rRNA gene profiles of middle ear fluid collected from children with acute otitis media. mSphere 2:00006–00017
108. Berger G (1989) Nature of spontaneous tympanic membrane perforation in acute otitis media in children. J Laryngol Otol 103:1150–1153
109. Torretta S, Marchisio P (2017) Otitis media in children: a proposal for a new nosological classification. Int J Pediatr Otorhinolaryngol. 93:174–175
110. Liu CM, Cosetti MK, Aziz M, Buchhagen JL, Contente-Cuomo TL, Price LB, Keim P, Lalwani AK (2011) The otologic microbiome a study of the bacterial microbiota in a pediatric patient with chronic serous otitis media using 16SrRNA gene-based pyrosequencing. Arch Otolaryngol Head Neck Surg. 137:664–668
111. Fagö-Olsen H, Dines LM, Sørensen CH, Jensen A (2019) The adenoids but not the palatine tonsils serve as a reservoir for bacteria associated with secretory otitis media in small. mSystems 4:e00169–e00218
112. Johnston J, Hoggard M, Biswas K, Astudillo-García C, Radcliff FJ, Mahadevan M, Douglas RG (2019) Pathogen reservoir hypothesis investigated by analyses of the adenotonsillar and middle ear microbiota. Int. J. Pediatr. Otorhinolaryngol. 118:103–109

113. Xu J, Dai W, Liang Q, Ren D (2020) The microbiomes of adenoid and middle ear in children with otitis media with effusion and hypertrophy from a tertiary hospital in China. Int. J. Pediatr. Otorhinolaryngol. 134:110058
114. Kim SK, Hong SJ, Pak KH, Hong SM (2019) Analysis of the microbiome in the adenoids of Korean children with otitis media with effusion. J. Int. Adv. Otol. 15:379–385
115. Walker RE, Walker CG, Camargo CA Jr, Bartley J, Flint D, Thompson JMD, Mitchell EA (2019) Nasal microbial composition and chronic otitis media with effusion: a case-control study. PLoS ONE 14:e0212473
116. Steele J, Sponseller J, Schmidt D, Cohen O, Tzipori S (2013) Hyperimmune bovine colostrum for treatment of GI infections: a review and update on *Clostridium difficile*. Hum. Vaccin Immunother. 9(7):1565–1568
117. Patıröglu T, Kondolot M (2013) The effect of bovine colostrum on viral upper respiratory tract infections in children with immunoglobulin A deficiency. Clin. Respir. J. 7(1):21–26
118. Ballard O, Morrow AL (2013) Human milk composition: nutrients and bioactive factors. Pediatr Clin. 60(1):49–74
119. Patel K, Rana R (2006) Pedimune in recurrent respiratory infection and diarrhoea—the Indian experience—the pride study. Indian J. Pediatr. 73(7):585–591
120. Rathe M, Müller K, Sangild PT, Husby S (2014) Clinical applications of bovine colostrum therapy: a systematic review. Nutr. Rev. 72(4):237–254
121. Alsayed A, Al-Doori A, Al-Dulaimi A, Alnaseri A, Abuhashish J, Aliasin K, Alfayoumi I (2020) Influences of bovine colostrum on nasal swab microbiome and viral upper respiratory tract infections – A case report. Respir Med Case Rep. 31:101189. https://doi.org/10.1016/j.rmcr.2020.101189. Epub 2020 Aug 19
122. Georas SN, Rezaee F (2014) Epithelial barrier function: at the frontline of asthma immunology and allergic airway inflammation. J Allergy Clin Immunol. 134:509–520
123. Soyka MB, Wawrzyniak P, Eiwegger T, Holzmann D, Treis A, Wanke K et al (2012) Defective epithelial barrier in chronic rhinosinusitis: the regulation of tight junctions by IFN-g and IL-4. Am Acad Allergy Asthma Immunol. 130:1087–1096
124. Mullish BH, Marchesi JR, McDonald JAK, Pass DA, Masetti G, Michael DR, Plummer S, Jack AA, Davies TS, Hughes TR, Wang D (2021) Probiotics reduce self-reported symptoms of upper respiratory tract infection in overweight and obese adults: should we be considering probiotics during viral pandemics? Gut Microbes. 13(1):1–9. https://doi.org/10.1080/19490976.2021.1900997

# Challenges in Understanding the Lung Microbiota 18

Olorunfemi R. Molehin, Olusola O. Elekofehinti, Adeniyi S. Ohunayo, and Oluwatosin A. Adetuyi


## Abstract

The human lung has been less explored from the microbiological point of view because of the low concentration of microorganisms in this area, physiological and anatomic barriers leading to sampling, isolation, and estimation difficulties. The latest research has found that the lung is not sterile contrary to some belief that the lung is a sterile environment although the respiratory tract is limiting in nutrients for the survival of microorganisms. However lack of microbial diversity has been studied to be associated with disease progression, it is clear that lifestyle such as smoking and underlying disease such as COPD has declined the microbial diversity of lungs. The sampling technique of the lung includes bronchoalveolar lavage (BAL), which is a very technical process that needs to be done with care for precision and accuracy. Many times, sputum has been used but often provides misleading and insufficient data due to contamination of pharyngeal microbiota. Quantity and type of DNA extraction protocol including their analysis are part of



O. R. Molehin (✉)
Department of Biochemistry, Faculty of Science, Ekiti State University, Ado-Ekiti, Nigeria
e-mail: olorunfemi.molehin@eksu.edu.ng

O. O. Elekofehinti
Bioinformatics and Molecular Biology Unit, Department of Biochemistry, Federal University of Technology, Akure, Nigeria
e-mail: ooelekofehinti@futa.edu.ng

A. S. Ohunayo
Department of Science Laboratory Technology, Faculty of Science, Ekiti State University, Ado-Ekiti, Nigeria
e-mail: adeniyi.ohunayo@eksu.edu.ng

O. A. Adetuyi
Department of Biochemistry, Osun State University, Osogbo, Nigeria
e-mail: oluwatosin.adetuyi@uniosun.edu.ng





G. Gupta et al. (eds.), *Microbiome in Inflammatory Lung Diseases*,
https://doi.org/10.1007/978-981-16-8957-4_18

the challenges facing the study of lung microbiome. Uncontrolled use of antimicrobials and lifestyle has contributed to the problem facing ease of studying lung microbiome.

**Keywords**

Microbiome · Lung · Bronchoalveolar lavage · COPD · Interleukin

## 18.1 Introduction

Some microorganisms have been studied to live peacefully with man. The role each group of microbes play in a particular niche cannot be under emphasized. Every particular habitat is unique in terms of microbes found in them, because of their diversity, the inability of some of them to be cultured on various media and reproducible results, it is important for studies to be genetic based. The microbiome is a term used to describe the totality of the genome of microorganisms living in or on a particular tissue or host; however, their presence could either be helpful or harmful. The estimation and evaluation of such microorganisms in their natural environments are sometimes difficult not because of their growth pattern or lack of analyzing devices but site or location of the natural environment to be explored [1].

Apparently, the estimation and determination of the roles of microorganism in a particular habitat are an interesting area of science, as interesting as it is, there are limiting factor faced in the evaluation and determination of specific functions of microbes in such an environment. However, colonization of the respiratory tract most importantly the lungs by microorganisms like bacteria, fungi, viruses may not necessarily give rise to a cytopathological effect as studies as shown that the presence of non-pathogenic microbes has contributed greatly to the general wellbeing of the body [2]. Colonization of the lungs by microorganisms is of lower microbial load but with higher diversity compared to other tissues and organs in the body; hence the microbial population of the lungs is of lower bioburden when compared to many other sites of the body, for example, the human digestive and genital tracts have more microbial population. The microbial load of the lungs has been an indication of pathological condition which leads to loss of diversity with increased colonization of some bacteria general reducing the concentration of some generals. The lung microbiome is difficult to study and characterize by conventional isolation and culturing techniques.

However, lifestyle, underlying disease, and misuse of antibiotics have been some of the problems in studying the lung microbiome [3]. Research has shown that the lung microbiome and its variations may influence the onset of diseases of the respiratory system therefore the biology of the lung microbiome has been able to differentiate between the non-pathogenic, pathogenic, opportunistic, and commensals. Metagenomics, 16S rRNA sequencing, and other tools such as DNA hybridization techniques have successful methods in studying lung microbiota. Such analyses are based on sample collection such could be sputum or bronchioalveolar

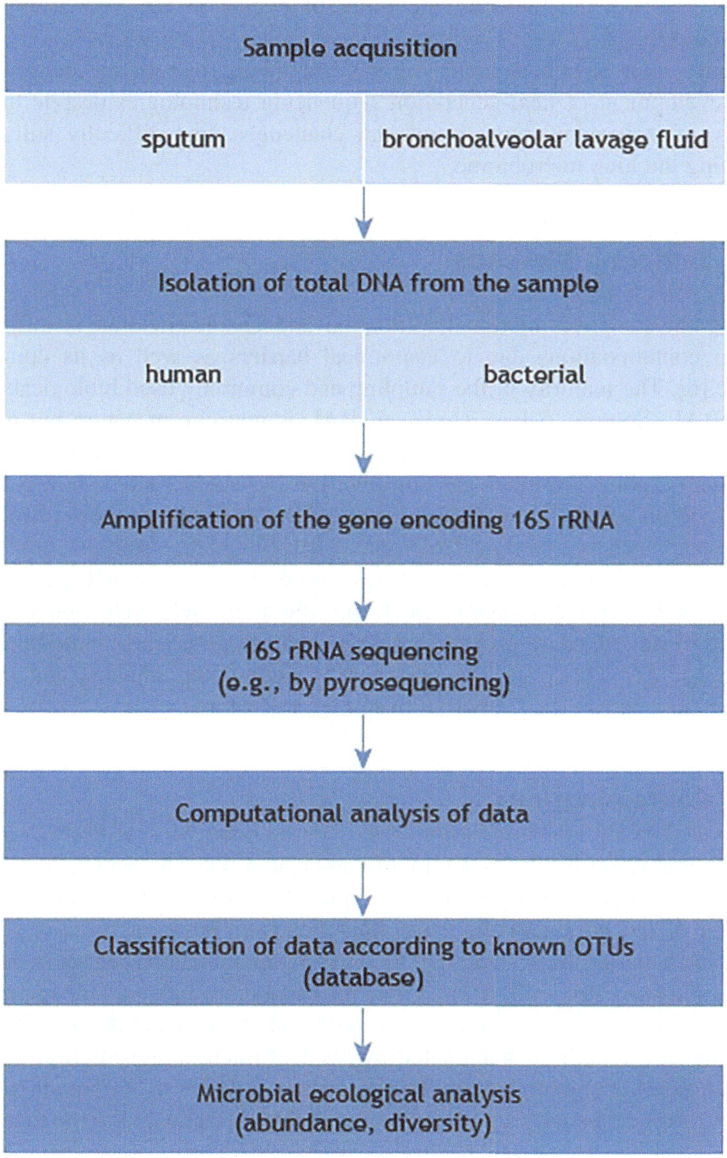


**Fig. 18.1** Flow chart of showing isolation and determination of lung microbiota [5]

lavage fluids followed by the isolation of the interested DNA (Bacteria, Fungi, Viruses, etc.) from the sample. Research on lung microbiome is influenced by changes with age, diet, lifestyle, and antibiotics among others [4]. The figure below is a flow chart showing the order of events in the estimation of lung microbiota (Fig. 18.1).

- Recently, there have been increased interests in lung microbiome studies through the development of next-generation sequencing technologies despite the availability of instrumentations, significant challenges, and difficulty still exist in studying the lung microbiome.

## 18.2 Sampling Methods

Sampling the lung for microbial estimation and characterization is difficult and prone to contaminations due to anatomical barriers as well as its complicated structure [6]. The majority of the sampling and commonly used biological samples include BAL, Sputum, Silver. However, BAL is invasive in nature but it reflects microbiome of the lung more precisely and accurately if only handled with care. On the contrary, sputum samples are non-invasive but studies have shown that they often provide inappropriate data as a result of contaminating microbes from neighboring tissues such as pharyngeal microbes. Microbiota of the sputum contains more genera when compared with BAL samples. Some of the genera include *Streptococcus*, *Veillonella*, and *Prevotella* which are about 2% relatively abundant when compared to either BAL or protected brush samples from the same individual having lung disease [7]. However, the presence of underlying ailments and depth of coughing can contribute to variation in sputum microbiota.

## 18.3 Contaminations

DNA contamination has been commonly associated with personnel, consumables, equipment, all which could be direct or indirect. However, due to the anatomical placement of the lung, contaminating microbes from adjacent or closely situated tissues can challenge the accuracy of the result. Contaminations can set in during the insertion and harvesting of BAL by bronchial catheters which pass the upper airway before assessing the lower lung. If this protocol is not carefully monitored and carried out, they can be contaminated by resident microorganisms from the upper airway or external contamination. The bacteria population of the mouth and other members of the oral cavity are often 2–3 log higher as compared to the lungs [8, 9].

## 18.4 Microbial Diversity

In time memorial studies, the diversity of microorganisms have been done solely through culturing. In the process, media (selective and differential) with appropriate temperature conditions have helped recover a tremendous range of organisms [10]. However, scientists noticed a lack of correlation in the estimation of no. of organisms that could be evaluated microscopically [11]. This problem leads to the

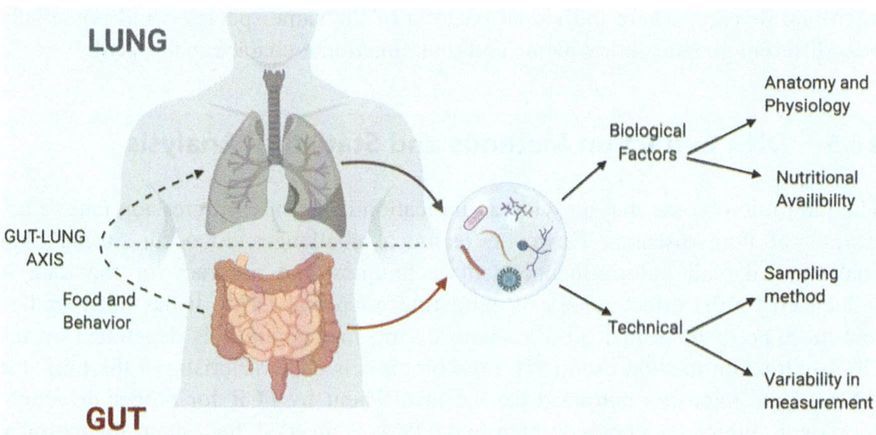


**Fig. 18.2** Biological and technical factors that contribute to the diversity of results in studies of the gut and lung microbiome [12]

inception of culture-independent techniques that are majorly based on the analysis of 16S rRNA genes. Despite the introduction of metagenomic, the analysis of the lung microbiome is still very much challenged due to limited nutritional availability. These features have led to fewer microbiota of the lung, with 103–106/g compared to the gut with 1011–1012/g (Fig. 18.2).

The microbial load of the lungs does not have the same concentration compared with other tracts and, therefore, the microbial diversity is influenced by many factors which include [13];

- Microbial proliferation as a result of micro-aspiration and inhalation of various microorganisms and direct mucous-dispersion.
- Microbial load reduction through coughing, mucociliary clearance, innate and adaptive immune responses.
- Bioavailability of nutrients and growth parameters.

The next-generation sequencing has shared more light to the study of microbial diversity of biological samples; low concentrated microbiome is particularly challenging because of their low density and they require the development of new scientific approaches. Low-biomass samples such as the human lungs are characterized by low density which in turn gives rise to low quantities of DNA. Low-biomass samples are prone to contaminations which affect the taxonomical analysis [14]. However, "noise" compromises legitimate bacterial signal originating from a biological specimen, makes it challenging to convert biological meaning from sequencing data [15]. Moreover, high microbial diversity is built on high no. of individuals for each single species capable of quick proliferation. The ability to proliferate quickly increases the diversification of microbes and expands the microbial density, where individual bacteria of the same species could potentially bear different genetic endowments and thus functional characteristics [16].

## 18.5 DNA Extraction Methods and Statistical Analysis

The lung microbiome may provide an indication of disease progression and pathogenesis of lung diseases. However, technical challenges do occur during their analysis. Like all culture-independent techniques, the accuracy of the data is influenced by the effectiveness of extracting microbial DNA. It has been studied that the type of microbial profile obtain during metagenomic is dependent on the DNA extraction method used [17]. Low biomass is a characteristic of the lung, the DNA of the microbes extracted can be insufficient by PCR for proper detection. Moreover, high concentrated microbial DNA is needed for better taxonomical estimation. It is important to develop more reliable and quality-based DNA extracting protocols from the lungs [18]. Although studies on low microbial biomass such as lungs are more prone to technical baises due to being overwhelmed by contaminants from backgroung DNA and do not routinely analyze negative controls from DNA extractions, even reporting statistically noteworthy taxa that overlap those observed in −ve controls. Hence, microbial contaminants in low microbial biomass samples can alter the relative abundance of the microbial communities under analysis which is often overlooked [17].

## 18.6 Use of Antibiotics

Antibiotics were quickly recognized as the most effective and life-saving medications for combating infectious infections, resulting in a significant reduction in morbidity and mortality. Antibiotic resistance emerged as a result of the careless and unthinking usage, misuse, and overuse of antibiotics (AR). AR is a severe health problem that poses a global threat to contemporary medicine and its achievements [19]. In addition, recent research has shed light on the potential influence of antibiotic use on the microbiome of the intestine. Antibiotics have been shown to reduce the diversity of microbiota in both adults and children [20]. The most common symptoms treated with antibiotics at drug stores in Ethiopia, Vietnam, China, India, Europe were respiratory tract problems, diarrhea, and wounds [21].

Irrational antibiotic use has been shown to induce the alteration of the lung microbiome and leads to the onset of pulmonary disorders. In a healthy rat lung, the predominant microorganisms are the proteobacteria and are followed closely by a lower percentage of some phyla such as Firmicutes, Bacteroidetes as well as Actinobacteria. The lungs are at a high risk of exposure to microorganisms from the environment, as well as the upper airway via microaspirations.

In a study done by Finn et al. [22], the effect of the usage of the antibiotic Levofloxacin on the lung microbiota of laboratory rats was investigated. The antibiotic understudy possessed a strong activity against most of the commensal bacteria in a healthy rat lung. The authors observed that while the lungs of untreated animals demonstrated a mixed bacteria flora, majorly belonging to the genus *Serratia*, the lungs of animals treated with the antibiotic comprised majorly of bacteria of the genus *Pantoea*.

Finn et al.[22] thus hypothesized that irrational usage of antibiotics affects the ecology of the microbiota via a reduction in bacteria diversity, further complementing the reports of Barfod et al. [23] that Vancomycin possesses the ability to preferentially disrupt murine lung microbiota.

## 18.7 The Life Style of the Respective Individual

However, in an era where bacteria are rapidly being identified as agents that aid the onset of chronic disorders such as malignancies and neurodegenerative disorders, of great importance is the need for understanding the negative effect and consequences of smoking on the microbiota in diseased conditions. Microbiome is a population of microorganisms that live in a specific habitat and includes bacteria, viruses, fungus, and protozoa, as well as their genes and genomes. The most complicated echo-system is the gut microbiome, with 10–100 trillion microbial populations, the majority of which are bacteria, followed by fungi and viruses [24]. Oral communities are the second type of community in the body of a human [25].

With the advent of the Human Microbiome Project, researchers have been able to know the constituents of the human microbiota employing methods that are culture-independent combined with next-generation DNA sequencing methods [25]. Based on standard culture methods, the once-dubbed sterile lung was discovered to contain a variety of microbiomes that differed in health and diseased situations [26]. Although the human microbiome is stable and has the tenacity to recover following disruption, its constituents are sensitive to various factors which include antibiotics, food, alcohol, as well as smoking. It is now obvious that the bacteria is not just a bystander in many clinical processes; its alteration is frequently a contributory factor or causative factor in pathophysiological processes [27]. Smoking influences the lung microbiome through changes in immunological balance, biofilm development, oxygen tension, as well as direct interaction with the bacteria it contains [26].

Highlighted below are some of the implications of smoking and how they influence the lung microbiota. For example, tobacco has negatively affected the human health and has been studied to cause some pathophysiologic changes that might lead to disease. Such have been implicated and linked with the presence of chemicals most of which are composed of free radicals, transition metals, pollutants, reactive oxygen species, and many other ingredients of tobacco forming a complex mixture of carcinogenic and toxigenic potentials [28]. *Pantoea agglomerans, Acinetobacter calcoaceticus*, and specific *Pseudomonadaceae* species such as *P. fluorescens* and *Stenotrophomonas maltophilia* were identified in fresh tobacco leaves, many other species were cultured from tobacco flakes and from particles of tobacco particles before the introduction of DNA sequencing techniques [28].

In the European Union, 15 different types of bacteria were found in cigarettes. Sapkota et al., discovered a wide range of bacteria in cigarettes, the commonly implicated includes *Pseudomonas aeruginosa, Acinetobacter, Clostridium, Klebsiella, Bacillus, Burkholderia*, as well as some genera from the soil and human commensals. As a result, smokers of cigarettes and tobacco may have other means of bacterial acquisition and colonization through this lifestyle. This maybe a process that leads to varied bacteria profiles of the lungs among smokers. Another reason by which addicted smokers may have a distinct bacterial load and population could be connected to decreased host cell defenses as a result of tobacco's immunosuppressive nature. Tobacco smoking has been shown to have a number of devastating effects on the peripheral immune response, including a decrease in natural killer cell activity and an increased susceptibility to infection.

Smoking changes macrophage and neutrophil function by increasing number of macrophages, neutrophils, eosinophils, and mast cells while reducing the number of airway dendritic cells [29]. Reduced bacterial-stimulated production of SO and SRE (e.g., TLR2) that is an integral expression that identifies and terminates many microbial structures, were found to elicit the immune response indicating defected phagocytosis process which hinders the clearance of antigen posed by bacteria [30]. As a result, smoking-related immunosuppression may allow for the colonization of new microorganisms. It is also likely that in a smoky atmosphere, certain taxa gain metabolic benefits such as biofilm development and enhanced epithelial adhesion. Cigarette smoke may promote the production of biofilms by certain bacteria [31]. The term "microenvironment" may also apply to the impact of smoking on specific microbiota members, such as dissolved oxygen concentration and acid production potentials. Because of the decreased dissolves oxygen demand, obligate anaerobes and microaerophilic are able to dominate the bacterial community [32]. Zhang et al. [33] concluded in their study that smoking affects the microbial diversity as well as microbial population of the LRT, indicating the need for studies of inflammation-induced via smoking to consider lung microbiota variation.

## 18.8 Biofilm

Biofilm is a complex community of microbial cells built through self-assembling of polymer matrix that protect and help them evade host defenses and antibiotics, allowing them to survive longer [34]. Aggregated biofilm complexity observed among *Streptococcus pneumoniae* and despite the antimicrobial potency of some cigarette ingredients other forms of smokes having pneumolysin have been found to condensate and it is expected to encourage propagation and attachment of microbial cells, both of which are critical antecedents of pneumococcal illness. According to Mutepe et al., similarly reported that smoking enhances *Staphylococcus aureus* biofilm formation and human cell adhesion. Smoke from cigarette has been proved to have bioactive effects on smoker's local microbiota predisposing them to respiratory infection [34]. This study has proven that cigarette smoke may increase biofilm formation through adaptation, attachment, and colonization of specific bacterial group in the human lungs.

## 18.9 Conclusion

The new era of scientific research through metagenomics will not only focus on soil, gut but also lung and every other sacred environments previously understudied. The human lung has been less studied due to it anatomical barrier. The position of the lung in the human body has made direct sampling very difficult. Despite the recent advancement in metagenomics, the lack of microbial density and loss of microbial diversity due to some underlying ailments are constantly challenging the ease of studying lung microbiome. It is important to increase the focus to lung microbiome just like how gut microbiota is currently studied. Finally, the knowledge on lung microbiome will not only give us an insight on the microbial population but will also create an awareness on the role of each individual genera is playing in this complex niche.

## References

1. Marsland BJ, Gollwitzer ES (2014) Host-microorganism interactions in lung diseases. Nat Rev Immunol 14(12):827–835. https://doi.org/10.1038/nri3769
2. Rogers GB, Shaw D, Marsh RL, Carroll MP, Serisier DJ, Bruce KD (2015) Respiratory microbiota addressing clinical questions, informing clinical practice. Thorax 70(1):74–81
3. Surette MG (2014) The cystic fibrosis lung microbiome. Ann Am Thorac Soc 11(1):S61–S65. https://doi.org/10.1513/AnnalsATS.201306-159MG
4. Dickson RP, Erb-Downward JR, Freeman CM, McCloskey L, Beck JM, Huffnagle GB, Curtis JL (2015) Spatial variation in the healthy human lung microbiome and the adapted island model of lung biogeography. Ann Am Thorac Soc 12:821–830
5. André NC, Felipe M, Silvia VC, Roberta KS, Rodrigo AA (2018) The pulmonary microbiome: challenges of a new paradigm. J Bras Pneumol 44(5):424–432. https://doi.org/10.1590/S1806-37562017000000209
6. Cox MJ, Ege MJ, von Mutius E (2019) The lung microbiome. European Respiratory Society, Lausanne. ISBN(electronic):978-1-84984-102-3
7. Hogan DA, Willger SD, Dolben EL, Hampton TH, Stanton BA, Morrison HG, Sogin ML, Czum J, Ashare A (2016) Analysis of lung microbiota in bronchoalveolar lavage, protected brush and sputum samples from subjects with mild-to-moderate cystic fibrosis lung disease. PLoS One 11:e0149998
8. Bassis CM, Erb-Downward JR, Dickson RP, Freeman CM, Schmidt TM, Young VB, Beck JM, Curtis JL, Huffnagle GB (2015) Analysis of the upper respiratory tract microbiotas as the source of the lung and gastric microbiotas in healthy individuals. MBio 6:e00037–e00015
9. Wu BG, Segal LN (2017) Lung microbiota and its impact on the mucosal immune phenotype. Microbiol Spectr. 5:BAD0005. https://doi.org/10.1128/microbiolspec.BAD-0005-2016
10. Lagier JC, Armougom F, Million M, Hugon P, Pagnier I, Robert C (2012) Microbial culturomics: paradigm shift in the human gut microbiome study. Clin Microbiol Infect 18: 1185–1193. https://doi.org/10.1111/1469-0691.12023

11. Staley JT, Konopka A (1985) Measurement of in situ activities of nonphotosynthetic microorganisms in aquatic and terrestrial habitats. Annu Rev Microbiol 39:321–346. https://doi.org/10.1146/annurev.mi.39.100185.001541
12. De Chang MD, Charles S, Dela Cruz MD (2019) Challenges in understanding lung microbiome: it is NOT like the gut microbiome. Respirology 25(3):244–245
13. Luigi S, Ioannis AC, Andrea B, Francesco I, Paolo L, Emanuele D, Skender T (2020) The human respiratory system and its microbiome at a glimpse. Biology 9(318):1–16. https://doi.org/10.3390/biology9100318
14. Salter SJ, Cox MJ, Turek EM, Calus ST, Cookson WO, Moffatt MF et al (2014) Reagent and laboratory contamination can critically impact sequence-based microbiome analyses. BMC Biol 12(1):87. https://doi.org/10.1186/s12915-014-0087-z
15. The Lancet Infectious Diseases (2018) Microbiome studies and "blue whales in the Himalayas". Lancet Infect Dis 18(9):925. https://doi.org/10.1016/S1473-3099(18)30503-6
16. Robbins RJ, Krishtalka L, Wooley JC (2016) Advances in biodiversity: metagenomics and the unveiling of biological dark matter. Stand Genomic Sci 11:69. https://doi.org/10.1186/s40793-016-0180-8
17. Pérez-Brocal V, Magne F, Ruiz-Ruiz S (2020) Optimized DNA extraction and purification method for characterization of bacterial and fungal communities in lung tissue samples. Sci Rep 10:17377. https://doi.org/10.1038/s41598-020-74137-2
18. Thomsen PF, Willerslev E (2015) Environmental DNA—an emerging tool in conservation for monitoring past and present biodiversity. Biol Conserv 183:4–18
19. Smith RA, M'ikanatha NM, Read AF (2015) Antibiotic resistance: a primer and call to action. Health Commun 30:309–314. https://doi.org/10.1080/10410236.2014.943634
20. Zaura E, Brandt BW, de Mattos MJT (2015) Same exposure but two radically different responses to antibiotics: resilience of the salivary microbiome versus long-term microbial shifts in feces. MBio 6. https://doi.org/10.1128/mBio.01693-15
21. Mao W, Vu H, Xie Z (2015) Systematic review on irrational use of medicines in China and Vietnam. PLoS One 10:1–16. https://doi.org/10.1371/journal.pone.0117710
22. Finn SMB, Scheuermann U, Holzknecht ZE et al (2019) The effect of levofloxacin on the lung microbiota of laboratory rats. Exp Lung Res 45:200–208. https://doi.org/10.1080/01902148.2019.1639225
23. Barfod KK, Vrankx K, Mirsepasi-Lauridsen HC et al (2015) The murine lung microbiome changes during lung inflammation and intranasal vancomycin treatment. Open Microbiol J 9(1): 167–179. https://doi.org/10.2174/1874285801509010167
24. Gill SR, Pop M, DeBoy RT et al (2006) Metagenomic analysis of the human distal gut microbiome. Science 312:1355–1359. https://doi.org/10.1126/science.1124234
25. Proctor LM (2011) The human microbiome project in 2011 and beyond. Cell Host Microbe 10: 287–291. https://doi.org/10.1016/j.chom.2011.10.001
26. Huang C, Shi G (2019) Smoking and microbiome in oral, airway, gut and some systemic diseases. J Transl Med 17:1–15. https://doi.org/10.1186/s12967-019-1971-7
27. Owyang C, Wu GD (2014) The gut microbiome in health and disease. Gastroenterology 146: 1433–1436. https://doi.org/10.1053/j.gastro.2014.03.032
28. Larsson L, Szponar B, Ridha B et al (2008) Identification of bacterial and fungal components in tobacco and tobacco smoke. Tob Induc Dis 4:4. https://doi.org/10.1186/1617-9625-4-4
29. Mehta H, Nazzal K, Sadikot RT (2008) Cigarette smoking and innate immunity. Inflamm Res 57:497–503. https://doi.org/10.1007/s00011-008-8078-6
30. Droemann D, Goldmann T, Tiedje T et al (2005) Toll-like receptor 2 expression is decreased on alveolar macrophages in cigarette smokers and COPD patients. Respir Res 6. https://doi.org/10.1186/1465-9921-6-68

31. Kulkarni R, Antala S, Wang A et al (2012) Cigarette smoke increases *Staphylococcus aureus* biofilm formation via oxidative stress. Infect Immun 80:3804–3811. https://doi.org/10.1128/IAI.00689-12
32. Mason MR, Preshaw PM, Nagaraja HN et al (2015) The subgingival microbiome of clinically healthy current and never smokers. ISME J 9:268–272. https://doi.org/10.1038/ismej.2014.114
33. Zhang R, Chen L, Cao L et al (2018) Effects of smoking on the lower respiratory tract microbiome in mice. Respir Res 19:253. https://doi.org/10.1186/s12931-018-0959-9
34. Oggioni MR, Trappetti C, Kadioglu A et al (2006) Switch from planktonic to sessile life: a major event in pneumococcal pathogenesis. Mol Microbiol 61:1196–1210. https://doi.org/10.1111/j.1365-2958.2006.05310.x

# Microbiome in Inflammatory Lung Diseases: Challenges and Future Prospects

# 19

Nitin Verma, Komal Thapa, and Kamal Dua

**Abstract**

The human microbiome is broadly recognized to have a crucial role in various bodily mechanisms that impact immunological homeostasis, inflammation, and metabolic activity. A recent study has revealed about the presence of commensal microbiome in mucosal surface of human body. The gut microbiome has extensively researched its role in controlling host metabolism or alterations in various disease states. There is a knowledge gap about the bacteria that live on mucosal surfaces. Unfortunately, there is currently a scarcity of scientific data on the involvement of lung microbiota in pulmonary illnesses. Previously lungs were considered as a sterile organ, and lung illnesses were usually accompanied mostly by bacterial pathogenesis. Emerging research suggests microbiomes in respiratory tracts both the upper and lower along with their importance in respiratory diseases. The current discusses the possibility of microbial disturbance about several lung illnesses in this book chapter. (e.g., Asthma, COPD, Pneumonia, Viral infection). Such disorders may be related to metabolic and biochemical stress. The book chapter also aims to emphasize the microbiome's function in lung illnesses using data from classical microbiology and microbiome literature.

N. Verma (✉) · K. Thapa
Chitkara University School of Pharmacy, Chitkara University, Baddi, Himachal Pradesh, India
e-mail: nitin.verma@chitkarauniversity.edu.in

K. Dua
Discipline of Pharmacy, Graduate School of Health, University of Technology, Sydney, NSW, Australia
e-mail: Kamal.Dua@uts.edu.au


G. Gupta et al. (eds.), *Microbiome in Inflammatory Lung Diseases*,
https://doi.org/10.1007/978-981-16-8957-4_19

**Keywords**

Lung microbiome · Commensal flora · Gut microbiome · Cystic fibrosis, COPD · Immune modulation · Microbial manipulations

## 19.1 Introduction

Microbes populate numerous mucosal areas of body such as lungs and gastrointestinal tract [1]. The primary function of the microbial biome is to maintain immune balance on diverse mucosal surfaces [2–4]. Food and lifestyle changes, age, and infections are external variables that have an impact on the commensal microbial ecology and produce adaptive changes in the host immune system. Although the current research, foreign gut infection boosts host systemic inflammation while eliminating healthy gut flora with possible pathogenic characteristics [4–7]. Initially, Bacterial pathogenesis was mainly produced by lungs illnesses [8, 9]. Emerging metagenomics and 16s ribosomal RNA (rRNA) investigations have shown the microbiome habitat in both upper and lower respiratory tracts and their critical involvement in respiratory disorders [10–12]. The human immune system influences microbiota, and the lungs have 50 times less microbiome than the oral cavity [13]. The gut microbiome has a significant effect on most illnesses and has received much attention in recent years, but little research has been done on the lung microbiome [9, 14]. One of the most studied lung disorders is cystic fibrosis and microbial dysbiosis [8, 15]. Research [16–27] has found the involvement of lung microbiome in allergic illness, asthma and chronic obstructive pulmonary disease (COPD) [16–27]. Some data hoarded demonstrates the direct and indirect function of gut microbiota in lung illnesses [27–31]. Clinical data suggest that gut microbiome modification benefits lung microbial homeostasis and, by extension, lung immunological homeostasis [31–35]. The function of lung microbiome homeostasis and lung morbidities and the considerations for lung microbiome analysis and the obstacles involved are discussed in this book chapter.

### 19.1.1 Interlink Between Inhabitants Microbial Flora and Mucosal Surfaces

The interaction of microbial flora in mucosal surfaces is of significant interest. Inflammatory bowel disease (IBD) has long been linked to lung illness and impairment [36, 37]. Chronic respiratory illness was described in six IBD patients who had no history of smoking. Bronchiectasis, chronic bronchitis, and obstructive pulmonary dysfunction were all present in these individuals. It has also been observed that patients with ulcerative colitis have decreased pulmonary function [38]. The carbon monoxide diffusion capacity, however, was not evaluated in this investigation. Another research [39] discovered a connection between Crohn's illness and decreased carbon monoxide diffusion ability. In one study, 314 Crohn's disease patients were shown to have an elevated risk of COPD [40]. The main risk factor for the development of COPD and Chron's disease and Ulcerative colitis Cigarette [41–43]. The smoke from cigarettes has the potential to alter the lung epithelium and the gut microbiota. A potential connection between illnesses and immunological responses to common commensal microorganisms has emerged. As a result, the data suggest that smoking disrupts the gut microbiota [44, 45].

### 19.1.2 Various Environmental Factors Associated with Microbial Status

#### 19.1.2.1 Smoking

Smoking destroys lung's epithelial layer and interruption of the gut hindrance have shown to prompt bacterial movement and systemic immune enactment [46–48]. This produces the sensible supposition that disruption of the lung epithelial hindrance could initiate the action of individuals from the lung microbiome [49]. A fascinating piece of supporting proof for this view was found by Rosas-Salazar et al. [50]. This examination utilized a lung carcinoma mice model infused with lung carcinoma cell. Mice exposed with cigarette smoke and aerosolized nontypeable Hemophilus influenza (NTHi) had a higher rate of metastatic growth than control mice injected simply with lung cancer cells. Furthermore, bacterial movement to acellular breakdown in the lungs tumors also happened under this joined therapy.

#### 19.1.2.2 Breast-Feeding

Breast milk plays a crucial functions in developing newborn microbiota [50]. First, the human bosom milk has its microbiome, bragging more than 200 species of microscopic organisms [51]. These commensals have been demonstrated to prevent harmful bacteria from colonizing and growing in the host. Second, prebiotic oligosaccharides are found in milk and are available for bacterial use in the gut. Human milk oligosaccharides also have antibacterial effects, limiting harmful bacteria adhesion and infection [52, 53]. Therefore breast milk plays a major role in the foundation of gut microbiome in babies. Indeed, breastfeeding care has been displayed to affect the advancement of the nasopharyngeal microbiome [54]. Furthermore, breastfeeding leads to microbial stability and early upper respiratory colonization of Corynebacterium, Moraxella, and Dolosigranulum [55].

#### 19.1.2.3 Antibiotics

The use of antibiotic has a crucial influence on microbial homeostasis and limits microbial diversity [56]. Anti-toxins are regularly recommended during pregnancy and can be found in mother's breast milk and in infant's blood after birth. There is two way effect of this on the microbiome. Firstly, microbes are killed by the mother before they are given to the offspring. Second, antibiotics in breast milk can disrupt the establishment and colonization of bacteria eliminated in the mother and were transmitted down to the child. In addition, according to research, antibiotic exposure throughout childhood can cause chronic inflammation and asthma [57, 58].

#### 19.1.2.4 Diet

Diet has a prominent effect on lung microbiome. While there is some evidence on how the maternal eating routine can impact the microbiome in a baby and might be identified with the advancement of child asthma, there are limited studies on the effect of diet on lung microbiome [59]. Anew report revealed that mice given treatment of vancomycin showed stable microbiome and metabolite profile with aggravated Th2 reactions. These creatures likewise showed more significant weakness to hypersensitive lung inflammation. This examination further revealed that gut dysbiosis could bother hypersensitive lung irritation through T cell-and DC-subordinate components hindered by bacterial short-chain unsaturated fats (SCFAs). Even though this investigation showed that lung illnesses might be influenced by gut microbiome alteration [60]. It is apparent from specific investigations that the gut microbiome impacts lung irritation and its pathology (asthma, COPD) anyway [61]. Other investigations have shown that dietary αtocopherol supplementation brings about inversion old enough instigated vulnerability to *Streptococcus pneumonia* lung disease [62–64].

#### 19.1.2.5 Pollution

Lung disorders such as COPD and asthma are directly affected by pollutants in our environment. For example, lung's exposure to smoke during cooking, consuming coal, and other biomass energies directly affects lung well-being [65]. The respiratory microbiome is assumed to be the first line of defense against toxins in the environment. Environmental contaminants can disrupt the established microbial community structure, causing unstability in pro-inflammatory as well as anti-oxidant conditions, resulting in disease [66]. Unfortunately, there is not much literature accessible in this field concerning the structure of the occupant microscopic organisms in the lung microbiome and its alteration against environmental pollutants [67].

## 19.2 Microbiome Pathogenic Role in Common Lung Disorders

### 19.2.1 Chronic Obstructive Pulmonary Disease (COPD)

COPD is a chronic and fatal lung illness. It provides a large epidemic now and has been expected to increase in this regard [68]. Exacerbations are rapid, brief worsening of the illness that causes a substantial but ephemeral loss in pulmonary and baseline function of lung regaining in about a month [69]. Conversely, baseline lung function is rarely decreased for an extended period of time or permanently [70]. The vicious circle paradigm [71] has also been used to characterize disease progression in COPD. Microorganisms cause inflammation and enhanced mucus secretion after first exposure. Itirritates the respiratory epithelium, impairing the innate immune responses. The weakened immune responses expose the lungs to further contamination through harmful germs, completing the vicious loop. H. influenza colonization and illness exacerbations are characterized due to the existence of that in the sufferer's sputum are defined as COPD [72]. P6is a H. influenzae membrane protein is mucin secretion has been reported to increase in epithelial cells of human and mice during COPD grown in vitro [73]. H. influenza is not found in between exacerbations [74]. Surprisingly, the similar H. influenzae strain was found in sputum samples from various exacerbation phases. This could suggest that infections stay primarily inside the host and proliferate throughout exacerbations, although when microorganisms reach unsafe levels during recurrences.

### 19.2.2 Cystic Fibrosis

A hereditary illness triggered through CFTR gene mutation is called Cystic fibrosis (CF) [75]. Symptoms CF causes in different organs across the body are affected; however the lungs seem to be the most severely affected. The presence of secretions within lungs and a higher rate of pulmonary bacterial infection are the disease's characteristics. These changes in function of lung, obviously, show a massive effect on microbiome of the lung. CF patients have long been known to suffer from lung infections caused by *S. aureus*, *H. influenza*, *B. cepacian*, and *P. aeruginosa* [76]. *M. abscesses*, *S. maltophilia*, MRSA, and *S. milleri* had actually been found to be a component of a CF lung microbiota. Lung transplantation is frequently required when bacterial infection induces lung tissue damage. P. aeruginosa invades transplanted tissue quickly; according to a recent study [77]. The invasive germs are likely to even have started inside the sinus prior moving onto other non-CF donor lungs. Cox et al. [78] investigated that in young and old CF, increasing age is related to decreased pulmonary performance and also reduced macrobiotic complexity, variety, and uniformity.

### 19.2.3 Asthma

According to hygiene theory, early exposure to bacteria can influence asthma and other disorders [79]. Asthma formation has also been associated with earlier antibiotic exposure. Antibiotic usage in pregnant women, as previously noted, can minimize bacterial colonization [80]. Forth to the bacterial endotoxin lipopolysaccharide (LPS) was shown in research to be preventive against asthma development. LPS exposure protects mice by increasing the production of the A20 protein that inhibits NF-kB stimulation. When asthmatic persons' lung microbiomes had been contrasted to healthy subjects, it was discovered that they had higher amounts of Proteobacteria but lower levels of Bacteriodetes [81].

### 19.2.4 Pneumonia (Idiopathic or Ventilator)

Our understanding of the pathogenesis of pneumonia has enhanced our understanding on the lung microbiome. The fundamental etiology of the disease is currently thought to be dysbiosis of the lung microbial flora rather than the introduction of pathogenic bacteria [82–84]. Children on mechanical ventilation are vulnerable to a variety of nosocomial infections called Ventilator-Associated Pneumonia (VAP): This increases\possibility of mortality, necessitates a longer hospital stay, and necessitates extensive rehabilitation [85]. The bacterial community in airway samples collected from ventilated children is pretty varied. The findings also reveal that after the diagnosis of VAP, the diversity declines, and the airways become rapidly dominated by pathogenic microorganisms. *Streptococcus* flora was found to be significantly higher in children with VAP compare to those without VAP in this research [86]. *Streptococcus pneumoniae* was shown to be the most significant causal bacterial pathogen for lower respiratory tract contaminations such as pneumonia for analysis of airway microbiome [87]. Environmental stressors, also immunological variations, disrupt the population's "balance," resulting in dysbiosis and, finally, infection [88]. One of the few studies that have done in-depth research compared geriatric pneumonia (with healthy older control individuals) and adult pneumonia in terms of changes in microbiota population during illness (with healthy adult controls). In pneumonia patients, the researchers discovered a reduction in gram-negative bacteria like Leptotrichia, Veillonella, and Prevotella and as well as the gram-positive species Parascardovia [89]. Another study found a significant connection between the microbiota of the tongue and pneumonia death in nursing care individuals [90].

### 19.2.5 Lung Cancer

The microbiome's significance to the growth of several malignancies has previously been investigated [91]. Latest studies have found an association in lung cancer and pulmonary microbiota [61, 92]. Studies have found a connection here in lung cancer and pulmonary microbiome [61, 92]. BAL was collected, and microbiota was investigated in recent research comparing individuals suffering from lung cancer with those suffering from benign lung masses [92]. The scientists discovered that Proteobacteria was the dominating phylum in this research. The Thermus genus was found more common in individuals with stage 3 or stage 4 malignancies. Legionella, a Proteobacteria genus, was prevalent in individuals who had metastases. Both of findings suggest, alterations in the microbiome could be utilized also as novel biomarker for the lung cancer.

### 19.2.6 Viral Infections of the Lung

Acute respiratory infections (ARIs) that can be seen between kids and adults are mostly produced through viruses [93]. ARI has been linked to children account for 40% with all healthcare-linked lung infections [94]. Most viral ARIs are caused by rhinoviruses (RV), "respiratory syncytial virus (RSV)"and influenza A virus (IAV), and "metapneumovirus (MPV)". As per the viewpoint pathophysiology of ARI as

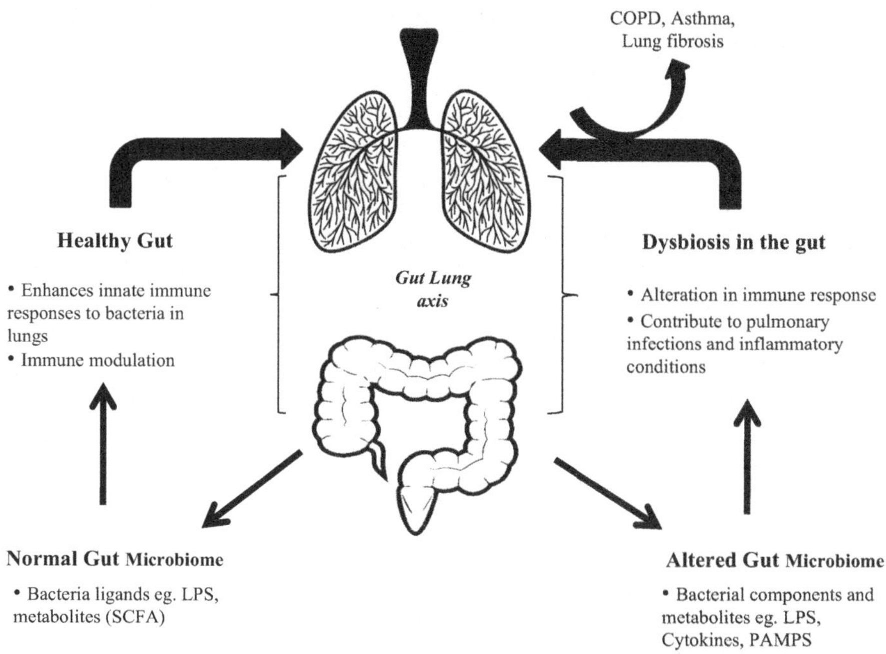


**Fig. 19.1** The gut-lung microbiome's function in the pathophysiology of different lung disorders

well as the role of microbial homeostasis with in lung, it is the only opportunity to explore cross-domain linkages or if the pulmonary microbiota may impact ARI in good or worse. RSV and IAV had lately been explored detailed during this setting. According to a recent study, Sublethal transmission with such a reduced variant of H5N1 IAV may alter bacterial makeup of a lower respiratory tract (LRT) [95]. The findings display a Bacillus to Lactobacillus shift inside the lung microbiota as well as gut microbiome reduction. However, it has been found that the pulmonary microbiome had enriched *Streptococcus* and reduced Pseudomonas [96]. Interestingly, Bartley et al. showed that time period microbial alterations result in unfavorable prognosis and IFN-1 activation as a result of IAV invasion, but that it might be prevented in proportion through upholding host's microbial configuration, in just this instance through calorie reduction [97]. Although Because the majority of ARI signs are related to increased inflammation, this stands to reason as virus infection-induced microbiome alteration could contribute into continuous loop of inflammatory response and abnormal immune responses, significantly increasing infection rate [97, 98]. Furthermore, Lung/or microbial gut regeneration may be used for therapeutic strategy to control systemically immune system responses to ARI virus-related causes (Fig. 19.1).

## 19.3 Lung Microbiome and Modulation of the Host Immune System

The GI microbial flora has been well recognized to have a significant function in controlling the inflammatory reaction of a host, particularly in the establishment of a Th17 reaction with in mucosal immune system [99]. Modifications within microbiota of young adults are related to close to zero inflammation in the lungs [99]. Microbial profiles linked with Th17 phenotypes in asthma patients have also been observed [100]. Adherents of the Proteobacteria taxa Bacillaceae, Pasteurellaceae, and Enterobacteriaceae, were shown to be linked to Th17 related genes expression in a research by Huang et al. In asthma patients, this Th17 inflammatory phenotype may constitute a separate route from the Th2 response [101]. However, Yadava et al. discovered, using elastase and LPS treatment led in a dysbiotic lung microbiota, that led to a significant enhancement in expression of IL-17A in an interventional animal model. This was attributable to expansion within phenotype of $\gamma\delta$ + T cells [102]. The mouse inflammatory phenotype was linked to airway alterations that are comparable to those seen in people with COPD. BAL from these animals exposed to LPS and elastase showed a reduction in variability and a rise in relative assemblages of Lactobacillus, Pseudomonas, and Chryseobacterium [103]. Using microbiota-depleted animals, The authors found the microbiota enhanced IL-17A release by $\gamma\delta$ + T-cells. The scientists demonstrated an elevation of the IL-17A immunological phenotype after shifting the optimized microbiome through Elastase/LPS infected animals as well as concomitant exposure with Elastase/LPS in antibiotic-infected mice. As a result, the lung microbiota seems to have an important role in immunological phenotypes, according to research.

Host with lung dysbiosis have been reported with upregulated IL-17 and Th17 inflammatory pathway. According to a latest study, invasion of microorganism in the epithelium of pIgR deficient mice resulted in impairement of mucosal immunity [104]. However there was no difference observed in terms of percentage between the pIgR deficient mice and wild-type mice which depicted that effects were because of microorganisms' invasion. NF-κB-mediated increased chemokine keratinocyte chemoattractant in BAL fluid was also observed in the BAL fluid of pIgR deficient mice [105]. Thus according to research, the lung microbiota has a significant impact. in modulation of inflammatory pathways and immune system within host.

### 19.3.1 The Respiratory Microbiome and Metabolism

The gut microbiota is essential in the host metabolic activities [106]. Because either pathogenic and commensal serve critical functions in metabolism. and understanding their roles may aid in the modulation of gut metabolism. The gut microbiota metabolizes various macromolecules, including phenol, bile acid, and choline metabolism [107, 108]. Bacterial function in the lung must be investigated further since certain bacteria perform harmful roles in the lung microbiome [109]. Recent research discovered metabolomic abnormalities in BAL of HIV-infected people compared with healthy and postulated that these differences were caused by microbiota changes [110]. A subsequent investigation conducted by a similar group discovered a connection between existences of pathogenic microorganisms inside the lungs with altered metabolite amounts [111]. Nocardioidaceae, Staphylococcaceae, and Caulobacteraceae were identified as the significant determinants to altered metabolite amounts in the research. Those microorganisms are notable due to their role in the pathogenesis of HIV-related pneumonia. Metabolism of bacteria has been shown to influence host pulmonary immunity. It was demonstrated in a landmark research that continuous expansion of the lung microbiome by oral genera results in a unique metabolic environment that promotes neutrophil-mediated and Th17 inflammation with inhibiting innate mechanisms [112, 113]. Several techniques have been researched to determine the influence of lung microbiota upon host metabolome. Garg et al. developed a technique for visualizing 3D human lungs and used it to map metabolomics and 16S rRNA gene sequencing information to individual particular related space. It also enabled a more accurate depiction of the relationship between microbial and molecular penetration [114].

### 19.3.2 Microbial Manipulations in the Treatment of Various Lung Disorders

The complex importance of intestinal microbiome has resulted in a growing number of clinical and experimental research examining gut microbiota modification as an intriguing therapeutic option for lung illnesses. Natural products, probiotics, prebiotics, and antibiotics addressing gut microbiota have also been studied in patients with lung illnesses.

#### 19.3.2.1 Probiotics and Prebiotics

Dietary supplements that have anti-inflammatory and immunomodulatory properties are called Probiotics and have been proven to decrease pulmonary exacerbations in cystic fibrosis patients [115]. Bifidobacterium longum, commonly known as BB536, is a versatile probiotic that has been demonstrated in a double-blind randomized experiment of Malaysian preschoolers kids 2–6 years. to attenuate upper respiratory illnesses with gut microbiota modifying characteristics [116]. As compared to the placebo group, BB536 therapy dramatically enhanced the population of the species Faecalibacterium, which is associated with anti-inflammation and immunomodulation. Many randomized clinical studies have demonstrated the probiotic supplements could be helpful in the diagnosis of respiratory relapse and inflammation of intestine in patients with "cystic fibrosis", although the information is incomplete and more elevated research will be required [117]. COPD patients receiving antibiotics for respiratory tract infection were given a multispecies probiotic or a placebo in a, placebo-controlled, randomized and double-blind trial [118]. Other herbal ingredient boosted the number of lactic acid-forming microorganisms Bifidobacterium and "Lactobacillus" spp. in asthmatic rat models

[119]. Lactobacillus is an important beneficial bacterium in cystic fibrosis. In the case of cystic fibrosis during randomized Clinical Study of affected children, Lactobacillus rhamnosus GG partially restored gut microbiome resulting in decreased microbial richness and intestinal inflammation [120]. In FVB/N mice with Pseudomonas aeruginosa pneumonia, oral treatment of Lactobacillus rhamnosus GG enhanced stomach penetrability and controlled inflammatory responses and spleen and colon homeostasis. The mechanism behind its protection is that it increases the expression of gut mucin, increases cell growth while decreasing pro-inflammatory cytokine expression [121]. A double-blind anticipated trial of a probiotic composition of Lactobacillus reuteri enhanced digestive function and reduced calprotectin concentrations in patients with cystic fibrosis [122]. The antibiotic streptomycin influenced the profile of gut microbiota pulmonary lymphocyte and hyper responsiveness of airway in mice model of cystic fibrosis. It decreased bacterial excess-growth in Cftr mice and mainly changed Lactobacillus levels [123].

Probiotics have been shown to improve the effectiveness of anti-tumor medications. L. acidophilus, for example, cisplatin has an anti-tumor effect and increases success rates in C57BL/6 mice having cancer [124]. During a research, Barnesiellaintestinihominisand Enterococcus hirae improved the cytotoxic activity of cyclophosphamide with developed lung cancer in mice by increasing the presence of IFN-producing T type cells in tumor regions [125].

Various probiotics have been shown to inhibit lung metastases via immune regulation. By boosting cytotoxic T and T helper cells, a probiotic-containing fermented milk preparation demonstrated substantial cytotoxicity on 4 T1 breast carcinoma cells and decreased metastasis to the lungs [126]. Additional research has resulted in cancer formation inhibition and decreased cancer extravasation and tumour vascularity, along with metastasis of the lung, due to the treatment of the BALB/c mice with fermented milk with L. casei CRL 431. The process was ascribed to changes in immunological response, such as an increase in CD8 + and CD4 + cells and a reduction in infiltrating macrophages [127].

In melanoma lung meta statics, the number of metastatic lung focuses was decreased by an α T17 immune-cell-dependent mechanism during commensal Microbiota transplantation [128, 129].

In addition, the potential therapeutic agent against lung cell cancer Bifidobacterium infant is, a recombinant probiotic bacterium, was presented. In C57BL/6 lung cancer mice, the tumor growth reduction and extended surviving period were shown using Bifidobacterium infants-mediated sFlt-1 gene-transference system [130].

#### 19.3.2.2 Supplementing Micronutrients to Regulate Gut Microbiota

Vitamin D insufficiency owing to good malabsorption frequently occurs in Cystic Fibrosis patients [131]. The treatment of vitamin D has been tested in the module of intestinal microbiota in cystic fibrosis. Vitamin D helps to keep the intestinal mucus layer intact and helps useful bacteria to fight with pathogenic bacteria [132]. Vitamin D's function is to improve intercellular interconnection by inhibiting intestinal epithelial apoptosis and by decreasing pro-inflammatory cytokines for example

IL-8 [132]. The intake of vitamin E vitamin C, beta-carotenes, riboflavins, and niacins is adversely linked to bacteroid intestines whereas vitamin E and beta-carotenes intakes are favourably linked to Firmicutes when cystic fibrosis occurs [133]. Consumption of such flavonoids can be linked with intestinal microbiota changes, as a clinical trial of adults with cystic fibrosis shows [134]. For example, ingestion of gallocatechin is positive for Actinomyces and Actinomycetacaceae genera, whereas ingestion is solely negative for Coriobacteria classes. It is likely to have a major influence on the therapy of cyst fibrosis [134, 135] on immunological function, inflammation and metabolism. A meal with 5% acidic oligosaccharide enhanced to reduce the degree of bacterial clearance following P. aeruginosa infections in mice [136]. Dietary treatments along with prebiotic and probiotic medications have been demonstrated to improve the symptoms of severe systemic inflammation in patients with cystic fibrosis, but additional investigation is needed to evaluate that impact [137, 138].

## 19.4 Challenges and Future Prospects

Collection and analysis of the sample are the main challenging aspects of the knowledge of lung microbiome in illnesses. The samples taken from the both the respiratory tract (upper and lower) and bronchial lavage in most lung microbiome trials are obtained. Respiratory infections and chronic lung illnesses alter air and intestinal flora composition. A bridge between these two different components has been documented concerning lung illnesses. This gut–lung axis allows endotoxins, cytokines, hormones, and microbial metabolites to enter the circulation. According to research, Alterations in gut microbiome is associated with alteration in immune response, inflammation, and development of lung injuries. Although current findings have shown that the intestinal microbiota is crucial in regulating immunological responses in lungs; hence, their metabolites can reach other organs via the circulation and exhibit anti-inflammatory and immune-regulating effects. The introduction of helpful bacteria from the gut to respiratory tract is an effective therapeutic strategy for various lung illnesses. The association between gut microbiota and lung diseases has yet to be investigated. Furthermore, fundamental perspectives into routes and mediators must be explored. On either side, treatment methods that can modify the gut microbiome, like antibiotics, prebiotics, and probiotics had been investigated in clinical and laboratory trials or have yielded promising results. They can improve immune responses by repairing microbiota dysbiosis. Despite the fact that microbiota modification as a treatment for inflammatory lung disorder seems restricted, antibiotic, probiotics, anti-inflammatory drugs, and nutrition may dramatically reduce illness recurrences. Over latest years, the broad application of culture-independent techniques like as 16s sequencing of rRNA gene and metagenomics have improved the research of lungs microbiota. Moreover, hence the need to improve and regulate techniques of extracting DNA across diverse types of samples while decoding the microbiota in order to minimize infection by products like DNA extraction kit and PCR master mix and in 16S rRNA sequencing of gene.

The primary treatment for asthma patients includes anti-inflammatory corticosteroid, that been usually given with conjunction to bronchodilators. While combination treatments relieve ailments, larger dosages cause lengthy negative side effects. As a result, taking pre- and probiotic supplementation, as well as corticosteroids and macrolide antibiotics may help to prevent asthma attacks by lowering hazardous viral contamination. Antibiotic therapy has been found to lower dangerous bacteria in COPD, but corticosteroids raised bacterial pathogens counts, demonstrating the efficacy of antibiotic therapy during COPD. In addition to standard treatment using long-acting muscarinic antagonists, inhaled corticosteroid, and long-acting -agonists, antibiotic and anti-inflammatory agents, anti-inflammatory and antibiotics agents aid in the management of COPD recurrences. Therefore, advance research is required to evaluate its impact upon the lung microbiota.

Pre-clinical and clinical research displays that microbiota is linked to cancer, and more research into the mechanism of such associations is required. The identification of particular bacteria such as *Capnocytophaga* and *Veillonella* in saliva samples in cancer patients using 16S sequencing can reveal changes in the lung cancer microbiota. Moreover, immunotherapy using monoclonal antibodies addressing immune antagonists like programmed cell death (PD)-1 with ligand PD-L1 is utilized as a treatment in several metastatic malignancies. There are no instances of patients utilizing particular therapies to address the microbiota in cancer, and that might be a diagnostic and therapeutic approach. Therefore, reversing dysbiosis and addressing the lung microbiota could be helpful in CLDs. By use of antibiotics and probiotics appears to be helpful, but strong effect evidence is missing, and treatment trials with translational studies and explanations are necessary to understand and implement beneficial therapies. It must next to be performed properly during clinical studies before being applied in medical care.

## References

1. Ferreira CM, Vieira AT, Vinolo MAR, Oliveira FA, Curi R, FDS M (2014) The central role of the gut microbiota in chronic inflammatory diseases. J Immunol Res 2014:689492
2. Power SE, O'Toole PW, Stanton C, Ross RP, Fitzgerald GF (2014) Intestinal microbiota, diet and health. Br J Nutr 111:387–402
3. McHardy IH, Goudarzi M, Tong M, Ruegger PM, Schwager E, Weger JR, Graeber TG, Sonnenburg JL, Horvath S, Huttenhower C, McGovern DP, Fornace AJ, Borneman J, Braun J (2013) Integrative analysis of the microbiome and metabolome of the human intestinal mucosal surface reveals exquisite inter-relationships. Microbiome 1:17
4. Jacobs JP, Braun J (2014) Immune and genetic gardening of the intestinal microbiome. FEBS Lett. https://doi.org/10.1016/j.febslet.2014.02.052
5. Kamada N, Chen GY, Inohara N, Núñez G (2013) Control of pathogens and pathobionts by the gut microbiota. Nat Immunol 14:685–690
6. Gómez-Hurtado I, Santacruz A, Peiró G, Zapater P, Gutiérrez A, Pérez-Mateo M, Sanz Y, Francés R (2011) Gut microbiota dysbiosis is associated with inflammation and bacterial translocation in mice with CCl4-induced fibrosis. PLoS One 6:e23037
7. Maier L, Diard M, Sellin ME, Chouffane E-S, Trautwein-Weidner K, Periaswamy B, Slack E, Dolowschiak T, Stecher B, Loverdo C, Regoes RR, Hardt W-D (2014) Granulocytes impose a

tight bottleneck upon the gut luminal pathogen population during *Salmonella typhimurium* colitis. PLoS Pathog 10:e1004557
8. Dickson R, Erb-Downward J, Huffnagle G (2013) The role of the bacterial microbiome in lung disease. Expert Rev Respir Med 7:245–257
9. Dickson RP, Huffnagle GB (2015) The lung microbiome: new principles for respiratory bacteriology in health and disease. PLoS Pathog 11:e1004923
10. Morris A, Beck JM, Schloss PD, Campbell TB, Crothers K, Curtis JL, Flores SC, Fontenot AP, Ghedin E, Huang L, Jablonski K, Kleerup E, Lynch SV, Sodergren E, Twigg H, Young VB, Bassis CM, Venkataraman A, Schmidt TM, Weinstock GM (2013) Comparison of the respiratory microbiome in healthy nonsmokers and smokers. Am J Respir Crit Care Med 187: 1067–1075
11. Bourzac K (2014) The bacterial tightrope. Nature 516:S14–S16
12. Hosgood HD III, Sapkota AR, Rothman N, Rohan T, Hu W, Xu J, Vermeulen R, He X, White JR, Wu G, Wei F (2014) The potential role of lung microbiota in lung cancer attributed to household coal burning exposures. Environ Mol Mutagen 55(8):643–651. https://doi.org/10.1002/em.21878
13. Scales BS, Dickson RP, Huffnagle GB (2016) A tale of two sites: how inflammation can reshape the microbiomes of the gut and lungs. J Leukoc Biol 100:1–8
14. Dwyer DNO, Dickson RP, Moore B (2016) The lung microbiome, immunity, and the pathogenesis of chronic lung disease. J Immunol 196:4839–4847
15. Stressmann FA, Rogers GB, Klem ER, Lilley AK, Donaldson SH, Daniels TW, Carroll MP, Patel N, Forbes B, Boucher RC, Wolfgang MC, Bruce KD (2011) Analysis of the bacterial communities present in lungs of patients with cystic fibrosis from American and British centers. J Clin Microbiol 49:281–291
16. Wills-karp M, Santeliz J, Karp CL (2001) Revisiting the hygiene hypothesis. Nat Rev Immunol 1:69–75
17. Risnes KR, Belanger K, Murk W, Bracken MB (2011) Antibiotic exposure by 6 months and asthma and allergy at 6 years: findings in a cohort of 1,401 US children. Am J Epidemiol 173: 310–318
18. McKeever TM, Lewis SA, Smith C, Collins J, Heatlie H, Frischer M, Hubbard R (2002) Early exposure to infections and antibiotics and the incidence of allergic disease: a birth cohort study with the west midlands general practice research database. J Allergy Clin Immunol 109:43–50
19. Erb-Downward JR, Thompson DL, Han MK, Freeman CM, McCloskey L, Schmidt LA, Young VB, Toews GB, Curtis JL, Sundaram B, Martinez FJ, Huffnagle GB (2011) Analysis of the lung microbiome in the "healthy" smoker and in COPD. PLoS One 6:e16384
20. Cabrera-Rubio R, Garcia-Nunez M, Seto L, Anto JM, Moya A, Monso E, Mira A (2012) Microbiome diversity in the bronchial tracts of patients with chronic obstructive pulmonary disease. J Clin Microbiol 50:3562–3568
21. Pragman AA, Kim HB, Reilly CS, Wendt C, Isaacson RE (2012) The lung microbiome in moderate and severe chronic obstructive pulmonary disease. PLoS One 7:e47305
22. Sze MA, Dimitriu PA, Hayashi S, Elliott WM, McDonough JE, Gosselink JV, Cooper J, Sin DD, Mohn WW, Hogge JC (2012) The lung tissue microbiome in chronic obstructive pulmonary disease. Am J Respir Crit Care Med 185:1073–1080
23. Huffnagle GB (2010) The microbiota and allergies/asthma. PLoS Pathog 6:1–3
24. Fagundes CT, Amaral FA, Vieira AT, Soares AC, Pinho V, Nicoli JR, Vieira LQ, Teixeira MM, Souza DG (2012) Transient TLR activation restores inflammatory response and ability to control pulmonary bacterial infection in germfree mice. J Immunol 188:1411–1420
25. Clarke TB, Davis KM, Lysenko ES, Zhou AY, Yu Y, Weiser JN (2010) Recognition of peptidoglycan from the microbiota by Nod1 enhances systemic innate immunity. Nat Med 16: 228–231
26. Bingula R, Filaire M, Radosevic-Robin N, Bey M, Berthon JY, Bernalier-Donadille A, Vasson MP, Filaire E (2017) Desired turbulence? Gut-lung axis, immunity, and lung cancer. J Oncol 2017:5035371

27. Bruzzese E, Callegari ML, Raia V, Viscovo S, Scotto R, Ferrari S, Morelli L, Buccigrossi V, Lo Vecchio A, Ruberto E, Guarino A (2014) Disrupted intestinal microbiota and intestinal inflammation in children with cystic fibrosis and its restoration with lactobacillus gg: a randomised clinical trial. PLoS One 9:1–12
28. Abrahamsson TR, Jakobsson HE, Andersson AF, Björkstén B, Engstrand L, Jenmalm MC (2012) Low diversity of the gut microbiota in infants with atopic eczema. J Allergy Clin Immunol 129:434–440. 440.e1–2
29. Sun Y, Cai Y, Huse SM, Knight R, Farmerie WG, Wang X, Mai V (2012) A large-scale benchmark study of existing algorithms for taxonomy-independent microbial community analysis. Brief Bioinform 13:107–121
30. Schloss PD, Westcott SL (2011) Assessing and improving methods used in operational taxonomic unit-based approaches for 16S rRNA gene sequence analysis. Appl Environ Microbiol 77:3219–3226
31. Liu Z, DeSantis TZ, Andersen GL, Knight R (2008) Accurate taxonomy assignments from 16S rRNA sequences produced by highly parallel pyrosequencers. Nucleic Acids Res 36:e120
32. Costello EEK, Lauber CCL, Hamady M, Fierer N, Gordon JI, Knight R (2009) Bacterial community variation in human body habitats across space and time. Science 326:1694–1697
33. Dickson RP, Erb-Downward JR, Prescott HC, Martinez FJ, Curtis JL, Lama VN, Huffnagle GB (2014) Cell-associated bacteria in the human lung microbiome. Microbiome 2:28
34. Beck JM, Schloss PD, Venkataraman A, Twigg H, Jablonski KA, Bushman FD, Campbell TB, Charlson ES, Collman RG, Crothers K, Curtis JL, Drews KL, Flores SC, Fontenot AP, Foulkes MA, Frank I, Ghedin E, Huang L, Lynch SV, Morris A, Palmer BE, Schmidt TM, Sodergren E, Weinstock GM, Young VB (2015) Multicenter comparison of lung and oral microbiomes of HIV-infected and HIV-uninfected individuals. Am J Respir Crit Care Med. 192:1335–1344
35. Kelly BJ, Imai I, Bittinger K, Laughlin A, Fuchs BD, Bushman FD, Collman RG (2016) Composition and dynamics of the respiratory tract microbiome in intubated patients. Microbiome 4:7
36. Charlson ES, Bittinger K, Haas AR, Fitzgerald AS, Frank I, Yadav A, Bushman FD, Collman RG (2011) Topographical continuity of bacterial populations in the healthy human respiratory tract. Am J Respir Crit Care Med 184:957–963
37. Hilty M, Burke C, Pedro H, Cardenas P, Bush A, Bossley C, Davies J, Ervine A, Poulter L, Pachter L, Moffatt MF, Cookson WOC (2010) Disordered microbial communities in asthmatic airways. PLoS One 5:e8578
38. Jovel J, Patterson J, Wang W, Hotte N, O'Keefe S, Mitchel T, Perry T, Kao D, Mason AL, Madsen KL, Wong GK-S (2016) Characterization of the gut microbiome using 16S or shotgun metagenomics. Front Microbiol 7:459
39. Janda JM, Abbott SL (2007) 16S rRNA gene sequencing for bacterial identification in the diagnostic laboratory: pluses, perils, and pitfalls. J Clin Microbiol 45:2761–2764
40. Charlson ES, Bittinger K, Chen J, Diamond JM, Li H, Collman RG, Bushman FD (2012) Assessing bacterial populations in the lung by replicate analysis of samples from the upper and lower respiratory tracts. PLoS One 7:e42786
41. Sethi S, Maloney J, Grove L, Wrona C, Berenson CS (2006) Airway inflammation and bronchial bacterial colonization in chronic obstructive pulmonary disease. Am J Respir Crit Care Med 173:991–998
42. Edwards FC, Truelove SC (1964) Course and prognosis of ulcerative colitis: part III complications. Gut 5(1):1
43. Kraft SC, Earle RH, Roesler M, Esterly JR (1976) Unexplained bronchopulmonary disease with inflammatory bowel disease. Arch Intern Med 136:454–459
44. Munck A, Murciano D, Pariente R, Cezard J, Navarro J (1995) Latent pulmonary function abnormalities in children with Crohn's disease. Eur Respir J 8:377–380
45. Jess T, Loftus EV, Harmsen WS, Zinsmeister AR, Tremaine WJ, Melton LJ, Munkholm P, Sandborn WJ (2006) Survival and cause specific mortality in patients with inflammatory bowel disease: a long term outcome study in Olmsted County, Minnesota, 1940–2004. Gut 55:1248–1254

46. Keely S, Talley NJ, Hansbro PM (2012) Pulmonary-intestinal cross-talk in mucosal inflammatory disease. Mucosal Immunol 5:7–18
47. Somerville KW, Logan RF, Edmond M, Langman MJ (1984) Smoking and Crohn's disease. Br Med J (Clin Res Ed) 289:954–956
48. Cosnes J (2004) Tobacco and IBD: relevance in the understanding of disease mechanisms and clinical practice. Best Pract Res Clin Gastroenterol 18:481–496
49. Biedermann L, Zeitz J, Mwinyi J, Sutter-Minder E, Rehman A, Ott SJ, Steurer-Stey C, Frei A, Frei P, Scharl M, Loessner MJ, Vavricka SR, Fried M, Schreiber S, Schuppler M, Rogler G (2013) Smoking cessation induces profound changes in the composition of the intestinal microbiota in humans. PLoS One 8:e59260
50. Rosas-Salazar C, Shilts MH, Tovchigrechko A, Schobel S, Chappell JD, Larkin EK, Shankar J, Yooseph S, Nelson KE, Halpin RA, Moore ML, Anderson LJ, Peebles RS, Das SR, Hartert TV (2016) Differences in the nasopharyngeal microbiome during acute respiratory tract infection with human rhinovirus and respiratory syncytial virus in infancy. J Infect Dis 214:1924–1928
51. Yu G, Gail MH, Consonni D, Carugno M, Humphrys M, Pesatori AC, Caporaso NE, Goedert JJ, Ravel J, Landi MT (2016) Characterizing human lung tissue microbiota and its relationship to epidemiological and clinical features. Genome Biol 17:163
52. Jones JG, Lawler P, Crawley JCW, Minty BD, Hulands G, Veall N (1980) Increased alveolar epithelial permeability in cigarette smokers. Lancet 315:66–68
53. Moazed F, Burnham EL, Vandivier RW, Kane CMO, Shyamsundar M, Hamid U, Abbott J, Thickett DR, Matthay MA, Mcauley DF, Calfee CS (2016) Cigarette smokers have exaggerated alveolar barrier disruption in response to lipopolysaccharide inhalation. Thorax 71:1130–1136
54. Kelly JR, Kennedy PJ, Cryan JF, Dinan TG (2015) Breaking down the barriers: the gut microbiome, intestinal permeability and stress-related psychiatric. Front Cell Neurosci. 9:392
55. Jungnickel C, Wonnenberg B, Karabiber O, Wolf A, Voss M, Wolf L, Honecker A, Kamyschnikow A, Herr C, Bals R, Beisswenger C (2015) Cigarette smoke-induced disruption of pulmonary barrier and bacterial translocation drive tumor-associated inflammation and growth. Am J Physiol Lung Cell Mol Physiol 309:L605–L613
56. Gallacher DJ, Kotecha S (2016) Respiratory microbiome of new-born infants. Front Pediatr 4: 10
57. Heikkila MP (2003) Inhibition of *Staphylococcus aureus* by the commensal bacteria of human milk. J Appl Microbiol 95:471–478
58. Jost T, Lacroix C, Braegger C, Chassard C (2015) Impact of human milk bacteria and oligosaccharides on neonatal gut microbiota establishment and gut health. Nutr Rev 73:426–437
59. Simon PM, Goode PL, Mobasseri A, Zopf D (1997) Inhibition of *Helicobacter pylori* binding to gastrointestinal epithelial cells by sialic acid-containing oligosaccharides. Infect Immun 65: 750–757
60. Biesbroek G, Tsivtsivadze E, Sanders EAM, Montijn R, Veenhoven RH, Keijser BJF, Bogaert D (2014) Early respiratory microbiota composition determines bacterial succession patterns and respiratory health in children. Am J Respir Crit Care Med 190:1283–1292
61. Zhao J, Murray S, LiPuma JJ, Costello EK, Yatsunenko T, Kostic AD, Karlsson FH, Yoshimoto S, Arthur JC, Atarashi K, Koren O, Jansson J, Holmes E, Li JV, Athanasiou T, Ashrafian H, Nicholson JK, Gajer P, Westermann AJ, Gorski SA, Vogel J, Sharma CM, Costello EK, Stagaman K, Dethlefsen L, Bohannan BJ, Relman DA, Lemon KP, Armitage GC, Relman DA, Fischbach MA, Dethlefsen L, Relman DA, Jernberg C, Lofmark S, Edlund C, Jansson JK, Jakobsson HE, Jernberg C, Lofmark S, Edlund C, Jansson JK, Dethlefsen L, Huse S, Sogin ML, Relman DA, Zhao J, Konstan MW, Wagener JS, VanDevanter DR, Filkins LM, Nick JA, Cox MJ, Klepac-Ceraj V, van der Gast CJ, VanDevanter DR, LiPuma JJ, Schloss PD, Quince C, Lanzen A, Davenport RJ, Turnbaugh PJ, Edgar RC, Haas BJ, Clemente JC, Quince C, Knight R, Huse SM, Welch DM, Morrison

HG, Sogin ML (2014) Modeling the impact of antibiotic exposure on human microbiota. Sci Rep 4:1694–1697

62. Gray LEK, O'Hely M, Ranganathan S, Sly PD, Vuillermin P (2017) The maternal diet, gut bacteria, and bacterial metabolites during pregnancy influence offspring asthma. Front Immunol 8:1–13
63. Cait A, Hughes MR, Antignano F, Cait J, Dimitriu PA, Maas KR, Reynolds LA, Hacker L, Mohr J, Finlay BB, Zaph C, McNagny KM, Mohn WW (2018) Microbiome-driven allergic lung inflammation is ameliorated by short-chain fatty acids. Mucosal Immunol 11:785–795
64. Sokolowska M, Frei R, Lunjani N, Akdis CA, O'Mahony L (2018) Microbiome and asthma. Asthma Res Pract 4:1
65. Bou Ghanem EN, Clark S, Du X, Wu D, Camilli A, Leong JM, Meydani SN (2015) The α-tocopherol form of vitamin E reverses age-associated susceptibility to *Streptococcus pneumoniae* lung infection by modulating pulmonary neutrophil recruitment. J Immunol 194:1090–1099
66. Bou Ghanem EN, Lee JN, Joma BH, Meydani SN, Leong JM, Panda A (2017) The alpha-tocopherol form of vitamin E boosts elastase activity of human PMNs and their ability to kill *Streptococcus pneumoniae*. Front Cell Infect Microbiol 7:1–9
67. Brown RL, Sequeira RP, Clarke TB (2017) The microbiota protects against respiratory infection via GM-CSF signaling. Nat Commun. 8:1512
68. Han MK, Postma D, Mannino DM, Giardino ND, Buist S, Curtis JL, Martinez FJ (2007) Gender and chronic obstructive pulmonary disease. Am J Respir Crit Care Med 176:1179–1184
69. Dickson RP, Martinez FJ, Huffnagle GB (2014) The role of the microbiome in exacerbations of chronic lung diseases. Lancet 384:691–702
70. Adar SD, Huffnagle GB, Curtis JL (2016) The respiratory microbiome: an underappreciated player in the human response to inhaled pollutants? Ann Epidemiol 26:355–359
71. Lopez AD, Shibuya K, Rao C, Mathers CD, Hansell AL, Held LS, Schmid V, Buist S (2006) Chronic obstructive pulmonary disease: current burden and future projections. Eur Respir J 27: 397–412
72. Wedzicha JA, Donaldson GC (2003) Exacerbations of chronic obstructive pulmonary disease. Respir Care 48:1204–1213. discussion 1213–5
73. Seemungal TAR, Donaldson GC, Bhowmik A, Jeffries DJ, Wedzicha JA (2000) Time course and recovery of exacerbations in patients with chronic obstructive pulmonary disease. Am J Respir Crit Care Med 161:1608–1613
74. Sethi S, Murphy TF (2008) Infection in the pathogenesis and course of chronic obstructive pulmonary disease. N Engl J Med 359:2355–2365
75. Murphy TF, Brauer AL, Schiffmacher AT, Sethi S (2004) Persistent colonization by *Haemophilus influenzae* in chronic obstructive pulmonary disease. Am J Respir Crit Care Med 170:266–272
76. Chen R, Lim JH, Jono H, Gu XX, Kim YS, Basbaum CB, Murphy TF, Li JD (2004) Nontypeable *Haemophilus influenzae* lipoprotein P6 induces MUC5AC mucin transcription via TLR2-TAK1-dependent p38 MAPK-AP1 and IKKβ-IκBα-NF-κB signaling pathways. Biochem Biophys Res Commun 324:1087–1094
77. Cutting GR (2015) Cystic fibrosis genetics: from molecular understanding to clinical application. Nat Rev Genet 16:45–56. HHS Public Access
78. Høiby N (1988) *Haemophilus influenzae*, *Staphylococcus aureus*, *Pseudomonas cepacia*, and *Pseudomonas aeruginosa* in patients with cystic fibrosis. Chest 94, 97S–102S
79. Beaume M, Köhler T, Greub G, Manuel O, Aubert J-D, Baerlocher L, Farinelli L, Buckling A, van Delden C, Achermann R, Amico P, Baumann P, Beldi G, Benden C, Berger C, Binet I, Bochud P-Y, Boely E, Bucher H, Bühler L, Carell T, Catana E, Chalandon Y, de Geest S, de Rougemont O, Dickenmann M, Duchosal M, Fehr T, Ferrari-Lacraz S, Garzoni C, Soccal PG, Giostra E, Golshayan D, Good D, Hadaya K, Halter J, Heim D, Hess C, Hillinger S, Hirsch HH, Hofbauer G, Huynh-Do U, Immer F, Klaghofer R, Koller M, Laesser B, Lehmann R,

Lovis C, Marti H-P, Martin PY, Martinolli L, Meylan P, Mohacsi P, Morard I, Morel P, Mueller U, Mueller NJ, Mueller-McKenna H, Müller A, Müller T, Müllhaupt B, Nadal D, Pascual M, Passweg J, Ziegler CP, Rick J, Roosnek E, Rosselet A, Rothlin S, Ruschitzka F, Schanz U, Schaub S, Seiler C, Stampf S, Steiger J, Stirnimann G, Toso C, Tsinalis D, Venetz J-P, Villard J, Wick M, Wilhelm M, Yerly P (2017) Rapid adaptation drives invasion of airway donor microbiota by pseudomonas after lung transplantation. Sci Rep 7:40309

80. Cox MJ, Allgaier M, Taylor B, Baek MS, Huang YJ, Daly RA, Karaoz U, Andersen GL, Brown R, Fujimura KE, Wu B, Tran D, Koff J, Kleinhenz ME, Nielson D, Brodie EL, Lynch SV (2010) Airway microbiota and pathogen abundance in age-stratified cystic fibrosis patients. PLoS One 5:e11044
81. Froidure A, Pilette C (2016) From the hygiene hypothesis to A20: the protective effect of endotoxins against asthma development. Clin Exp Allergy 46:192–193
82. Foglia E, Meier MD, Elward A (2007) Ventilator-associated pneumonia in neonatal and pediatric intensive care unit patients. Clin Microbiol Rev 20:409–425
83. Mourani PM, Sontag MK (2017) Ventilator-associated pneumonia in critically ill children: a new paradigm. Pediatr Clin N Am 64:1039–1056
84. File TM (2003) Community-acquired pneumonia. Lancet 362:1991–2001
85. Bogaert D, Keijser B, Huse S, Rossen J, Veenhoven R, van Gils E, Bruin J, Montijn R, Bonten M, Sanders E (2011) Variability and diversity of nasopharyngeal microbiota in children: a metagenomic analysis. PLoS One 6:e17035
86. De Steenhuijsen Piters WAA, Huijskens EGW, Wyllie AL, Biesbroek G, Van Den Bergh MR, Veenhoven RH, Wang X, Trzcinski K, Bonten MJ, Rossen JWA, Sanders EAM, Bogaert D (2016) Dysbiosis of upper respiratory tract microbiota in elderly pneumonia patients. ISME J 10:97–108
87. Kageyama S, Takeshita T, Furuta M, Tomioka M, Asakawa M, Suma S, Takeuchi K, Shibata Y, Iwasa Y, Yamashita Y, Newman A (2017) Relationships of variations in the tongue microbiota and pneumonia mortality in nursing home residents. J Gerontol Ser A 00:1–6
88. Vogtmann E, Goedert JJ (2016) Epidemiologic studies of the human microbiome and cancer. Br J Cancer 114:237–242
89. Lee SH, Sung JY, Yong D, Chun J, Kim SY, Song JH, Chung KS, Kim EY, Jung JY, Kang YA, Kim YS, Kim SK, Chang J, Park MS (2016) Characterization of microbiome in bronchoalveolar lavage fluid of patients with lung cancer comparing with benign mass like lesions. Lung Cancer 102:89–95
90. Zilberman-Schapira G, Zmora N, Itav S, Bashiardes S, Elinav H, Elinav E (2016) The gut microbiome in human immunodeficiency virus infection. BMC Med 14:83
91. Twigg HL, Knox KS, Zhou J, Crothers KA, Nelson DE, Toh E, Day RB, Lin H, Gao X, Dong Q, Mi D, Katz BP, Sodergren E, Weinstock GM (2016) Effect of advanced HIV infection on the respiratory microbiome. Am J Respir Crit Care Med 194:226–235
92. Pichon M, Lina B, Josset L (2017) Impact of the respiratory microbiome on host responses to respiratory viral infection. Vaccine 5:40
93. Harris J-AS (2006) Infection control in pediatric extended care facilities. Infect Control Hosp Epidemiol 27:598–603
94. Yildiz S, Mazel-Sanchez B, Kandasamy M, Manicassamy B, Schmolke M (2018) Influenza A virus infection impacts systemic microbiota dynamics and causes quantitative enteric dysbiosis. Microbiome 6:1–17
95. dos Borges LGA, Giongo A, de Pereira LM, Trindade FJ, Gregianini TS, Campos FS, Ghedin E, da Veiga ABG (2018) Comparison of the nasopharynx microbiome between influenza and non-influenza cases of severe acute respiratory infections: a pilot study. Heal Sci Reports:e47
96. Bartley JM, Zhou X, Kuchel GA, Weinstock GM, Haynes L (2017) Impact of age, caloric restriction, and influenza infection on mouse gut microbiome: an exploratory study of the role of age-related microbiome changes on influenza responses. Front Immunol 8:1–11

97. Ederveen THA, Ferwerda G, Ahout IM, Vissers M, de Groot R, Boekhorst J, Timmerman HM, Huynen MA, van Hijum SAFT, de Jonge MI (2018) Haemophilus is overrepresented in the nasopharynx of infants hospitalized with RSV infection and associated with increased viral load and enhanced mucosal CXCL8 responses. Microbiome 6:1–13
98. Ivanov II, de Frutos RL, Manel N, Yoshinaga K, Rifkin DB, Sartor RB, Finlay BB, Littman DR (2008) Specific microbiota direct the differentiation of IL-17-producing T-helper cells in the mucosa of the small intestine. Cell Host Microbe 4:337–349
99. Kumar P, Monin L, Castillo P, Elsegeiny W, Horne W, Eddens T, Vikram A, Good M, Schoenborn AA, Bibby K, Montelaro RC, Metzger DW, Gulati AS, Kolls JK (2016) Intestinal Interleukin-17 receptor signaling mediates reciprocal control of the gut microbiota and autoimmune inflammation. Immunity 44:659–671
100. Segal LN, Alekseyenko A, Clemente JC, Berger K, Goldring R, Rom WN, Blaser MJ, Weiden MD (2014) Enrichment of lung microbiome with Supraglotic microbes is associated with increased pulmonary inflammation. Ann Am Thorac Soc 11:S71–S71
101. Huang YJ, Nariya S, Harris JM, Lynch SV, Choy DF, Arron JR, Boushey H (2015) The airway microbiome in patients with severe asthma: associations with disease features and severity. J Allergy Clin Immunol 136:874–884
102. Yadava K, Pattaroni C, Sichelstiel AK, Trompette A, Gollwitzer ES, Salami O, Von Garnier C, Nicod LP, Marsland BJ (2016) Microbiota promotes chronic pulmonary inflammation by enhancing IL-17A and autoantibodies. Am J Respir Crit Care Med 193:975–987
103. Richmond BW, Brucker RM, Han W, Du RH, Zhang Y, Cheng DS, Gleaves L, Abdolrasulnia R, Polosukhina D, Clark PE, Bordenstein SR, Blackwell TS, Polosukhin VV (2016) Airway bacteria drive a progressive COPD-like phenotype in mice with polymeric immunoglobulin receptor deficiency. Nat Commun 7:1–12
104. Gomez-Arango LF, Barrett HL, McIntyre HD, Callaway LK, Morrison M, Nitert MD, Tremellen A, Tobin J, Wilkinson S, McSweeney C, O'Rourke P, Lingwood B, Kang A, Shanahan E, Fukuma N, Angel N, Foxcroft K (2016) Connections between the gut microbiome and metabolic hormones in early pregnancy in overweight and obese women. Diabetes 65:2214–2223
105. Le Chatelier E, Nielsen T, Qin J, Prifti E, Hildebrand F, Falony G, Almeida M, Arumugam M, Batto JM, Kennedy S, Leonard P, Li J, Burgdorf K, Grarup N, Jorgensen T, Brandslund I, Nielsen HB, Juncker AS, Bertalan M, Levenez F, Pons N, Rasmussen S, Sunagawa S, Tap J, Tims S, Zoetendal EG, Brunak S, Clement K, Dore J, Kleerebezem M, Kristiansen K, Renault P, Sicheritz-Ponten T, de Vos WM, Zucker JD, Raes J, Hansen T, Bork P, Wang J, Ehrlich SD, Pedersen O (2013) Richness of human gut microbiome correlates with metabolic markers. Nature 500:541–546
106. Devaraj S, Hemarajata P, Versalovic J (2013) The human gut microbiome and body metabolism: implications for obesity and diabetes. Clin Chem 59:617–628
107. Velasquez MT, Ramezani A, Manal A, Raj DS (2016) Trimethylamine N-oxide: the good, the bad and the unknown. Toxins (Basel) 8:326
108. Horáčková Š, Plocková M, Demnerová K (2017) Importance of microbial defence systems to bile salts and mechanisms of serum cholesterol reduction. Biotechnol Adv. 36:682–690
109. Duda-Chodak A, Tarko T, Satora P, Sroka P (2015) Interaction of dietary compounds, especially polyphenols, with the intestinal microbiota: a review. Eur J Nutr 54:325–341
110. Davies PL, Spiller OB, Beeton ML, Maxwell NC, Remold-O'Donnell E, Kotecha S (2010) Relationship of proteinases and proteinase inhibitors with microbial presence in chronic lung disease of prematurity. Thorax 65:246–251
111. Cribbs SK, Park Y, Guidot DM, Martin GS, Brown LA, Lennox J, Jones DP (2014) Metabolomics of bronchoalveolar lavage differentiate healthy HIV-1-infected subjects from controls. AIDS Res Hum Retrovir 30:579–585
112. Cribbs SK, Uppal K, Li S, Jones DP, Huang L, Tipton L, Fitch A, Greenblatt RM, Kingsley L, Guidot DM, Ghedin E, Morris A (2016) Correlation of the lung microbiota with metabolic profiles in bronchoalveolar lavage fluid in HIV infection. Microbiome 4:3

113. Segal LN, Clemente JC, Tsay JJ, Koralov SB, Keller C, Wu BG, Li Y, Shen N, Ghedin E, Morris A, Diaz P, Huang L, Wikoff WR, Ubeda C, Artacho A, Weiden MD (2016) Enrichment of the lung microbiome with oral taxa is associated with lung inflammation of a Th17 phenotype. Nat Microbiol 1:1–24
114. Garg N, Wang M, Hyde E, da Silva RR, Melnik AV, Protsyuk I, Bouslimani A, Lim YW, Wong R, Humphrey G, Ackermann G, Spivey T, Brouha SS, Bandeira N, Lin GY, Rohwer F, Conrad DJ, Alexandrov T, Knight R, Dorrestein PC (2017) Three-dimensional microbiome and metabolome cartography of a diseased human lung. Cell Host Microbe 22:705–716.e4
115. Jafari SA, Mehdizadeh-Hakkak A, Kianifar HR, Hebrani P, Ahanchianm H, Abbasnejad E (2013) Effects of probiotics on quality of life in children with cystic fibrosis; a randomized controlled trial. Iran J Pediatr 23:669–674
116. Lau AS, Yanagisawa N, Hor YY, Lew LC, Ong JS, Chuah LO, Lee YY, Choi SB, Rashid F, Wahid N, Sugahara H (2018) *Bifidobacterium longum* BB536 alleviated upper respiratory illnesses and modulated gut microbiota profiles in Malaysian pre-school children. Benef Microbes 9, 61–70. doi: https://doi.org/10.3920/BM2017.0063
117. Nikniaz Z, Nikniaz L, Bilan N, Somi MH, Faramarzi E (2017) Does probiotic supplementation affect pulmonary exacerbation and intestinal inflammation in cystic fibrosis: a systematic review of randomized clinical trials. World J Pediatr 13:307–313. https://doi.org/10.1007/s12519-017-0033-6
118. Koning CJ, Jonkers D, Smidt H, Rombouts F, Pennings HJ, Wouters E, Stobberingh E, Stockbrügger R (2010) The effect of a multispecies probiotic on the composition of the faecal microbiota and bowel habits in chronic obstructive pulmonary disease patients treated with antibiotics. Br J Nutr 103:1452–1460. https://doi.org/10.1017/S0007114509993497
119. Kong YH, Qi SHI, Na HAN, Zhang L, Zhang YY, Gao TX, Chen CHEN, Li YL (2016) Structural modulation of gut microbiota in rats with allergic bronchial asthma treated with recuperating lung decoction. Biomed Environ Sci 29(8):574–583
120. Bruzzese E, Callegari ML, Raia V, Viscovo S, Scotto R, Ferrari S, Morelli L, Buccigrossi V, Lo Vecchio A, Ruberto E, Guarino A (2014) Disrupted intestinal microbiota and intestinal inflammation in children with cystic fibrosis and its restoration with *Lactobacillus* GG: a randomised clinical trial. PLoS One 9(2):e87796
121. Khailova L, Baird CH, Rush AA, Barnes C & Wischmeyer PE (2017) *Lactobacillus rhamnosus* GG treatment improves intestinal permeability and modulates inflammatory response and homeostasis of spleen and colon in experimental model of *Pseudomonas aeruginosa* pneumonia. Clin Nutr 36, 1549–1557. doi: https://doi.org/10.1016/j.clnu.2016.09.025
122. Del Campo R, Garriga M, Perez-Aragon A, Guallarte P, Lamas A, Maiz L, Bayón C, Roy G, Cantón R, Zamora J, Baquero F (2014) Improvement of digestive health and reduction in proteobacterial populations in the gut microbiota of cystic fibrosis patients using a *Lactobacillus reuteri* probiotic preparation: a double blind prospective study. J Cyst Fibros 13:716–722. https://doi.org/10.1016/j.jcf.2014.02.007
123. Bazett M, Bergeron M, Haston CK (2016) Streptomycin treatment alters the intestinal microbiome, pulmonary T cell profile and airway hyperresponsiveness in a cystic fibrosis mouse model. Sci Rep 6:19189. https://doi.org/10.1038/srep19189
124. Gui QF, Lu HF, Zhang CX, Xu ZR, Yang YH (2015) Well balanced commensal microbiota contributes to anti-cancer response in a lung cancer mouse model. Genet Mol Res 14:5642–5651. https://doi.org/10.4238/2015.May.25.16
125. Dasari S, Kathera C, Janardhan A, Praveen Kumar A, Viswanath B (2017) Surfacing role of probiotics in cancer prophylaxis and therapy: a systematic review. Clin Nutr 36:1465–1472. https://doi.org/10.1016/j.clnu.2016.11.017
126. Zitvogel L, Daillere R, Roberti MP, Routy B, Kroemer G (2017) Anticancer effects of the microbiome and its products. Nat Rev Microbiol 15:465–478. https://doi.org/10.1038/nrmicro.2017.44

127. Aragon F, Carino S, Perdigon G, de Moreno de LeBlanc A (2015) Inhibition of growth and metastasis of breast cancer in mice by milk fermented with *Lactobacillus casei* CRL 431. J Immunother 38:185–196. https://doi.org/10.1097/CJI.0000000000000079
128. Tanasienko OA, Cheremshenko NL, Titova GP, Potebnya MG, Gavrilenko MM, Nagorna SS, Kovalenko NK (2005) Elevation of the efficacy of antitumor vaccine prepared on the base of lectines from *B. subtilis* B-7025 upon its combined application with probiotics in vivo. Exp Oncol 27:336–338
129. Cheng M, Qian L, Shen G, Bian G, Xu T, Xu W, Shen G, Hu S (2014) Microbiota modulate tumoral immune surveillance in lung through a gammadeltaT17 immune cell-dependent mechanism. Cancer Res 74:4030–4041. https://doi.org/10.1158/0008-5472.CAN-13-2462
130. Zhu H, Li Z, Mao S, Ma B, Zhou S, Deng L, Liu T, Cui D, Zhao Y, He J, Yi C (2011) Antitumor effect of sFlt-1 gene therapy system mediated by *Bifidobacterium infantis* on Lewis lung cancer in mice. Cancer Gene Ther 18:884–896. https://doi.org/10.1038/cgt.2011.57
131. Sexauer WP, Hadeh A, Ohman-Strickland PA, Zanni RL, Varlotta L, Holsclaw D, Fiel S, Graff GR, Atlas A, Bisberg D, Hadjiliadis D (2015) Vitamin D deficiency is associated with pulmonary dysfunction in cystic fibrosis. J Cyst Fibros 14:497–506. https://doi.org/10.1016/j.jcf.2014.12.006
132. Kanhere M, Chassaing B, Gewirtz AT, Tangpricha V (2018) Role of vitamin D on gut microbiota in cystic fibrosis. J Steroid Biochem Mol Biol 175:82–87. https://doi.org/10.1016/j.jsbmb.2016.11.001
133. Li L, Krause L, Somerset S (2017) Associations between micronutrient intakes and gut microbiota in a group of adults with cystic fibrosis. Clin Nutr 36:1097–1104. https://doi.org/10.1016/j.clnu.2016.06.029
134. Li L, Somerset S (2018) Associations between flavonoid intakes and gut microbiota in a group of adults with cystic fibrosis. Nutrients 10:E1264. https://doi.org/10.3390/nu10091264
135. Bernard H, Desseyn JL, Bartke N, Kleinjans L, Stahl B, Belzer C, Knol J, Gottrand F, Husson MO (2015) Dietary pectin-derived acidic oligosaccharides improve the pulmonary bacterial clearance of Pseudomonas aeruginosa lung infection in mice by modulating intestinal microbiota and immunity. J Infect Dis 211:156–165. https://doi.org/10.1093/infdis/jiu391
136. Li L, Somerset S (2014) The clinical significance of the gut microbiota in cystic fibrosis and the potential for dietary therapies. Clin Nutr 33:571–580. https://doi.org/10.1016/j.clnu.2014.04.004
137. Mohamed-Hussein AA, Mohamed NA, Ibrahim M-EA (2007) Changes in pulmonary function in patients with ulcerative colitis. Respir Med 101:977–982
138. Shah BR, Li B, Al Sabbah H, Xu W, Mráz J (2020) Effects of prebiotic dietary fibers and probiotics on human health: with special focus on recent advancement in their encapsulated formulations. Trends Food Sci Technol 102:178–192. 10.1016%2Fj.tifs.2020.06.010

# Microbiota Targeted Via Nanotechnology for Lung Cancer Therapy: Challenges and Future Perspectives

20

Monika Yadav and Anita Kamra Verma

**Abstract**

Lung cancer is the foremost reason of cancer associated deaths globally and poses a great threat to human health. It has become progressively clear that we live in a symbiotic association with microbes within us. Human microbiota have been linked with normal physiology and function, reports have proved that microbiome may have a significant share in lung cancer progression as well as advancement, and communications among microbial inhabitants have the ability to impact disease, signifying that microbiome would be an emergent target in cancer management. Traditional approaches to modulate the microbiome (for instance, probiotics, antibiotics) have been revealed to increase efficiency of treatments of cancer in few cases although complications like collateral harm to the commensal microbiome and reliability of these strategies inspire works for evolving new approaches specially devised for the cancer–microbiome interface. Microbiota modulation can be employed as an adjuvant to conventional cancer strategies like immunotherapy and chemotherapy. Considering the achievement of nanoparticles in renovating cancer diagnostics and management, nanoparticles mediated approaches that are able to modulate interactions between microbiota and tumor microenvironment (TME) have the aptitude to offer novel approaches for cancer management. Opportunities at the interface of cancer, microbiome, and nanotechnology are immense. Here, we will emphasize main adaptable areas for using nanotechnology towards influencing the microbiome for cancer treatments, provide overview on current research advances about microbiome in lung cancer and discuss future challenges in this emerging area.

M. Yadav · A. K. Verma (✉)
Nano-Biotech Lab, Kirori Mal College, University of Delhi, Delhi, India
e-mail: akverma@kmc.du.ac.in





G. Gupta et al. (eds.), *Microbiome in Inflammatory Lung Diseases*,
https://doi.org/10.1007/978-981-16-8957-4_20

**Keywords**

Lung cancer · Microbiome · Nanotechnology · Microbiome–cancer interface · Immunotherapy

## 20.1 Introduction

Lung cancer is a heterogenic condition and the most collective as well as the most recurrent reason of cancer associated mortalities globally both in women and men [1]. Annually, with over 1.8 million cases of lung cancer are reported with1.6 million casualties and 5 year- survival rates fluctuating within 4–17%, subject to cancer stage and national differences [2]. It is commonly divided into two types comprising NSCLC accounting for 80–90% of lung cancers and SCLC that has been reducing in incidence in various nations over the past 20 years [1]. Various treatments options for lung cancer comprise chemotherapy, immunotherapy, surgery, targeted therapy, and radiotherapy. The cause of high mortality rate is that mostly diagnosis happens at a very advanced stage and thus gets poor assistance from limited treatment options [3]. Alternatively, multi-organ metastasis and relapse in both pre- and post-medication are crucial reasons of cancer related deaths without successful treatment. So, there are rising crisis as well as public requirement in investigating the oncogenesis and novel therapeutic strategies for this fatal disease. Lung cancer is extensively contemplated to be a complex disease initiated by communications amid environmental circumstances and host [4]. Among various ecological harmful circumstances, microbes have a crucial role in sustaining micro-ecological equilibrium and controlling immune responses of the host to multi-therapies. Though the non-diseased lungs were believed as a sterile ecosystem but currently with the advancement of high throughput next generation sequencing (NGS) technologies it has been found that in lung tissues some microbial species existed that influences the balance between pathogenesis and health in the lung microenvironment [5]. It has been revealed that particular bacteria and the equilibrium of bacteria in lung microbiome can cause oncogenesis via several pathways such as inflammation initiated by bacterial infection, immune response modulation triggered by dysbiosis or bacteria-derived carcinogens [6]. The primarily genera existing in lungs include *Pseudomonas, Prevotella*, *Veillonella*, *Streptococcus* and phyla *Proteobacteria, Firmicutes,* and *Bacteroides*. Currently investigations about the lung microbiota and critical breakthroughs in the association of microbiota with lung diseases are rising quickly. It has been supposed that advanced knowledge of this interrelation will offer innovative perceptions into the lung diseases pathogenesis. Some oncogenic *Escherichia coli* strains imitate swelling and are precisely magnified at inflammation sites, the microbiota composition can be altered by inflammation that consecutively stimulates carcinogenesis [7]. Notably, the microbiome role in cancer expends beyond both the host gut microbiome and the primary tumor site, as particular cancer causing bacteria are frequently discovered in distal tumor sites like liver metastases and lymph nodes metastases [7]. The recent advancement in microbiome understanding has formed huge curiosity in evolving strategies for microbiome modulation to eliminate lung cancer triggering bacteria or advances cancer therapies [8]. Existing strategies for modulation of microbiome such as prebiotics, antibiotics, diet modifications, and fecal microbiota transplants lack selectivity in attaining aimed modulation because they were not initially established with the TME in mind and might not interrelate with microbes that are unreachable via oral route of administration. As such, new strategies for microbiome interference headed towards lung cancer treatment should be able to (1) cross the intricate microenvironment (including the microbiome, the TME, and the tumor–microbiome interface barriers), (2) precisely intervene with the accountable molecular signaling pathways, (3) must be performing outside the primary tumor sites (for instance, metastases to lymph nodes). Nanoparticles are employed to possibly encounter these various necessities as they are proficient to connect across macroscopic and molecular length scales a crucial prerequisite for communications with macroscopic tumors, bacteria, and small molecule metabolites. Although the nanotechnology application for modulation of microbiome for lung cancer management is still naive but seminal work present that emphasizes its immense possibilities and capabilities. These capabilities first depend on categorizing which bacterial species are dangerous in cancer instigation and advancement verses useful for treatment of lung cancer [9]. In this chapter, we concise and assessed the recent advancement of the interactions and fundamental mechanism among microbiome and lung cancer. Similarly, we also conferred the possibilities about the oncogenesis and medicinal applications of microbiota in lung cancer. Furthermore, we discussed how the various properties of nanotechnology make them exclusively suitable for microbiome interference. We too emphasize seminal work where nanotechnologies have been exploited to upgrade cancer treatment via modulation of microbiome or intervention with bacteria-derived carcinogens. We also emphasize an evolving area of exploration where exclusive roles of the microbe communities, microbiome, or microbes may be exploited as a source of motivation for refining conventional delivery of nanoparticle. Conclusively, we share our assessment on the contests and stance of nanotechnology applications towards lung cancer treatment through microbiome interferences.

## 20.2 Gut and Lung Microbiome

Microbiome is defined by protozoa, fungi, bacteria, virus and their associated genome and metabolites [10]. Presently, more and more attempts have been engrossed on understanding the interaction of microbiome with the human body, mainly gut microbiota that is believed as an overlooked organ intervening host homeostasis through intricate processes. The emerging research on gut microbiome is entirely assisted by the quick advancement and application of metagenomics, bioinformatics, and increased throughput molecular technologies [11]. Until now lungs were considered as sterile organs, but recent investigations supported with high throughput sequencing technologies questioned the ancient philosophy

[12]. Nowadays, it is mostly documented that instabilities in microbiota influence various diseases of the lung. Lung microbiome included virus, fungi, bacteria which are resultant from inhaling process of oropharynx, nasopharynx, environmental air exchange, and mucosal secretions [5]. In healthy human body, fungi like *Candida*, *Penicillium*, and *Aspergillus* coexist with bacteria of the genus, *Haemophilus*, *Veillonella*, *Streptococcus* as well as *Propionibacterium* but do not initiate infection of lungs [10]. It is not unexpected while studying several unfamiliar communications in other tissues. All the multiple unknown communications between immunity, metabolism, and microbiome, in microbial niche of other tissues influence several lung pathogenesis of lung cancer, cystic fibrosis, and asthma [13–15]. Indeed, the human body is an energetically stable integrity and microorganisms in several body sites can directly interrelate with each other comprising dispersal of mucosal membrane, digestive and respiratory activities, and indirectly through metabolites in systematic circulation, cytokine and inflammatory substances that displayed the potential microbial interaction among the oral cavities, gut, and lungs. The modifications of native lung microbiome populations generally rest on three phases that can be briefed as microbial growth rates, eradication, and migration under the situation of health and diseases [5, 16]. Some investigators have conveyed that the prime source of lung microbiome is oral microbiome after an established and authenticated observation (via secretions in oral cavities, micro-aerosols mucosal dispersion, and swallowing) [16]. Both the gut and respiratory tract can interconnect with each other through biological activities comprising inhalation and micro-aspiration. The pH and temperature in gastro-intestinal tract are comparatively stable and also migration of microbes is single directional as well as continuously amended by complex chemical and physical conditions. Conversely, lungs regularly exchange gas with outer atmosphere to sustain ample reserve of microbiota and oxygen. Additionally, in upper respiratory tract there is no gradient variety of temperature, pressure, and physical barrier that could offer bidirectional circumstances for lung inhabitant microbial migration and energetic fluctuations [17, 18]. In spite of dissimilar alterations in micro-anatomic characteristics, population and composition dynamics in lung and gut microbiome, both organs have a comparable homeostasis and some biological features like co-evolution and interaction with immune cells, mucosal-immune system, constant exposure to outer environment and microbiota maturation process. Interestingly, cumulative clinical findings showed that various diseases of the lungs were more possibly to progress in patients having gastro-intestinal ailments [19]. These remarkable findings direct us to reexamine whether the microbial communication network actually exists and modifies host vulnerability to either external or internal pathogenic aspects. Therefore, an innovative hypothesis was suggested as "Microbiota-Gut-Lung axis" found on varied as well as intricate gut and lung microbiota networks proven founded on several long-term epidemiology studies. Nevertheless, the mechanism underlying the "Microbiota-Gut-Lung axis" stays indefinable and stronger proofs are still obligatory to kindle the lamp.

### 20.2.1 Microbiome and Tissue Homeostasis

The lung and gut microbiome is evolving as a crucial regulator of lung cancer progress and management. The human beings during their entire lifetime have a close mutual beneficial association with the microbes and use as a provider of beneficiary vital molecules and shield against external invasions. Human microbiome is biological communal of pathogenic, symbiotic, and commensal micro-organisms comprising bacteria, viruses, protozoa, and fungi. The humans developed microbiota following birth via direct surface contact with mother and through vertical transmission inheritance, soon afterwards that its evolution begins in response to ecological circumstances such as exposure to environment, drugs, and diet, accumulating in an adult microbiome by the age of two [20]. Since bacterial composition of microbiota has been the most abundant within commensal microbiota, majority of studies are focused on them. Undeniably, it is approximated that a lone human being holds above one billion bacteria and majority of them are commensals, therefore, total microbiome is approximated to comprise 0.2 kg of weight [21]. Composition and size of the microbiome display a comparatively temporal solidity but extensive interpersonal disparity in a single individual based on anatomical site. Therefore, before using it to design efficient microbiota transplant practices or as a biomarker for cancer therapies, that are key existing contests in various pathologies comprising lung cancer, the personal composition of microbiome must be taken into consideration. Among other purposes, microbiota normalizes tissue homeostasis and host immunity. Thus, the same micro-organisms which are favorable for human health, in particular conditions can stimulate the growth of cancer [6, 22]. The eubiosis is acclimatized by various circumstances such as antibiotic exposure, lifestyle, genetic background, chronic infections, environmental and hereditary factors, and diseases. All these circumstances can promote the agitation of the equilibrium of microbial community composition, a condition identified as dysbiosis. This condition can be minor and sequential restoring after the elimination of unfavorable stimuli. But, sometimes this imbalance can be chronified changing tissue homeostasis and thus causing conditions such as cancer [23]. Indirect consequences of microbial constituents associated with a dysregulated inflammatory IR or direct effects on cell alteration have been observed to be engrossed in carcinogenesis [24].

## 20.3 Role of Microbiome in Cancers

Cancers are commonly believed to be multi-factorial pathological progression wherein healthy cells start multiplying in an un-programmed way causing the blocking of autophagy, apoptosis, DNA damage, and inflammation. There are cumulative pathogenic and commensal micro-organisms present in the humans with stated carcinogenic effects and most of them are notably epidemiologically associated with carcinogenesis [25]. The findings confirmed the close association between respiratory tract and microbial communities. The instigations of

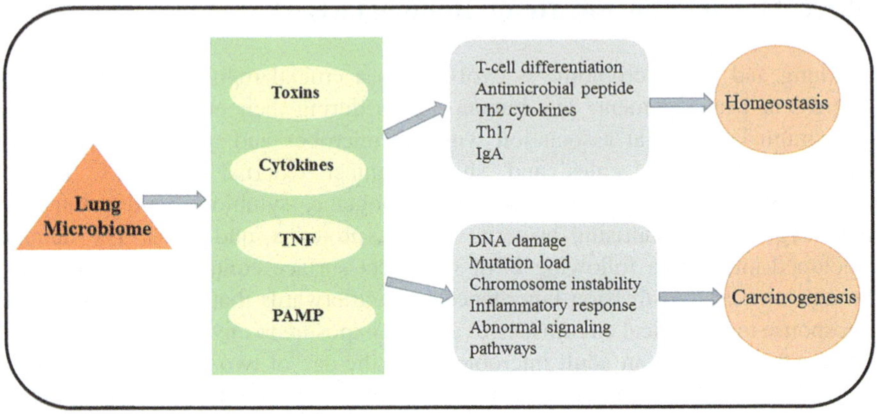


**Fig. 20.1** Lung microbiota have a dual function in sustaining homeostasis and stimulating carcinogenesis in distinctive circumstances. Lung microbiome can prompt carcinogenesis through enhancing mutation load, DNA damage, chromosome instability, inflammatory response alterations, and instigations of abnormal signaling pathway via multiple cytokines and bacterial toxins production. Alternatively, the progression of colonization and evolution of lungs inhabitant microbiome also contributes in the evolution of lung and supporting host homeostasis as well as convening vulnerability to lung ailments during difficult external environment exposure

surface-bound tumors are frequently related with the destruction of host mucosal-immune barrier. Once the mucosal exterior is impaired and if the damage cannot be restored in time, the microenvironment of the commensal microbiome and original tissue will be reassembled. Else, this injury will keep intensifying and causes repetitive inflammation that could prompt cancer at the end. It is promising that to communicate with the tumor immune microenvironment in the enduring co-existence the microbiota situated in the intra-tumor or surface-bound tumor, exploit tumor derivative carbon sources and additional nutrients [26, 27]. In collective, cumulative findings have proved that the host vulnerability to carcinogenic reasons can be altered by dysbiosis of commensal microbial communities. Metabolites or genotoxic toxins generated by bacteria could impair host-DNA directly as well as trigger genomic instability through natural killer immune receptors, reactive nitrogen or oxygen species that lead to cancer like distinctive modifications when increasing injury consequence prevails the host self-restore ability [28]. These reports projecting the carcinogenesis advancement can offer an innovative understanding to support us identify the method from normal tissue to pre-cancerous lesions and to progressive lung cancer. Additionally, lung microbiota comprising virus or bacterial infection can possibly enter epithelial cells of the airways initiating the wound remedial cascade in chronic pathogenic stimuli or prompting host immune response [29]. It is further possible that lung microbiota could have a dual function in sustaining body constancy and stimulating cancer (Fig. 20.1). Inclusive, these cohort findings or epidemiological examinations, particularly whether the microbiota associated with chronic diseases of the lung contribute in the instigation of lung cancer, still expect additional research. Emergent investigations have also outstretched fascinations about associations between microbiome and lung cancer by high throughput sequencing and epidemiological investigation.

## 20.4 Nano-Bioengineering and Bio-nanotools: Types and Sub-Types

Recently, several nanotechnology derivative tool such as nano-biodevices and nano-biocarriers have earned remarkable scientific attention in tackling concerns of ca lung treatment. The most extensively investigated bio-nano carriers for cancer treatments can be briefed as: Cell and cell membrane-derivative nanocarriers, microbiotic nanocarriers, nano-biodevices, and ligand-conjugated nanocarriers (Fig. 20.2).

### 20.4.1 Cells and Cell Membrane-Derived Nanocarriers

Cells and cell membrane based nanocarriers are also popularizing due to their distinct characteristics represented as enhanced biocompatibility, biodegradable

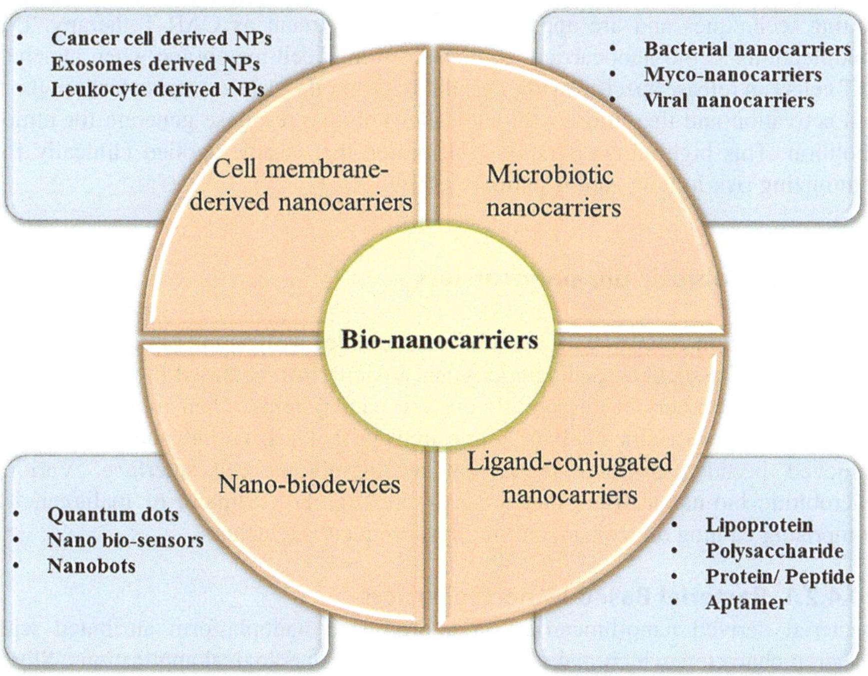


**Fig. 20.2** Bio-nanocarriers and nano-biodevices for oncological applications in lung cancer

nature, and tailoring capability for novel targeting strategies against tumor growth. This diverse group includes whole cell based nano-biocarriers, specialized micro-vesicles or exosomes based nanocarriers, etc. The source for these bio-nanocarriers is derived from biological system in the form of desired cells of our interest (leukocytes, cancer cells) and specialized component (vesicles, exosomes) derived from them. Such therapeutics can effectively induce multiple actions at target site with self stealthing ability. Tumor specific macrophages exhibit protumor characters under the influence from cancer and other immune cells at tumor specific site. Targeting tumor is not very efficient due to suppressed antitumor immune response and elevated tumor-interstitial pressure, presence of irregular vasculature and intense stroma population at tumor site. Designing and fabrication of such macrophages at target site as engineered bio-nanocarriers can overcome conventional hindrance, facilitate intra-tumoral infiltration and localization of various cargos like endotoxins, lipoproteins, antibodies, complement and polysaccharides at site of action [30, 31]. Similarly, dendritic cells (DC) founded bio-nanocarriers are promising for cancer immunotherapy and can also be utilized in designing cancer specific vaccines. This approach can come up with potential outcomes due to presence of diverse surface receptors and directly employed in antitumor immune response generation. Its cross talk with other immune cells can also direct the same with enhance cellular uptake, antigen processing, presentation, and activation of other immune cells at tumor specific site [30]. Another important immune cells are lymphocytes such as T cells, these cells were primarily been engineered with gene editing techniques and are applied for cancer treatment as CAR-T therapy. But modifications as bio-nanocarriers composed from T cell membrane when attached to T cells can release specific cytokines like IL-5 and IL-21 in autocrine fashion for T cell activation and therefore, cytotoxic killer immune response generate for tumor abolition. This highlights effective contribution that can be applied clinically for minimizing risk in lung cancer patients [32].

## 20.4.2 Microbiotic Bio-nanocarriers

The complication coupled with microbiome (complex association of bacteria, fungi) in the form of associated pathogenicity and toxicity has increased the considerable interest of researchers in microbiotic derived nanosystems. Their role as a carrier/vector system for specified delivery of desired agent at target tumor site can be easily achieved because of their interaction at microbiome–cell interface. Various microbiotic bio-nanocarriers that are explored for the treatment of malignancies comprising ca lung are considered in the subsequent segments:

### 20.4.2.1 Bacterial Based Nano-Biocarriers

Bacterial derived nano-biocarriers area privileged nanoplatform attributed with bacterial characters which make them exclusive for oncological applications. Slight modification in them can attenuate pathogenicity, tumor specific target capability and predetermined drug expression along with flexible payload can do wonder for tumor regression. Diverse variety includes nano-engineered bacterial carriers, bacterial specific minicells, bacterial ghosts, bacterial derived polymers (for instance, cellulose, ε-poly-l-lysine, hyaluronic acid, alginate, and poly-l-lactic acid) and magnetosomes that are proved to be utilized in oncological applications [33].

#### 20.4.2.2 Bacterial Minicells

Bacterial minicells are enucleated bioengineered nanocarrier system with approximate diameter of 400 nm. They are known for their surface interaction with tumors cells because of the presence of bi-specific antibodies with O-polysaccharide linker group presented on minicells and tumor specific receptors that are expressed on tumor cells. These minicells can effectively be loaded with desired anticancer drugs (doxorubicin cisplatin, paclitaxel), interfering siRNA or short hairpin loop shRNA for gene targeted therapy against tumor growth including lung cancer too. Advantage of this nano-biocarriers is its bio-stability that can be maintained for long time period [33, 34].

#### 20.4.2.3 Magnetosomes

Magnetosomes are referred as nanoparticles coated with organic membrane within magneto tactic bacteria, their response in presence of applied magnetic field appears to have magneto-sensitive nature. Its role as nano-biocarriers can be employed for lung theranostics where it can act as dynamic player for lung cancer detection similar to MRI reporter gene MagA that act as tracing device for MRI generation and can also be applied for therapeutic treatment with multiple applications as magnetic iron mediated lung cancer management [35]. Study led by Maruyama and colleagues has utilized the potential of nano-engineered biomagnetites isolated as bio-magnetic particles from bacterium *Magnetospirillum magneticum* for monitoring the mutations that occurred at EGFR gene in NSCLC [36]. Improving on to this, further magnetic manipulation with functional proteins such as protein G (*Streptococcus*) and AMB-1 using gene fusion techniques has also enhanced the detection of lung cancer cells [37]. Similarly, in lung cancer cell line A549 presence of magnetosomes has initiated apoptosis and reduced viability of lung cancer cells with significant decrease in Bcl-2 expression [38]. Further, magnetosomes encapsulated with anticancer drug doxorubicin and transferrin are also known to enhance the antitumor efficacy of designed nano-biocarriers in hepatocellular carcinoma, this reveals its potential role for oncological applications [39].

#### 20.4.2.4 Bacterial Ghosts

Bacterial ghosts as nano-biocarriers are preserved envelopes of bacterial structures with no cellular content in it making them ideal for treatment. The desired modification can be added to the surface of nano-biocarriers in the form of native or recombinant antigens or DNA that are required for stimulation of specific antitumor immune response against growing tumor. Further, it can also be incorporated with desired payload (anticancer drug, antioxidant, fluorescent agents) for enhanced effectiveness. Advantages of such nano-biocarrier systems are large scale production, easy isolation of desired variant, reduced risk of pathogenicity, and effective lyophilization for long-term storage. Such characteristics make them superb, for lung cancer treatment [40, 41].

#### 20.4.2.5 Bacterial Polymer Derived Nanocarriers

Bacterial polymer derived nanocarriers are known for their wide range of surface modification added to existing nanocarriers in the form of bio-polymers, this includes polysaccharides (dextran, bacterial starch, alginate, and hyaluronic acid); polyamides (poly γ-glutamic acid); polyesters (polyhydroxy alkanoates and polythioesters); polyphenols (lignin). These modified bacterial polymer hybrids perform diverse action like drug delivery, biosensor fabrication, imaging, thus their presence has raised the promising prospects allied to it for lung cancer management and related effectiveness in other cancer cases too [18].

#### 20.4.2.6 Fungal Based Nano-biocarriers

Researchers have also focused on fungal based nano-biocarriers. Their development with added advancement in the form of fungal polymer component (β-glucan, chitin, and chitosan) coated on nanocarriers for its better utilization in wide oncological treatment is identified with accelerating trends [42]. Lung and other related cancer studies have been reported with enhanced efficacy when these nano-biocarriers were encapsulated with anticancer drugs. Further, production of biogenic metal nanoparticles such as silver or zinc nanoparticles is also reported with promising anticancer potential for lung cancer treatment. These nano-biocarriers are designed with slight modification where nanoparticles are conjugated with yeast cells or vice versa, synthesizing other similar sub-types; therefore, highlights its progressive demands as a cost effective alternative for present microbiome associated counterparts for lung malignancies [43].

### 20.4.3 Novel Nano-bio-devices

Nano-bio-devices are equipment, tools, or their constituent engineered via multidisciplinary study in bio-nanotechnology for several biomedical and clinical therapies [18]. There are four different types of nano-biodevices founded on the material of production such as tissue based, immune-sensors, DNA-based, and microbial based nano-bio-devices. Further, nano-devices exploiting the purpose of membrane proteins have been improved for varied oncological uses. The advancement in bio-nanotechnology leads to use of numerous nanomaterials like nano-dots, nano-balls, nano-walls, nano-pillars, and nano-tubes for the construction of nano-biodevices like nano-chips, nano-wire arrays, nano-robots, and nano-biosensors that have massive oncological applications [18, 44].

### 20.4.4 Ligand-Conjugated Nanocarriers

Bio-nanoparticles wherein the traditional nanocarriers are altered for assisting cancer targeting using bioinspired, natural, bioengineered biomolecules are considered under this group of bio-nanoparticles. To oblige the tenacity of effective tumor aiming bio-nanocarriers can be either of the following sub-types: nano-reservoir (surface coating with biomolecule), conjugate or nano-matrix (entire nano-system is made-up of biomolecule). In most of the situations, the biomolecules have some anticancer properties in themselves. Many innovative biomolecules which are investigated upon for assisting the drive of active tumor aiming via bio-nanotechnology comprise aptamers and aptasensors, proteins and peptides, lipoproteins, carbohydrates, and polysaccharides. The simple nanoparticles that could be altered with any of biomolecules for different oncological applications are polymeric nanoparticles, lipid nanoparticles, lipid-polymeric hybrid nanoparticles, carbon nanostructures, inorganic nanoparticles, and quantum dots [18].

Lipoproteins are an assemblage of nanoparticles constituted of neutral lipids, phospholipids, and apo-proteins. They have a specific biodegradation, bio-transport, and bio-synthetic pathway that mark them a striking nanocarrier for focused applications. Nevertheless, it has been described that only the LDL and HDL are appropriate for oncological treatments owing to their exclusive alpha-helical protein intercalation. Other lipoproteins have large diameter and curvature that subjects them to water and makes them unbalanced [45]. More recently, innovative bio-nanocarriers conjugated with lipoprotein as their targeting ligand are being studied for the cancer management. The various alteration approaches that are used for the assembly of such bio-nanocarriers comprise reconstitution-facilitated core loading, non-covalent surface loading, and covalent modification of phospholipids or proteins [45, 46]. Protein and peptide bio-nanocarriers are the extremely popular types of bio-hybrids due to their superior biodegradability and biocompatibility than synthetic nanocarriers. As these are the most ubiquitous biomolecules they have multi-modal responsibilities in the treatment of several diseases and also have therapeutic and drug delivery purposes along with being used as biomarkers in lung cancer. Though, peptides are being favored over proteins for aiming purposes, as proteins are difficult to conjugate to nanoparticles, have low in-vivo bioavailability and high molecular weights. In recent times, most commonly used peptides for lung cancer aiming comprising somatostatin peptide sequence, iRGD peptide sequence, cell-penetrating peptides, and peptides targeting EGFR, bombesin peptide, fibroblast growth factor peptide (tbFGF) [47]. Hatakeyama et al. have developed an innovative 7-mer peptide labeled as "I-peptide" which imitate carbohydrates and prevent carbohydrate–facilitated cell localization for the management of several types of cancers comprising lung cancer [48]. Recently, several carbohydrate and polysaccharide-based nanocarriers have developed as a notable platform for onco-targeting. The lungs are rich in macrophages and have various GLUT1 and mannose receptors present on their cell surface that facilitate the transport and endocytosis of carbohydrate and polysaccharide altered nanocarriers into the tumor site [49]. Aptamers are the class of oligonucleotides with unique 3D conformations that have been fabricated with the blend of systematic ligand evolution and in-vitro selection procedures to bestow extreme selectivity and likeness for particular target [50]. With attainment of promptness in bio-engineering the aptamers are being researched for their oncological applications. The procedure of bio-engineering and aptamer fabrication is greatly assisted by an in-vitro selection procedure recognized as SELEX. The details of all the above discussed bio-nanocarriers are referred to in the next section. Aptamers owing to their considerable molecular detection aptitude are being utilized for modifying nanoparticles and engineering gadgets such as aptasensors. Aptasensors are bio-sensors engineered exploiting particular aptamers [50].

## 20.5 Bio-Nano Carriers for Clinical Management of Lung Cancer

Besides numerous undeterred cancer related questions, the existing situation necessities an urgent realist strategy. Bio-nanotechnology developed as a result of the convergent evolution of biotechnology with nanotechnology in oncological sciences to propose numerous rewards while overpowering their specific downsides. Bio-nanotechnology has supported discovery of drugs and advancement along with treatment of several types of malignancies comprising Ca lung. The applications of all the above discussed bio-nanocarriers in the perspective of lung cancer treatment have been briefed upon in Table 20.1.

## 20.6 Conclusion and Future Perspectives

The triple communication between host, environment, and microbiome sustains lung homeostasis in healthy functioning. Undeniably, microbiome was proved to be participating in several disease conditions instigation and advancement, but the intricate mechanisms are still unknown. Communally, there are still certain key glitches in this field first, several reports prove the part of gut microbiota in numerous lung diseases but there are methodological contests concerning to the classification of the low biomass-lung microbiota by the NSG. Secondly, due to the poorer richness and dearth of fully characterized reference genomes, the function of microbial components other than bacteria like virus and fungi is mostly unknown in lung cancer. Third, microbiome role in the development of lung cancer has fascinated more responsiveness to the communication among microbiome and tumor immune microenvironment, but lung tumor microenvironment derivative microbiome was also described to perform straight on the tumor tissue and the microbial biomarkers in initial stage lung cancer are still requisite to be investigated. However, with numerous optimistic outcomes, the application of principles of bio-nanotechnology and nano-bioscience has offered an enormous prospective to redefine the existing oncological development. A complete fleet of diverse forms of bio-altered nanosystems is being explored to overcome several problems in the way

**Table 20.1** Bio-nanotools for lung cancer treatment

| Onco-modality | Bio-nanotools | Findings | Ref. |
|---|---|---|---|
| Diagnostics | Nano-device based on nano-porous glass-integrated volumetric chip | The ELISA-based detection device is surface-modified for the rapid detection of lung cancer biomarkers like CYFRA21-1, CEA, and SCCA with extreme sensitivity | [51] |
| | Gadolinium doped-carbon 11 choline-Lenvatinib (GdCo@Ln) nanoparticles | GdCo@ln nanoparticles had clinically assisted PET imaging of lung cancer and considerably enhanced the survival in patients | [52] |
| | Bacteriophage-T4 nano-probe Labeled with Alexa Fluor 546 and Cy3 fluorescent dye | The nano-probes were detected to give fluorescent signal improvement of-90%. They had great intracellular stability and can be employed as a molecular nano-probe for cell imaging and flow cytometry | [53] |
| Secondary prophylaxis/ metastasis prevention | PEGylated liposome polycation DNA complex for siRNA-mediated tumor targeting | The nano-biohybrid enhanced the tumor localization and siRNA-mediated gene silencing (3-times higher) *via* downregulation of surviving in lung cancer. The treatment in H-460 cells was testified to have effective antitumor activity as revealed by 90% apoptosis, that was 4-times higher than the non-targeted nano-hybrids treatment groups | [54] |
| Chemotherapy | Doxorubicin nanoparticles conjugated with GE-11 peptide | GE11 peptide targets the EGFR cell receptor over lung cancer cells with high specificity. Studies on A549 cells shown that the liposomes coated with 10% GE-11 have high antitumoral activity and 2.6-times lowered IC50 values as compared to the non-targeted liposomes. EGFR-mediated cellular uptake was considerable from fluorescent microscopy and flowcytometry | [55] |
| | Anti-carbonic anhydrase IX antibody and cell-penetrating peptide (CPP33) dual-ligand altered triptolide-loaded liposomes (A-CPP-TL-LP) | The A-CPP-TL-LP confirmed high in-vitro cytotoxicity and effective tumor permeability in the C-IX expressing 3 D tumor spheroids. In-vivo pharmacodynamic studies in orthotopic mice model of lung tumor stated n systemic toxicity after pulmonary administration | [56] |

(continued)

**Table 20.1** (continued)

| Onco-modality | Bio-nanotools | Findings | Ref. |
|---|---|---|---|
| Phototherapy | Porphyrin high density lipoprotein (P-HDL) nanoparticles | P-HDL mediated photo therapy to target scavenger receptor class B type I (SR-BI) is a novel strategy towards photo thermal abolition of tumors. Their radiation of P-HDL nanoparticles at 671 nm laser showed higher therapeutic effectiveness in H-460 cells. The in-vivo study in the lung tumor model exhibited 73% cell apoptosis with no signs of cytotoxicity to normal neighboring tissues | [57] |
| | Platelet membrane (PLM) cloaked hollow nanoparticles of bismuth selenide (HNBS) for ICG delivery | PLM have high tumor permeability, protracted systemic circulation and prohibited non-targeted drug release. The HNBS had high ICG loading and great stability under hyperthermia demonstrating an effective means of tumor management | [58] |
| Immunotherapy | T cell labeled with gold nanoparticles and CT imaging for immunotherapy | The nano-biohybrids exhibited high tumor site accumulation and cancer cell tracking. Substantial tumor regression and higher release of cytokines were evident from proliferation assay | [59] |
| | MicroRNA 125b-encapsulated hyaluronic acid-PEI-nanoparticles targeted to TAM (HA-miR-NPs) | Intra-peritoneal administration of the HA-miR-NPs confirmed 300-times improved iNOS (Mlbiomarker) to Arg-l (M2biomarker) ratio and about 6-times higher Ml toM2 macrophage ratio when compared to the control group. | [60] |
| Gene therapy | CDC 20 siRNA-encapsulated cationic-liposomes | The amphiphilic cationic-liposomes encapsulated with synthetic CDC20 exhibited tumor growth inhibition by arresting cell cycle at the G2/M phase and prevent lung metastasis inC57BL/6 J metastatic lung cancer mice model | [61] |
| | CD44-targeted lipid-modified Hyaluronic acid-modified SSB/PLKl siRNA self-assembly nanosystems (HA@siRNA) | The cy3-loaded nano-system was examined to undergo high cellular endocytosis. The SSB/PLKl siRNA-encapsulated nano-system exhibited CD44-specific gene knock-down in the tumor initiating stem cells and primary lung cancer cells | [62] |

(continued)

**Table 20.1** (continued)

| Onco-modality | Bio-nanotools | Findings | Ref. |
|---|---|---|---|
| Combination therapy | Photo thermally active bioinspired lipoprotein (BL-NPs) nanoparticles | Administration of BL-NPs to solid lung tumors disturbed the tumor stromal cells and extra cellular matrix assisting tumor priming to the secondary BL-NPs nanoparticles. With this approach, about 27-times higher tumor accessibility, 4.27-fold higher tumor infiltration, and 97.4% higher anti-metastatic effects were detected on consequent administration of the BL-NPs | [63] |
| | Indocyanine green (ICG) and imiquimod (IQ) co-loaded poly (lactic-co-glycolic) acid (PLGA) nano-biohybrid | Toll-like receptor agonist (IQ), photo therapeutic agent (ICG) in combination with the checkpoint-blockade by CTLA4 confirmed greater check point blockade and anticancer activity in mice tumor models | [64] |
| Theranostics | Folate-functionalized polyethyleneimine passivated-reducible carbon dots(F@P-CDs) loaded with siRNAs (EGFRandcyclinBl) | F@P-CDs exhibited an extremely precise intracellular siRNA payload release escorted by blue photoluminescence upon irradiation at 360 nm in the acidic intracellular microenvironment. The in-vitro cytotoxicity assessment exhibited extremely targeted delivery ability and high biocompatibility of nano-biohybrid. The multi-functionality of nano-biohybrids offers an assuring theranostic tool for real-time monitoring and treatment of several types of cancers comprising lung cancer | [65] |
| | Octreotide-decorated honokiol and epirubicin-loaded liposomes (0-HNE-LP) | The 0-HNE-LP were demonstrated to have high in-vitro cytotoxicity in Lewis lung carcinoma cells (LC). Molecular signaling including VE-Cadherin, caspase3, MMP-2, FAK and PI3K were altered by 0-HNE-LP to display high anticancer activity. The 0-HNE-LP observed high in-vivo safety and efficiency in the LC cell-induced mice tumor model | [66] |

of onco-targeting. These upcoming nano-tools have been believed to direct their perfectionism upon the approaching diagnostic, protective, theranostic, and therapeutic characteristics of numerous forms of neoplasms comprising lung cancer. Several innovative nanocarriers have been significantly investigated to explain many disputes encountered by traditional approaches such as drug resistance non-specific drug targeting, detection of cancer advancement and metastasis, off-target side-effects and poor tumor bioavailability. Though, the projecting importance of microbiome, biotechnology and nanotechnology in the oncological perspective has derived a magnificent view for inhibition and management of Ca lung, it was commonly acknowledged that expansion of this field is restricted and needs extra multidisciplinary and extensive research.

## References


1. Bray F, Ferlay J, Soerjomataram I, Siegel RL, Torre LA, Jemal A (2020) Global cancer statistics 2018: GLOBOCAN estimates of incidence and mortality worldwide for 36 cancers in 185 countries (vol 68, pg 394, 2018). CA Cancer J Clin 70(4):313–313
2. Hirsch FR, Scagliotti GV, Mulshine JL, Kwon R, Curran WJ Jr, Wu YL, Paz-Ares L (2017) Lung cancer: current therapies and new targeted treatments. Lancet 389(10066):299–311
3. The Lancet (2019) Lung cancer: some progress, but still a lot more to do. Lancet. https://doi.org/10.1016/S0140-6736(19)32795-3
4. Brandi G, Frega G (2019) Microbiota: overview and implication in immunotherapy-based cancer treatments. Int J Mol Sci 20(11):2699
5. Dickson RP, Huffnagle GB (2015) The lung microbiome: new principles for respiratory bacteriology in health and disease. PLoS Pathog 11(7):e1004923
6. Helmink BA, Khan MW, Hermann A, Gopalakrishnan V, Wargo JA (2019) The microbiome, cancer, and cancer therapy. Nat Med 25(3):377–388
7. Song W, Anselmo AC, Huang L (2019) Nanotechnology intervention of the microbiome for cancer therapy. Nat Nanotechnol 14(12):1093–1103
8. McQuade JL, Daniel CR, Helmink BA, Wargo JA (2019) Modulating the microbiome to improve therapeutic response in cancer. Lancet Oncol 20(2):e77–e91
9. Gopalakrishnan V, Helmink BA, Spencer CN, Reuben A, Wargo JA (2018) The influence of the gut microbiome on cancer, immunity, and cancer immunotherapy. Cancer Cell 33(4): 570–580
10. Quigley EM (2017) Microbiota-brain-gut axis and neurodegenerative diseases. Curr Neurol Neurosci Rep 17(12):1–9
11. Claesson MJ, O'Toole PW (2010) Evaluating the latest high-throughput molecular techniques for the exploration of microbial gut communities. Gut Microbes 1(4):277–278
12. Mao Q, Jiang F, Yin R, Wang J, Xia W, Dong G, Ma W, Yang Y, Xu L, Hu J (2018) Interplay between the lung microbiome and lung cancer. Cancer Lett 415:40–48
13. Invernizzi R, Molyneaux PL (2019) The contribution of infection and the respiratory microbiome in acute exacerbations of idiopathic pulmonary fibrosis. Eur Respir Rev 28(152): 190045
14. Huang D, Su X, Yuan M, Zhang S, He J, Deng Q, Qiu W, Dong H, Cai S (2019) The characterization of lung microbiome in lung cancer patients with different clinicopathology. Am J Cancer Res 9(9):2047
15. Durack J, Huang YJ, Nariya S, Christian LS, Ansel KM, Beigelman A, Castro M, Dyer AM, Israel E, Kraft M, Martin RJ, Mauger DT, Rosenberg SR, King TS, White SR, Denlinger LC, Holguin F, Lazarus SC, Lugogo N, Peters SP, Smith LJ, Wechsler ME, Lynch SV, Boushey HA (2018) Bacterial biogeography of adult airways in atopic asthma. Microbiome 6(1):1–16

16. Bassis CM, Erb-Downward JR, Dickson RP, Freeman CM, Schmidt TM, Young VB, Beck JM, Curtis JL, Huffnagle GB (2015) Analysis of the upper respiratory tract microbiotas as the source of the lung and gastric microbiotas in healthy individuals. MBio 6(2):e00037–e00015
17. Sampaio-Maia B, Caldas IM, Pereira ML, Pérez-Mongiovi D, Araujo R (2016) The oral microbiome in health and its implication in oral and systemic diseases. Adv Appl Microbiol 97:171–210
18. Liu NN, Ma Q, Ge Y, Yi CX, Wei LQ, Tan JC, Chu Q, Li JQ, Zhang P, Wang H (2020) Microbiome dysbiosis in lung cancer: from composition to therapy. NPJ Precis Oncol 4(1):1–12
19. Wang H, Liu JS, Peng SH, Deng XY, Zhu DM, Javidiparsijani S, Wang GR, Li DQ, Li LX, Wang YC, Luo JM (2013) Gut-lung crosstalk in pulmonary involvement with inflammatory bowel diseases. World J Gastroenterol: WJG 19(40):6794
20. Samuelson DR, Welsh DA, Shellito JE (2015) Regulation of lung immunity and host defense by the intestinal microbiota. Front Microbiol 6:1085
21. Sender R, Fuchs S, Milo R (2016) Revised estimates for the number of human and bacteria cells in the body. PLoS Biol 14(8):e1002533
22. Bhatt AP, Redinbo MR, Bultman SJ (2017) The role of the microbiome in cancer development and therapy. CA Cancer J Clin 67(4):326–344
23. Moya A, Ferrer M (2016) Functional redundancy-induced stability of gut microbiota subjected to disturbance. Trends Microbiol 24(5):402–413
24. Ramírez-Labrada AG, Isla D, Artal A, Arias M, Rezusta A, Pardo J, Gálvez EM (2020) The influence of lung microbiota on lung carcinogenesis, immunity, and immunotherapy. Trends Cancer 6(2):86–97
25. de Martel C, Georges D, Bray F, Ferlay J, Clifford GM (2020) Global burden of cancer attributable to infections in 2018: a worldwide incidence analysis. Lancet Glob Health 8(2): e180–e190
26. Rooks MG, Garrett WS (2016) Gut microbiota, metabolites and host immunity. Nat Rev Immunol 16(6):341–352
27. Nejman D, Livyatan I, Fuks G, Gavert N, Zwang Y, Geller LT, Rotter-Maskowitz A, Weiser R, Mallel G, Gigi E, Meltser A, Douglas GM, Kamer I, Gopalakrishnan V, Dadosh T, Levin-Zaidman S, Avnet S, Atlan T, Cooper ZA, Arora R, Cogdill AP, MAW K, Ologun G, Bussi Y, Weinberger A, Lotan-Pompan M, Golani O, Perry G, Rokah M, Bahar-Shany K, Rozeman EA, Blank CU, Ronai A, Shaoul R, Amit A, Dorfman T, Kremer R, Cohen ZR, Harnof S, Siegal T, Yehuda-Shnaidman E, Gal-Yam EN, Shapira H, Baldini N, MGI L, Ben-Nun A, Kaufman B, Nissan A, Golan T, Dadiani M, Levanon K, Bar J, Yust-Katz S, Barshack I, Peeper DS, Raz DJ, Segal E, Wargo JA, Sandbank J, Shental N, Straussman R (2020) The human tumor microbiome is composed of tumor type–specific intracellular bacteria. Science 368(6494): 973–980
28. Espinoza JL, Minami M (2018) Sensing bacterial-induced DNA damaging effects via natural killer group 2 member D immune receptor: from dysbiosis to autoimmunity and carcinogenesis. Front Immunol 9:52
29. Mathieu E, Escribano-Vazquez U, Descamps D, Cherbuy C, Langella P, Riffault S, Remot A, Thomas M (2018) Paradigms of lung microbiota functions in health and disease, particularly, in asthma. Front Physiol 9:1168
30. Huang Y, Gao X, Chen J (2018) Leukocyte-derived biomimetic nanoparticulate drug delivery systems for cancer therapy. Acta Pharm Sin B 8(1):4–13
31. Pierigè F, Serafini S, Rossi L, Magnani M (2008) Cell-based drug delivery. Adv Drug Deliv Rev 60(2):286–295
32. Fesnak AD, June CH, Levine BL (2016) Engineered T cells: the promise and challenges of cancer immunotherapy. Nat Rev Cancer 16(9):566–581
33. Nguyen HN, Romero Jovel S, Nguyen THK (2017) Nanosized minicells generated by lactic acid bacteria for drug delivery. J Nanomaterials. 2017:6847297
34. Developers Himanshu M (2013) "Minicells" safely deliver targeted drugs. Cancer 3:5

35. Guan X, Yang B, Xie M, Ban DK, Zhao X, Lal R, Zhang F (2019) MRI reporter gene MagA suppresses transferrin receptor and maps Fe2+ dependent lung cancer. Nanomedicine 21: 102064
36. Maruyama K, Takeyama H, Mori T, Ohshima K, Ogura SI, Mochizuki T, Matsunaga T (2007) Detection of epidermal growth factor receptor (EGFR) mutations in non-small cell lung cancer (NSCLC) using a fully automated system with a nano-scale engineered biomagnetite. Biosens Bioelectron 22(9–10):2282–2288
37. Maeda Y, Yoshino T, Matsunaga T (2009) Novel nanocomposites consisting of in vivo-biotinylated bacterial magnetic particles and quantum dots for magnetic separation and fluorescent labeling of cancer cells. J Mater Chem 19(35):6361–6366
38. Novakova ZV, Gasparova I, Krajciova L, Molcan M, Varga I, Timko M, Danisovic L (2017) Effect of magnetosomes on cell proliferation, apoptosis induction and expression of Bcl-2 in the human lung cancer cell line A549. Biologia 72(5):554–560
39. Wang J, Geng Y, Zhang Y, Wang X, Liu J, Basit A, Miao T, Liu W, Jiang W (2019) Bacterial magnetosomes loaded with doxorubicin and transferrin improve targeted therapy of hepatocellular carcinoma. Nano 3(3):284
40. Kraśko JA, Žilionytė K, Darinskas A, Strioga M, Rjabceva S, Zalutsky I, Derevyanko M, Kulchitsky V, Lubitz W, Kudela P, Miseikyte-Kaubriene E, Karaman O, Didenko H, Potebnya H, Chekhun V, Pašukonienė V (2017) Bacterial ghosts as adjuvants in syngeneic tumour cell lysate-based anticancer vaccination in a murine lung carcinoma model. Oncol Rep 37(1):171–178
41. Hajam IA, Dar PA, Won G, Lee JH (2017) Bacterial ghosts as adjuvants: mechanisms and potential. Vet Res 48(1):1–13
42. Roudi R, Mohammadi SR, Roudbary M, Mohsenzadegan M (2017) Lung cancer and β-glucans: review of potential therapeutic applications. Investig New Drugs 35(4):509–517
43. Majeed S, Abdullah MS, Dash GK, Ansari MT, Nanda A (2016) Biochemical synthesis of silver nanoparticles using filamentous fungi Penicillium decumbens (MTCC-2494) and its efficacy against A-549 lung cancer cell line. Chin J Nat Med 14(8):615–620
44. Kashimura Y, Oshima A, Sumitomo K (2016) Fabrication of nanobiodevices that utilize the function of membrane proteins. NTT Tech Rev 1–14
45. Ng KK, Lovell JF, Zheng G (2011) Lipoprotein-inspired nanoparticles for cancer theranostics. Acc Chem Res 44(10):1105–1113
46. Chaudhary J, Bower J, Corbin IR (2019) Lipoprotein drug delivery vehicles for cancer: rationale and reason. Int J Mol Sci 20(24):6327
47. Raha S, Paunesku T, Woloschak G (2011) Peptide-mediated cancer targeting of nanoconjugates. Wiley Interdiscip Rev Nanomed Nanobiotechnol 3(3):269–281
48. Hatakeyama S, Sugihara K, Shibata TK, Nakayama J, Akama TO, Tamura N, Wong SM, Bobkov AA, Takano Y, Ohyama C, Fukuda M, Fukuda MN (2011) Targeted drug delivery to tumor vasculature by a carbohydrate mimetic peptide. Proc Natl Acad Sci 108(49): 19587–19592
49. Chen F, Huang G, Huang H (2020) Sugar ligand-mediated drug delivery. Future Med Chem 12(2):161–171
50. Zavyalova E, Kopylov A (2018) DNA aptamer-based molecular nanoconstructions and nanodevices for diagnostics and therapy. In: Nanostructures for the engineering of cells, tissues and organs. William Andrew Publishing, Norwich, NY, pp 249–290
51. Li Y, Xuan J, Song Y, Qi W, He B, Wang P, Qin L (2016) Nanoporous glass integrated in volumetric bar-chart chip for point-of-care diagnostics of non-small cell lung cancer. ACS Nano 10(1):1640–1647
52. Zhou X, Ling K, Liu M, Zhang X, Ding J, Dong Y, Liang Z, Li J, Zhang J (2019) Targeted delivery of cisplatin-derived nanoprecursors via a biomimetic yeast microcapsule for tumor therapy by the oral route. Theranostics 9(22):6568
53. Robertson KL, Soto CM, Archer MJ, Odoemene O, Liu JL (2011) Engineered T4 viral nanoparticles for cellular imaging and flow cytometry. Bioconjug Chem 22(4):595–604

54. LI, S. D., & Huang, L. (2006) Surface-modified lpd nanoparticles for tumor targeting. Ann N Y Acad Sci 1082(1):1–8
55. Cheng L, Huang FZ, Cheng LF, Zhu YQ, Hu Q, Li L, Wei L, Chen DW (2014) GE11-modified liposomes for non-small cell lung cancer targeting: preparation, ex vitro and in vivo evaluation. Int J Nanomedicine 9:921
56. Lin C, Zhang X, Chen H, Bian Z, Zhang G, Riaz MK, Tyagi D, Lin G, Zhang Y, Wang J, Lu A, Yang Z (2018) Dual-ligand modified liposomes provide effective local targeted delivery of lung-cancer drug by antibody and tumor lineage-homing cell-penetrating peptide. Drug Deliv 25(1):256–266
57. Ujiie H, Ding L, Fan R, Kato T, Lee D, Fujino K, Kinoshita T, Lee CY, Waddell TK, Keshavjee S, Wilson BC, Zheng G, Chen J, Yasufuku K (2019) Porphyrin–high-density lipoprotein: a novel photosensitizing nanoparticle for lung cancer therapy. Ann Thorac Surg 107(2):369–377
58. Ding K, Zheng C, Sun L, Liu X, Yin Y, Wang L (2020) NIR light-induced tumor phototherapy using ICG delivery system based on platelet-membrane-camouflaged hollow bismuth selenide nanoparticles. Chin Chem Lett 31(5):1168–1172
59. Meir R, Shamalov K, Betzer O, Motiei M, Horovitz-Fried M, Yehuda R, Popovtzer A, Popovtzer R, Cohen CJ (2015) Nanomedicine for cancer immunotherapy: tracking cancer-specific T-cells in vivo with gold nanoparticles and CT imaging. ACS Nano 9(6):6363–6372
60. Parayath NN, Parikh A, Amiji MM (2018) Repolarization of tumor-associated macrophages in a genetically engineered nonsmall cell lung cancer model by intraperitoneal administration of hyaluronic acid-based nanoparticles encapsulating microRNA-125b. Nano Lett 18(6): 3571–3579
61. Mukherjee A, Bhattacharyya J, Sagar MV, Chaudhuri A (2013) Liposomally encapsulated CDC20 siRNA inhibits both solid melanoma tumor growth and spontaneous growth of intravenously injected melanoma cells on mouse lung. Drug Deliv Transl Res 3(3):224–234
62. Ganesh S, Iyer AK, Morrissey DV, Amiji MM (2013) Hyaluronic acid based self-assembling nanosystems for CD44 target mediated siRNA delivery to solid tumors. Biomaterials 34(13): 3489–3502
63. Tan T, Hu H, Wang H, Li J, Wang Z, Wang J, Wang S, Zhang Z, Li Y (2019) Bioinspired lipoproteins-mediated photothermia remodels tumor stroma to improve cancer cell accessibility of second nanoparticles. Nat Commun 10(1):1–17
64. Chen Q, Xu L, Liang C, Wang C, Peng R, Liu Z (2016) Photothermal therapy with immune-adjuvant nanoparticles together with checkpoint blockade for effective cancer immunotherapy. Nat Commun 7(1):1–13
65. Wu YF, Wu HC, Kuan CH, Lin CJ, Wang LW, Chang CW, Wang TW (2016) Multi-functionalized carbon dots as theranosticnanoagent for gene delivery in lung cancer therapy. Sci Rep 6(1):1–12
66. Song XL, Ju RJ, Xiao Y, Wang X, Liu S, Fu M, Liu JJ, Gu LY, Li XT, Cheng L (2017) Application of multifunctional targeting epirubicin liposomes in the treatment of non-small-cell lung cancer. Int J Nanomedicine 12:7433

# Correction to: Interplay of Microbiome, Inflammation, and Immunity in Inflammatory Lung Diseases

Hitesh Malhotra, Anjoo Kamboj, Peeyush Kaushik, and Rupesh K. Gautam

**Correction to:**
**Chapter 4 in: G. Gupta et al. (eds.),**
***Microbiome in Inflammatory Lung Diseases,***
**https://doi.org/10.1007/978-981-16-8957-4_4**

This chapter was inadvertently published with incorrect affiliations which have now been corrected as below:

H. Malhotra · P. Kaushik
Guru Gobind Singh College of Pharmacy, Yamunanagar, India
A. Kamboj
Chandigarh College of Pharmacy, Mohali, India
R. K. Gautam (✉)
MM School of Pharmacy, MM University, Sadopur-Ambala, India

---

The updated version of this chapter can be found at
https://doi.org/10.1007/978-981-16-8957-4_4



G. Gupta et al. (eds.), *Microbiome in Inflammatory Lung Diseases*,
https://doi.org/10.1007/978-981-16-8957-4_21

The manufacturer's authorised representative in the EU is Springer Nature Customer Service Centre GmbH, Europaplatz 3, 69115 Heidelberg, Germany. If you have any concerns regarding our products, please contact ProductSafety@springernature.com

Printed and bound by CPI Group (UK) Ltd, Croydon, CR0 4YY
30/09/2026
02225324-0015